Electrical Power Systems

IEEE Press
445 Hoes Lane, PO Box 1331
Piscataway, NJ 08855-1331

Editorial Board
John B. Anderson, *Editor in Chief*

R. S. Blicq	J. D. Irwin	J. M. F. Moura
M. Eden	S. V. Kartalopoulos	I. Peden
D. M. Etter	P. Laplante	E. Sánchez-Sinencio
G. F. Hoffnagle	A. J. Laub	L. Shaw
R. F. Hoyt	M. Lightner	D. J. Wells

Dudley R. Kay, *Director of Book Publishing*
Carrie Briggs, *Administrative Assistant*
Lisa S. Mizrahi, *Review and Publicity Coordinator*

Valerie Zaborski, *Production Editor*

IEEE Power Systems Engineering Series

Dr. Paul M. Anderson, Series Editor
Power Math Associates, Inc.

Series Editorial Advisory Committee

Dr. Roy Billinton
University of Saskatchewan

Dr. Atif S. Debs
Georgia Institute of Technology

Dr. M. El-Hawary
Technical University of
 Nova Scotia

Mr. Richard G. Farmer
Arizona Public
 Service Company

Dr. Charles A. Gross
Auburn University

Dr. G. T. Heydt
Purdue University

Dr. George Karady
Arizona State University

Dr. Donald W. Novotny
University of Wisconsin

Dr. A. G. Phadke
Virginia Polytechnic and
 State University

Dr. Chanan Singh
Texas A & M University

Dr. E. Keith Stanek
University of Missouri-Rolla

Dr. J. E. Van Ness
Northwestern University

Electrical Power Systems
Revised Printing

Design and Analysis

Mohamed E. El-Hawary
Technical University of Nova Scotia

A JOHN WILEY & SONS, INC., PUBLICATION

IEEE Press Power Systems Engineering Series
Dr. Paul M. Anderson, Series Editor

The Institute of Electrical and Electronics Engineers, Inc., New York

This book may be purchased at a discount from the publisher when ordered in bulk quantities. For more information contact:

IEEE PRESS Marketing
Attn: Special Sales
P.O. Box 1331
445 Hoes Lane
Piscataway, NJ 08855-1331
Fax: (908) 981-8062

Revised printing © 1995 by the Institute of Electrical and Electronics Engineers, Inc.
345 East 47th Street, New York, NY 10017-2394

© 1983 by Reston Publishing Company

This is the IEEE revised printing of a book previously published by Reston Publishing Company under the title *Electrical Power Systems: Design and Analysis*.

No part of this publication may be reproduced, stored in a retrieval system or transmitted in any form or by any means, electronic, mechanical, photocopying, recording, scanning or otherwise, except as permitted under Sections 107 or 108 of the 1976 United States Copyright Act, without either the prior written permission of the Publisher, or authorization through payment of the appropriate per-copy fee to the Copyright Clearance Center, 222 Rosewood Drive, Danvers, MA 01923, (978) 750-8400, fax (978) 750-4470. Requests to the Publisher for permission should be addressed to the Permissions Department, John Wiley & Sons, Inc., 111 River Street, Hoboken, NJ 07030, (201) 748-6011, fax (201) 748-6008.

Printed in the United States of America

10 9 8 7 6 5 4 3 2

ISBN 0-7803-1140-X

IEEE Order Number: PC5606

Library of Congress Cataloging-in-Publication Data

El-Hawary, M. E.
    Electrical power systems : design and analysis / Mohamed E. El
-Hawary. — Rev. printing.
      p. cm. — (IEEE Press power systems engineering series)
    Includes bibliographical references and index.
    ISBN 0-7803-1140-X
    1. Electric power systems—Design and construction   I. Title.
II. Series.
TK1001.E54   1995   94-44628
621.31—dc20   CIP

They are all plain to him that understandeth and right to them to find knowledge. Receive my instruction and not silver and knowledge rather than choice gold. For wisdom is better than rubies, and all the things that may be desired are not to be compared to it. I wisdom dwell with prudence, and find out knowledge of witty inventions.

Proverbs 8:9–12

Read: In the name of thy Lord Who Createth, Createth man from a clot. Read: And thy Lord is the Most Bounteous, Who teacheth by the pen, Teacheth man which he knew not.

Coran 96:1–5

Contents

Preface   xi

Chapter I Introduction   1

1.1   The Development of Electric Power Systems, **1**
1.2   Outline of the Text, **7**

Chapter II Some Basic Principles   11

2.1   Introduction, **11**
2.2   Power Concepts, **11**
2.3   Three Phase Systems, **18**
2.4   Power System Representation, **31**
      Some Solved Problems, **35**
      Problems, **39**

Chapter III Power Generation and the Synchronous Machine   43

3.1   Introduction, **43**
3.2   The Synchronous Machine: Preliminaries, **44**
3.3   Fields in a Synchronous Machine, **48**
3.4   A Simple Equivalent Circuit, **52**
3.5   Open-Circuit and Short-Circuit Characteristics, **55**
3.6   Principal Steady-State Characteristics, **59**
3.7   Power-Angle Characteristics and the Infinite Bus Concept, **63**
3.8   Static Stability Limit Curves, **68**
3.9   Accounting for Saliency, **71**
3.10   Salient-Pole Machine Power Angle Characteristics, **75**
      Some Solved Problems, **79**
      Problems, **85**

Chapter IV The Transmission Subsystem   91

4.1 Introduction, **91**
4.2 Electric Transmission Line Parameters, **92**
4.3 Line Inductance, **96**
4.4 Line Capacitance, **129**
4.5 Two-Port Networks, **154**
4.6 Transmission Line Models, **164**
     Some Solved Problems, **183**
     Appendix 4-A, **192**
     Appendix 4-B, **193**
     Appendix 4-C, **193**
     Problems, **197**

Chapter V The Load Subsystem   217

5.1 Introduction, **217**
5.2 General Theory of Transformer Operation, **218**
5.3 Transformer Connections, **232**
5.4 Three-Phase Induction Motors, **256**
     Some Solved Problems, **266**
     Problems, **277**

Chapter VI Analysis of Interconnected Systems   283

6.1 Introduction, **283**
6.2 Reduction of Interconnected Systems, **284**
6.3 The Per Unit System, **296**
6.4 Network Nodal Admittance Formulation, **299**
6.5 The General Form of the Load-Flow Equations, **304**
6.6 The Load-Flow Problem, **310**
6.7 Getting Started, **315**
6.8 Newton-Raphson Method, **318**
6.9 The Newton-Raphson Method for Load-Flow Solution, **328**
     Some Solved Problems, **338**
     Problems, **345**

Chapter VII High-Voltage Direct-Current Transmission   355

7.1 Introduction, **355**
7.2 Main Applications of HVDC, **358**
7.3 HVDC Converters, **360**

7.4   Classifications of Direct-Current Links, **363**
7.5   Some Advantages of HVDC Transmission, **365**
7.6   Some Economic Considerations, **370**
7.7   Converter Circuits: Configurations and Properties, **376**
7.8   Analysis of the Three-Phase Bridge Converter, **396**
7.9   Inversion in Three-Phase Bridge Converter, **413**
7.10 HVDC Link and Converter Control Characteristics, **416**
7.11 Analysis of HVDC Link Performance, **420**
      Some Solved Problems, **427**
      Problems, **459**

Chapter VIII Faults on Electric Energy Systems   469

8.1   Introduction, **469**
8.2   Transients During a Balanced Fault, **470**
8.3   The Method of Symmetrical Components, **474**
8.4   Sequence Networks, **483**
8.5   Line-to-Ground Fault, **501**
8.6   Double Line-to-Ground Fault, **507**
8.7   Line-to-Line Fault, **512**
8.8   The Balanced Three-Phase Fault, **515**
      Some Solved Problems, **518**
      Problems, **536**

Chapter IX System Protection   541

9.1   Introduction, **541**
9.2   Productive Relays, **544**
9.3   The X-R Diagram, **555**
9.4   Relay Comparators, **559**
9.5   Generator Protection, **567**
9.6   Transformer Protection, **575**
9.7   Bus Bar Protection, **582**
9.8   Transmission Line Overcurrent Protection, **587**
9.9   Pilot-Wire Feeder Protection, **598**
9.10 Distance Protection, **601**
9.11 Power Line Carrier Protection, **604**
9.12 Computer Relaying, **608**
      Some Solved Problems, **612**
      Problems, **621**

Chapter X Power System Stability   625

10.1 Introduction, **625**
10.2 The Swing Equation, **626**
10.3 Electric Power Relations, **632**
10.4 Concepts in Transient Stability, **641**
10.5 A Method for Stability Assessment, **646**
10.6 Improving System Stability, **657**
  Some Solved Problems, **658**
  Problems, **692**

Chapter XI Optimal Operation of Electric Power Systems   701

11.1 Introduction, **701**
11.2 Modeling of Fuel Costs for Thermal Generation, **702**
11.3 Optimal Operation of an All-Thermal System: Equal Incremental Cost Loading, **708**
11.4 Accounting for Transmission Losses, **716**
11.5 Optimal Operation of an All-Thermal System, Including Losses, **722**
11.6 Optimal Operation of Hydrothermal Systems, **738**
  Some Solved Problems, **744**
  Problems, **775**

Index   785

Preface

This book is intended to provide an introduction to a number of important topics in engineering. The book's audience consists mainly of post-secondary electrical engineering students as well as practicing engineers interested in learning the fundamental concepts of power systems analysis and related designs. Background requirements include a basic electric circuit course and some mathematical notions from algebra and calculus.

The text material is arranged in a format which is aimed at furthering the readers' understanding by providing ample practical examples within the text to illustrate the concepts discussed. In addition, each chapter contains a section that offers additional solved problems that serve to illustrate the interrelation between the concepts discussed in the chapter from a system's point of view.

The text treats first models of the major components of modern day electric power systems. Thus, chapters three through five provide detailed discussions of synchronous machines, transmission lines, transformers and the induction motor which is a major system load component.

Chapter six deals with analysis of interconnected systems with major emphasis on load flow analysis. Chapter seven is intended to present—in a reasonable amount of detail—elements of high voltage, direct current transmission which are becoming increasingly important.

Chapter eight details analysis problems in systems with fault conditions. This is followed in Chapter nine by a treatment of system protection.

Chapter ten is devoted to transient stability problems at an introductory level. The final chapter on optimal economic operation of power systems provides a comprehensive yet simple introduction to that important area. Each of the chapters is concluded by a section of problems for drill purposes. It is assumed that the reader has access to a modest computing facility such as a programmable calculator.

I am indebted to my many students who have contributed immensely to the development of this text, in particular students at Memorial University of Newfoundland and the Technical University of Nova Scotia, who took great interest in this project. To my colleagues and friends from electric utilities and the academe alike, my sincere appreciation for their contributions.

The drafting of the manuscript involved the patient and able typing work done by many at Memorial University and lately at the Technical University of Nova Scotia. Many thanks to Mrs. Minnie Ewing, Ms. Marilyn Tiller, Ms. Brenda Young, Mrs. Ethil Pitt, and Ms. Valerie Blundell of Memorial and Frances Julian of the Technical University of Nova Scotia. Margaret McNeily of Kennett Square, Pennsylvania, skillfully copyedited the original manuscript. I gratefully appreciate her help. Also, my thanks to Dan McCauley of Reston Publishing Company for his work on this book. Finally, the patience and understanding of my wife Ferial, and children are appreciated.

M. E. El-Hawary

Halifax, Nova Scotia
February, 1982

CHAPTER I

Introduction

The purpose of this chapter is twofold. We first provide a brief perspective on the development of electric power systems. This is not intended to be a detailed historical review, but rather it uses historical landmarks as a background to highlight the features and structure of the modern power systems. Following this we offer an outline of the text material.

1.1 THE DEVELOPMENT OF ELECTRIC POWER SYSTEMS

Electric power is one major industry that has shaped and contributed to the progress and technological advances of mankind over the past century. It is not surprising then that the growth of electric energy consumption in the world has been nothing but phenomenal. In the United States, for example, electric energy sales have grown to well over 400 times in the period between the turn of the century and the early 1970s. This growth rate was 50 times as much as the growth rate in all other energy forms used during the same period.

Edison Electric Illuminating Company of New York pioneered the central station electric power generation by the opening of the Pearl Street station in 1881. This station had a capacity of four 250-hp boilers supplying steam to six engine-dynamo sets. Edison's system used a 110-V dc underground distribution network with copper conductors insulated with a jute wrapping. The *low voltage of the circuit* limited the service area of a central station, and consequently central stations proliferated throughout metropolitan areas.

The invention of the transformer, then known as the "inductorium," made ac systems possible. The first practical ac distribution system in the United States was installed by W. Stanley at Great Barrington, Massachusetts, in 1866 for Westinghouse, who acquired the American rights to the transformer from its British inventors Gaulard and Gibbs. Early ac distribution utilized 1000-V overhead lines.

By 1895, Philadelphia had about twenty electric companies with distribution systems operating at 100-V and 500-V two-wire dc and 220-V three-wire dc; single-phase, two-phase, and three-phase ac; with frequencies of 60, 66, 125, and 133 cycles per second; and feeders at 1000–1200 V and 2000–2400 V.

The consolidation of electric companies enabled the realization of economies of scale in generating facilities, the introduction of a certain degree of equipment standardization, and the utilization of the load diversity between areas. Generating unit sizes of up to 1300 MW are in service, an era that was started by the 1973 Cumberland Station of the Tennessee Valley Authority. A major generating station is shown in Figure 1-1, with the turbine-generator hall shown in Figure 1-2.

Underground distribution at voltages up to 5 kV was made possible by the development of rubber-base insulated cables and paper-insulated, lead-covered cables in the early 1900s. Since that time higher distribution voltages have been necessitated by load growth that would otherwise overload low-voltage circuits and by the requirement to transmit large blocks of power over great distances. Common distribution voltages in today's system are in 5-, 15-, 25-, 35-, and 69-kV voltage classes.

The growth in size of power plants and in the higher voltage equipment was accompanied by interconnections of the generating facilities. These interconnections decreased the probability of service interruptions, made the utilization of the most economical units possible, and decreased the total reserve capacity required to meet equipment-forced outages. This growth was also accompanied by the use of sophisticated analysis tools such as the network analyzer shown in Figure 1-3. Central control of the interconnected systems was introduced for reasons of economy and safety. Figure 1-4 shows the control room in a system control center. The advent of the load dispatcher heralded the dawn of power systems engineering, whose

Figure 1-1. Site for a Major Nuclear Generating Station.
(*Courtesy Ontario Hydro*)

Figure 1-2. Turbine-Generator Hall in a Major Generating Station.
(*Courtesy Ontario Hydro*)

Figure 1-3. A Network Analyzer Facility.
(Courtesy Ontario Hydro)

Figure 1-4. Inside a Power System Control Center.
(*Courtesy Ontario Hydro*)

objective is to provide the best system to meet the load demand reliably, safely, and economically, utilizing state-of-the-art computer facilities.

Extra high voltage (EHV) has become the dominant factor in the transmission of electric power over great distances. By 1896, an 11-kV three-phase line was transmitting 10 MW from Niagara Falls to Buffalo over a distance of 20 miles. Today, transmission voltages of 230 kV (see Figure 1-5), 287 kV, 345 kV, 500 kV, 735 kV, and 765 kV are commonplace, with the first 1100-kV line scheduled for energization in the early 1990s. A prototype 1200-kV transmission tower is shown in Figure 1-6. The trend is motivated by the economy of scale due to the higher transmission capacities possible, more efficient use of right-of-way, lower transmission losses, and reduced environmental impact.

The preference for ac was first challenged in 1954 when the Swedish State Power Board energized the 60-mile, 100-kV dc submarine cable utilizing U. Lamm's Mercury Arc valves at the sending and receiving ends of the world's first high-voltage direct current (HVDC) link connecting the Baltic island of Gotland and the Swedish mainland. Today numerous installations with voltages up to 800-kV dc have become operational around the globe. Solid-state technology advances have also enabled the use of the silicon-controlled rectifiers (SCR) or thyristor for HVDC applications since the late 1960s. Whenever cable transmission is required (underwater or in a metropolitan area), HVDC is more economically attractive than ac.

Protecting isolated systems has been a relatively simple task, which is carried out using overcurrent directional relays with selectivity being obtained by time grading. High-speed relays have been developed to meet the increased short-circuit currents due to the larger size units and the complex interconnections.

1.2 OUTLINE OF THE TEXT

Chapter 2 lays the foundations for the development in the rest of the book. The intention of the discussion offered here is to provide a brief review of fundamentals including electric circuit analysis and some mathematical background, to make the treatment self-contained. A student with an introductory electric circuit background may safely omit this chapter.

Chapters 3, 4, and 5 are sequentially structured to follow the flow of electric energy from conversion to utilization. Thus Chapter 3 treats the synchronous machine from an operational modeling point of view. Emphasis here is on performance characteristics of importance to the electric power specialist. Chapter 4 provides a comprehensive treatment of EHV transmission lines starting from parameter evaluation for different circuit

Figure 1-5. Transmission Towers for a 230-kV Line.
(*Courtesy Ontario Hydro*)

Figure 1-6. A Prototype 1200-kV Transmission Line Tower.
(*Courtesy U.S. Department of Energy, Bonneville Power Administration*)

and conductor configurations. Various transmission line performance modeling approaches are covered along with a unique section on the errors involved when using simplified models over the more elaborate ones. Chapter 5 is entitled "The System Load" and deals with the power transformer as well as control and instrument transformers in addition to induction motor models as the latter is a major load component. A brief discussion of load modeling philosophy is given at the end of the chapter.

Chapter 6 treats interconnected system analysis and covers aspects of network reduction, per unit systems, and the load flow problem. A comprehensive treatment of high-voltage direct-current transmission is given in Chapter 7. Here again emphasis is placed on analysis and control aspects that should be of interest to the electric power systems specialist.

Faults on electric energy systems are considered in Chapter 8. Here we start with the transient phenomenon of a symmetrical short circuit, followed by a treatment of unbalanced and balanced faults. Realizing the crucial part that system protection plays in maintaining service integrity is the basis for Chapter 9. Here an introduction to this important area is given. The transient stability problem is treated in Chapter 10 from an introductory point of view. Chapter 11 introduces the subject of economic dispatch under the title "Optimal Operation of Electric Power Systems." The treatment covers thermal systems where losses are neglected, followed by a case including losses. The chapter is concluded by an introduction to hydrothermal dispatch.

The text of each chapter includes a number of examples that illustrate the concepts discussed. Following each chapter there is a set of solved problems that involves, in many instances, increased sophistication, and it helps to bring together the overall thrust of the concepts and techniques treated. The student should have, then, no difficulty in dealing with the drill problems included at the end of each chapter.

CHAPTER II

Some Basic Principles

2.1 INTRODUCTION

The intention of this chapter is to lay the groundwork for the study of electric energy systems. This is done by developing some basic tools involving concepts, definitions, and some procedures fundamental to electric energy systems. The chapter can be considered as simply a review of topics that will be utilized throughout this work. We start by introducing the principal electrical quantities that we will deal with.

2.2 POWER CONCEPTS

The electric power systems specialist is in many instances more concerned with electric power in the circuit rather than the currents. As the power into an element is basically the product of voltage across and current through it, it seems reasonable to swap the current for power without losing any information in describing the phenomenon. In treating sinusoidal steady-state behavior of circuits, some further definitions are necessary. To illustrate the concepts, we will use a cosine representation of the waveforms.

Consider impedance element $Z = Z\underline{/\phi}$. For a sinusoidal voltage, $v(t)$ is given by

$$v(t) = V_m \cos \omega t$$

The instantaneous current in the circuit is

$$i(t) = I_m \cos(\omega t - \phi)$$

where

$$I_m = V_m/|Z|$$

The instantaneous power is thus given by

$$p(t) = v(t)i(t) = V_m I_m [\cos(\omega t)\cos(\omega t - \phi)]$$

Using the trigonometric identity

$$\cos \alpha \cos \beta = \tfrac{1}{2}[\cos(\alpha - \beta) + \cos(\alpha + \beta)]$$

we can write the instantaneous power as

$$p(t) = \frac{V_m I_m}{2}[\cos \phi + \cos(2\omega t - \phi)]$$

The average power p_{av} is seen to be

$$p_{av} = \frac{V_m I_m}{2} \cos \phi \qquad (2.1)$$

Since the average of $\cos(2\omega t - \phi)$ is zero, through 1 cycle, this term therefore contributes nothing to the average of p.

It is more convenient to use the effective (rms) values of voltage and current than the maximum values. Substituting $V_m = \sqrt{2}(V_{rms})$, and $I_m = \sqrt{2}(I_{rms})$, we get

$$p_{av} = V_{rms} I_{rms} \cos \phi \qquad (2.2)$$

Thus the power entering any network is the product of the effective values of terminal voltage and current and the cosine of the phase angle ϕ which is called the *power factor* (PF). This applies to sinusoidal voltages and currents only. For a purely resistive load, $\cos \phi = 1$, and the current in the circuit is fully engaged in conveying power from the source to the load resistance. When reactance as well as resistance are present, a component of the current in the circuit is engaged in conveying the energy that is periodically stored in and discharged from the reactance. This stored energy, being shuttled to and from the magnetic field of an inductance or the electric field of a capacitance, adds to the current in the circuit but does not add to the average power.

The average power in a circuit is called *active power*, and the power that supplies the stored energy in reactive elements is called *reactive power*. Active power is P, and the reactive power, designated Q, are thus*

$$P = VI\cos\phi \qquad (2.3)$$
$$Q = VI\sin\phi \qquad (2.4)$$

In both equations, V and I are rms values of terminal voltage and current, and ϕ is the phase angle by which the current lags the voltage.

Both P and Q are of the same dimension, that is in watts. However, to emphasize the fact that the Q represents the nonactive power, it is measured in reactive voltampere units (var). Larger and more practical units are kilovars and megavars, related to the basic unit by

$$1 \text{ Mvar} = 10^3 \text{ kvar} = 10^6 \text{ var}$$

Figure 2-1 shows the time variation of the various variables discussed in this treatment.

Assume that V, $V\cos\phi$, and $V\sin\phi$, all shown in Figure 2-2, are each multiplied by I, the rms value of current. When the components of voltage $V\cos\phi$ and $V\sin\phi$ are multiplied by current, they become P and Q respectively. Similarly, if I, $I\cos\phi$, and $I\sin\phi$ are each multiplied by V, they become VI, P, and Q respectively. This defines a power triangle.

We define a quantity called the *complex or apparent power*, designated S, of which P and Q are components. By definition,

$$\begin{aligned} S &= P + jQ \\ &= VI\cos\phi + jVI\sin\phi \\ &= VI(\cos\phi + j\sin\phi) \end{aligned}$$

Using Euler's identity, we thus have

$$S = VIe^{j\phi}$$

or

$$S = VI\underline{/\phi}$$

If we introduce the conjugate current defined by the asterisk (*)

$$I^* = |I|\underline{/\phi}$$

---

*If we write the instantaneous power as

$$p(t) = V_{\text{rms}}I_{\text{rms}}[\cos\phi(1+\cos 2\omega t)] + V_{\text{rms}}I_{\text{rms}}\sin\phi\sin 2\omega t$$

then it is seen that

$$p(t) = P(1+\cos 2\omega t) + Q\sin 2\omega t$$

Thus P and Q are the average power and the amplitude of the pulsating power respectively.

14 Some Basic Principles

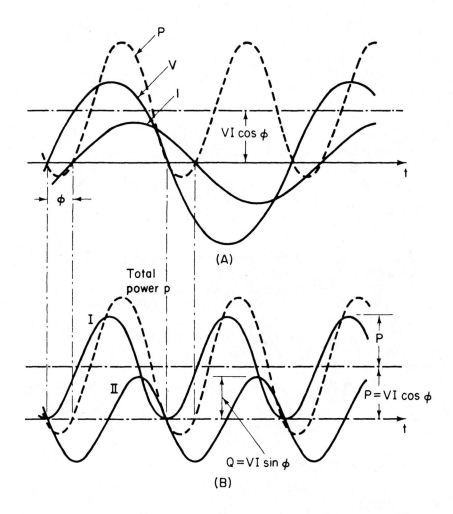

Figure 2-1. Voltage, Current and Power in a Single-Phase Circuit.

it becomes immediately obvious that an equivalent definition of complex or apparent power is

$$S = VI^* \qquad (2.5)$$

We can write the complex power in two alternative forms by using the

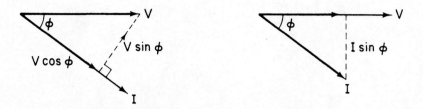

Figure 2-2. Phasor Diagrams Leading to Power Triangles.

relationships
$$V = ZI \quad \text{and} \quad I = YV$$
This leads to
$$S = ZII^* = Z|I|^2 \tag{2.6}$$
or
$$S = VY^*V^* = Y^*|V|^2 \tag{2.7}$$

Consider the series circuit shown in Figure 2-3. Here the applied voltage is equal to the sum of the voltage drops:
$$V = I(Z_1 + Z_2 + \cdots + Z_n)$$
Multiplying both sides of this relation by I^* results in
$$S = VI^* = II^*(Z_1 + Z_2 + \cdots + Z_n)$$
or
$$S = \sum_{i=1}^{n} S_i \tag{2.8}$$
with
$$S_i = |I|^2 Z_i \tag{2.9}$$
being the individual element's complex power. Equation (2.8) is known as the summation rule for complex powers. The summation rule also applies to parallel circuits. The use of the summation rule and concepts of complex power may prove advantageous in solving problems of power system analysis.

The phasor diagrams shown in Figure 2-2 can be converted into complex power diagrams by simply following the definitions relating complex power to voltage and current. Consider the situation with an inductive circuit in which the current lags the voltage by the angle ϕ. The conjugate of the current will be in the first quadrant in the complex plane as shown in

16 Some Basic Principles

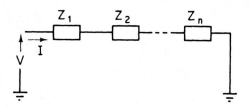

Figure 2-3. Series Circuit.

Figure 2-4(a). Multiplying the phasors by **V**, we obtain the complex power diagram shown in Figure 2-4(b). Inspection of the diagram as well as the previous development leads to a relation for the power factor of the circuit:

$$\cos\phi = \frac{P}{|S|}$$

Example 2.1

Consider the circuit composed of a series R-L branch in parallel with capacitance with the following parameters:

$$R = 0.5 \text{ ohms}$$
$$X_L = 0.8 \text{ ohms}$$
$$B_c = 0.6 \text{ siemens}$$

Assume

$$V = 100\underline{/0}\text{ V}$$

Calculate the input current and the active, reactive, and apparent power into the circuit.

Solution

The current into the R-L branch is given by

$$I_Z = \frac{100}{0.5 + j0.8} = 106.00\underline{/-57.99°}\text{ A}$$

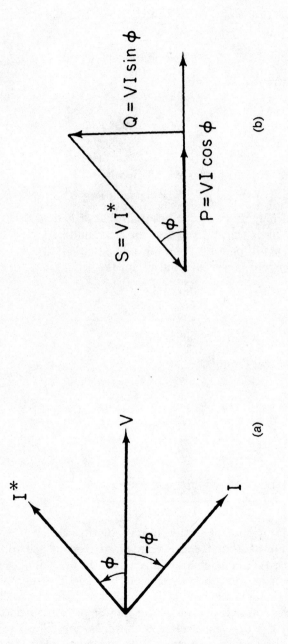

Figure 2-4. Complex Power Diagram.

The power factor (PF) of the R-L branch is
$$\mathrm{PF}_Z = \cos \phi_Z = \cos 57.99°$$
$$= 0.53$$

The current into the capacitance is
$$I_c = j(0.6)(100) = 60\underline{/90°} \text{ A}$$

The input current I_t is
$$I_t = I_c + I_Z$$
$$= 106.00\underline{/-57.99°} + 60\underline{/90°}$$
$$= 63.64\underline{/-28.01°}$$

The power factor (PF) of the overall circuit is
$$\mathrm{PF}_t = \cos \phi_t = \cos 28.01° = 0.88$$

Note that the magnitude of I_t is less than that of I_Z, and that $\cos \phi$ is higher than $\cos \phi_z$. This is the effect of the capacitor, and its action is called *power factor correction* in power system terminology.

The apparent power into the circuit is
$$S_t = VI_t^*$$
$$= (100\underline{/0})(63.64)\underline{/28.01°}$$
$$= 6364.00\underline{/28.01°} \text{ VA}$$

In rectangular coordinates we get
$$S_t = 5617.98 + j2988.76$$

Thus the active and reactive powers are:
$$P_t = 5617.98 \text{ W}$$
$$Q_t = 2988.76 \text{ var}$$

2.3 THREE-PHASE SYSTEMS

The major portion of all the electric power presently used is generated, transmitted, and distributed using balanced three-phase voltage systems. The single-phase voltage sources referred to in the preceding section originate in many instances as part of a three-phase system. Three-phase operation is preferable to single-phase because a three-phase winding makes more efficient use of generator copper and iron. Power flow in single-phase circuits was shown in the previous section to be pulsating. This drawback is not present in a three-phase system as will be shown later. Also three-phase motors start more conveniently and, having constant torque, run more

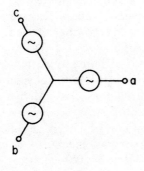

(a)

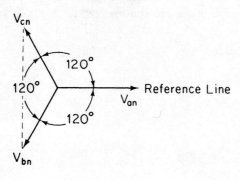

(b)

Figure 2-5. A Y-Connected Three-Phase System and the Corresponding Phasor Diagram.

satisfactorily than single-phase motors. However, the complications of additional phases are not compensated for by the slight increase of operating efficiency when polyphase systems other than three-phase are used.

A balanced three-phase voltage system is composed of three single-phase voltages having the same magnitude and frequency but time-displaced from one another by 120°. Figure 2-5(a) shows a schematic representation where the three single-phase voltage sources appear in a Y connection; a Δ configuration is also possible. A phasor diagram showing each of the phase voltages is also given in Figure 2-5(b). As the phasors revolve at the angular frequency ω with respect to the reference line in the counterclockwise (positive) direction, the positive maximum value first occurs for phase a and then in succession for phases b and c. Stated in a different way, to an observer in the phasor space, the voltage of phase a arrives first followed by that of b and then that of c. For this reason the three-phase voltage of Figure 2-5 is said to have the phase sequence abc (*order* or *phase sequence* or *rotation* are all synonymous terms). This is important for certain applications. For example, in three-phase induction motors, the phase sequence determines whether the motor turns clockwise or counterclockwise.

With very few exceptions, synchronous generators (commonly referred to as *alternators*) are three-phase machines. For the production of a set of three voltages phase-displaced by 120 electrical degrees in time, it follows that a minimum of three coils phase-displaced 120 electrical degrees in space must be used. An elementary three-phase two-pole machine with one coil per phase is shown in Figure 2-6.

We find it convenient for clarity of the presentation to consider representing each coil as a separate generator. An immediate extension of the single-phase circuits discussed above would be to carry the power from

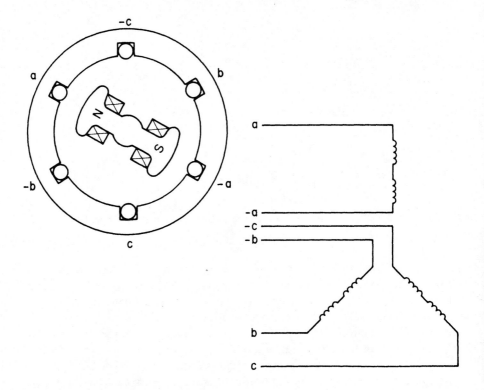

Figure 2-6. An Elementary Three-Phase Two-Pole Machine.

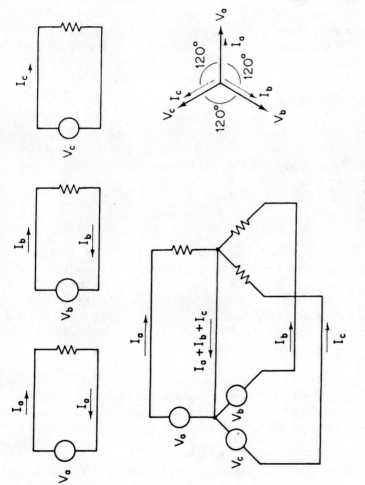

Figure 2-7. A Three-Phase System.

the three generators along six wires. However, for the sake of economy, instead of having a return wire from each load to each generator, a single wire is used for the return of all three. The current in the return wire will be $I_a + I_b + I_c$; and for a balanced load, these will cancel out as may be seen by inspecting the phasor diagram in Figure 2-7. If the load is unbalanced, the return current will still be small compared to either I_a, I_b, or I_c. Thus the return wire could be made smaller than the other three. This connection is known as a four-wire three-phase system. It is desirable for safety and system protection to have a connection from the electrical system to ground. A logical point for grounding is the generator neutral point, the junction of the Y.

Current and Voltage Relations

Balanced three-phase systems can be studied using techniques developed for single-phase circuits. The arrangement of the three single-phase voltages into a Y or a Δ configuration requires some modification in dealing with the overall system.

Y Connection

With reference to Figure 2-8, the common terminal n is called the *neutral* or *star (Y) point*. The voltages appearing between any two of the line terminals a, b, and c have different relationships in magnitude and phase to the voltages appearing between any one line terminal and the neutral point n. The set of voltages V_{ab}, V_{bc}, and V_{ca} are called the *line voltages*, and the set of voltages V_{an}, V_{bn}, and V_{cn} are referred to as the *phase voltages*. Analysis of phasor diagrams provides the required relationships.

The effective values of the phase voltages are shown in Figure 2-8 as V_{an}, V_{bn}, and V_{cn}. Each has the same magnitude, and each is displaced 120° from the other two phasors. To obtain the magnitude and phase angle of the line voltage from a to b (i.e., V_{ab}), we apply Kirchhoff's voltage law:

$$V_{ab} = V_{an} + V_{nb} \tag{2.10}$$

This equation states that the voltage existing from a to b is equal to the voltage from a to n (i.e., V_{an}) plus the voltage from n to b. Thus Eq. (2.10) can be rewritten as

$$V_{ab} = V_{an} - V_{bn} \tag{2.11}$$

Since for a balanced system, each phase voltage has the same magnitude, let us set

$$|V_{an}| = |V_{bn}| = |V_{cn}| = V_p \tag{2.12}$$

2.3 Three-Phase Systems

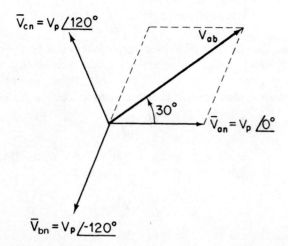

Figure 2-8. Illustrating the Phase and Magnitude Relations Between the Phase and Line Voltage of a Y Connection.

where V_p denotes the effective magnitude of the phase voltage. Accordingly we may write

$$V_{an} = V_p \underline{/0°} \qquad (2.13)$$

$$V_{bn} = V_p \underline{/-120°} \qquad (2.14)$$

$$V_{cn} = V_p \underline{/-240°} = V_p \underline{/120°} \qquad (2.15)$$

Substituting Eqs. (2.13) and (2.14) in Eq. (2.11) yields

$$V_{ab} = V_p \left(1 - 1 \underline{/-120°}\right)$$
$$= \sqrt{3}\, V_p \underline{/30°} \qquad (2.16)$$

Similarly we obtain

$$V_{bc} = \sqrt{3}\, V_p \underline{/-90°} \qquad (2.17)$$

$$V_{ca} = \sqrt{3}\, V_p \underline{/150°} \qquad (2.18)$$

The expressions obtained above for the line voltages show that they constitute a balanced three-phase voltage system whose magnitudes are $\sqrt{3}$

times the phase voltages. Thus we write

$$V_L = \sqrt{3}\, V_p \tag{2.19}$$

A current flowing out of a line terminal a (or b or c) is the same as that flowing through the phase source voltage appearing between terminals n and a (or n and b, or n and c). We can thus conclude that for a Y-connected three-phase source, the line current equals the phase current. Thus

$$I_L = I_p \tag{2.20}$$

In the above equation, I_L denotes the effective value of the line current and I_p denotes the effective value for the phase current.

Δ Connection

We consider now the case when the three single-phase sources are rearranged to form a three-phase Δ connection as shown in Figure 2-9. It is clear from inspection of the circuit shown that the line and phase voltages have the same magnitude:

$$|V_L| = |V_p| \tag{2.21}$$

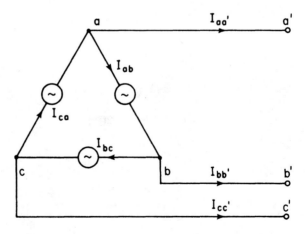

Figure 2-9. A Δ-Connected Three-Phase Source.

The phase and line currents, however, are not identical, and the relationship between them can be obtained using Kirchhoff's current law at one of the line terminals.

In a manner similar to that adopted for the Y-connected source, let us consider the phasor diagram shown in Figure 2-10. Assume the phase currents to be

$$I_{ab} = I_p \underline{/0}$$
$$I_{bc} = I_p \underline{/-120°}$$
$$I_{ca} = I_p \underline{/120°}$$

The current that flows in the line joining a to a' is denoted $I_{aa'}$ and is given by

$$I_{aa'} = I_{ca} - I_{ab}$$

As a result, we have

$$I_{aa'} = I_p \left[1\underline{/120°} - 1\underline{/0} \right]$$

which simplifies to

$$I_{aa'} = \sqrt{3}\, I_p \underline{/150°}$$

Similarly,

$$I_{bb'} = \sqrt{3}\, I_p \underline{/30°}$$
$$I_{cc'} = \sqrt{3}\, I_p \underline{/-90°}$$

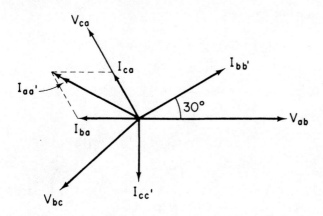

Figure 2-10. Illustrating Relation Between Phase and Line Currents in a Δ Connection.

Note that a set of balanced three phase currents yields a corresponding set of balanced line currents that are $\sqrt{3}$ times the phase values:

$$I_L = \sqrt{3}\, I_p \tag{2.22}$$

where I_L denotes the magnitude of any of the three line currents.

Power Relationships

Assume that the three-phase generator is supplying a balanced load with the three sinusoidal phase voltages:

$$v_a(t) = \sqrt{2}\, V_p \sin \omega t$$
$$v_b(t) = \sqrt{2}\, V_p \sin(\omega t - 120°)$$
$$v_c(t) = \sqrt{2}\, V_p \sin(\omega t + 120°)$$

With the currents given by

$$i_a(t) = \sqrt{2}\, I_p \sin(\omega t - \phi)$$
$$i_b(t) = \sqrt{2}\, I_p \sin(\omega t - 120° - \phi)$$
$$i_c(t) = \sqrt{2}\, I_p \sin(\omega t + 120° - \phi)$$

where ϕ is the phase angle between the current and voltage in each phase. The total power in the load is

$$p_{3\phi}(t) = v_a(t) i_a(t) + v_b(t) i_b(t) + v_c(t) i_c(t)$$

This turns out to be

$$p_{3\phi}(t) = 2V_p I_p [\sin(\omega t)\sin(\omega t - \phi)$$
$$+ \sin(\omega t - 120)\sin(\omega t - 120 - \phi)$$
$$+ \sin(\omega t + 120)\sin(\omega t + 120 - \phi)]$$

Using a trigonometric identity, we get

$$p_{3\phi}(t) = V_p I_p \{3\cos\phi - [\cos(2\omega t - \phi) + \cos(2\omega t - 240 - \phi)$$
$$+ \cos(2\omega t + 240 - \phi)]\}$$

Note that the last three terms in the above equation are the reactive power terms and they add up to zero. Thus we obtain

$$p_{3\phi}(t) = 3 V_p I_p \cos\phi \tag{2.23}$$

When referring to the voltage level of a three-phase system, one invariably understands the line voltages. From the above discussion the relationship between the line and phase voltages in a Y-connected system is

$$|V_L| = \sqrt{3}|V|$$

The power equation thus reads in terms of line quantities:

$$p_{3\phi} = \sqrt{3}|V_L||I_L|\cos\phi \tag{2.24}$$

We note that the total instantaneous power is constant, having a magnitude of three times the real power per phase. We may be tempted to assume that the reactive power is of no importance in a three-phase system since the Q terms cancel out. However, this situation is analogous to the summation of balanced three-phase currents and voltages that also cancel out. Although the sum cancels out, these quantities are still very much in evidence in each phase. We thus extend the concept of complex or apparent power (S) to three-phase systems by defining

$$S_{3\phi} = 3V_p I_p^* \tag{2.25}$$

where the active power and reactive power are obtained from

$$S_{3\phi} = P_{3\phi} + jQ_{3\phi}$$

as

$$P_{3\phi} = 3|V_p||I_p|\cos\phi \tag{2.26}$$

$$Q_{3\phi} = 3|V_p||I_p|\sin\phi \tag{2.27}$$

In terms of line values, we can assert that

$$S_{3\phi} = \sqrt{3}\, V_L I_L^* \tag{2.28}$$

and

$$P_{3\phi} = \sqrt{3}|V_L||I_L|\cos\phi \tag{2.29}$$

$$Q_{3\phi} = \sqrt{3}|V_L||I_L|\sin\phi \tag{2.30}$$

In specifying rated values for power system apparatus and equipment such as generators, transformers, circuit breakers, etc., we use the magnitude of the apparent power $S_{3\phi}$ as well as line voltage for specification values. In specifying three-phase motor loads, we use the horsepower output rating and voltage. To convert from horsepower to watts, recall that

$$1 \text{ hp} = 746 \text{ W}$$

Since the horsepower is a mechanical output, the electrical input will be somewhat higher due to the losses in the energy conversion process (more on this in Chapter 5). The efficiency of a process η is defined by

$$\eta = \frac{P_{\text{out}}}{P_{\text{in}}}$$

The value of efficiency should be used when converting a mechanical load to an equivalent electrical representation.

Example 2.2

A Y-connected, balanced three-phase load consisting of three impedances of $10/30°$ ohms each as shown in Figure 2-11 is supplied with the balanced line-to-neutral voltages:

$$V_{an} = 220/0 \text{ V}$$

$$V_{bn} = 220/240° \text{ V}$$

$$V_{cn} = 220/120° \text{ V}$$

A. Calculate the phasor currents in each line.
B. Calculate the line-to-line phasor voltages.
C. Calculate the total active and reactive power supplied to the load.

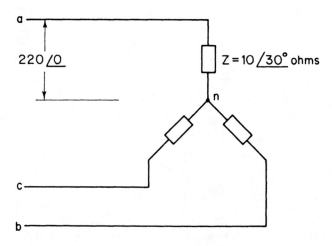

Figure 2-11. Load Connection for Example 2.2.

Solution

A. The phase currents are obtained as

$$I_{an} = \frac{220}{10\angle 30} = 22\angle -30° \text{ A}$$

$$I_{bn} = \frac{220\angle 240}{10\angle 30} = 22\angle 210° \text{ A}$$

$$I_{cn} = \frac{220\angle 120}{10\angle 30} = 22\angle 90° \text{ A}$$

B. The line-to-line voltages are obtained as

$$V_{ab} = V_{an} - V_{bn}$$
$$= 220\angle 0 - 220\angle 240°$$
$$= 220\sqrt{3} \angle 30°$$
$$V_{bc} = 220\sqrt{3}\angle 30 - 120 = 220\sqrt{3}\angle -90°$$
$$V_{ca} = 220\sqrt{3}\angle -210°$$

C. The apparent power into phase a is given by

$$S_a = V_{an} I_{an}^*$$
$$= (220)(22)\angle 30°$$
$$= 4840\angle 30° \text{ VA}$$

The total apparent power is three times the phase value:

$$S_t = 4840 \times 3\angle 30° = 14520.00\angle 30° \text{ VA}$$
$$= 12574.69 + j7260.00$$

Thus

$$P_t = 12574.69 \text{ W}$$
$$Q_t = 7260.00 \text{ var}$$

Example 2.3

Repeat Example 2.2 as if the same three impedances were connected in a Δ connection.

Solution

From Example 2.2 we have

$$V_{ab} = 220\sqrt{3}\,\underline{/30°}$$
$$V_{bc} = 220\sqrt{3}\,\underline{/-90°}$$
$$V_{ca} = 220\sqrt{3}\,\underline{/-210°}$$

The currents in each of the impedances are

$$I_{ab} = \frac{220\sqrt{3}\,\underline{/30°}}{10\,\underline{/30}} = 22\sqrt{3}\,\underline{/0°}$$

$$I_{bc} = 22\sqrt{3}\,\underline{/-120°}$$
$$I_{ca} = 22\sqrt{3}\,\underline{/120°}$$

The line currents are obtained with reference to Figure 2-12 as

$$I_a = I_{ab} - I_{ca}$$
$$= 22\sqrt{3}\,\underline{/0} - 22\sqrt{3}\,\underline{/-120°}$$
$$= 66\,\underline{/30°}$$
$$I_b = I_{bc} - I_{ab}$$
$$= 66\,\underline{/-90°}$$
$$I_c = I_{ca} - I_{bc}$$
$$= 66\,\underline{/-210°}$$

The apparent power in the impedance between a and b is

$$S_{ab} = V_{ab}I_{ab}^*$$
$$= \left(220\sqrt{3}\,\underline{/30°}\right)\left(22\sqrt{3}\,\underline{/0}\right)$$
$$= 14520\,\underline{/30°}$$

The total three-phase power is then

$$S_t = 43560\,\underline{/30°}$$
$$= 37724.04 + j21780.00$$

As a result,

$$P_t = 37724.04 \text{ W}$$
$$Q_t = 21780.00 \text{ var}$$

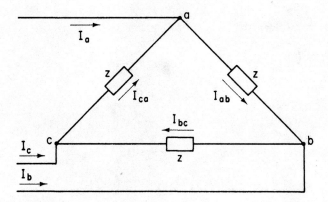

Figure 2-12. Load Connection for Example 2.3.

2.4 POWER SYSTEM REPRESENTATION

A major portion of the modern power system utilizes three-phase ac circuits and devices. It is clear that a detailed representation of each of the three phases in the system is cumbersome and can also obscure information about the system. A balanced three-phase system is solved as a single-phase circuit made of one line and the neutral return; thus a simpler representation would involve retaining one line to represent the three phases and omitting the neutral. Standard symbols are used to indicate the various components. A transmission line is represented by a single line between two ends. The simplified diagram is called the *single-line diagram*.

The one-line diagram summarizes the relevant information about the system for the particular problem studied. For example, relays and circuit breakers are not important when dealing with a normal state problem. However, when fault conditions are considered, the location of relays and circuit breakers is important and is thus included in the single-line diagram.

The International Electrotechnical Commission (IEC), the American National Standards Institute (ANSI), and the Institute of Electrical and

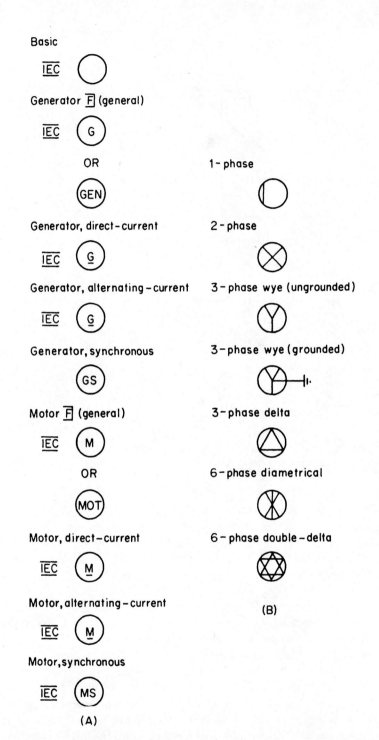

Figure 2-13. Symbols for Rotating Machines (A) and Their Winding Connections (B).

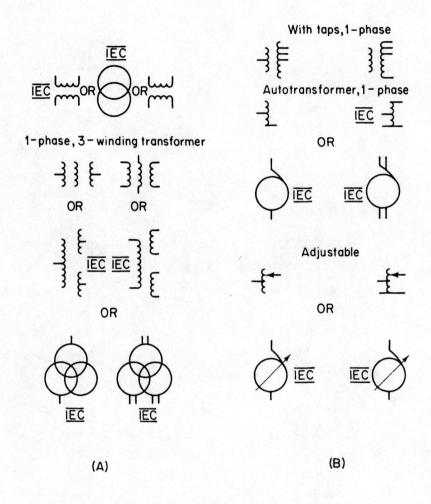

Figure 2-14. (A) Transformer Symbols. (B) Symbols for Single-Phase Transformers.

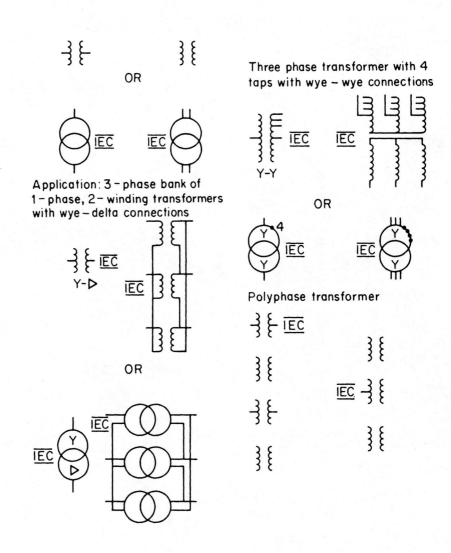

Figure 2-15. Symbols for Three-Phase Transformers.

Electronics Engineers (IEEE) have published a set of standard symbols for electrical diagrams. A basic symbol for a rotating machine is a circle. Figure 2-13(a) shows rotating machine symbols. If the winding connection is desired, the connection symbols may be shown in the basic circle using the representations given in Figure 2-13(b). The symbols commonly used for transformer representation are given in Figure 2-14(a). The two-circle symbol is the symbol to be used on schematics for equipment having international usage according to IEC. Figure 2-14(b) shows symbols for a number of single-phase transformers, and Figure 2-15 shows both single-line symbols and three-line symbols for three-phase transformers.

SOME SOLVED PROBLEMS

Problem 2-A-1

In the circuit shown in Figure 2-16, the source phasor voltage is $V = 30\underline{/15°}$. Determine the phasor currents I_2 and I_3 and the impedance Z_2. Assume that I_1 is equal to five A.

Solution

The voltage V_2 is given by

$$V_2 = V - (1)I_1$$
$$= 30\underline{/15°} - 5\underline{/0}$$
$$= 25.20\underline{/17.94°}$$

The current I_3 is thus

$$I_3 = \frac{V_2}{10} = 2.52\underline{/17.94°}$$

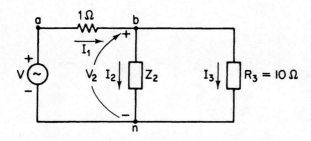

Figure 2-16. Circuit for Problem 2-A-1.

36 Some Basic Principles

The current I_2 is obtained using KCL. Thus
$$I_2 = I_1 - I_3$$
$$= 2.72\underline{/-16.61°}$$

Finally we have
$$Z_2 = \frac{V_2}{I_2} = 9.28\underline{/34.56°} = 7.64 + j5.26 \text{ ohms}$$

Problem 2-A-2

For the circuit of Problem 2-A-1, calculate the apparent power produced by the source and the individual apparent powers consumed by the 1-ohm resistor, the impedance Z_2, and the resistance R_3. Show that conservation of power holds true.

Solution

The apparent power produced by the source is
$$S_s = VI_1^* = (30\underline{/15°})(5\underline{/0})$$
$$= 150\underline{/15°} \text{ VA}$$
$$= 144.89 + j38.82 \text{ VA}$$

The apparent power taken by the 1-ohm resistor is
$$S_1 = V_{ab}I_1^* = 1(I_1)(I_1^*)$$
$$= |I_1|^2 = 25 \text{ W}$$

The apparent power taken by the impedance Z_2 is
$$S_2 = V_2 I_2^* = (25.20\underline{/17.94})(2.72\underline{/16.61°})$$
$$= 68.44\underline{/34.56°}$$
$$= 56.37 + j38.82 \text{ VA}$$

The apparent power taken by the resistor R_3 is
$$S_3 = V_2 I_3^* = (25.20\underline{/17.94})(2.52\underline{/-17.94})$$
$$= 63.52 \text{ W}$$

The total power consumed is
$$S_t = S_1 + S_2 + S_3 = 144.89 + j38.82 \text{ VA}$$

This is equal to the source apparent power, which proves the principle of conservation of power.

Problem 2-A-3

A three-phase transmission link is rated 100 kVA at 2300 V. When operating at rated load, the total resistive and reactive voltage drops in the link are, respectively, 2.4 and 3.6 percent of the rated voltage. Determine the input power and power factor when the link delivers 60 kW at 0.8 PF lagging at 2300 V.

Solution

The active voltage drop per phase is

$$\Delta V_r = IR = \frac{0.024(2300)}{\sqrt{3}}$$

The reactive drop is

$$\Delta V_x = IX = \frac{0.036(2300)}{\sqrt{3}}$$

But the rated current is

$$I = \frac{100 \times 10^3}{2300\sqrt{3}} = 25.1 \text{ A}$$

As a result,

$$R = 1.27 \text{ ohms}$$
$$X = 1.90 \text{ ohms}$$

For a load of 60 kW at 0.8 PF lagging, the phase current is

$$I_l = \frac{60 \times 10^3}{(2300\sqrt{3})(0.8)} = 18.83 \text{ A}$$

The active and reactive powers consumed by the link are thus

$$P_{lk} = 3I_l^2 R = 1350.43 \text{ W}$$
$$Q_{lk} = 3I_l^2 X = 2020.32 \text{ Var}$$

As a result, the apparent power consumed by the link is

$$S_{lk} = 1350.43 + j2020.32$$

The apparent load power is

$$S_l = \frac{60 \times 10^3}{0.8} \underline{/\cos^{-1} 0.8}$$
$$= 60,000 + j45,000 \text{ VA}$$

Thus the total apparent power is now obtained as

$$S_t = 61350.43 + j47020.32$$
$$= 77296.74 \underline{/37.47°}$$

Some Basic Principles

As a result

$$\cos\phi_t = \cos 37.47°$$
$$= 0.79 \text{ lagging}$$
$$P_t = 61.35043 \text{ kW}$$

Problem 2-A-4

A 60-hp, three-phase, 440-V induction motor operates at 0.8 PF lagging.

A. Find the active, reactive, and apparent power consumed per phase.

B. Suppose the motor is supplied from a 440-V source through a feeder whose impedance is $0.5 + j0.3$ ohm per phase. Calculate the voltage at the motor side, the source power factor, and the efficiency of transmission.

Solution

The active power is

$$P_r = 60 \times 746 = 44{,}760.00 \text{ W}$$

$$\cos\phi_r = 0.8$$

$$S_r = \frac{P_r}{\cos\phi_r}\underline{/\phi_r} = 55950\underline{/36.87°}$$

$$Q_r = S_r \sin\phi_r = 33570 \text{ V}_{ar}$$

Observe that we are given the sending end voltage V_s in magnitude value. Referring to a phasor diagram with I as reference, we can write for phase values,

$$|V_s|^2 = (V_r\cos\phi_r + IR)^2 + (V_r\sin\phi_r + IX)^2$$

A few manipulations yield

$$|V_s|^2 = |V_r|^2 + 2(P_{r\phi}R + Q_{r\phi}X) + |Z|^2\left(\frac{P_{r\phi}}{V_r\cos\phi_r}\right)^2$$

Thus substituting the given values, we get

$$\left(\frac{440}{\sqrt{3}}\right)^2 = |V_r|^2 + 2\left[\left(\frac{44{,}760}{3}\right)(0.5) + \left(\frac{33{,}570.00}{3}\right)(0.3)\right]$$

$$+ 0.34\left[\frac{44{,}760}{3(0.8)|V_r|}\right]^2$$

or
$$|V_r|^4 - 42899.33|V_r|^2 + 1.1826 \times 10^8 = 0$$

Solving the quadratic, we obtain
$$|V_r|^2 = 39938.27$$
or
$$|V_r| = 199.85 \text{ V}$$

We can now calculate the line current:
$$I = \frac{44{,}760}{\sqrt{3}\,(346.14)(0.8)} = 93.32 \text{ A}$$

The active and reactive power absorbed by the link is
$$S_l = 3(I^2R + jI^2X)$$
$$= 13062.93 + j7837.76 \text{ VA}$$

The sending end apparent power is thus
$$S_s = S_r + S_l$$
$$= 57822.87 + j41407.87$$
$$= 71.12028 \times 10^3 \underline{/35.61°}$$
$$P_s = 57.82287 \times 10^3 \text{ W}$$
$$\cos \phi_s = 0.81$$

The efficiency is thus obtained as
$$\eta = \frac{P_r}{P_s} = 0.7741$$

PROBLEMS

Problem 2-B-1

Find the phase currents I_A, I_B, and I_C as well as the neutral current I_n for the three-phase network with an unbalanced load as shown in Figure 2-17. Assume

$$V_{An} = 100/\underline{0}$$

$$V_{Bn} = 100/\underline{-120°}$$

$$V_{Cn} = 100/\underline{120°}$$

Problem 2-B-2

Calculate the apparent power consumed by the load of Problem 2-B-1.

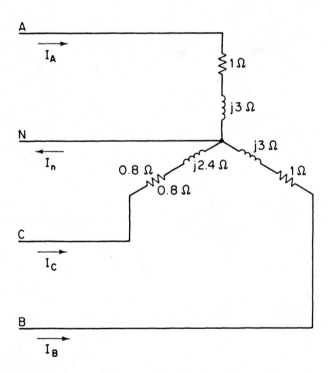

Figure 2-17. Three-Phase Load for Problem 2-B-1.

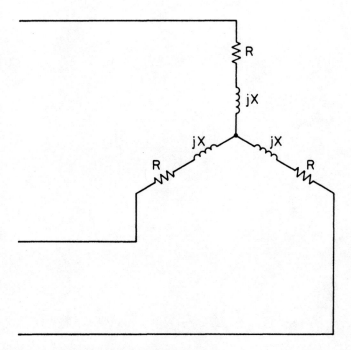

Figure 2-18. Circuit Model for Problem 2-B-3.

Problem 2-B-3

A 60-hp, three-phase, 440-V induction motor operates at 0.75 PF lagging. Find the active, reactive, and apparent power consumed per phase. Find the values of R and jX if the motor is modeled as shown in Figure 2-18.

Problem 2-B-4

Repeat Problem 2-B-3 if the motor's efficiency is 85%.

Problem 2-B-5

Repeat Problem 2-B-4 if the PF is 0.7 lagging.

CHAPTER III

Power Generation and the Synchronous Machine

3.1 INTRODUCTION

On October 28, 1831, Michael Faraday experimented with a revolving copper plate mounted on a horizontal brass axle and placed in the short gap between the pole pieces of a powerful magnet. Two contacts to a galvanometer were made—the first one on the periphery and the second one on the axle. Powerful currents were produced and lasted as long as the plate revolved. The news of Faraday's achievements quickly led to a large number of inventions. Notable among these was Benjamin G. Lamme's invention of the synchronous converter and the rotary condenser. Lamme also provided the electrical design of the 5000-hp generator, which inaugurated the hydroelectric power development at Niagara Falls in 1895. This signaled the dawn of the age of major electric power systems.

The daily functions of today's civilization depend on electric power, the bulk of which is produced by electric utility systems. The backbone of such a system is a number of generating stations operating in parallel. At each station there may be several synchronous generators operating in parallel. Synchronous machines represent the largest single-unit electric machine in production. Generators with power ratings of several hundred to over a thousand megavoltamperes (MVA) are fairly common in many utility systems. A synchronous machine provides a reliable and efficient means for energy conversion.

The operation of a synchronous generator is (like all other electromechanical energy conversion devices) based on Faraday's law of electromagnetic induction. The term *synchronous* refers to the fact that this type of machine operates at constant speed and frequency under steady-state conditions. Synchronous machines are equally capable of operating as motors, in which case the electric energy supplied at the armature terminals of the unit is converted into mechanical form.

For a meaningful study of electric power systems, we start by discussing the synchronous machine. Our attention will be limited to the fundamental models of the machine describing its steady-state balanced three-phase sinusoidal operational behavior. Analytical methods of examining the performance of polyphase synchronous machines will also be developed in this chapter.

3.2 THE SYNCHRONOUS MACHINE: PRELIMINARIES

The armature winding of a synchronous machine is on the stator, and the field winding is on the rotor as shown in Figure 3-1. The field is excited by the direct current that is conducted through carbon brushes bearing on slip (or collector) rings. The dc source is called the *exciter* and is often mounted on the same shaft as the synchronous machine. Various excitation systems with ac exciters and solid-state rectifiers are used with large turbine generators. The main advantages of these systems include the elimination of cooling and maintenance problems associated with slip rings, commutators, and brushes. The pole faces are shaped such that the radial distribution of the air-gap flux density B is approximately sinusoidal as shown in Figure 3-2.

The armature winding will include many coils. One coil is shown in Figure 3-1 and has two coil sides (a and $-a$) placed in diametrically opposite slots on the inner periphery of the stator with conductors parallel to the shaft of the machine. The rotor is turned at a constant speed by a mechanical power source connected to its shaft. As a result, the flux waveform sweeps by the coil sides a and $-a$. The induced voltage in the coil is a sinusoidal time function. It is evident that for each revolution of the

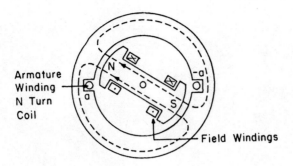

Figure 3-1. Simplified Sketch of a Synchronous Machine.

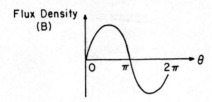

Figure 3-2. Space Distribution of Flux Density in a Synchronous Generator.

two poles, the coil voltage passes through a complete cycle of values. The frequency of the voltage in cycles per second (hertz) is the same as the rotor speed in revolutions per second. Thus a two-pole synchronous machine must revolve at 3600 r/min to produce a 60-Hz voltage.

P-Pole Machines

Many synchronous machines have more than two poles. A P-pole machine is one with P poles. As an example we consider an elementary single-phase four-pole generator shown in Figure 3-3. There are two complete cycles in the flux distribution around the periphery as shown in Figure 3-4. The armature winding in this case consists of two coils (a_1, $-a_1$, and a_2, $-a_2$) connected in series. The generated voltage goes through two complete cycles per revolution of the rotor, and thus the frequency f in hertz is twice the speed in revolutions per second. In general, the coil voltage of a machine with P-poles passes through a complete cycle every

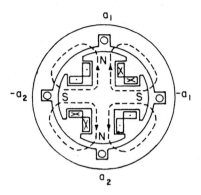

Figure 3-3. Four-Pole Synchronous Machine.

time a pair of poles sweeps by, or $P/2$ times for each revolution. The frequency f is therefore given by

$$f = \frac{P}{2}\left(\frac{n}{60}\right) \qquad (3.1)$$

where n is the shaft speed in revolutions per minute (r/min).

In treating P-pole synchronous machines, it is more convenient to express angles in electrical degrees rather than in the more familiar mechanical units. Here we conceptually concentrate on a single pair of poles and recognize that the conditions associated with any other pair are simply repetitions of those of the pair under consideration. A full cycle of generated

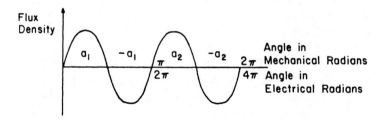

Figure 3-4. Space Distribution of Flux Density in a Four-Pole Synchronous Machine.

voltage will be described when the rotor of a four-pole machine has turned 180 mechanical degrees. This cycle represents 360 electrical degrees in the voltage wave. Extension of this argument to a P-pole machine leads to the conclusion that

$$\theta_e = \left(\frac{P}{2}\right)\theta_m$$

where θ_e and θ_m denote angles in electrical and mechanical degrees respectively.

Cylindrical vs. Salient-Pole Construction

Machines like the ones illustrated in Figures 3-1 and 3-3 have rotors with *salient* poles. There is another type of rotor, which is shown in Figure 3-5. The machine with such a rotor is called a *cylindrical rotor* or *nonsalient*-pole machine. The choice between the two designs (salient or nonsalient) for a specific application depends on the proposed prime mover. For hydroelectric generation, a salient-pole construction is employed. This is because hydraulic turbines run at relatively low speeds, and in this case a large number of poles is required to produce the desired frequency as indicated by Eq. (3.1). On the other hand, steam and gas turbines perform better at relatively high speeds, and two- or four-pole cylindrical rotor turboalternators are used in this case. This will avoid the use of protruding parts on the rotor, which at high speeds will give rise to dangerous mechanical stresses.

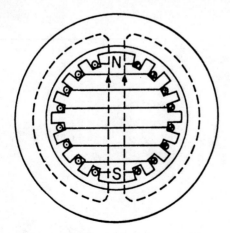

Figure 3-5. A Cylindrical Rotor Two-Pole Machine.

3.3 FIELDS IN A SYNCHRONOUS MACHINE

An understanding of the nature of the magnetic field produced by a polyphase winding is necessary for the analysis of polyphase ac machines. We will consider a two-pole, three-phase machine. The windings of the individual phases are displaced by 120 electrical degrees in space. This is shown in Figure 3-6. The magnetomotive forces developed in the air gap due to currents in the windings will also be displaced 120 electrical degrees in space. Assuming sinusoidal, balanced three-phase operation, the phase currents are displaced by 120 electrical degrees in time. The instantaneous values of the currents are

$$i_a = I_m \cos \omega t \tag{3.2}$$
$$i_b = I_m \cos(\omega t - 120°) \tag{3.3}$$
$$i_c = I_m \cos(\omega t - 240°) \tag{3.4}$$

where I_m is the maximum value of the current, and the time origin is arbitrarily taken as the instant when the phase a current is a positive maximum. The phase sequence is assumed to be abc. The instantaneous currents are shown in Figure 3-7.

The magnetomotive force (MMF) of each phase is proportional to the corresponding current. Accordingly, for the maximum current I_m, the time maximum of the MMF is

$$F_{\max} = K I_m$$

where K is a constant of proportionality that depends on the winding

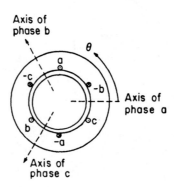

Figure 3-6. Simplified Two-Pole, Three-Phase Stator Winding.

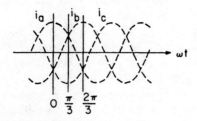

Figure 3-7. Instantaneous Three-Phase Currents.

distribution and the number of series turns in the winding per phase. We thus have

$$A_{a(p)} = F_{max}\cos \omega t \qquad (3.5)$$

$$A_{b(p)} = F_{max}\cos(\omega t - 120°) \qquad (3.6)$$

$$A_{c(p)} = F_{max}\cos(\omega t - 240°) \qquad (3.7)$$

where $A_{a(p)}$ is the amplitude of the MMF component wave at time t.

At time t, all three phases contribute to the air-gap MMF at a point P (whose spatial angle is θ). We thus have the resultant MMF given by

$$A_p = A_{a(p)}\cos\theta + A_{b(p)}\cos(\theta - 120°) + A_{c(p)}\cos(\theta - 240°) \qquad (3.8)$$

Using Eqs. (3.5) to (3.7), we have

$$A_p = F_{max}[\cos\theta\cos\omega t + \cos(\theta - 120°)\cos(\omega t - 120°) \\ + \cos(\theta - 240°)\cos(\omega t - 240°)] \qquad (3.9)$$

Equation (3.9) can be simplified using the following trigonometric identity:

$$\cos\alpha\cos\beta = \tfrac{1}{2}[\cos(\alpha - \beta) + \cos(\alpha + \beta)]$$

As a result, we have

$$A_p = \tfrac{1}{2}[\cos(\theta - \omega t) + \cos(\theta + \omega t) \\ + \cos(\theta - \omega t) + \cos(\theta + \omega t - 240°) \\ + \cos(\theta - \omega t) + \cos(\theta + \omega t - 480°)]F_{max}$$

The three cosine terms involving $(\theta + \omega t)$, $(\theta + \omega t - 240°)$, and $(\theta + \omega t - 480°)$ are three equal sinusoidal waves displaced in phase by 120° with a zero sum. Therefore

$$A_p = \tfrac{3}{2}[F_{max}\cos(\theta - \omega t)] \qquad (3.10)$$

The wave of Eq. (3.10) depends on the spatial position θ as well as time. The angle ωt provides rotation of the entire wave around the air gap at the constant angular velocity ω. At time t_1, the wave is a sinusoid with its positive peak displaced ωt_1 from the point P (at θ); at a later instant (t_2), the wave has its positive peak displaced ωt_2 from the same point. We thus see that a polyphase winding excited by balanced polyphase currents produces the same effect as a permanent magnet rotating within the stator.

The MMF wave created by the three-phase armature current in a synchronous machine is commonly called *armature-reaction* MMF. This MMF wave rotates at synchronous speed and is directly opposite to phase a at the instant when phase a has its maximum current ($t=0$). The dc field winding produces a sinusoid F with an axis 90° ahead of the axis of phase a in accordance with Faraday's law.

The resultant magnetic field in the machine is the sum of the two contributions from the field and armature reaction. Figure 3-8 shows a sketch of the armature and field windings of a cylindrical rotor generator. The space MMF produced by the field winding is shown by the sinusoid F. This is shown for the specific instant when the electromotive force (EMF) of phase a due to excitation has its maximum value. The time rate of change of flux linkages with phase a is a maximum under these conditions, and thus the axis of the field is 90° ahead of phase a. The armature-reaction wave is shown as the sinusoid A in the figure. This is drawn opposite phase a because at this instant both I_a and the EMF of the field E_f (also called excitation voltage) have their maximum value. The resultant magnetic field in the machine is denoted R and is obtained by graphically adding the F and A waves.

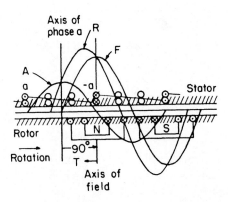

Figure 3-8. Spatial MMF Waves in a Cylindrical Rotor Synchronous Generator.

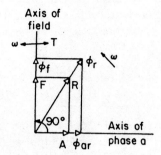

Figure 3-9. A Space Phasor Diagram for Armature Current in Phase with Excitation Voltage.

Sinusoids can conveniently be handled using phasor methods. We can thus perform the addition of the A and F waves using phasor notation.

Figure 3-9 shows a space phasor diagram where the fluxes ϕ_f (due to the field), ϕ_{ar} (due to armature reaction), and ϕ_r (the resultant flux) are represented. It is clear that under the assumption of a uniform air gap and no saturation, these are proportional to the MMF waves F, A, and R.

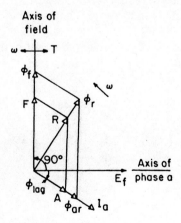

Figure 3-10. A Space Phasor Diagram for Armature Current Lagging the Excitation Voltage.

respectively. The figure is drawn for the case when the armature current is in phase with the excitation voltage. The situation for the case when the armature current lags the excitation voltage E_f is shown in Figure 3-10.

3.4 A SIMPLE EQUIVALENT CIRCUIT

The operation of a synchronous machine with cylindrical rotor can be conveniently analyzed if the effect of the armature-reaction flux is represented by an inductive reactance. The basis for this is shown in Figure 3-11, where the phasor diagram of component fluxes and corresponding voltages is given. The field flux ϕ_f is added to the armature-reaction flux ϕ_{ar} to yield the resultant air-gap flux ϕ_r. The armature-reaction flux ϕ_{ar} is in phase with the armature current I_a. The excitation voltage E_f is generated by the field flux, and E_f lags ϕ_f by 90°. Similarly, E_{ar} and E_r are generated by ϕ_{ar} and ϕ_r respectively, with each of the voltages lagging the flux causing it by 90°.

If we introduce the constant of proportionality x_ϕ that relates the rms values of E_{ar} and I_a, we can write

$$E_{ar} = -jx_\phi I_a \qquad (3.12)$$

where the $-j$ underscores the 90° lagging effect. We therefore have

$$E_r = E_f - jx_\phi I_a \qquad (3.13)$$

An equivalent circuit based on Eq. (3.13) is given in Figure 3-12. We thus conclude that the inductive reactance x_ϕ accounts for the armature-reaction effects. This reactance is known as the *magnetizing reactance* of the machine.

The terminal voltage of the machine denoted by V_t is obtained as the difference between the air-gap voltage E_r and the voltage drops in the

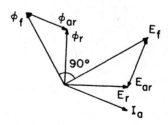

Figure 3-11. Phasor Diagram for Fluxes and Resulting Voltages in a Synchronous Machine.

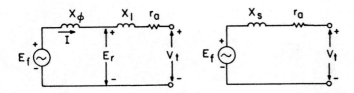

Figure 3-12. Two Equivalent Circuits for the Synchronous Machine.

armature resistance r_a, and the leakage-reactance x_l. Here x_l accounts for the effects of leakage flux as well as space harmonic field effects not accounted for by x_ϕ. A simple impedance commonly known as the *synchronous impedance* Z_s results from combining x_ϕ, x_l, and r_a according to

$$Z_s = r_a + jX_s \tag{3.14}$$

The synchronous reactance X_s is clearly given by

$$X_s = x_l + x_\phi \tag{3.15}$$

It is emphasized here that the above model applies to an unsaturated cylindrical rotor machine supplying balanced polyphase currents to its load. Our voltage relationship is now given by

$$E_f = V_t + I_a Z_s \tag{3.16}$$

Example 3-1

A 5-kVA, 220-V, 60-Hz, six-pole, Y-connected synchronous generator has a leakage reactance per phase of 0.78 ohms and negligible armature resistance. The armature-reaction EMF for this machine is related to the armature current by

$$E_{ar} = -j16.88(I_a)$$

Assume that the generated EMF is related to field current by

$$E_f = 25 I_f$$

A. Compute the field current required to establish rated voltage across the terminals of a unity power factor load that draws rated generator armature current.

B. Determine the field current needed to provide rated terminal voltage to a load that draws 125 percent of rated current at 0.8 PF lagging.

Solution

The rated current is given by

$$I_a = \frac{5 \times 10^3}{\sqrt{3} \times 220} = 13.12 \text{ A}$$

The phase value of terminal voltage is

$$V_t = \frac{220}{\sqrt{3}} = 127.02 \text{ V}$$

With reference to the equivalent circuit of Figure 3-12, we have

A.
$$E_r = V_t + jI_a x_l$$
$$= 127.02\underline{/0} + (13.12\underline{/0})(0.78\underline{/90°})$$
$$= 127.43\underline{/4.61°}$$
$$E_{ar} = -j(16.88)(13.12) = 221.47\underline{/-90°}$$

The required field excitation voltage E_f is therefore,

$$E_f = E_r - E_{ar}$$
$$= 127.43\underline{/4.61°} - 221.47\underline{/-90°}$$
$$= 264.24\underline{/61.27°} \text{ V}$$

Consequently, using the given field voltage versus current relation,

$$I_f = \frac{E_f}{25} = 10.5696 \text{A}$$

B. With conditions given, we have

$$I_a = 13.12 \times 1.25 = 16.40 \text{ A}$$
$$E_r = 127.02\underline{/0} + (16.4\underline{/-36.87°})(0.78\underline{/90°})$$
$$= 135.08\underline{/4.34°} \text{ V}$$
$$E_{ar} = -j(16.88)(16.40\underline{/-36.87°})$$
$$= 276.83\underline{/-126.87°} \text{ V}$$
$$E_f = E_r - E_{ar}$$
$$= 135.08\underline{/4.34°} - 276.83\underline{/-126.87°}$$
$$= 379.69\underline{/37.61°} \text{ V}$$

We therefore calculate the required field current as

$$I_f = \frac{379.69}{25} = 15.19 \text{ A}$$

3.5 OPEN-CIRCUIT AND SHORT-CIRCUIT CHARACTERISTICS

Open-circuit and short-circuit test results on synchronous generators are of importance to appropriately account for saturation effects as well as for the determination of machine constants. The open-circuit characteristic is a curve of the armature terminal voltage on open circuit as a function of the field excitation with the machine running at synchronous speed. An experimental setup for the test is shown in Figure 3-13.

As the test name implies, in a short-circuit test the armature terminals of the synchronous machine are short-circuited through ammeters, and the field current is gradually increased until a maximum safe value is reached. This is shown in Figure 3-14. An open-circuit characteristic and a short-circuit characteristic are shown in Figure 3-15. The short-circuit armature current is directly proportional to the field current up to almost 150 percent of rated armature current.

Determination of Synchronous Reactance

The synchronous reactance X_s can be determined on the basis of Eq. (3.16) provided that terminal voltage V_t, generated voltage E_f, and the corresponding current phasor are available. We thus have the synchronous

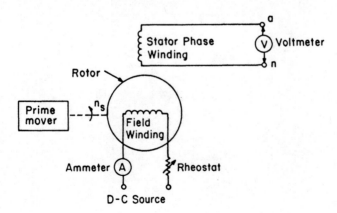

Figure 3-13. Experimental Setup for the Open-Circuit Characteristic of a Synchronous Machine.

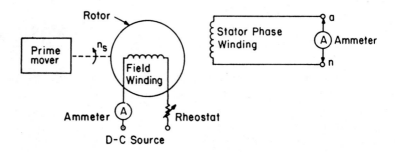

Figure 3-14. Experimental Setup for Determining the Short-Circuit Characteristics of a Synchronous Machine.

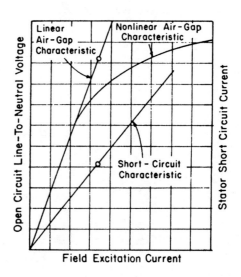

Figure 3-15. Typical Open-Circuit and Short-Circuit Characteristics of a Synchronous Machine.

3.5 Open-Circuit and Short-Circuit Characteristics

impedance:

$$Z_s = \frac{E_f - V_t}{I_a} \quad (3.17)$$

The imaginary part of Z_s is the required synchronous reactance.

A conveniently fast alternate method makes use of the short-circuit and open-circuit characteristics of the machine. To understand this, recall that for a short-circuit condition,

$$V_t = 0$$

Consequently, Eq. (3.17) yields, for this condition,

$$Z_s = \frac{E_f}{I_{a_{sc}}}$$

If we assume, as is usual, that the armature resistance is negligible, then we assert that

$$X_s = \frac{E_f}{I_{a_{sc}}} \quad (3.18)$$

The unsaturated value of the synchronous reactance can be obtained using E_f from the air-gap line in Figure 3-15—corresponding to $I_{a_{sc}}$. For operation near rated terminal voltage, we assume that the machine is equivalent to an unsaturated one with a straight-line magnetization curve through the origin and the rated voltage point on the open-circuit characteristic as shown by the dashed line in Figure 3-16. Accordingly,

$$X_s = \frac{V_t}{I'_{a_{sc}}}$$

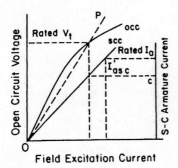

Figure 3-16. Defining the Synchronous Reactance.

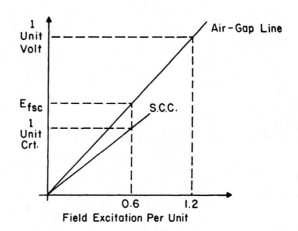

Figure 3-17. Geometry of Characteristics for Example 3-2.

Example 3-2

The short-circuit characteristic of a synchronous generator is such that rated armature current is obtained by 0.6 per-unit* (p.u.) excitation. Rated voltage on the air-gap line is obtained by 1.2 per-unit excitation. Find the value of the unsaturated synchronous reactance.

Solution

The geometry of the problem is shown in Figure 3-17. From similarity of triangles, we deduce that

$$\frac{E_{f_{sc}}}{0.6} = \frac{1}{1.2}$$

Consequently,

$$E_{f_{sc}} = 0.5 \text{ p.u.}$$

From which we calculate

$$X_s = \frac{E_{f_{sc}}}{I_a} = \frac{0.5}{1} = 0.5 \text{ p.u.}$$

---

*The numerical per-unit (p.u.) value of a quantity is its ratio to a chosen base quantity of the same dimension. This topic will be discussed in detail in Chapter 6.

3.6 PRINCIPAL STEADY-STATE CHARACTERISTICS

We now discuss some operating performance characteristics of importance in practical applications of the synchronous machine. These describe the interrelations among terminal voltage, field current, armature current, and power factor.

We begin by considering a synchronous generator delivering power to a constant power factor load at a constant frequency. A *compounding curve* shows the variation of the field current required to maintain rated terminal voltage with the load. Typical compounding curves for various power factors are shown in Figure 3-18. The computation of points on the curve follows easily as a consequence of applying Eq. (3.16). Figure 3-19 shows phasor diagram representations for three different power factors. The following example illustrates the ideas involved.

Example 3-3

A 9375-kVA, three-phase, Y-connected, 13,800-V (line-to-line), two-pole, 60-Hz turbine generator has an armature resistance of 0.064 ohms per phase and a synchronous reactance of 1.79 ohms per phase. Find the full load generated voltage per phase at:

A. Unity power factor.
B. A power factor of 0.8 lagging.
C. A power factor of 0.8 leading.

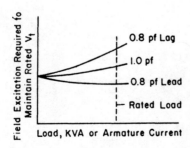

Figure 3-18. Synchronous-Machine Compounding Curves.

60 Power Generation and the Synchronous Machine

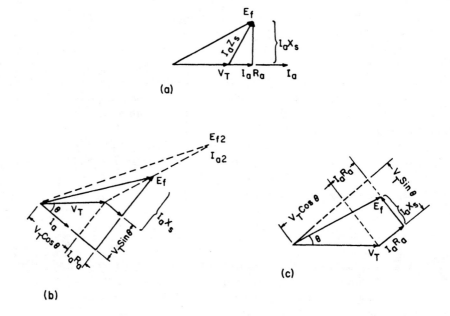

Figure 3-19. Phasor Diagrams for a Synchronous Machine Operating at Different Power Factors are: a) Unity PF Loads, b) Lagging PF Loads, c) Leading PF Loads.

Solution

The magnitude of full load current is obtained as

$$I_a = \frac{9375 \times 10^3}{\sqrt{3} \times 13,800} = 392.22 \text{ A}$$

The terminal voltage per phase is taken as reference

$$V_t = \frac{13,800}{\sqrt{3}} = 7967.43 \underline{/0} \text{ V}$$

3.6 Principal Steady-State Characteristics

The synchronous impedance is obtained as

$$Z_s = r_a + jX_s$$
$$= 0.064 + j1.79$$
$$= 1.7911 \underline{/87.95°} \text{ ohms per phase}$$

The generated voltage per phase is obtained using Eq. (3.16) as follows:

A. For the unity power factor: $\phi = 0$.

$$I_a = 392.22 \underline{/0} \text{ A}$$
$$E_f = 7967.43 + (392.22 \underline{/0})(1.7911 \underline{/87.95°})$$
$$= 8023.31 \underline{/5.02°} \text{ V}$$

B. For a power factor of 0.8 lagging: $\phi = -36.87°$.

$$I_a = 392.22 \underline{/-36.87°} \text{ A}$$
$$E_f = 7967.43 + (392.22 \underline{/-36.87°})(1.7911 \underline{/87.95°})$$
$$= 8426.51 \underline{/3.72°} \text{ V}$$

C. For a power factor of 0.8 leading: $\phi = +36.87°$.

$$I_a = 392.22 \underline{/36.87°} \text{ A}$$
$$E_f = 7967.43 + (392.22 \underline{/36.87°})(1.7911 \underline{/87.95°})$$
$$= 7588.22 \underline{/4.36°} \text{ V}$$

In Figure 3-20 we show the variation of the machine terminal voltage with armature current for different power factors and fixed field current. Each of the curves is given for a field current corresponding to producing full load (rated) armature current at rated terminal voltage.

Important characteristics of the synchronous machine are given by the reactive-capability curves. These give the maximum reactive power loadings corresponding to various active power loadings for rated voltage operation. Armature heating constraints govern the machine for power factors from rated to unity. Field heating represents the constraints for lower power factors. Figure 3-21 shows a typical set of curves for a large turbine generator.

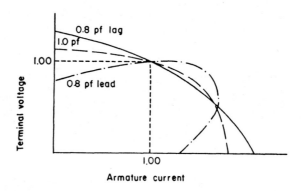

Figure 3-20. Constant Field Current Voltampere Characteristic of Synchronous Machine.

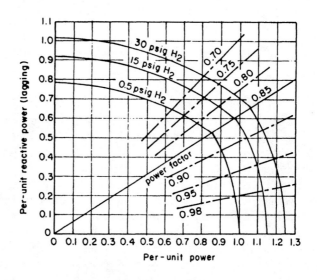

Figure 3-21. Generator Reactive-Capability Curves.

3.7 POWER-ANGLE CHARACTERISTICS AND THE INFINITE BUS CONCEPT

Consider the simple circuit shown in Figure 3-22. The impedance Z connects the sending end, whose voltage is E and receiving end, with voltage V. Let us assume that in polar form we have

$$E = E\underline{/\delta}$$
$$V = V\underline{/0}$$
$$Z = Z\underline{/\psi}$$

We therefore conclude that the current I is given by

$$I = \frac{E - V}{Z}$$

This reduces to

$$I = \frac{E}{Z}\underline{/\delta - \psi} - \frac{V}{Z}\underline{/-\psi} \qquad (3.19)$$

The complex power S_1 at the sending end is given by

$$S_1^* = E^*I$$

Similarly, the complex power S_2 at the receiving end is

$$S_2^* = V^*I$$

Using Eq. (3.19), we thus have

$$S_1^* = \frac{E^2}{Z}\underline{/-\psi} - \frac{EV}{Z}\underline{/-\psi - \delta} \qquad (3.20)$$

$$S_2^* = \frac{EV}{Z}\underline{/\delta - \psi} - \frac{V^2}{Z}\underline{/-\psi} \qquad (3.21)$$

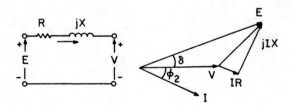

Figure 3-22. Equivalent Circuit and Phasor Diagram for a Simple Link.

Recall that
$$S^* = P - jQ$$
We thus have the power equations:
$$P_1 = \frac{E^2}{Z}\cos(-\psi) - \frac{EV}{Z}\cos(\psi+\delta) \tag{3.22}$$
$$Q_1 = \frac{E^2}{Z}\sin(\psi) - \frac{EV}{Z}\sin(\psi+\delta) \tag{3.23}$$
$$P_2 = \frac{EV}{Z}\cos(\delta-\psi) - \frac{V^2}{Z}\cos\psi \tag{3.24}$$
$$Q_2 = \frac{EV}{Z}\sin(\psi-\delta) - \frac{V^2}{Z}\sin\psi \tag{3.25}$$

An important case is when the resistance is negligible; then
$$\psi = 90° \tag{3.26}$$
$$Z = X \tag{3.27}$$

Here we have Eqs. (3.22)–(3.25) reducing to
$$P_1 = P_2 = \frac{EV}{X}\sin\delta \tag{3.28}$$
$$Q_1 = \frac{E^2 - EV\cos\delta}{X} \tag{3.29}$$
$$Q_2 = \frac{EV\cos\delta - V^2}{X} \tag{3.30}$$

In large-scale power systems, a three-phase synchronous machine is paralleled through an equivalent system reactance (X_e) to the network which has a high generating capacity relative to any single unit. We often refer to the network or system as an *infinite bus* when a change in input mechanical power or in field excitation to the unit does not cause an appreciable change in system frequency or terminal voltage. Figure 3-23 shows such a situation, where V is the infinite bus voltage.

The foregoing analysis shows that in the present case we have for power transfer,
$$P = P_{max}\sin\delta \tag{3.31}$$
with
$$P_{max} = \frac{EV}{X_t} \tag{3.32}$$
and
$$X_t = X_s + X_e \tag{3.33}$$

It is clear that if an attempt were made to advance δ further than 90° (corresponding to maximum power transfer) by increasing the mechanical

3.7 Power-Angle Characteristics and the Infinite Bus Concept

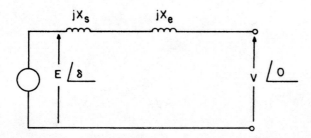

Figure 3-23. A Synchronous Machine Connected to an Infinite Bus.

power input, the electrical power output would decrease from the P_{max} point. Therefore the angle δ increases further as the machine accelerates. This drives the machine and system apart electrically. The value P_{max} is called the *steady-state stability limit* or *pull-out power*. We will consider the following example, which illustrates the utility of the above.

Example 3-4

A synchronous generator with a synchronous reactance of 1.3 p.u. is connected to an infinite bus whose voltage is one p.u. through an equivalent reactance of 0.2 p.u. The maximum permissible output is 1.2 p.u.

A. Compute the excitation voltage E.
B. The power output is gradually reduced to 0.7 p.u. with fixed field excitation. Find the new current and power angle δ.

Solution

A. The total reactance is
$$X_t = 1.3 + 0.2 = 1.5$$

Thus we have
$$1.2 = \frac{EV}{X_t}$$
$$= \frac{(E)(1)}{1.5}$$

Therefore,
$$E = 1.8 \text{ p.u.}$$

B. We have for any angle δ,
$$P = P_{max} \sin \delta$$

Therefore,
$$0.7 = 1.2 \sin \delta$$
This results in
$$\delta = 35.69°$$
The current is
$$I = \frac{E - V}{jX_t}$$
Substituting the given values, we obtain
$$I = \frac{1.8 \underline{/35.69°} - 1.0}{j1.5}$$
$$= 0.7648 \underline{/-23.75°} \text{ A}$$

Reactive Power Generation

Inspection of Eq. (3.30) reveals that the generator produces reactive power ($Q_2 > 0$) if
$$E \cos \delta > V$$
In this case the generator appears to the network as a capacitor. This condition applies for high magnitude E, and the machine is said to be overexcited. On the other hand, the machine is underexcited if it consumes reactive power ($Q_2 < 0$). Here we have
$$E \cos \delta < V$$
Figure 3-24 shows phasor diagrams for both cases. The overexcited synchronous machine is normally employed to provide synchronous condenser action, where usually no real load is carried by the machine ($\delta = 0$). In this case we have
$$Q_2 = \frac{V(E - V)}{X} \qquad (3.34)$$
Control of reactive power generation is carried out by simply changing E, by varying the dc excitation. An example will help underline the use of these concepts.

Example 3-5

Compute the reactive power generated by the machine of Example 3-4 under the conditions in part (b). If the machine is required to generate a reactive power of 0.4 p.u. while supplying the same active power by changing the field excitation, find the new excitation voltage and power angle δ.

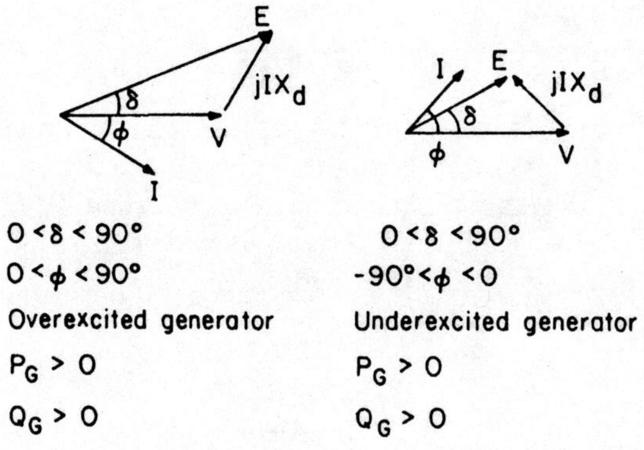

3.7 Power-Angle Characteristics and the Infinite Bus Concept

$0 < \delta < 90°$
$0 < \phi < 90°$
Overexcited generator
$P_G > 0$
$Q_G > 0$

$0 < \delta < 90°$
$-90° < \phi < 0$
Underexcited generator
$P_G > 0$
$Q_G > 0$

Figure 3-24. Phasor Diagrams for Overexcited and Underexcited Synchronous Machines.

Solution

The reactive power generated is obtained according to Eq. (3.30) as

$$Q_2 = \frac{1(1.8\cos 35.69 - 1)}{1.5} = 0.308$$

With a new excitation voltage and stated active and reactive powers, we have using Eq. (3.28) and (3.30)

$$0.7 = \frac{(E)(1)}{(1.5)}\sin\delta$$

$$0.4 = \frac{1(E\cos\delta - 1)}{1.5}$$

We thus obtain

$$\tan\delta = \frac{(1.5)(0.7)}{(1.6)}$$

$$\delta = 33.27°$$

From the above we get

$$E = \frac{(1.5)(0.7)}{\sin(33.27)} = 1.91$$

3.8 STATIC STABILITY LIMIT CURVES

Let us consider a machine with synchronous reactance X_s connected to an infinite bus of voltage V through the reactance X_e as shown in Figure 3-25. We wish to determine the relationship between P, Q, V_t, and δ as V and E are allowed to vary. The active power P and reactive power Q are obtained on the basis of Eqs. (3.28) and (3.30):

$$P = \frac{EV_t \sin(\delta - \theta)}{X_s} \qquad (3.35)$$

$$Q = \frac{EV_t \cos(\delta - \theta) - V_t^2}{X_s} \qquad (3.36)$$

In order to eliminate E from the above relationships, we note that two additional expressions for P can be obtained since active power losses are not present. The first is based on the transfer from node 1 to node 3 across the total reactance X_t. The second is based on the transfer between nodes 2 and 3 across the reactance X_e. We therefore have

$$P = \frac{EV \sin \delta}{X_t} = \frac{V_t V \sin \theta}{X_e} \qquad (3.37)$$

where

$$X_t = X_s + X_e$$

Thus we conclude that

$$E = V_t \frac{X_t \sin \theta}{X_e \sin \delta} \qquad (3.38)$$

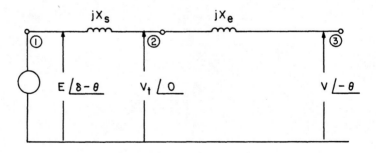

Figure 3-25. Equivalent Circuit for a Synchronous Machine Connected through an External Network to an Infinite Bus.

3.8 Static Stability Limit Curves

Using Eq. (3.38) in Eqs. (3.35) and (3.36), we get

$$P = \frac{V_t^2 X_t}{X_s X_e \sin \delta}[\sin\theta \sin(\delta - \theta)] \qquad (3.39)$$

$$Q + \frac{V_t^2}{X_s} = \frac{V_t^2 X_t}{X_s X_e \sin \delta}[\sin\theta \cos(\delta - \theta)] \qquad (3.40)$$

To eliminate θ we will recall the following two trigonometric identities:

$$\sin\alpha \sin\beta = \tfrac{1}{2}[\cos(\alpha - \beta) - \cos(\alpha + \beta)] \qquad (3.41)$$
$$\sin\alpha \cos\beta = \tfrac{1}{2}[\sin(\alpha + \beta) + \sin(\alpha - \beta)] \qquad (3.42)$$

Applying Eq. (3.41) to Eq. (3.39), we obtain

$$P = \frac{V_t^2 X_t}{2 X_s X_e \sin \delta}[\cos(\delta - 2\theta) - \cos\delta]$$

Moreover, Eq. (3.42) applied to Eq. (3.40) gives

$$Q + \frac{V_t^2}{X_s} = \frac{V_t^2 X_t}{2 X_s X_e \sin \delta}[\sin\delta - \sin(\delta - 2\theta)]$$

Rearranging we get

$$P + \frac{V_t^2 X_t}{2 X_s X_e \tan \delta} = \frac{V_t^2 X_t}{2 X_s X_e \sin \delta}\cos(\delta - 2\theta) \qquad (3.43)$$

$$Q - \frac{V_t^2 (X_s - X_e)}{2 X_s X_e} = \frac{-V_t^2 X_t}{2 X_s X_e \sin \delta}\sin(\delta - 2\theta) \qquad (3.44)$$

Squaring both sides of Eqs. (3.43) and (3.44), and adding, we obtain the desired result:

$$\left(P + \frac{V_t^2 X_t}{2 X_s X_e \tan \delta}\right)^2 + \left[Q - \frac{V_t^2(X_s - X_e)}{2 X_s X_e}\right]^2$$
$$= \left(\frac{V_t^2 X_t}{2 X_s X_e \sin \delta}\right)^2 \qquad (3.45)$$

Equation (3.45) indicates that the locus of P and Q delivered by the machine is a circle with center at (P_0, Q_0) and radius R, where

$$P_0 = \frac{-V_t^2 X_t}{2 X_s X_e \tan \delta} \qquad (3.46)$$

$$Q_0 = \frac{V_t^2}{2}\left(\frac{1}{X_e} - \frac{1}{X_s}\right) \qquad (3.47)$$

$$R_0 = \frac{V_t^2}{2 \sin \delta}\left(\frac{1}{X_s} + \frac{1}{X_e}\right) \qquad (3.48)$$

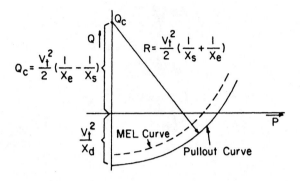

Figure 3-26. Static Stability Limit Curve for a Synchronous Machine.

The static stability limit curve for the machine is obtained from Eq. (3.45) by setting δ to 90°. Hence we have

$$P^2 + \left[Q - \frac{V_t^2}{2}\left(\frac{1}{X_e} - \frac{1}{X_s}\right)\right]^2 = \left[\frac{V_t^2}{2}\left(\frac{1}{X_s} + \frac{1}{X_e}\right)\right]^2 \qquad (3.49)$$

The static stability limit curve is commonly referred to as the *pull-out curve* of P and Q and will determine the minimum permissible output var for output watts and terminal voltage specifications. Figure 3.26 shows such a curve.

Example 3-6

Given a generator and a system with reactances of $X_s = 1.2$ and $X_e = 0.2$, both on a 100-MVA base. Assume a generator terminal voltage of 0.95 p.u. Infinite bus voltage is unknown. Find the minimum permissible output var for the p.u. output watts varying from zero to one in steps of 0.25.

Solution

The given parameters specify the static stability limit curve equation as

$$P^2 + (Q - 1.88)^2 = 6.93$$

For $P = 0$, we have

$$Q = -0.752$$

Similarly, we get

$$\text{For } P = 0.25, \quad Q = -0.740$$
$$\text{For } P = 0.5, \quad Q = -0.704$$
$$\text{For } P = 0.75, \quad Q = -0.643$$
$$\text{For } P = 1.00, \quad Q = -0.555$$

3.9 ACCOUNTING FOR SALIENCY

The presence of protruding field poles in a salient-pole machine introduces nonuniformity of the magnetic reluctance of the air gap. The reluctance along the polar axis is appreciably less than that along the interpolar axis. We often refer to the polar axis as the *direct axis* and the interpolar as the *quadrature axis*. This effect can be taken into account by resolving the armature current I_a into two components, one in time phase and the other in time quadrature with the excitation voltage as shown in Figure 3-27. The component I_d of the armature current is along the direct axis (the axis of the field poles), and the component I_q is along the quadrature axis.

Let us consider the effect of the direct-axis component alone. With I_d lagging the excitation EMF E_f by 90°, the resulting armature-reaction flux ϕ_{ad} is directly opposite the field poles as shown in Figure 3-28. The effect of the quadrature-axis component is to produce an armature-reaction flux ϕ_{aq},

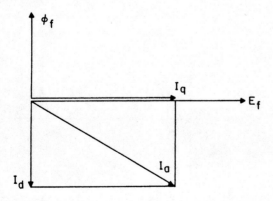

Figure 3-27. Resolution of Armature Current in Two Components.

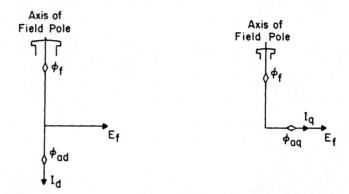

Figure 3-28. Direct-Axis and Quadrature-Axis Air-Gap Fluxes in a Salient-Pole Synchronous Machine.

which is in the quadrature-axis direction as shown in Figure 3-28. The phasor diagram with both components present is shown in Figure 3-29.

We recall that in the cylindrical rotor machine case, we employed the synchronous reactance x_s to account for the armature-reaction EMF in an equivalent circuit. The same argument can be extended to the salient-pole case. With each of the components currents I_d and I_q, we associate component synchronous-reactance voltage drops, $jI_d x_d$ and $jI_q x_q$ respectively. The direct-axis synchronous reactance x_d and the quadrature-axis synchronous

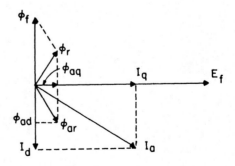

Figure 3-29. Phasor Diagram for a Salient-Pole Synchronous Machine.

reactance x_q are given by

$$x_d = x_l + x_{\phi d}$$
$$x_q = x_l + x_{\phi q}$$

where x_l is the armature leakage reactance and is assumed to be the same for direct-axis and quadrature-axis currents. The direct-axis and quadrature-axis magnetizing reactances $x_{\phi d}$ and $x_{\phi q}$ account for the inductive effects of the respective armature-reaction flux. Figure 3-30 shows a phasor diagram implementing the result.

$$E_f = V_t + I_a r_a + jI_d x_d + jI_q x_q \tag{3.50}$$

In many instances, the power factor angle Φ at the machine terminals is explicitly known rather than the internal power factor angle $(\phi + \delta)$, which is required for the resolution of I_a into its direct-axis and quadrature-axis components. We can circumvent this difficulty by recalling that in phasor notation,

$$I_a = I_q + I_d \tag{3.51}$$

Substitution of Eq. (3.51) into Eq. (3.50) for I_q and rearranging, we obtain

$$E_f = V_t + I_a(r_a + jx_q) + jI_d(x_d - x_q) \tag{3.52}$$

Let us define

$$E_f' = V_t + I_a(r_a + jx_q) \tag{3.53}$$

E_f' as defined is in the same direction as E_f since jI_d is also along the same direction. Our procedure then is to obtain E_f' as given by Eq. (3.53) and then obtain the component I_d based on the phase angle of E_f'. Finally, we find E_f as a result of

$$E_f = E_f' + jI_d(x_d - x_q) \tag{3.54}$$

This is shown in Figure 3-31. An example is taken up at this point to illustrate the procedure.

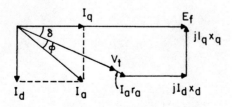

Figure 3-30. Phasor Diagram for a Synchronous Machine.

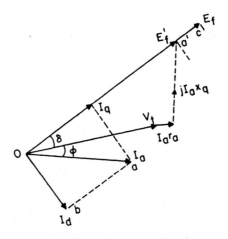

Figure 3-31. A Modified Phasor Diagram for a Salient-Pole Synchronous Machine.

Example 3-7

A 5-kVA, 220-V, Y-connected, three-phase, salient-pole synchronous generator is used to supply power to a unity PF load. The direct-axis synchronous reactance is 12 ohms and the quadrature-axis synchronous reactance is 7 ohms. Assume that rated current is delivered to the load at rated voltage and that armature resistance is negligible. Compute the excitation voltage and the power angle.

Solution

$$V_t = 127.02 \text{ V}$$

$$I_a = \frac{5 \times 10^3}{220\sqrt{3}} = 13.12 \text{ A}.$$

We calculate

$$E'_f = V_t + jI_a x_q$$
$$= 127.02 + j(13.12)(7) = 156.75 \underline{/35.87°}$$

Moreover,

$$I_d = I_a \sin 35.87 = 7.69 \text{ A}$$
$$|E_f| = |E'_f| + |I_d(x_d - x_q)|$$
$$= 156.75 + 7.69(12 - 7) = 195.20 \text{ V}$$
$$\delta = 35.87°$$

Figure 3.32 pertains to this example.

3.10 Salient-Pole Machine Power Angle Characteristics

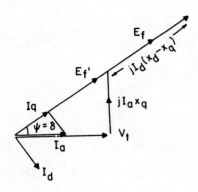

Figure 3-32. Phasor Diagram for Example 3-7.

TABLE 3-1
Typical per-Unit Values of Machine Reactances

	2-Pole Turbine Generators	4-Pole Turbine Generators	Salient-Pole Generators	Condensers (air-cooled)
x_d	0.85–1.45	1.00–1.45	0.6–1.5	1.25–2.20
x_q	0.92–1.42	0.92–1.42	0.4–0.8	0.95–1.3

Note: Machine kVA rating as base.

Table 3-1 gives typical ranges for values of x_d and x_q for synchronous machines. Note that x_q is less than x_d because of the greater reluctance of the air gap in the quadrature axis.

3.10 SALIENT-POLE MACHINE POWER ANGLE CHARACTERISTICS

The power angle characteristics for a salient-pole machine connected to an infinite bus of voltage V through a series reactance of x_e can be arrived at by considering the phasor diagram shown in Figure 3-33. The active power delivered to the bus is

$$P = (I_d \sin \delta + I_q \cos \delta) V \tag{3.55}$$

Similarly, the delivered reactive power Q is

$$Q = (I_d \cos \delta - I_q \sin \delta) V \tag{3.56}$$

To eliminate I_d and I_q, we need the following identities obtained from

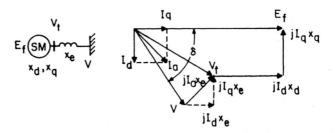

Figure 3-33. A Salient-Pole Machine Connected to an Infinite Bus through an External Impedance.

inspection of the phasor diagram:

$$I_d = \frac{E_f - V\cos\delta}{X_d} \tag{3.57}$$

$$I_q = \frac{V\sin\delta}{X_q} \tag{3.58}$$

where

$$X_d = x_d + x_e \tag{3.59}$$

$$X_q = x_q + x_e \tag{3.60}$$

Substitution of Eqs. (3.57) and (3.58) into Eqs. (3.55) and (3.56) yields

$$P = \frac{VE_f}{X_d}\sin\delta + \frac{V^2}{2}\left(\frac{1}{X_q} - \frac{1}{X_d}\right)\sin 2\delta \tag{3.61}$$

$$Q = \frac{VE_f}{X_d}\cos\delta - V^2\left(\frac{\cos^2\delta}{X_d} + \frac{\sin^2\delta}{X_q}\right) \tag{3.62}$$

Equations (3.61) and (3.62) contain six quantities—the two variables P and δ and the four parameters E_f, V, X_d, and X_q—and can be written in many different ways. The following form illustrates the effect of saliency. Define P_d and Q_d as

$$P_d = \frac{VE_f}{X_d}\sin\delta \tag{3.63}$$

and

$$Q_d = \frac{VE_f}{X_d}\cos\delta - \frac{V^2}{X_d} \tag{3.64}$$

The above equations give the active and reactive power generated by a

3.10 Salient-Pole Machine Power Angle Characteristics

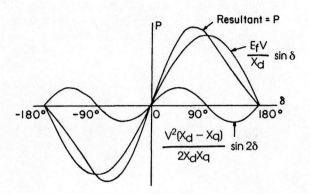

Figure 3-34. Power Angle Characteristics of a Salient-Pole Synchronous Machine.

round rotor machine with synchronous reactance X_d. We thus have

$$P = P_d + \frac{V^2}{2}\left(\frac{1}{X_q} - \frac{1}{X_d}\right)\sin 2\delta \qquad (3.65)$$

$$Q = Q_d - V^2\left(\frac{1}{X_q} - \frac{1}{X_d}\right)\sin^2\delta \qquad (3.66)$$

The second term in the above two equations introduces the effect of salient poles, and in the power equation the term corresponds to reluctance torque. Note that if $X_d = X_q$, as in a uniform air-gap machine, the second terms in both equations are zero. Figure 3-34 shows the power angle characteristics of a typical salient-pole machine.

The pull-out power and power angle δ for the salient-pole machine can be obtained by solving equation (3.67) requiring the partial derivative of P with respect to δ to be equal to zero.

$$\frac{\partial P}{\partial \delta} = 0 \qquad (3.67)$$

The actual value of pull-out power can be shown to be higher than that obtained assuming nonsaliency.

Example 3-8

A salient-pole synchronous machine is connected to an infinite bus through a link with reactance of 0.2 p.u. The direct-axis and quadrature-axis reactances of the machine are 0.9 and 0.65 p.u. respectively. The excitation voltage is 1.3 p.u., and the voltage of the infinite bus is maintained at 1 p.u. For a power angle of 30°, compute the active and reactive power supplied to the bus.

Solution

The active power formula is given by Eq. (3.61). Our data are used to calculate X_d and X_q as

$$X_d = x_d + x_e = 0.9 + 0.2 = 1.1$$
$$X_q = x_q + x_e = 0.65 + 0.2 = 0.85$$

Therefore,

$$P = \frac{(1.3)(1)}{1.1} \sin 30° + \frac{1}{2}\left(\frac{1}{0.85} - \frac{1}{1.1}\right) \sin 60°$$
$$= 0.7067 \text{ p.u.}$$

Similarly, the reactive power is obtained using Eq. (3.62) as:

$$Q = \frac{(1.3)(1)}{1.1} \cos 30° - \left(\frac{\cos^2 30°}{1.1} + \frac{\sin^2 30°}{0.85}\right)$$
$$= 0.0475 \text{ p.u.}$$

Example 3-9

A synchronous machine is supplied from a constant-voltage source. At no-load, the motor armature current is found to be negligible when the excitation is 1.0 per unit. The per-unit motor constants are $x_d = 1.0$ and $x_q = 0.6$.

   A. If the machine loses synchronism when the angle between the quadrature axis and the terminal voltage phasor direction is 60 electrical degrees, what is the per-unit excitation at pull-out?
   B. What is the load on the machine at pull-out? Assume the same excitations as in part (a).

Solution

$$P = \frac{VE_f}{X_d} \sin \delta + \frac{V^2}{2 X_d X_q}(X_d - X_q) \sin 2\delta$$
$$= \frac{E_f(1)}{1} \sin \delta + \frac{1}{2 \times 0.6}(1 - 0.6) \sin 2\delta$$
$$= E_f \sin \delta + \frac{1}{3} \sin 2\delta$$

For pull-out power we have

$$\frac{\partial P}{\partial \delta} = E_f \cos \delta + \frac{2}{3} \cos 2\delta = 0$$

The pull-out angle is

$$\delta_m = 60°$$

Hence we obtain

$$E_f = \frac{2}{3} \text{ p.u.}$$

Consequently the pull-out load is found to be

$$P = \frac{2}{3} \sin 60° + \frac{1}{3} \sin 120° = 0.866 \text{ p.u.}$$

SOME SOLVED PROBLEMS

Problem 3-A-1

A 180-kVA, 440-V, 300 r/min, 60-Hz, three-phase, Y-connected cylindrical rotor synchronous generator has the following particulars: $r_a = 0$ and $x_l = 0.296$ ohms. The air-gap line is described by $E = 17 I_f$, expressed per phase. The short-circuit characteristic is described by $I_{sc} = 10.75 I_f$.

A. Find the value of the unsaturated synchronous reactance.
B. Find the value of the armature reaction reactance.
C. Find the value of the field current needed to yield rated terminal voltage at rated current for a 0.8 lagging PF load.

Solution

A. Using the air-gap line and short-circuit characteristics given, we obtain

$$X_s = \frac{E_{f sc}}{I_{a\,sc}} = \frac{17 I_f}{10.75 I_f} = 1.58 \text{ ohms}$$

B. The armature reaction reactance is obtained as

$$X_\phi = X_s - X_l = 1.58 - 0.296$$
$$= 1.285 \text{ ohms}$$

C.
$$E_f = V_t + j I_a X_s$$
$$= \frac{440}{\sqrt{3}} + (236.19 \underline{/-36.87})(1.58 \underline{/90°}) = 563.8 \underline{/32°} \text{ V}$$

Therefore,
$$I_f = \frac{563.8}{17} = 33.17 \text{ A}$$

Problem 3-A-2

A 40,000-kVA, 14,000-V, Y-connected alternator has negligible armature resistance. Some pertinent data are as follows:

- Short-circuit characteristic: $I_a = 7I_f$
- Air-gap line, volts per phase: $E = 33I_f$
- Open-circuit characteristic volts per phase:

$$E = \frac{21{,}300 I_f}{430 + I_f}$$

Find both the unsaturated and saturated synchronous reactances for this machine.

Solution

The unsaturated synchronous reactance is obtained as

$$X_s = \frac{33}{7} = 4.71 \text{ ohms per phase}$$

To calculate the saturated synchronous reactance we need the field current corresponding to rated terminal voltage from the open-circuit characteristics. This is obtained as the solution to

$$\frac{14{,}000}{\sqrt{3}} = \frac{21{,}300 I_f}{430 + I_f}$$

The result is

$$I_f = 262.966 \text{ A}$$

From the short-circuit characteristic,

$$I_{a_{sc}} = 7I_f = 1840.763 \text{ A}$$

We thus conclude that

$$X_s = \frac{\frac{14{,}000}{\sqrt{3}}}{1840.763} = 4.39 \text{ ohms per phase}$$

Problem 3-A-3

For the machine of Problem 3-A-1 it is required to compute the field current needed to provide rated voltage when rated current is delivered to the load at:

A. A PF of 0.8 leading.
B. A PF of 0.8 lagging.

Assume E_{ar} at rated current corresponds to 305 equivalent field volts.

Solution

$$I_{rated} = \frac{180 \times 10^3}{\sqrt{3} \times 440} = 236.19 \text{ A}$$

A. For a load with 0.8 PF leading, we have

$$E_r = V_t + I_a(jx_l)$$
$$= \frac{440}{\sqrt{3}} + (236.19 \underline{/36.87°})(0.296 \underline{/90°})$$
$$= 219.34 \underline{/14.77°}$$

E_{ar} lags I_a by 90°. Therefore,

$$E_{ar} = 305 \underline{/-53.13°}$$
$$E_f = E_r - E_{ar}$$
$$= 219.34 \underline{/14.77°} - 305 \underline{/-53.13°}$$
$$= 301.33 \underline{/84.46°} \text{ V}$$
$$I_f = \frac{301.33}{17} = 17.73 \text{ A}$$

B. For a load with 0.8 PF lagging, we have

$$E_r = \frac{440}{\sqrt{3}} + (236.19 \underline{/-36.87°})(0.296 \underline{/90°})$$
$$= 301.22 \underline{/10.70°}$$
$$E_{ar} = 305 \underline{/-126.87°}$$
$$E_f = E_r - E_{ar}$$
$$= 301.22 \underline{/10.70°} - 305 \underline{/-126.87°}$$
$$= 565.14 \underline{/32.05°}$$
$$I_f = \frac{565.14}{17} = 33.24 \text{ A}$$

Problem 3-A-4

The synchronous reactance of a cylindrical rotor machine is 1.2 p.u. The machine is connected to an infinite bus whose voltage is 1 p.u. through an equivalent reactance of 0.3 p.u. For a power output of 0.7 p.u., the power angle is found to be 30°.

A. Find the excitation voltage E_f and the pull-out power.
B. For the same power output the power angle is to be reduced to 25°. Find the value of the reduced equivalent reactance connecting the machine to the bus to achieve this. What would be the new pull-out power?

Solution

A.
$$X_t = 1.5 \text{ p.u.}$$
$$P = P_{max} \sin \delta$$
$$0.7 = P_{max} \sin(30°)$$

Hence
$$P_{max} = 1.4 \text{ p.u.}$$

Thus
$$1.4 = \frac{(E_f)(1)}{1.5}$$
$$E_f = 2.1$$

B.
$$0.7 = P_{max} \sin(25°)$$
$$P_{max} = 1.66$$
$$1.66 = \frac{E_f V}{X_{new}} = \frac{(2.1)(1)}{X_{new}}$$
$$X_{new} = \frac{2.1}{1.66} = 1.27 \text{ p.u.}$$
$$x_e = 1.27 - 1.2 = 0.07 \text{ p.u.}$$

Problem 3-A-5

A cylindrical rotor machine is supplying a load of 0.8 PF lagging at an infinite bus. The ratio of the excitation voltage to the infinite bus voltage is found to be 1.25. Compute the power angle δ.

Solution

For a power factor of 0.8 we have

$$\frac{Q}{P} = \tan\phi = 0.75$$

Using the active and reactive power formulae, (3.28) and (3.30) we have

$$\frac{Q}{P} = \frac{\cos\delta - \frac{V}{E}}{\sin\delta}$$

$$0.75 = \frac{\cos\delta - 0.8}{\sin\delta}$$

Cross-multiplying we have

$$0.8 + 0.75\sin\delta = \cos\delta$$

Using

$$\cos^2\delta = 1 - \sin^2\delta$$

We get

$$\left[(0.75)^2 + 1\right]\sin^2\delta + (2)(0.8)(0.75)\sin\delta + \left[(0.8)^2 - 1\right] = 0$$

Consequently,

$$\sin\delta = 0.23$$
$$\delta = 13.34°$$

Problem 3-A-6

The apparent power delivered by a cylindrical rotor synchronous machine to an infinite bus is 1.2 p.u. The excitation voltage is 1.3 p.u. and the power angle is 20°. Compute the synchronous reactance of the machine, given that the infinite bus voltage is 1 p.u.

Solution

We have that the apparent power S is given by

$$S^2 = P^2 + Q^2$$

Using the formulae of active and reactive power (Eq. 3.28 and 3.30) we have by squaring and adding,

$$S^2 = \left(\frac{V^2}{X^2}\right)(E^2 + V^2 - 2EV\cos\delta)$$

$$(1.2)^2 = \left(\frac{1}{X^2}\right)[1.69 + 1 - (2)(1.3)(1)\cos 20°]$$

From which

$$X = 0.414 \text{ p.u.}$$

Problem 3-A-7

The synchronous reactance of a cylindrical rotor machine is 0.9 p.u. The machine is connected to an infinite bus through two parallel identical transmission links with reactance of 0.6 p.u. each. The excitation voltage is 1.5 p.u., and the machine is supplying a load of 0.8 p.u.

A. Compute the power angle δ for the given conditions.

B. If one link is opened with the excitation voltage maintained at 1.5 p.u., find the new power angle to supply the same load as in part (a).

Solution

A. The reactance of the two lines in parallel is

$$X_{e_1} = \frac{(0.6)(0.6)}{0.6+0.6} = 0.3 \text{ p.u.}$$

The total reactance between the source and bus is thus

$$X_{t_1} = X_s + X_{e_1} = 0.9 + 0.3 = 1.2 \text{ p.u.}$$

The power angle is found from

$$P = \frac{E_f V}{X_{t_1}} \sin \delta_1$$

$$0.8 = \frac{(1.5)(1)}{1.2} \sin \delta_1$$

$$\delta_1 = 39.79°$$

B. With only one line in service,

$$X_{e_2} = 0.6$$

$$X_{t_2} = X_s + X_{e_2} = 0.9 + 0.6 = 1.5 \text{ p.u.}$$

The new power angle is found from

$$0.8 = \frac{(1.5)(1)}{1.5} \sin \delta_2$$

$$\delta_2 = 53.13°$$

We note that the power angle is increased with the opening of one line.

Problem 3-A-8

A salient pole machine supplies a load of 1.2 p.u. at unity power factor to an infinite bus whose voltage is maintained at 1.05 p.u. The machine excitation voltage is computed to be 1.4 p.u. when the power angle is 25°. Evaluate the direct-axis and quadrature-axis synchronous reactances.

Solution

The active and reactive power equations for the salient-pole machine are given by Eq. (3.61) and (3.62). Using the given data we obtain

$$1.2 = \frac{(1.4)(1.05)}{X_d}\sin 25° + \frac{(1.05)^2}{2}\left(\frac{1}{X_q} - \frac{1}{X_d}\right)\sin 50°$$

$$0 = \frac{(1.4)(1.05)}{X_d}\cos 25° - (1.05)^2\left(\frac{\cos^2 25°}{X_d} + \frac{\sin^2 25}{X_q}\right)$$

These reduce to

$$0.19897\left(\frac{1}{X_d}\right) + 0.4223\left(\frac{1}{X_q}\right) = 1.2$$

$$0.4267\left(\frac{1}{X_d}\right) - 0.1969\left(\frac{1}{X_q}\right) = 0$$

Solving the above two equations we obtain

$$\frac{1}{X_d} = 1.077 \quad \text{and} \quad \frac{1}{X_q} = 2.334$$

The required reactances are thus obtained as

$$X_d = 0.9284 \quad \text{and} \quad X_q = 0.4284$$

PROBLEMS

Problem 3-B-1

A 10 MVA, 13.8 kV, 60 Hz, two-pole, Y connected three phase alternator has an armature winding resistance of 0.07 ohms per phase and a leakage reactance of 1.9 ohms per phase. The armature reaction EMF for the machine is related to the armature current by

$$E_{ar} = -j19.91 I_a$$

Assume that the generated EMF is related to the field current by

$$E_f = 60 I_f$$

A. Compute the field current required to establish rated voltage across the terminals of a load when rated armature current is delivered at 0.8 PF lagging.

B. Compute the field current needed to provide rated terminal voltage to a load that draws 100 per cent of rated current at 0.85 PF lagging.

Problem 3-B-2

A 5 kVA, 220-V, 60 Hz, six pole, Y connected synchronous generator has a leakage reactance per phase of 0.8 ohms and negligible armature resistance. The air-gap line is described by $E = 18I_f$ and the short circuit characteristic is described by $I_{sc} = 6I_f$.

A. Find the value of the unsaturated synchronous reactance.
B. Find the value of armature reaction reactance.
C. Find the value of the field current needed to yield rated terminal voltage at rated current for a unity PF load.

Problem 3-B-3

A 9375 kVA, 13.800 kV, 60 Hz, two pole, Y-connected synchronous generator is delivering rated current at rated voltage and unity PF. Find the armature resistance and synchronous reactance given that the field excitation voltage is 11935.44 V and leads the terminal voltage by an angle 47.96°.

Problem 3-B-4

The magnitude of the field excitation voltage for the generator of Problem (3-B-3) is maintained constant at the value specified above. Find the terminal voltage when the generator is delivering rated current at 0.8 PF lagging.

Problem 3-B-5

A 1,250 kVA, three-phase, Y-connected, 4160 V, ten-pole, 60 Hz synchronous generator has an armature resistance of 0.126 ohms per phase and a synchronous reactance of 3 ohms per phase. Find the full load generated voltage per phase at 0.8 PF lagging.

Problem 3-B-6

A 180 kVA, three-phase, Y-connected, 440 V, 60 Hz synchronous generator has a synchronous reactance of 1.6 ohms and a negligible armature resistance. Find the full load generated voltage per phase at 0.8 PF lagging.

Problem 3-B-7

For the generator of (3-B-2); find the maximum power output if the terminal voltage and field excitation voltage are kept constant at the values defined in Problem 3-B-2. If the power output is gradually reduced to 3 kW with fixed field excitation find the new current and power angle δ.

Problem 3-B-8

Repeat Problem 3-B-7 for the generator of Problem 3-B-6 with power output reduced to 150 kW.

Problem 3-B-9

For the generator of Problems 3-B-7 compute the reactive power generated by the machine. If the machine is required to generate a reactive power of 600 Var while supplying the same active power by changing the field excitation, find the new excitation voltage and power angle.

Problem 3-B-10

Repeat Problem 3-B-9 for the generator operating under the conditions of Problem 3-B-8. Assume the new reactive power to be 90 kVar.

Problem 3-B-11

The synchronous reactance of a cylindrical rotor synchronous generator is 0.9 p.u. If the machine is delivering active power of 1 p.u. to an infinite bus whose voltage is 1 p.u. at unity PF, calculate the excitation voltage and the power angle.

Problem 3-B-12

A cylindrical rotor machine is delivering active power of 0.8 p.u. and reactive power of 0.6 p.u. at a terminal voltage of 1 p.u. If the power angle is 20°, compute the excitation voltage and the machine's synchronous reactance.

Problem 3-B-13

A cylindrical rotor machine is delivering active power of 0.8 p.u. and reactive power of 0.6 p.u. when the excitation voltage is 1.2 p.u. and the pwer angle is 25°. Find the terminal voltage and synchronous reactance of the machine.

Problem 3-B-14

The synchronous reactance of a cylindrical rotor machine is 0.8 p.u. The machine is connected to an infinite bus through two parallel identical transmission links with reactance of 0.4 p.u. each. The excitation voltage is 1.4 p.u. and the machine is supplying a load of 0.8 p.u.

A. Compute the power angle δ for the outlined conditions.
B. If one link is opened with the excitation voltage maintained at 1.4 p.u. Find the new power angle to supply the same load as in "a".

Problem 3-B-15

The synchronous reactance of a cylindrical rotor generator is 1 p.u. and its terminal voltage is 1 p.u. when connected to an infinite bus through a reactance 0.4 p.u. Find the minimum permissible output vars for zero output active power and unity output active power.

Problem 3-B-16

The reactances x_d and x_q of a salient-pole synchronous generator are 0.95 and 0.7 per unit, respectively. The armature resistance is negligible. The generator delivers rated kVA at unity PF and rated terminal voltage. Calculate the excitation voltage.

Problem 3-B-17

The machine of problem (3-B-16) is connected to an infinite bus through a link with reactance of 0.2 p.u. The excitation voltage is 1.3 p.u. and the infinite bus voltage is maintained at 1 p.u. For a power angle of 25°, compute the active and reactive power supplied to the bus.

Problem 3-B-18

The reactances x_d and x_q of a salient-pole synchronous generator are 1.00 and 0.6 per unit respectively. The excitation voltage is 1.77 p.u. and the infinite bus voltage is maintained at 1 p.u. For a power angle of 19.4°, compute the active and reactive power supplied to the bus.

Problem 3-B-19

For the machine of Problem 3-B-18, assume that the active power supplied to the bus is 0.8 p.u. compute the power angle and the reactive power supplied to the bus. (Hint: assume $\cos \delta \simeq 1$ for an approximation)

Problem 3-B-20

For the machine of Problem 3-B-18, assume that the reactive power supplied to the bus is 0.6 p.u. Compute the power angle and the active power supplied to the bus.

Problem 3-B-21

A salient pole machine supplies a load of 1.2 p.u. at unity PF to an infinite bus. The direct axis and quadrature axis synchronous reactances are

$$x_d = 0.9283 \qquad x_q = 0.4284$$

The power angle δ is 25°. Evaluate the excitation and terminal voltages.

Problem 3-B-22

The condition for pull-out for a salient-pole machine is given by Eq. (3-67), with P as given in Eq. (3-65). Show that the angle δ at which pull-out occurs satisfies

$$\cos^2\delta + A\cos\delta - 0.5 = 0$$

with

$$A = E_f X_q / 2V[X_d - X_q]$$

Show also that if $\dfrac{2}{A^2} < 1$, then approximately we have

$$\cos\delta_m \simeq \frac{V[X_d - X_q]}{E_f X_q}$$

and that

$$P_{max} = \left(\frac{VE_f}{X_d}\right) + \left(\frac{V^3}{E_f}\right)\left(\frac{1}{X_d}\right)\left(\frac{X_d}{X_q} - 1\right)^2$$

Problem 3-B-23

For the machine of Problem (3-B-18), find the exact pull-out angle and power. Can we apply the approximate formulae of Problem 3-B-22 to this machine?

Problem 3-B-24

A salient-pole machine has the following particulars

$$X_d = 1.1 \quad X_q = 0.9$$
$$E_f = 1.2 \quad V = 1$$

Calculate the exact value of the pull-out angle and compare with the approximate result using the expression of Problem 3-B-22.

CHAPTER IV

The Transmission Subsystem

4.1 INTRODUCTION

The electric energy produced at generating stations is transported over high-voltage transmission lines to utilization points. In the early days (until 1917), electric systems were operated as isolated systems with only point-to-point transmission at voltages that are considered low by today's standards. Operating voltages increased rapidly from the 3300-V level used in the Willamette-Portland line (1890) to the 11-kV level used to transmit nearly 10 MW from Niagara Falls to Buffalo, N.Y., 20 miles away, in 1896. Two 287-kV circuits were completed in 1936 to transmit a block of 240 MW over a distance of 266 miles from the Hoover Dam across the desert to the outskirts of Los Angeles. The first 345-kV line grew out of a test program by the American Electric Power (AEP) system that started in 1946. This line was completed in 1953, and it ushered in the beginning of a 345-kV system that AEP placed to overlay its extensive 138-kV transmission. During the same period the Swedish State Power Board established a 400-kV system between its northern hydroplants and its southern load centers, which was placed in operation in 1952.

The 345-kV system established the principle of the use of bundle conductors, the V-configuration of insulator strings (to restrain swings), and the use of aluminum in line structures.

The first 500-kV line was energized in 1964 to tie a minemouth station in West Virginia to load centers in the eastern part of the state. One reason for the preference of this voltage level over the 345-kV level was that upgrading from 230-kV to 345-kV represented a gain of only 140 percent compared to a 400 percent gain when using the 500-kV level. Hydro Quebec inaugurated its 735-kV, 375-mile line in the same year. A line voltage of 765 kV was introduced into service by AEP in 1969. The 1980s witnessed the introduction of higher voltage levels in the Bonneville Power Administration's (BPA) 1100-kV transmission system.

The trend toward higher voltages is mainly motivated by the resulting increased line capacity while reducing line losses per unit of power transmitted. The reduction in losses is significant and is an important aspect of energy conservation. Better use of land is a benefit of the larger capacity. This can be illustrated by a comparison of the right-of-way width of 56 m required for an 1100-kV line with a capacity of 10,000 MW, to 76 m required for two double-circuit 500-kV lines to transmit the same capacity of 10,000 MW.

The purpose of this chapter is to develop a fundamental understanding of transmission line modeling and performance analysis. This is done for the major configurations in service. We begin by discussing the parameters of a transmission line.

4.2 ELECTRIC TRANSMISSION LINE PARAMETERS

An electric transmission line is modeled using four parameters that affect its performance characteristics. The four parameters are the series resistance, series inductance, shunt capacitance, and shunt conductance. The line resistance and inductive reactance are of importance in many problems of interest. For some studies it is possible to omit the shunt capacitance and conductance and thus simplify the equivalent circuit considerably.

We deal here with aspects of determining these parameters on the basis of line length, type of conductor used, and the spacing of the conductors as they are mounted on the supporting structure. We will start by a discussion of the nature of conductors and introduce some common terminology.

A wire or combination of wires not insulated from one another is called a *conductor*. A stranded conductor is composed of a group of wires, usually twisted or braided together.

TABLE 4-1
American Wire Gage versus Diameters in Mils

AWG	4/0	3/0	2/0	1/0	1	2	3	4	5	6
Diameter, mils	460	409.6	364.8	324.9	289.3	257.6	229.4	204.3	181.9	162.0
AWG	7	8	9	10	20	30	40	50		
Diameter, mils	144.3	128.5	114.4	101.9	32.0	10.0	3.1	1.0		

Wire sizes have been indicated commercially in terms of gage numbers for many years. Present practice calls for specifying wire sizes in terms of their diameters in mils (unit of length, 1/1000th of an inch). The cross-sectional area is given in circular mils. A circular mil is the area of a circle of 1 mil in diameter. This circle has an area of $(\pi/4)(1)$ mil^2 or 0.7854 mil^2. The American wire gage, usually abbreviated AWG, is based on a simple geometric progression. The diameter of No. 0000 is defined as 0.46 in., and of No. 36 as 0.005 in. There are 38 sizes between these two; thus the ratio of any diameter to the diameter of the next greater number is

$$n_A = \left(\frac{0.46}{0.005}\right)^{1/39} = 1.1229322$$

Observing that $n_A^6 = 2.005$ leads us to conclude that the diameter is doubled for a difference of six gage numbers. Table 4-1 gives a selection of AWG versus conductor diameters in mils for reference purposes. For conductors of sizes larger than 4/0, circular mils are used in North American practice.

In a concentrically stranded conductor, each successive layer contains six more wires than the preceding one. There are two basic constructions: the one-wire core and the three-wire core. The total number of wires (N) in a conductor with n layers over the core is given by

$$N = 3n(n+1) + 1 \quad \text{for 1-wire core}$$
$$N = 3n(n+2) + 3 \quad \text{for 3-wire core}$$

The wire size d in a stranded conductor with total conductor area A circular mils and N wires is

$$d = \left(\frac{A}{N}\right)^{1/2} \text{ mils}$$

Types of Conductors and Conductor Materials

Phase conductors in EHV-UHV transmission systems employ aluminum conductors and aluminum or steel conductors for overhead ground wires. Many types of cables are available. These include:

A. Aluminum Conductors
   There are five designs in common use:
   1. Homogeneous designs: These are denoted as All-Aluminum-Conductors (AAC) or All-Aluminum-Alloy Conductors (AAAC).
   2. Composite designs: These are essentially aluminum-conductor-steel-reinforced conductors (ACSR) with steel core material.
   3. Expanded ACSR: These use solid aluminum strand with a steel core. Expansion is by open helices of aluminum wire, flexible concentric tubes, or combinations of aluminum wires and fibrous ropes.
   4. Aluminum-clad conductor (Alumoweld).
   5. Aluminum-coated conductors.

B. Steel Conductors
   Galvanized steel conductors with various thicknesses of zinc coatings are used.

Line Resistance

The resistance of the conductor is the most important cause of power loss in a power line. Direct-current resistance is given by the familiar formula:

$$R_{dc} = \frac{\rho l}{A} \text{ ohms}$$

where

ρ = resistivity of conductor
l = length
A = cross-sectional area

Any consistent set of units may be used in the calculation of resistance. In the SI system of units, ρ is expressed in ohm-meters, length in meters, and area in square meters. A system commonly used by power systems engineers expresses resistivity in ohms circular mils per foot, length in feet, and area in circular mils.

Table 4-2 gives the value of ρ for several materials used in power systems networks. The resistance of the conductor is obtained at 20°C when ρ given in the table is used. The resistance of a conductor at any other temperature may be obtained from

$$R_2 = R_1[1 + \alpha(T_2 - T_1)]$$

Here R_2 is the resistance at temperature T_2, and R_1 is the resistance at

TABLE 4-2
Resistivity and Temperature Coefficient of Conductor Materials

Material	Resistivity (ρ) at 20°C		Temperature Coefficient (α) at 20°C
	Micro-ohm cm	Ohms circular mils per ft	
Aluminum	2.83	17.0	0.0039
Brass	6.4–8.4	38–51	0.002
Bronze	13–18	78–108	0.0005
Copper			
Hard drawn	1.77	10.62	0.00382
Annealed	1.72	10.37	0.00393
Iron	10	60	0.0050
Silver	1.59	9.6	0.0038
Sodium	4.3	26	0.0044
Steel	12–88	72–530	0.001–0.005

temperature T_1. The variations of resistance with temperature are usually unimportant (for example, 17 percent increase in copper resistance for a temperature change from 0°C to 40°C).

There are certain limitations in the use of this equation for calculating the resistance of transmission line conductors:

1. A slight error is introduced when the conductor is stranded rather than solid. This is because the individual strands are slightly longer than the length of the cable itself.

2. When ac flows in a conductor, the current is not distributed uniformly over the conductor cross-sectional area. This is called *skin effect* and is a result of the nonuniform flux distribution in the conductor. This increases the resistance of the conductor by reducing the effective cross section of the conductor through which the current flows. Manufacturer-supplied conductor tables give the resistance at commercial frequencies of 25, 50, and 60 Hz.

3. The resistance of magnetic conductors varies with current magnitude. The flux and therefore the magnetic losses inside the conductor depend on the current magnitude. Tables on magnetic conductors such as ACSR (Aluminum Cable, Steel Reinforced) include resistance tabulations at two current-carrying levels to show this effect.

4. In a transmission line there is a nonuniformity of current distribution in addition to that caused by skin effect. In a two-wire line, fewer lines of flux link the elements nearest each other on opposite sides of the line than link the elements farther apart. Thus the near sides will have lower inductance than elements in the far sides. The result is a higher current density in the elements of adjacent

conductors nearest each other than in the elements farther apart. The effective resistance is increased by the nonuniformity of current distribution. The phenomenon is known as *proximity effect*. It is present for three-phase as well as single-phase circuits. For the usual spacing of overhead lines at 60 Hz, the proximity effect is neglected.

4.3 LINE INDUCTANCE

For normal line designs, the inductive reactance is by far the most dominating impedance element. To develop expressions for the inductance of three-phase transmission lines, it is first necessary to develop a few concepts that greatly simplify the problem. We choose the following approach.

The internal inductance of a cylindrical conductor is derived first in Appendix 4-A. This is followed by a derivation of the flux linkages between two points outside the conductor in Appendix 4-B. With these two expressions at hand, the single-phase two-wire line case is considered. The generalization to a multiconductor configuration is obtained as an immediate consequence of these results.

The calculation of the inductances of balanced three-phase single-circuit lines is shown using the method of inductive voltage drops. The necessity of line transposition becomes evident from this discussion, and hence the case of transposed lines is treated. Bundle-conductor line inductances are considered, and some advantages of their use are treated.

Inductance of a Single-Phase Two-Wire Line

The inductance of a simple two-wire line consisting of two solid cylindrical conductors of radii r_1 and r_2 shown in Figure 4-1 can be obtained using a step-by-step approach. The inductance of the circuit due to the current in conductor 1 is the sum of contributions from flux linkages internal and external to the conductor. In Appendix 4-A we conclude that the inductance due to internal flux is

$$L_i = \tfrac{1}{2}(10^{-7}) \text{ henries/meter} \quad (4.1)$$

The inductance due to external flux linkages is shown in Appendix 4-B to be

$$L_{P1,P2} = (2 \times 10^{-7}) ln\left(\frac{D_2}{D_1}\right) \quad (4.2)$$

Substituting $D_2 = D$ and $D_1 = r_1$, we get for the external contribution:

$$L_{1\,\text{ext}} = (2 \times 10^{-7}) ln\left(\frac{D}{r_1}\right) \quad (4.3)$$

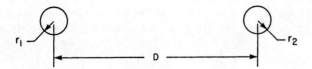

Figure 4-1. Single-Phase Two-Wire Line Configuration.

The total inductance of the circuit due to the current in conductor 1 only is therefore

$$L_1 = (2 \times 10^{-7})\left[\frac{1}{4} + \ln\left(\frac{D}{r_1}\right)\right] \text{ henries/meter} \quad (4.4)$$

Similarly, the inductance due to current in conductor 2 is

$$L_2 = (2 \times 10^{-7})\left[\frac{1}{4} + \ln\left(\frac{D}{r_2}\right)\right] \text{ henries/meter} \quad (4.5)$$

Thus L_1 and L_2 are the phase inductances. For the complete circuit we have

$$L_t = L_1 + L_2 \quad (4.6)$$

Thus

$$L_t = (2 \times 10^{-7})\left[\frac{1}{2} + \ln\left(\frac{D}{r_1}\right) + \ln\left(\frac{D}{r_2}\right)\right] \quad (4.7)$$

A more concise form of the inductance expression may be obtained if we observe that

$$\ln e^{1/4} = \frac{1}{4}$$

so that

$$L_1 = (2 \times 10^{-7}) \ln\left(\frac{D}{r_1'}\right) \quad (4.8)$$

$$L_2 = (2 \times 10^{-7}) \ln\left(\frac{D}{r_2'}\right) \quad (4.9)$$

$$L_t = (4 \times 10^{-7}) \ln\left(\frac{D}{\sqrt{r_1' r_2'}}\right) \quad (4.10)$$

where

$$r_i' = r_i e^{-1/4}$$
$$= 0.7788 r_i \quad (4.11)$$

Basically, we have omitted the internal flux term while compensating for it by using an adjusted value for the radius of the conductor. The quantity r' is commonly referred to as the solid conductor's *geometric mean radius* (GMR).

We can employ the inductive voltage drop equations to arrive at the same conclusions as follows:

$$V_1 = j\omega(L_{11}I_1 + L_{12}I_2) \tag{4.12}$$

$$V_2 = j\omega(L_{12}I_1 + L_{22}I_2) \tag{4.13}$$

where V_1 and V_2 are the voltage drops per unit length along conductors 1 and 2 respectively. The self-inductances L_{11} and L_{22} correspond to conductor geometries:

$$L_{11} = (2 \times 10^{-7}) \ln\left(\frac{1}{r_1'}\right) \tag{4.14}$$

$$L_{22} = (2 \times 10^{-7}) \ln\left(\frac{1}{r_2'}\right) \tag{4.15}$$

The mutual inductance L_{12} corresponds to the conductor separation D. Thus

$$L_{12} = (2 \times 10^{-7}) \ln\left(\frac{1}{D}\right) \tag{4.16}$$

Now we have

$$I_2 = -I_1$$

The complete circuit's voltage drop is

$$V_1 - V_2 = j\omega(L_{11} + L_{22} - 2L_{12})I_1 \tag{4.17}$$

In terms of the geometric configuration, we have

$$V = V_1 - V_2$$

$$V = j\omega(2 \times 10^{-7})\left[\ln\left(\frac{1}{r_1'}\right) + \ln\left(\frac{1}{r_2'}\right) - 2\ln\left(\frac{1}{D}\right)\right]I_1$$

$$= j\omega(4 \times 10^{-7}) \ln\left(\frac{D}{\sqrt{r_1'r_2'}}\right)$$

Thus

$$L_t = (4 \times 10^{-7}) \ln\left(\frac{D}{\sqrt{r_1'r_2'}}\right) \tag{4.18}$$

where

$$L_t = L_{11} + L_{22} - 2L_{12}$$

4.3 Line Inductance

We recognize this as the inductance of two series-connected magnetically coupled coils, each with self-inductance L_{11} and L_{22} respectively, and having a mutual inductance L_{12}.

The phase inductance expressions given in Eqs. (4.8) and (4.9) can be obtained from the voltage drop equations as follows:

$$V_1 = j\omega(2 \times 10^{-7})\left[I_1 \ln\left(\frac{1}{r_1'}\right) + I_2 \ln\left(\frac{1}{D}\right)\right]$$

But

$$I_2 = -I_1$$

Thus

$$V_1 = j\omega(2 \times 10^{-7})\left[I_1 \ln\left(\frac{D}{r_1'}\right)\right]$$

In terms of phase inductance we have

$$V_1 = j\omega L_1 I_1$$

Thus for phase one,

$$L_1 = (2 \times 10^{-7}) \ln\left(\frac{D}{r_1'}\right) \text{ henries/meter} \quad (4.19)$$

Similarly, for phase two,

$$L_2 = (2 \times 10^{-7}) \ln\left(\frac{D}{r_2'}\right) \text{ henries/meter} \quad (4.20)$$

Normally we have identical line conductors.

In practice, we deal with the inductive reactance of the line per phase per mile and use the logarithm to the base 10. Performing this conversion, we obtain

$$X = k \log \frac{D}{r'} \text{ ohms per conductor per mile} \quad (4.21)$$

where

$$k = 4.657 \times 10^{-3} f$$
$$= 0.2794 \text{ at } 60 \text{ Hz} \quad (4.22)$$

assuming identical line conductors.

Expanding the logarithm in the expression of Eq. (4.21), we get

$$X = k \log D + k \log \frac{1}{r'} \quad (4.23)$$

The first term is called X_d and the second is X_a. Thus

$$X_d = k \log D \text{ inductive reactance spacing factor}$$
$$\text{in ohms per mile} \qquad (4.24)$$

$$X_a = k \log \frac{1}{r'} \text{ inductive reactance at 1-ft}$$
$$\text{spacing in ohms per mile} \qquad (4.25)$$

Factors X_a and X_d may be obtained from tables available in many handbooks.

Example 4-1

Find the inductive reactance per mile per phase for a single-phase line with phase separation of 20 ft and conductor radius of 0.06677 ft.

Solution

We first find r', as follows:

$$\begin{aligned} r' &= re^{-1/4} \\ &= (0.06677)(0.7788) \\ &= 0.052 \text{ ft} \end{aligned}$$

We therefore calculate

$$X_a = 0.2794 \log \frac{1}{0.052}$$
$$= 0.35875$$
$$X_d = 0.2794 \log 20$$
$$= 0.36351$$
$$X = X_a + X_d = 0.72226 \text{ ohms per mile}$$

Bundle Conductors

At voltages above 230 kV (extra high voltage) and with circuits with only one conductor per phase, the corona effect becomes more excessive. Associated with this phenomenon is a power loss as well as interference with communication links. Corona is the direct result of high-voltage gradient at the conductor surface. The gradient can be reduced considerably by using more than one conductor per phase. The conductors are in close proximity compared with the spacing between phases. A line such as this is called a *bundle-conductor line*. The bundle consists of two or more conductors (subconductors) arranged on the perimeter of a circle called the *bundle circle* as shown in Figure 4-2. Another important advantage of bundling is the attendant reduction in line reactances, both series and shunt. The analysis of bundle-conductor lines is a specific case of the general multiconductor configuration problem that we treat next.

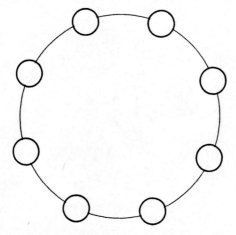

Figure 4-2. Bundle Conductor.

General Multiconductor Configurations

The complex problem of finding the inductances in a multiconductor configuration is of fundamental importance. The results of such consideration prove useful in the cases of stranded conductors, multicircuit lines, and bundle-conductor lines.

Let us consider a group of n conductors as shown in Figure 4-3 where the sum of all currents in the conductors is zero. The voltage drops per unit length for the conductors are given by the generalization of the expression in Eqs. (4.12) and (4.13) as

$$\begin{aligned} V_1 &= j\omega(L_{11}I_1 + L_{12}I_2 + \cdots + L_{1n}I_n) \\ V_2 &= j\omega(L_{12}I_1 + L_{22}I_2 + \cdots + L_{2n}I_n) \\ &\vdots \\ V_n &= j\omega(L_{1n}I_1 + L_{2n}I_2 + \cdots + L_{nn}I_n) \end{aligned} \quad (4.26)$$

The apparent self- and mutual inductances L_{jj} and L_{jk} are given by

$$L_{jj} = (2 \times 10^{-7}) \ln\left(\frac{1}{r'_j}\right) \text{ henries/meter} \quad (4.27)$$

102 The Transmission Subsystem

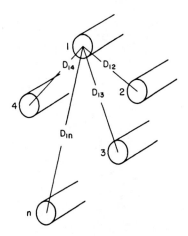

Figure 4-3. A Multiconductor Configuration.

$$L_{kj} = (2 \times 10^{-7}) \ln\left(\frac{1}{D_{kj}}\right) \text{ henries/meter} \qquad (4.28)$$

The above expressions form the basis for the evaluation of line inductances in practice.

Inductance of a Multiconductor Single-Phase Line

As an example of the application of Eq. (4.26), let us consider a single-phase line composed of two bundle conductors as shown in Figure 4-4. Each conductor forming one side of the line is shown as an arbitrary arrangement of a number of conductors. We require that these conductors share the current equally. Conductor A is composed of N_1 identical subconductors, each of which carries the current (I/N_1). Conductor B, the return circuit, is composed of N_2 identical filaments, each of which carries the current $(-I/N_2)$.

We write the expression of voltage drop per unit length of conductor 1 as

$$V_1 = j\omega(2 \times 10^{-7})\left[\left(\frac{I}{N_1}\right)\sum_{j=1}^{N_1} \ln\left(\frac{1}{D_{1j}}\right) - \left(\frac{I}{N_2}\right)\sum_{j=N_1+1}^{N_1+N_2} \ln\left(\frac{1}{D_{1j}}\right)\right] \qquad (4.29)$$

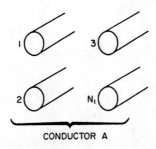

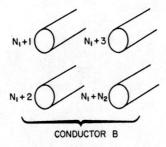

Figure 4-4. Single-Phase Bundle-Conductor Line.

But

$$\frac{1}{N_1}\sum_{j=1}^{N_1} \ln\left(\frac{1}{D_{1j}}\right) = \ln\left\{\frac{1}{\left[\prod_{j=1}^{N_1}(D_{1j})\right]^{1/N_1}}\right\}$$

So that

$$V_1 = j\omega(2\times 10^{-7})I\ln\left\{\frac{\left[\prod_{j=N_1+1}^{N_1+N_2}(D_{1j})\right]^{1/N_2}}{\left[\prod_{j=1}^{N_1}(D_{1j})\right]^{1/N_1}}\right\} \quad (4.30)$$

Here $D_{11} = r_1'$ is used to make the expressions more compact and the symbol $\prod$ denotes the product operation.

Let us define D_{m_1} and D_{s_1} by

$$D_{m_1} = \left[\prod_{j=N_1+1}^{N_1+N_2} (D_{1j}) \right]^{1/N_2} \quad (4.31)$$

$$D_{s_1} = \left[\prod_{j=1}^{N_1} (D_{1j}) \right]^{1/N_1} \quad (4.32)$$

The first expression is the N_2th root of the product of N_2 terms made of all the distances between subconductor 1 and all other subconductors in phase B. The second is the N_1th root of the product of N_1 terms made of all the distances between subconductor 1 and all other subconductors in phase A. According to these definitions, we have

$$V_1 = j\omega(2 \times 10^{-7}) I \ln\left(\frac{D_{m_1}}{D_{s_1}}\right)$$

We can now generalize to the case of the ith subconductor in phase A to obtain

$$V_i = j\omega(2 \times 10^{-7}) I \ln\left(\frac{D_{m_i}}{D_{s_i}}\right) \quad (4.33)$$

where

$$D_{m_i} = \left[\prod_{j=N_1+1}^{N_1+N_2} (D_{ij}) \right]^{1/N_2} \quad (4.34)$$

$$D_{s_i} = \left[\prod_{j=1}^{N_1} (D_{ij}) \right]^{1/N_1} \quad (4.35)$$

The inductance L_i for the ith subconductor is obtained using the relation

$$V_i = j\omega(L_i I_i) \quad (4.36)$$

Thus we have

$$L_i = \frac{I}{I_i}(2 \times 10^{-7}) \ln\left(\frac{D_{m_i}}{D_{s_i}}\right) \quad (4.37)$$

The N_1 subconductors on conductor A are connected in parallel. Hence its equivalent inductance can be obtained as the parallel combination of the N_1 inductances L_i. Thus

$$\frac{1}{L_A} = \sum_{i=1}^{N_1} \left(\frac{1}{L_i}\right) \quad (4.38)$$

Using Eq. (4.38) to evaluate L_A directly may be tedious. A good approximation results from a simple averaging calculation as shown below.

$$L_{av} = \left(\frac{1}{N_1}\right) \sum_{i=1}^{N_1} (L_i) \qquad (4.39)$$

Thus the equivalent inductance L_A of the parallel combination of N_1 subconductors of inductance L_{av} each is

$$L_A = \frac{L_{av}}{N_1} \qquad (4.40)$$

Thus the line inductance for phase A is

$$L_A = (2 \times 10^{-7}) \ln \frac{\left[\prod_{i=1}^{N_1} (D_{m_i})\right]^{1/N_1}}{\left[\prod_{i=1}^{N_1} (D_{s_i})\right]^{1/N_1}} \qquad (4.41)$$

We can now define

$$D_m = \left[\prod_{i=1}^{N_1} (D_{m_i})\right]^{1/N_1} \qquad (4.42)$$

$$D_{s_A} = \left[\prod_{i=1}^{N_1} (D_{s_i})\right]^{1/N_1} \qquad (4.43)$$

or in an expanded form,

$$D_m = \left[\prod_{i=1}^{N_1} \left(\prod_{j=N_1+1}^{N_1+N_2} (D_{ij})\right)\right]^{(1/N_1 N_2)} \qquad (4.44)$$

$$D_{s_A} = \left[\prod_{i=1}^{N_1} \left(\prod_{j=1}^{N_1} (D_{ij})\right)\right]^{(1/N_1^2)} \qquad (4.45)$$

Let us note that D_m is the $N_1 N_2$th root of the $N_1 N_2$ terms, which are the product of the distances from the N_1 subconductors of conductor A to all the N_2 subconductors of conductor B. This is called the *mutual geometric mean distance* (GMD).

D_{s_A} as defined is the N_1^2th root of N_1^2 terms, which are the product of the distances between all subconductors of conductor A. This is called the *self geometric mean distance*. Thus we rewrite Eq. (4.41) as

$$L_A = (2 \times 10^{-7}) \ln\left(\frac{D_m}{D_{s_A}}\right) \text{ henries/meter} \qquad (4.46)$$

and

$$L_B = (2 \times 10^{-7}) \ln\left(\frac{D_m}{D_{s_B}}\right) \text{ henries/meter} \qquad (4.47)$$

The self geometric mean distance D_s is commonly referred to as the GMR. In common practice we refer to D_s as the bundle conductor's GMR.

Assuming that the two conductors are identical, then

$$L_A = L_B = L$$

with

$$L = (2 \times 10^{-7}) \ln\left(\frac{\text{GMD}}{\text{GMR}}\right) \text{ henries/meter} \qquad (4.48)$$

where

$$\text{GMD} = \left[\prod_{i=1}^{N} \prod_{j=N+1}^{2N} (D_{ij})\right]^{1/N^2} \qquad (4.49)$$

$$\text{GMR} = \left[\prod_{i=1}^{N} \prod_{j=1}^{N} (D_{ij})\right]^{1/N^2} \qquad (4.50)$$

The concepts of the geometric mean distance and geometric mean radius enable us to deal with multiconductor configurations in much the same manner as those with solid conductor systems. In fact, we can state that our inductance expression is a generalization of the one obtained for the single-phase solid conductor case for which the inductance expression is given by Eq. (4.48).

Example 4-2

Consider a 375-kV, single-phase line with bundle conductors as shown in Figure 4-5. The phase separation D_1 is 12.19 m, and the subconductor spacing is $S = 45.72$ cm. The subconductor diameter is 4.577 cm. Calculate the line inductance by applying Eq. (4.48).

Solution

We have four subconductors; thus

$$N_1 = N_2 = 2$$

The geometric mean distance is therefore

$$D_m = \left[\prod_{i=1}^{2} \left(\prod_{j=3}^{4} (D_{ij})\right)\right]^{1/4}$$

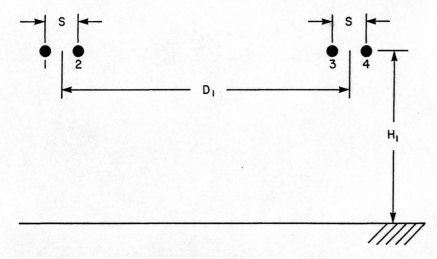

Figure 4-5. A 375-kV Single-Phase Bundle-Conductor Line.

Expanding the products, we have

$$D_m = \left[\prod_{i=1}^{2} (D_{i3}D_{i4}) \right]^{1/4}$$
$$= (D_{13}D_{14}D_{23}D_{24})^{1/4}$$

From the geometry of Figure 4-5,

$$D_{13} = D_{24} = D_1 = 12.19 \text{ m}$$
$$D_{14} = D_1 + S = 12.19 + 0.4572 = 12.6472 \text{ m}$$
$$D_{23} = D_1 - S = 12.19 - 0.4572 = 11.7328 \text{ m}$$

We thus have

$$D_m = \left[(12.19)^2 (12.6472)(11.7328) \right]^{1/4}$$
$$= 12.1857 \text{ m}$$

The geometric mean radius is

$$D_s = \left[\prod_{i=1}^{2} \prod_{j=1}^{2} (D_{ij}) \right]^{1/4}$$
$$= \left[\prod_{i=1}^{2} (D_{i1}D_{i2}) \right]^{1/4}$$
$$= (D_{11}D_{12}D_{21}D_{22})^{1/4}$$

Observe that
$$D_{11} = D_{22} = r'_1 = 0.7788 r_1$$
$$= (0.7788)\left(\frac{4.577}{2} \times 10^{-2}\right)$$
$$= 1.7823 \times 10^{-2} \text{ m}$$

Also
$$D_{12} = D_{21} = S = 0.4572 \text{ m}$$

Therefore,
$$D_s = \left[(1.7824 \times 10^{-2})^2 (0.4572)^2\right]^{1/4}$$
$$= 9.027 \times 10^{-2} \text{ m}$$

As a result of the above, we obtain the following result using Eq. (4.48):
$$L = (2 \times 10^{-7}) \ln\left(\frac{12.1857}{9.027 \times 10^{-2}}\right)$$
$$= 0.981 \times 10^{-6} \text{ henries/meter}$$

Inductance of a Single-Phase Symmetrical Bundle-Conductor Line

Consider a symmetrical bundle with N subconductors arranged in a circle of radius A. The angle between two subconductors is $2\pi/N$. The arrangement is shown in Figure 4-6. Considering subconductor 1, the distances $D_{12}, D_{13}, \ldots, D_{1n}$ are easily seen to be given by

$$D_{12} = 2A \sin\left(\frac{\pi}{N}\right)$$
$$D_{13} = 2A \sin\left(\frac{2\pi}{N}\right)$$
$$\vdots$$
$$D_{1N} = 2A \sin\left[\frac{(N-1)\pi}{N}\right]$$
(4.51)

The current in each of the subconductors is (I/N) for phase A and $(-I/N)$ for phase B.

The voltage drop per unit length of subconductor 1 is

$$V_1 = j\omega(2 \times 10^{-7})\left\{\frac{I}{N}\left[\ln\frac{1}{r'_1} + \ln\frac{1}{D_{12}} + \ln\frac{1}{D_{13}} + \cdots + \ln\frac{1}{D_{1N}}\right]\right.$$
$$\left. - \frac{I}{N}\left[\ln\frac{1}{D_{1(N+1)}} + \ln\frac{1}{D_{1(N+2)}} + \cdots + \ln\frac{1}{D_{1(2N)}}\right]\right\}$$

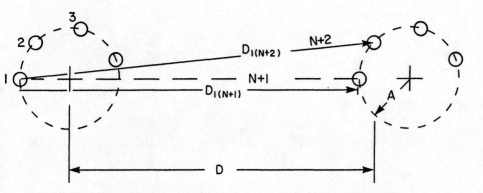

Figure 4-6. Single-Phase Symmetrical Bundle-Conductor Circuit.

As a result,

$$V_1 = j\omega(2 \times 10^{-7}) \frac{I}{N} \ln\left[\frac{D_{1(N+1)} D_{1(N+2)} \cdots D_{1(2N)}}{r_1' D_{12} D_{13} \cdots D_{1N}}\right]$$

We have

$$D_{12} D_{13} \cdots D_{1N} = (A)^{N-1} \left\{ \left[2\sin\left(\frac{\pi}{N}\right)\right]\left[2\sin\left(\frac{2\pi}{N}\right)\right] \right.$$
$$\left. \cdots \left[2\sin\left(\frac{(N-1)\pi}{N}\right)\right] \right\}$$
$$= N(A)^{N-1}$$

where we utilize the trigonometric identity:

$$\left[2\sin\left(\frac{\pi}{N}\right)\right]\left[2\sin\left(\frac{2\pi}{N}\right)\right] \cdots \left[2\sin\left(\frac{(N-1)\pi}{N}\right)\right] = N$$

We define the geometric mean distance (GMD) by

$$\text{GMD} = \left\{ \left[D_{1(N+1)}\right]\left[D_{1(N+2)}\right] \cdots \left[D_{1(2N)}\right] \right\}^{1/N}$$

$$V_1 = j\omega(2 \times 10^{-7})\left(\frac{I}{N}\right) \ln\left[\frac{(\text{GMD})^N}{Nr_1'(A)^{N-1}}\right] \qquad (4.52)$$

which reduces to

$$V_1 = j\omega(2 \times 10^{-7}) I \ln\left\{\frac{\text{GMD}}{\left[Nr_1'(A)^{N-1}\right]^{1/N}}\right\}$$

As a result, we obtain

$$L = (2 \times 10^{-7}) \ln\left\{\frac{\text{GMD}}{\left[Nr_1'(A)^{N-1}\right]^{1/N}}\right\} \text{ henries/meter} \qquad (4.53)$$

Let us observe that practically the distances $D_{1(N+1)}$, $D_{1(N+2)}, \ldots,$ are all almost equal in value to the distance D between the bundle centers. As a result,

$$\text{GMD} \cong D \qquad (4.54)$$

We also note that the geometric mean radius in this case is

$$\text{GMR} = \left[Nr'(A)^{N-1}\right]^{1/N} \qquad (4.55)$$

We thus have the same form as derived before for the solid conductor case:

$$L = (2 \times 10^{-7}) \ln\left(\frac{\text{GMD}}{\text{GMR}}\right) \qquad (4.56)$$

In many instances, the subconductor spacing S in the bundle circle is given. It is easy to find the radius A using the formula

$$S = 2A \sin\left(\frac{\pi}{N}\right) \qquad (4.57)$$

which is a consequence of the geometry of the bundle as shown in Figure 4-7.

Example 4-3

Let us apply the formulae just derived to the bundle conductor line of Example 4-2. Here we have the approximation

$$\text{GMD} = 12.19 \text{ m}$$

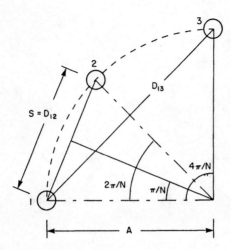

Figure 4-7. Conductor Geometry.

Clearly the subconductor spacing S is the same as twice the bundle radius A; thus

$$A = \frac{0.4572}{2} = 0.2286$$

The subconductor radius is

$$r = \frac{0.04577}{2}$$
$$r' = (e^{-1/4})r$$
$$= 0.0178 \text{ m}$$

The GMR is therefore given by

$$\text{GMR} = \left[2(0.0178)(0.2286)^1\right]^{1/2}$$
$$= 0.0903 \text{ m}$$

As a result,

$$L = (2 \times 10^{-7}) \ln\left(\frac{12.19}{0.0903}\right)$$
$$= 0.9811 \times 10^{-6} \text{ henries/meter}$$

which is just about identical to the result obtained earlier.

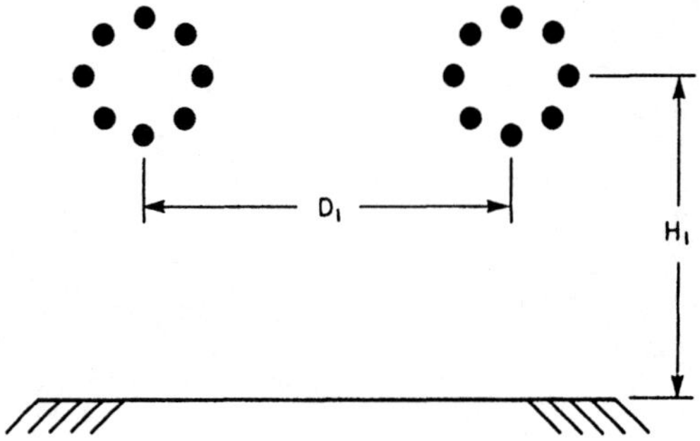

Figure 4-8. 1000-kV Single-Phase Bundle-Conductor Line.

Example 4-4

Figure 4-8 shows a 1000-kV, single-phase, bundle-conductor line with eight subconductors per phase. The phase spacing is $D_1 = 16.76$ m, and the subconductor spacing is $S = 45.72$ cm. Each subconductor has a diameter of 4.572 cm. Calculate the line inductance.

Solution

We first evaluate the bundle radius A. Thus

$$0.4572 = 2A \sin\left(\frac{\pi}{8}\right)$$

Therefore

$$A = 0.5974 \text{ m}$$

Assume that the following practical approximation holds:

$$\text{GMD} = D_1 = 16.76 \text{ m}$$

The subconductor's geometric mean radius is

$$r_1' = 0.7788\left(\frac{4.572}{2} \times 10^{-2}\right)$$

$$= 1.7803 \times 10^{-2} \text{ m}$$

Thus we have

$$L = (2 \times 10^{-7}) \ln\left\{\frac{\text{GMD}}{\left[Nr_1'(A)^{N-1}\right]^{1/N}}\right\}$$

$$= (2 \times 10^{-7}) \ln\left\{\frac{16.76}{\left[(8)(1.7803 \times 10^{-2})(0.5974)^7\right]^{1/8}}\right\}$$

The result of the above calculation is

$$L = 7.027 \times 10^{-7} \text{ henries/meter}$$

Inductance of a Balanced Three-Phase Single-Circuit Line

We consider a three-phase line whose phase conductors have the general arrangement shown in Figure 4-9. We use the voltage drop per unit length concept. This is a consequence of Faraday's law. In engineering practice we have a preference for this method. In our three-phase system, we can write

$$V_1 = j\omega(L_{11}I_1 + L_{12}I_2 + L_{13}I_3)$$
$$V_2 = j\omega(L_{12}I_1 + L_{22}I_2 + L_{23}I_3)$$
$$V_3 = j\omega(L_{13}I_1 + L_{23}I_2 + L_{33}I_3)$$

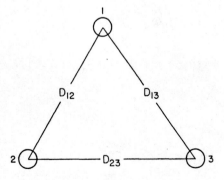

Figure 4-9. A Balanced Three-Phase Line.

114 The Transmission Subsystem

Here we generalize the expressions of Eqs. (4.14) and (4.16) to give

$$L_{ii} = (2 \times 10^{-7}) \ln\left(\frac{1}{r'_i}\right) \qquad (4.58)$$

$$L_{kj} = (2 \times 10^{-7}) \ln\left(\frac{1}{D_{kj}}\right) \qquad (4.59)$$

We now substitute for the inductances in the voltage drops equations to obtain the following:

$$V'_1 = I_1 \ln\left(\frac{1}{r'_1}\right) + I_2 \ln\left(\frac{1}{D_{12}}\right) + I_3 \ln\left(\frac{1}{D_{13}}\right)$$

$$V'_2 = I_1 \ln\left(\frac{1}{D_{12}}\right) + I_2 \ln\left(\frac{1}{r'_2}\right) + I_3 \ln\left(\frac{1}{D_{23}}\right)$$

$$V'_3 = I_1 \ln\left(\frac{1}{D_{13}}\right) + I_2 \ln\left(\frac{1}{D_{23}}\right) + I_3 \ln\left(\frac{1}{r'_3}\right)$$

Here

$$V'_i = \frac{V_i}{j\omega(2 \times 10^{-7})}$$

We next use the condition of balanced operation to eliminate one current from each equation. Thus

$$I_1 + I_2 + I_3 = 0$$

The result is

$$V'_1 = I_1 \ln\left(\frac{D_{13}}{r'_1}\right) + I_2 \ln\left(\frac{D_{13}}{D_{12}}\right)$$

$$V'_2 = I_1 \ln\left(\frac{D_{23}}{D_{12}}\right) + I_2 \ln\left(\frac{D_{23}}{r'_2}\right)$$

$$V'_3 = I_2 \ln\left(\frac{D_{13}}{D_{23}}\right) + I_3 \ln\left(\frac{D_{13}}{r'_3}\right) \qquad (4.60)$$

We note that for this general case, the voltage drop in phase one, for example, depends on the current in phase two in addition to its dependence on I_1. Thus the voltage drops will not be a balanced system. This situation is undesirable.

Consider the case of equilaterally spaced conductors generally referred to as the *delta* configuration; that is

$$D_{12} = D_{13} = D_{23} = D$$
$$r'_1 = r'_2 = r'_3 = r'$$

Recall that

$$ln\, 1 = 0$$

It is clear that the voltage drops will thus be given by

$$V'_1 = I_1 ln\left(\frac{D}{r'}\right)$$
$$V'_2 = I_2 ln\left(\frac{D}{r'}\right)$$
$$V'_3 = I_3 ln\left(\frac{D}{r'}\right) \qquad (4.61)$$

and in this case the voltage drops will form a balanced system.

Consider the so often called H-type configuration. The conductors are in one horizontal plane as shown in Figure 4-10. The distances between conductors are thus

$$D_{12} = D_{23} = D$$
$$D_{13} = 2D$$

and the voltage drops are given by

$$V'_1 = I_1 ln\left(\frac{2D}{r'}\right) + I_2 ln\, 2$$
$$V'_2 = I_2 ln\left(\frac{D}{r'}\right)$$
$$V'_3 = I_2 ln\, 2 + I_3 ln\left(\frac{2D}{r'}\right) \qquad (4.62)$$

We note that only conductor two has a voltage drop proportional to its current.

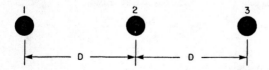

Figure 4-10. H-Type Line.

Transposition of Line Conductors

The equilateral triangular spacing configuration is not the only configuration commonly used in practice. Thus the need exists for equalizing the mutual inductances. One means for doing this is to construct transpositions or rotations of overhead line wires. A transposition is a physical rotation of the conductors, arranged so that each conductor is moved to occupy the next physical position in a regular sequence such as *a-b-c*, *b-c-a*, *c-a-b*, etc. Such a transposition arrangement is shown in Figure 4-11. If a section of line is divided into three segments of equal length separated by rotations, we say that the line is "completely transposed."

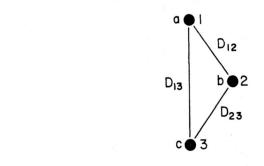

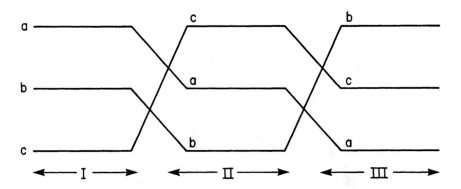

Figure 4-11. Transposed Line.

Consider a completely transposed three-phase line. In segment I, the voltage drops are

$$V'_{a_I} = I_a \ln\left(\frac{D_{13}}{r'}\right) + I_b \ln\left(\frac{D_{13}}{D_{12}}\right)$$

$$V'_{b_I} = I_a \ln\left(\frac{D_{23}}{D_{12}}\right) + I_b \ln\left(\frac{D_{23}}{r'}\right)$$

$$V'_{c_I} = I_b \ln\left(\frac{D_{13}}{D_{23}}\right) + I_c \ln\left(\frac{D_{13}}{r'}\right) \quad (4.63)$$

This is a direct application of Eq. (4.60) for equal conductor diameters. In segment II, phase a occupies physical position 2, phase b occupies physical position 3, and phase c occupies physical position 1. Therefore,

$$V'_{a_{II}} = I_c \ln\left(\frac{D_{23}}{D_{12}}\right) + I_a \ln\left(\frac{D_{23}}{r'}\right)$$

$$V'_{b_{II}} = I_a \ln\left(\frac{D_{13}}{D_{23}}\right) + I_b \ln\left(\frac{D_{13}}{r'}\right)$$

$$V'_{c_{II}} = I_c \ln\left(\frac{D_{13}}{r'}\right) + I_a \ln\left(\frac{D_{13}}{D_{12}}\right) \quad (4.64)$$

Similarly, for segment III, we obtain

$$V'_{a_{III}} = I_c \ln\left(\frac{D_{13}}{D_{23}}\right) + I_a \ln\left(\frac{D_{13}}{r'}\right)$$

$$V'_{b_{III}} = I_b \ln\left(\frac{D_{13}}{r'}\right) + I_c \ln\left(\frac{D_{13}}{D_{12}}\right)$$

$$V'_{c_{III}} = I_b \ln\left(\frac{D_{23}}{D_{12}}\right) + I_c \ln\left(\frac{D_{23}}{r'}\right) \quad (4.65)$$

The total drop for a completely transposed unit length of the line is given by

$$V'_a = \tfrac{1}{3}(V'_{a_I} + V'_{a_{II}} + V'_{a_{III}})$$
$$V'_b = \tfrac{1}{3}(V'_{b_I} + V'_{b_{II}} + V'_{b_{III}})$$
$$V'_c = \tfrac{1}{3}(V'_{c_I} + V'_{c_{II}} + V'_{c_{III}})$$

These yield

$$V'_a = I_a \ln \frac{(D_{12}D_{13}D_{23})^{1/3}}{r'}$$

$$V'_b = I_b \ln \frac{(D_{12}D_{13}D_{23})^{1/3}}{r'}$$

$$V'_c = I_c \ln \frac{(D_{12}D_{13}D_{23})^{1/3}}{r'} \quad (4.66)$$

Thus by completely transposing a line, the mutual inductance terms disappear, and the voltage drops are proportional to the current in each phase.

It is evidently clear that the phase inductance L in the present case is given by

$$L = (2 \times 10^{-7}) \ln \frac{(D_{12}D_{13}D_{23})^{1/3}}{r'} \text{ henries/meter} \quad (4.67)$$

This again is of the same form as the expression for the inductance in the case of a single-phase two-wire system. Defining the geometric mean distance GMD as

$$\text{GMD} = (D_{12}D_{13}D_{23})^{1/3} \quad (4.68)$$

and the geometric mean radius GMR as

$$\text{GMR} = r' \quad (4.69)$$

we attain

$$L = (2 \times 10^{-7}) \ln \left(\frac{\text{GMD}}{\text{GMR}}\right) \text{ henries/meter} \quad (4.70)$$

Again, we can obtain the inductive reactance per conductor per mile as the sum

$$X = X_a + X_d \quad (4.71)$$

where

$$X_a = k \log \frac{1}{\text{GMR}} \quad (4.72)$$

$$X_d = k \log \text{GMD} \quad (4.73)$$

with $k = 0.2794$ at 60 Hz, as before.

Example 4-5

Calculate the inductance per phase of the 345-kV three-phase solid conductor line shown in Figure 4-12. Assume that the conductor diameter is 4.475 cm and the phase separation D_1 is 7.92 m. Assume that the line is transposed.

Solution

The geometric mean distance is given by

$$\text{GMD} = [D_1 D_1 (2D_1)]^{1/3}$$
$$= 1.2599 D_1$$
$$= 9.9786 \text{ m}$$

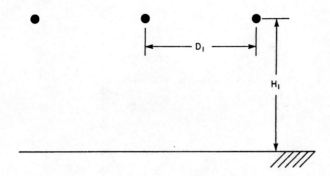

Figure 4-12. A 345-kV Three-Phase Line.

The geometric mean radius is

$$r' = (e^{-1/4})\frac{4.475 \times 10^{-2}}{2}$$
$$= 0.0174 \text{ m}$$

Therefore

$$L = (2 \times 10^{-7}) \ln\left(\frac{9.9786}{0.0174}\right)$$
$$= 1.27 \times 10^{-6} \text{ henries/meter}$$

Inductance of Multiconductor Three-Phase Systems

Let us consider a single-circuit, three-phase system with multiconductor-configured phase conductors as shown in Figure 4-13. Let us assume equal current distribution in the phase subconductors and complete transposition. We can combine the concepts developed earlier to arrive at the phase inductance for the system. The result is the following inductance expression:

$$L = (2 \times 10^{-7}) \ln\left(\frac{\text{GMD}}{\text{GMR}}\right) \qquad (4.74)$$

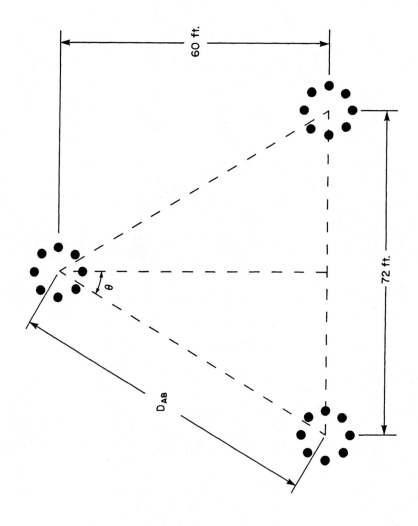

Figure 4-13. Multiconductor Single-Circuit Three-Phase Line.

4.3 Line Inductance

In this case the geometric mean distance is given by

$$\text{GMD} = (D_{AB} D_{BC} D_{CA})^{1/3} \tag{4.75}$$

where D_{AB}, D_{BC}, and D_{CA} are the distances between phase centers. The geometric mean radius (GMR) is obtained using the same expression as that for the single-phase system. Thus

$$\text{GMR} = \left[\prod_{i=1}^{N} (D_{si}) \right]^{1/N} \tag{4.76}$$

For the case of symmetrical bundle conductors, we have

$$\text{GMR} = \left[Nr'(A)^{N-1} \right]^{1/N} \tag{4.77}$$

It is clear that the inductive reactance per mile per phase X_L in the case of a three-phase, bundle-conductor line can be obtained using

$$X_L = X_a + X_d \tag{4.78}$$

where as before for 60 Hz operation,

$$X_a = 0.2794 \log \frac{1}{\text{GMR}} \tag{4.79}$$

$$X_d = 0.2794 \log \text{GMD} \tag{4.80}$$

The GMD and GMR are defined by Eqs. (4.75) and (4.77).

Example 4-6

An 1100-kV, three-phase line has an eight subconductor-bundle delta arrangement with a 42 in. diameter. The subconductors are ACSR 84/19 (Chukar) with $r' = 0.0534$ ft. The horizontal phase separation is 72 ft, and the vertical separation is 60 ft. Calculate the inductive reactance of the line in ohms per mile per phase.

Solution

From the geometry of the phase arrangement, we have

$$\tan \theta = \frac{36}{60}$$

$$\theta = 30.96°$$

$$D_{AB} = \frac{60}{\cos 30.96°}$$

$$= 69.97 \text{ ft}$$

Thus

$$\text{GMD} = [(69.97)(69.97)(72)]^{1/3} = 70.64 \text{ ft}$$

For Chukar we have $r' = 0.0534$ ft. The bundle particulars are $N = 8$ and

$A = (42/2)$ in. Therefore,

$$\text{GMR} = \left[8(0.0534)\left(\frac{21}{12}\right)^7\right]^{1/8}$$
$$= 1.4672 \text{ ft}$$

Thus

$$X_a = 0.2794 \log \frac{1}{1.4672}$$
$$= -0.0465$$
$$X_d = 0.2794 \log 70.64$$
$$= 0.5166$$

As a result,

$$X_L = X_a + X_d = 0.4701 \text{ ohms per mile}$$

Inductance of Three-Phase, Double-Circuit Lines

A three-phase, double-circuit line is essentially two three-phase circuits connected in parallel. Normal practice calls for identical construction for the two circuits. If the two circuits are widely separated, then we can

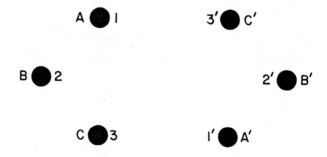

Figure 4-14. Double-Circuit Conductors' Relative Positions in Segment *I* of Transposition.

obtain the line reactance as simply half that of one single-circuit line. For the situation where the two circuits are on the same tower, the above approach may not produce results of sufficient accuracy. The error introduced is mainly due to neglecting the effect of mutual inductance between the two circuits. We derive here a simple but more accurate expression for calculating the reactance of double-circuit lines.

Consider a three-phase, double-circuit line with full line transposition such that in segment I, the relative phase positions are as shown in Figure 4-14. The relative phase positions for segments II and III are shown in Figures 4-15 and 4-16.

The phase inductance in this case can be obtained using the voltage drops concept. For phase A in segment I, we have

$$V'_{A_I} = \frac{l}{3}\left[\frac{I_A}{2}\ln\left(\frac{1}{r'}\right) + \frac{I_B}{2}\ln\left(\frac{1}{D_{12}}\right) + \frac{I_C}{2}\ln\left(\frac{1}{D_{13}}\right)\right.$$

$$\left. + \frac{I_A}{2}\ln\left(\frac{1}{D_{11'}}\right) + \frac{I_B}{2}\ln\left(\frac{1}{D_{12'}}\right) + \frac{I_C}{2}\ln\left(\frac{1}{D_{13'}}\right)\right]$$

(4.81)

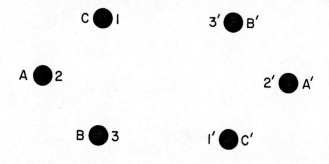

Figure 4-15. Double-Circuit Conductors' Relative Positions in Segment *II* of Transposition.

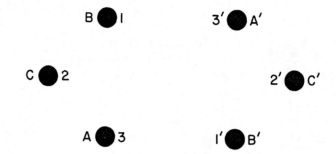

Figure 4-16. Double-Circuit Conductors' Relative Positions in Segment *III* of Transposition.

For phase A' we have for segment I,

$$V'_{A_I} = \frac{l}{3}\left[\frac{I_A}{2}\ln\left(\frac{1}{r'}\right) + \frac{I_B}{2}\ln\left(\frac{1}{D_{1'2}}\right) + \frac{I_C}{2}\ln\left(\frac{1}{D_{31'}}\right)\right.$$
$$\left. + \frac{I_A}{2}\ln\left(\frac{1}{D_{11'}}\right) + \frac{I_B}{2}\ln\left(\frac{1}{D_{1'2'}}\right) + \frac{I_C}{2}\ln\left(\frac{1}{D_{1'3'}}\right)\right] \quad (4.82)$$

In segment II, we have

$$V'_{A_{II}} = \frac{l}{3}\left[\frac{I_A}{2}\ln\left(\frac{1}{r'}\right) + \frac{I_B}{2}\ln\left(\frac{1}{D_{23}}\right) + \frac{I_C}{2}\ln\left(\frac{1}{D_{12}}\right)\right.$$
$$\left. + \frac{I_A}{2}\ln\left(\frac{1}{D_{22'}}\right) + \frac{I_B}{2}\ln\left(\frac{1}{D_{23'}}\right) + \frac{I_C}{2}\ln\left(\frac{1}{D_{21'}}\right)\right] \quad (4.83)$$

$$V'_{A'_{II}} = \frac{l}{3}\left[\frac{I_A}{2}\ln\left(\frac{1}{r'}\right) + \frac{I_B}{2}\ln\left(\frac{1}{D_{32'}}\right) + \frac{I_C}{2}\ln\left(\frac{1}{D_{12'}}\right)\right.$$
$$\left. + \frac{I_A}{2}\ln\left(\frac{1}{D_{22'}}\right) + \frac{I_B}{2}\ln\left(\frac{1}{D_{2'3'}}\right) + \frac{I_C}{2}\ln\left(\frac{1}{D_{1'2'}}\right)\right] \quad (4.84)$$

In segment III, we have

$$V'_{A_{III}} = \frac{l}{3}\left[\frac{I_A}{2}\ln\left(\frac{1}{r'}\right) + \frac{I_B}{2}\ln\left(\frac{1}{D_{31}}\right) + \frac{I_C}{2}\ln\left(\frac{1}{D_{23}}\right)\right.$$
$$\left. + \frac{I_A}{2}\ln\left(\frac{1}{D_{33'}}\right) + \frac{I_B}{2}\ln\left(\frac{1}{D_{31'}}\right) + \frac{I_C}{2}\ln\left(\frac{1}{D_{32'}}\right)\right] \quad (4.85)$$

$$V'_{A'_{III}} = \frac{l}{3}\left[\frac{I_A}{2}\ln\left(\frac{1}{r'}\right) + \frac{I_B}{2}\ln\left(\frac{1}{D_{13'}}\right) + \frac{I_C}{2}\ln\left(\frac{1}{D_{23'}}\right)\right.$$
$$\left. + \frac{I_A}{2}\ln\left(\frac{1}{D_{33'}}\right) + \frac{I_B}{2}\ln\left(\frac{1}{D_{3'1'}}\right) + \frac{I_C}{2}\ln\left(\frac{1}{D_{2'3'}}\right)\right] \quad (4.86)$$

The total voltage drops along phases A and A' are therefore obtained using

$$V'_A = V'_{A_I} + V'_{A_{II}} + V'_{A_{III}}$$

and

$$V'_{A'} = V'_{A'_I} + V'_{A'_{II}} + V'_{A'_{III}}$$

Since phase A' and A are in parallel, we take the average voltage drop; thus

$$V'_A = V'_{A'} = \frac{V'_A + V'_{A'}}{2}$$

We also utilize the balanced three-phase condition:

$$I_A + I_B + I_c = 0$$

After some algebra, we get

$$V'_A = \frac{lI_A}{12}\ln\left[\frac{(D_{12}D_{1'2'}D_{12'}D_{1'2})(D_{13}D_{1'3'}D_{13'}D_{1'3})(D_{23}D_{2'3'}D_{2'3}D_{23'})}{(r')^6(D_{11'}^2 D_{22'}^2 D_{33'}^2)}\right] \quad (4.87)$$

From the above expression we conclude that the inductance per phase per unit length is given by

$$L = (2 \times 10^{-7})\ln\left(\frac{\text{GMD}}{\text{GMR}}\right) \quad (4.88)$$

where the double-circuit geometric mean distance is given by

$$\text{GMD} = \left(D_{AB_{eq}} D_{BC_{eq}} D_{AC_{eq}}\right)^{1/3} \quad (4.89)$$

with mean distances defined by

$$D_{AB_{eq}} = (D_{12}D_{1'2'}D_{12'}D_{1'2})^{1/4}$$
$$D_{BC_{eq}} = (D_{23}D_{2'3'}D_{2'3}D_{23'})^{1/4}$$
$$D_{AC_{eq}} = (D_{13}D_{1'3'}D_{13'}D_{1'3})^{1/4} \quad (4.90)$$

where subscript eq. refers to equivalent spacing. The GMR is

$$\text{GMR} = [(\text{GMR}_A)(\text{GMR}_B)(\text{GMR}_C)]^{1/3} \qquad (4.91)$$

with phase GMR's defined by

$$\begin{aligned}\text{GMR}_A &= [r'(D_{11'})]^{1/2} \\ \text{GMR}_B &= [r'(D_{22'})]^{1/2} \\ \text{GMR}_C &= [r'(D_{33'})]^{1/2}\end{aligned} \qquad (4.92)$$

We see from the above development that the same methodology adopted for the single-circuit case can be utilized for the double-circuit case.

Example 4-7

Calculate the inductance per phase for the 345-kV, three-phase, double-circuit line whose phase conductors have a GMR of 0.0588 ft, with the horizontal conductor configuration as shown in Figure 4-17.

Solution

We use Eq. (4.90):

$$D_{AB_{eq}} = [(25)(25)(50)(100)]^{1/4}$$
$$= 42.04 \text{ ft}$$
$$D_{BC_{eq}} = [(25)(25)(50)(100)]^{1/4}$$
$$= 42.04 \text{ ft}$$
$$D_{AC_{eq}} = [(50)(50)(125)(25)]^{1/4}$$
$$= 52.87 \text{ ft}$$

As a result,

$$\text{GMD} = [(42.04)(42.04)(52.87)]^{1/3}$$
$$= 45.381 \text{ ft}$$

The equivalent GMR is obtained using Eq. (4.92) as

$$r_{eq} = [(0.0588)^3 (75)^3]^{1/6}$$
$$= 2.1 \text{ ft}$$

As a result,

$$L = (2 \times 10^{-7}) \ln\left(\frac{45.381}{2.1}\right)$$
$$= 0.61463 \times 10^{-6} \text{ henries/meter}$$

Let us calculate the inductance of one circuit. Here we have

$$D_{eq} = [(25)(25)(50)]^{1/3}$$
$$= 31.5 \text{ ft}$$
$$r' = 0.0588$$

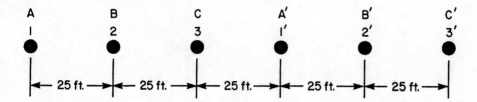

Figure 4-17. Configuration for Example 4-7.

Consequently,

$$L_s = (2 \times 10^{-7}) ln\left(\frac{31.5}{0.0588}\right)$$

$$= 1.257 \times 10^{-6} \text{ henries/meter}$$

The inductance of the double-circuit is obtained as

$$L = \frac{L_s}{2}$$

$$= 0.62835 \times 10^{-6} \text{ henries/meter}$$

Note that the relative error involved in this case is just below 3 percent. The relative error is computed as

$$\frac{0.61463 - 0.62835}{0.61463} = -0.0223$$

Double-Circuit, Bundle-Conductor Lines

A double-circuit line may use bundle conductors. In this case, the same method indicated for the case of double-circuit, single-conductor lines is used with r' replaced by the bundle's GMR in the calculation of the overall GMR. The following example illustrates the procedure.

Example 4-8

Find the inductance of the 345-kV, double-circuit line shown in Figure 4-18. Assume that the GMR for each subconductor is 0.0587 ft. Bundle spacing is 18 in.

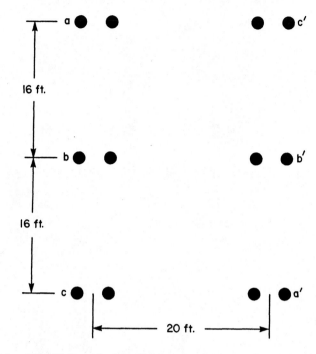

Figure 4-18. 345-kV, Double-Circuit Line.

Solution

$$N = 2$$
$$S = 18 \text{ in.}$$
$$A = 9 \text{ in.} = 0.75 \text{ ft}$$

We have

$$\text{GMR}_c = 0.0587 \text{ ft}$$

Thus

$$\text{GMR}_b = [(2)(0.0587)(0.75)]^{1/2}$$
$$= 0.29673 \text{ ft}$$

As a result:

$$r'_A = \left[(\text{GMR}_b)\sqrt{(20)^2 + (32)^2}\right]^{1/2}$$
$$= 3.3463 \text{ ft}$$
$$r'_B = [(\text{GMR}_b)(20)]^{1/2}$$
$$= 2.4361 \text{ ft}$$
$$r'_C = r'_A = 3.3463 \text{ ft}$$

Hence,

$$\text{GMR}_p = (r'_A r'_B r'_c)^{1/3}$$
$$= 3.0103 \text{ ft}$$
$$D_{AB_{eq}} = \left[16\sqrt{(16)^2 + (20)^2}\right]^{1/2}$$
$$= 20.24 \text{ ft}$$
$$D_{BC_{eq}} = \left[16\sqrt{(16)^2 + (20)^2}\right]^{1/2}$$
$$= 20.24 \text{ ft}$$
$$D_{AC_{eq}} = [(20)(32)]^{1/2}$$
$$= 25.30 \text{ ft}$$
$$\text{GMD} = [(20.24)(20.24)(25.30)]^{1/3}$$
$$= 21.80 \text{ ft}$$

The inductance is thus given by

$$L = (2 \times 10^{-7}) \ln\left(\frac{21.8}{3.0103}\right)$$
$$= 3.9602 \times 10^{-7} \text{ henries/meter}$$

4.4 LINE CAPACITANCE

We have discussed in the previous sections two line parameters that constitute the series impedance of the transmission line. The line inductance normally dominates the series resistance and determines the power transmission capacity of the line. There are two other line-parameters whose effects can be appreciable for high transmission voltages and line length. The line's shunt admittance consists of the first parameter represented by the conductance (g) and the second parameter represented by the capacitive susceptance (b). The conductance of a line is usually not a major factor since it is dominated by the capacitive susceptance $b = \omega C$. The line capacitance is a leakage (or charging) path for the ac line currents.

The capacitance of a transmission line is the consequence of the potential differences between the conductors themselves as well as potential differences between the conductors and ground. Charges on conductors arise, and the capacitance is the charge per unit potential difference. Since we are dealing with alternating voltages, we would expect that the charges on the conductors are also alternating (i.e., time varying). The time variation of the charges results in currents called the *line-charging currents*. In this section it is our intent to study the evaluation of line capacitance for a number of conductor configurations. The format of this section is similar to the previous one.

Capacitance of Single-Phase Line

Consider a single-phase, two-wire line of infinite length with conductor radii of r_1 and r_2 and separation D as shown in Figure 4-19. In Appendix 4-C we show that the potential at an arbitrary point P at distances r_a and r_b from A and B respectively is given by

$$V_p = \frac{q}{2\pi\varepsilon_0} ln\left(\frac{r_b}{r_a}\right) \quad (4.93)$$

where q is the charge density in coulombs per unit length.

The potential V_A on the conductor A of radius r_1 is therefore obtained by setting $r_a = r_1$ and $r_b = D$ to yield

$$V_A = \frac{q}{2\pi\varepsilon_0} ln\left(\frac{D}{r_1}\right) \quad (4.94)$$

Likewise for conductor B of radius r_2, we have

$$V_B = \frac{q}{2\pi\varepsilon_0} ln\left(\frac{r_2}{D}\right) \quad (4.95)$$

The potential difference between the two conductors is therefore

$$V_{AB} = V_A - V_B = \frac{q}{\pi\varepsilon_0} ln\left(\frac{D}{\sqrt{r_1 r_2}}\right) \quad (4.96)$$

The capacitance between the two conductors is defined as the charge on one conductor per unit of potential difference between the two conductors. As a result,

$$C_{AB} = \frac{q}{V_{AB}} = \frac{\pi\varepsilon_0}{ln\left(\frac{D}{\sqrt{r_1 r_2}}\right)} \quad \text{farads per meter} \quad (4.97)$$

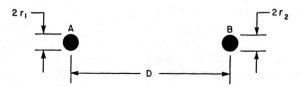

Figure 4-19. Single-Phase, Two-Wire Line.

4.4 Line Capacitance

If $r_1 = r_2 = r$, we have

$$C_{AB} = \frac{\pi \varepsilon_0}{\ln\left(\dfrac{D}{r}\right)} \tag{4.98}$$

Converting to microfarads (μF) per mile and changing the base of the logarithmic term, we have

$$C_{AB} = \frac{0.0388}{2\log\left(\dfrac{D}{r}\right)} \quad \mu\text{F per mile} \tag{4.99}$$

Equation (4.99) gives the line-to-line capacitance between the conductors. The capacitance to neutral for conductor A is defined as

$$C_{AN} = \frac{q}{V_A} = \frac{2\pi \varepsilon_0}{\ln\left(\dfrac{D}{r_1}\right)} \tag{4.100}$$

Likewise, observing that the charge on conductor B is $-q$, we have

$$C_{BN} = \frac{-q}{V_B} = \frac{2\pi \varepsilon_0}{\ln\left(\dfrac{D}{r_2}\right)} \tag{4.101}$$

For $r_1 = r_2$, we have

$$C_{AN} = C_{BN} = \frac{2\pi \varepsilon_0}{\ln\left(\dfrac{D}{r}\right)} \tag{4.102}$$

Observe that

$$C_{AN} = C_{BN} = 2 C_{AB} \tag{4.103}$$

This is consistent with the perception depicted in Figure 4-20 of the capacitance between the two lines C_{AB} as a series combination of C_{AN} and C_{BN}.

In terms of μF per mile, we have

$$C_{AN} = \frac{0.0388}{\log\dfrac{D}{r}} \quad \mu\text{F per mile to neutral} \tag{4.104}$$

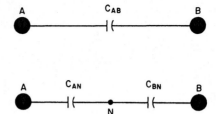

Figure 4-20. Capacitance between Lines as Series Combination of Capacitances to Neutral.

The capacitive reactance X_c is given by

$$X_c = \frac{1}{2\pi f C} = k' \log \frac{D}{r} \quad \text{ohms} \cdot \text{mile to neutral} \quad (4.105)$$

where

$$k' = \frac{4.1 \times 10^6}{f} \quad (4.106)$$

Expanding the logarithm, we have

$$X_c = k' \log D + k' \log \frac{1}{r} \quad (4.107)$$

The first term is called $X_{d'}$, the capacitive reactance spacing factor, and the second is called $X_{a'}$, the capacitive reactance at 1-ft spacing.

$$X_{d'} = k' \log D \quad (4.108)$$
$$X_{a'} = k' \log \frac{1}{r} \quad (4.109)$$
$$X_c = X_{d'} + X_{a'} \quad (4.110)$$

The last relationships are very similar to those for the inductance case. One difference that should be noted is that the conductor radius for the capacitance formula is the actual outside radius of the conductor and not the modified value r'.

Example 4-9

Find the capacitive reactance in ohms · mile per phase for a single-phase line with phase separation of 20 ft and conductor radius of 0.06677 ft for 60-Hz operation.

Solution

Note that this line is the same as that of Example 4-1. We have for $f = 60$ Hz:

$$k' = \frac{4.1 \times 10^6}{f} = 0.06833 \times 10^6$$

We calculate

$$X_{d'} = k' \log 20$$
$$= 88.904 \times 10^3$$
$$X_{a'} = k' \log \frac{1}{0.06677}$$
$$= 80.32 \times 10^3$$

As a result,

$$X_c = X_{d'} + X_{a'}$$
$$X_c = 169.224 \times 10^3 \text{ ohms} \cdot \text{mile to neutral}$$

Including the Effect of Earth

The effect of the presence of ground should be accounted for if the conductors are not high enough above the ground. This can be done using the theory of image charges. These are imaginary charges of the same magnitude as the physical charges but of opposite sign and are situated below the ground at a distance equal to that between the physical charge and ground. The situation is shown in Figure 4-21. Observe that the potential at ground due to the charge and its image is zero, which is consistent with the usual assumption that ground is a plane of zero potential.

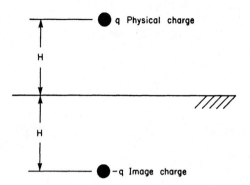

Figure 4-21. Image Charge Concept.

General Multiconductor Configurations

As we have seen in the case of problems of calculating inductances, the case of multiconductors is of fundamental importance. The basis for results presented here is the formula for potential derived in Appendix 4-C. This is given by

$$V_P = \frac{q}{2\pi\varepsilon_0} ln\left(\frac{1}{r_a}\right) + \frac{-q}{2\pi\varepsilon_0} ln\left(\frac{1}{r_b}\right) \qquad (4.111)$$

Considering a system of n parallel and very long conductors with charges $q_1, q_2, \ldots, q_n$ respectively, we can state that the potential at point P having distances $r_1, r_2, \ldots, r_n$ to the conductor as shown in Figure 4-22 is given by

$$V_P = \frac{q_1}{2\pi\varepsilon_0} ln\left(\frac{1}{r_1}\right) + \frac{q_2}{2\pi\varepsilon_0} ln\left(\frac{1}{r_2}\right) + \cdots + \frac{q_n}{2\pi\varepsilon_0} ln\left(\frac{1}{r_n}\right) \qquad (4.112)$$

This is a simple extension of the two-conductor case.

If we consider the same n parallel long conductors and wish to account for the presence of ground, we make use of the theory of images. As a result, we will have n images charges $-q_1, -q_2, \ldots, -q_n$ situated below the ground

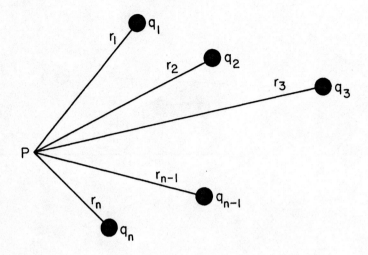

Figure 4-22. A Multiconductor Configuration.

at distance $r_1^-, r_2^-, \ldots, r_n^-$ from P. This is shown in Figure 4-23. The potential at P is therefore

$$V_P = \frac{q_1}{2\pi\varepsilon_0} \ln\left(\frac{1}{r_1}\right) + \frac{q_2}{2\pi\varepsilon_0} \ln\left(\frac{1}{r_2}\right) + \cdots + \frac{q_n}{2\pi\varepsilon_0} \ln\left(\frac{1}{r_n}\right)$$

$$+ \frac{-q_1}{2\pi\varepsilon_0} \ln\left(\frac{1}{r_1^-}\right) + \frac{-q_2}{2\pi\varepsilon_0} \ln\frac{1}{r_2^-} + \cdots + \frac{-q_n}{2\pi\varepsilon_0} \ln\left(\frac{1}{r_n^-}\right) \quad (4.113)$$

The above reduces to

$$V_P = \frac{q_1}{2\pi\varepsilon_0} \ln\left(\frac{r_1^-}{r_1}\right) + \frac{q_2}{2\pi\varepsilon_0} \ln\left(\frac{r_2^-}{r_2}\right) + \cdots + \frac{q_n}{2\pi\varepsilon_0} \ln\left(\frac{r_n^-}{r_n}\right) \quad (4.114)$$

The use of this relationship in finding the capacitance for many systems will be treated next.

136 The Transmission Subsystem

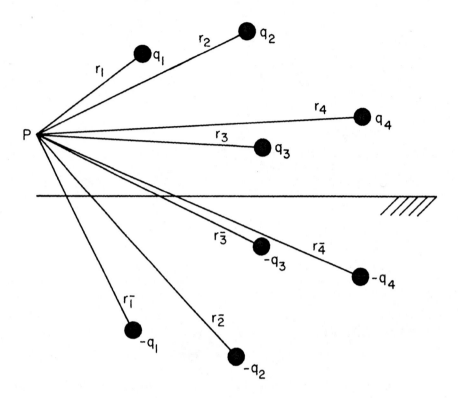

Figure 4-23. A Multiconductor Configuration Accounting for Ground Effect.

Capacitance of a Single-Phase Line Considering the Effect of Ground

Consider a single-phase line with conductors A and B as before. To account for ground effects, we introduce the image conductors A' and B'. The situation is shown in Figure 4-24.

The voltage of phase A is given according to Eq. (4.114) by

$$V_A = \frac{q}{2\pi\varepsilon_0} \ln\left(\frac{H}{r}\right) + \frac{-q}{2\pi\varepsilon_0} \ln\left(\frac{H_{AB'}}{D}\right) \qquad (4.115)$$

4.4 Line Capacitance   137

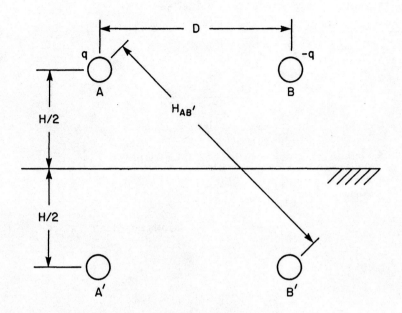

Figure 4-24. Single-Phase Line and Its Image.

The above reduces to

$$V_A = \frac{q}{2\pi\varepsilon_0} \ln\left(\frac{H}{r} \cdot \frac{D}{H_{AB'}}\right) \quad (4.116)$$

The voltage of phase B is

$$V_B = \frac{q}{2\pi\varepsilon_0} \ln\left(\frac{H_{AB'}}{D}\right) + \frac{-q}{2\pi\varepsilon_0} \ln\left(\frac{H}{r}\right) \quad (4.117)$$

which reduces to

$$V_B = \frac{q}{2\pi\varepsilon_0} \ln\left(\frac{H_{AB'}}{D} \cdot \frac{r}{H}\right) \quad (4.118)$$

The voltage difference is thus

$$V_{AB} = V_A - V_B$$
$$= \frac{q}{\pi\varepsilon_0} \ln\left(\frac{H}{r} \cdot \frac{D}{H_{AB'}}\right) \quad (4.119)$$

138 The Transmission Subsystem

The capacitance between the two conductors is thus

$$C_{AB} = \frac{q}{V_{AB}} \qquad (4.120)$$

or

$$C_{AB} = \frac{\pi \varepsilon_0}{\ln\left(\dfrac{D}{r} \cdot \dfrac{H}{H_{AB'}}\right)} \qquad (4.121)$$

The capacitance to neutral is obtained using

$$C_{AN} = \frac{q}{V_A}$$

$$= \frac{2\pi \varepsilon_0}{\ln\left(\dfrac{D}{r} \cdot \dfrac{H}{H_{AB'}}\right)} \quad \text{farads per meter} \qquad (4.122)$$

Observe that again

$$C_{AB} = \frac{C_{AN}}{2}$$

Let us examine the effect of including ground on the capacitance for a single-phase line in the following example.

Example 4-10

Find the capacitance to neutral for a single-phase line with phase separation of 20 ft and conductor radius of 0.06677 ft. Assume the height of the conductor above ground is 80 ft.

Solution

We have

$D = 20$ ft
$r = 0.06677$ ft
$H = 160$ ft

As a result,

$$H_{AB'} = \sqrt{(160)^2 + (20)^2} = 161.2452 \text{ ft}$$

Therefore we have

$$C_{AN_1} = \frac{2\pi\varepsilon_0}{ln\left(\frac{20}{0.06677} \cdot \frac{160}{161.2452}\right)}$$

$$= \frac{2\pi\varepsilon_0}{5.6945} \quad \text{farads per meter}$$

If we neglect earth effect, we have

$$C_{AN_2} = \frac{2\pi\varepsilon_0}{ln\left(\frac{20}{0.06677}\right)}$$

$$= \frac{2\pi\varepsilon_0}{5.7022} \quad \text{farads per meter}$$

The relative error involved if we neglect earth effect is:

$$\frac{C_{AN_1} - C_{AN_2}}{C_{AN_1}} = 0.00136$$

which is clearly less than 1 percent.

Capacitance of a Single-Circuit, Three-Phase Line

We consider the case of a three-phase line with conductors not equilaterally spaced. We assume that the line is transposed and as a result can assume that the capacitance to neutral in each phase is equal to the average value. This approach provides us with results of sufficient accuracy for our purposes. This configuration is shown in Figure 4-25.

The potential of conductor A in the first segment is

$$V_{A_I} = \left[\frac{q_a}{2\pi\varepsilon_0}ln\left(\frac{1}{r}\right) + \frac{q_b}{2\pi\varepsilon_0}ln\left(\frac{1}{D_{12}}\right) + \frac{q_c}{2\pi\varepsilon_0}ln\left(\frac{1}{D_{13}}\right)\right] \quad (4.123)$$

In segment II, conductor A is in position 2, conductor B is in position 3, and conductor C is in position 1. Thus,

$$V_{A_{II}} = \frac{1}{2\pi\varepsilon_0}\left[q_a ln\left(\frac{1}{r}\right) + q_b ln\left(\frac{1}{D_{23}}\right) + q_c ln\left(\frac{1}{D_{12}}\right)\right] \quad (4.124)$$

Similarly, in segment III,

$$V_{A_{III}} = \frac{1}{2\pi\varepsilon_0}\left[q_a ln\left(\frac{1}{r}\right) + q_b ln\left(\frac{1}{D_{13}}\right) + q_c ln\left(\frac{1}{D_{23}}\right)\right] \quad (4.125)$$

The average potential is

$$V_A = \tfrac{1}{3}(V_{A_I} + V_{A_{II}} + V_{A_{III}}) \quad (4.126)$$

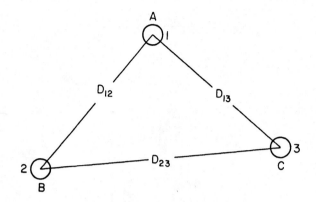

Figure 4-25. Three-Phase Line with General Spacing.

This combined with the three-phase balanced condition

$$q_a + q_b + q_c = 0$$

will result in

$$V_A = \frac{q_a}{2\pi\varepsilon_0} \ln\left[\frac{(D_{12}D_{23}D_{13})^{1/3}}{r}\right] \quad (4.127)$$

The capacitance to neutral is therefore given by

$$C_{AN} = \frac{q_a}{V_A}$$

$$= \frac{2\pi\varepsilon_0}{\ln\left(\frac{D_{eq}}{r}\right)} \quad (4.128)$$

where

$$D_{eq} = \sqrt[3]{D_{12}D_{23}D_{13}} \quad (4.129)$$

Observe that D_{eq} is the same as the geometric mean distance obtained in the case of inductance. Moreover, we have the same expression for the capacitance as that for the single-phase line. Thus

$$C_{AN} = \frac{2\pi\varepsilon_0}{\ln\left(\frac{\text{GMD}}{r}\right)} \quad \text{farad per meter} \quad (4.130)$$

If we account for the influence of earth, we come up with a slightly modified expression for the capacitance. Consider the same three-phase line with the attendant image line shown in Figure 4-26. The line is assumed to be transposed. As a result, the average phase A voltage will be given by

$$V_A = \frac{1}{3(2\pi\varepsilon_0)}\left\{\left[q_a\ln\left(\frac{H_1}{r}\right) + q_b\ln\left(\frac{H_{12}}{D_{12}}\right) + q_c\ln\left(\frac{H_{13}}{D_{13}}\right)\right]\right.$$
$$+ \left[q_a\ln\left(\frac{H_2}{r}\right) + q_b\ln\left(\frac{H_{23}}{D_{23}}\right) + q_c\ln\left(\frac{H_{12}}{D_{12}}\right)\right]$$
$$\left. + \left[q_a\ln\left(\frac{H_3}{r}\right) + q_b\ln\left(\frac{H_{13}}{D_{13}}\right) + q_c\ln\left(\frac{H_{23}}{D_{23}}\right)\right]\right\} \quad (4.131)$$

Each of the above square brackets corresponds to a segment of the line.

With our usual assumption of a balanced three-phase operation

$$q_a + q_b + q_c = 0$$

we get

$$V_A = \frac{q_a}{3(2\pi\varepsilon_0)}\ln\left[\frac{(D_{12}D_{23}D_{13})(H_1H_2H_3)}{r^3(H_{12}H_{13}H_{23})}\right] \quad (4.132)$$

From the above,

$$C_{AN} = \frac{2\pi\varepsilon_0}{\ln\left[\frac{D_{eq}}{r}\left(\frac{H_1H_2H_3}{H_{12}H_{13}H_{23}}\right)^{1/3}\right]}$$

or

$$C_{AN} = \frac{2\pi\varepsilon_0}{\ln\left(\frac{D_{eq}}{r}\right) + \ln\left(\frac{H_1H_2H_3}{H_{12}H_{13}H_{23}}\right)^{1/3}} \quad (4.133)$$

We define the mean distances

$$H_s = (H_1H_2H_3)^{1/3} \quad (4.134)$$
$$H_m = (H_{12}H_{23}H_{13})^{1/3} \quad (4.135)$$

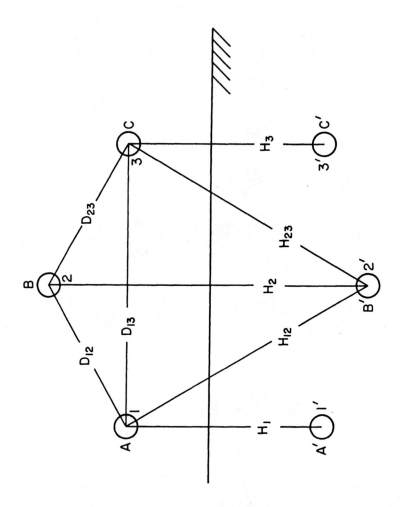

Figure 4-26. Three-Phase Line with Ground Effect Included.

4.4 Line Capacitance

Then our capacitance expression reduces to

$$C_{AN} = \frac{2\pi\varepsilon_0}{\ln\left(\frac{D_{eq}}{r}\right) - \ln\left(\frac{H_m}{H_s}\right)} \qquad (4.136)$$

We can thus conclude that including the effect of ground will give a higher value for the capacitance than that obtained by neglecting the ground effect.

Example 4-11

Find the capacitance to neutral for the single-circuit, three-phase, 345-kV line with conductors having an outside diameter of 1.063 in. with phase configuration as shown in Figure 4-27.

Solution

$$\text{GMD} = [(23.5)(23.5)(47)]^{1/3}$$
$$= 29.61 \text{ ft}$$
$$r = \frac{1.063}{(2)(12)} = 0.0443 \text{ ft}$$
$$C_{AN} = \frac{2\pi\varepsilon_0}{\ln\left(\frac{\text{GMD}}{r}\right)}$$
$$= 8.5404 \times 10^{-12} \quad \text{farads per meter}$$

Example 4-12

Calculate the capacitance to neutral for the line of Example 4-11, including the effect of earth, assuming the height of the conductors is 50 ft.

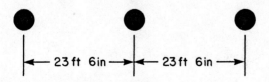

Figure 4-27. Conductor Layout for Example 4-11.

Solution

We have from Figure 4-28:

$$H_1 = H_2 = H_3 = 2 \times 50 = 100 \text{ ft}$$

$$H_{12} = H_{23} = \sqrt{(23.5)^2 + (100)^2} = 102.72$$

$$H_{13} = \sqrt{(47)^2 + (100)^2} = 110.49$$

$$ln\left(\frac{H_s}{H_m}\right) = ln\left[\frac{(100)(100)(100)}{(102.72)(102.72)(110.49)}\right]^{1/3}$$

$$= -0.0512$$

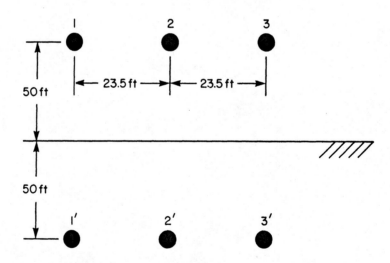

Figure 4-28. Configuration for Example 4-12.

Thus

$$C_{AN} = \frac{1}{(18 \times 10^9)(6.505 - 0.0512)}$$
$$= 8.6082 \times 10^{-12} \quad \text{farads per meter}$$

Capacitance of Double-Circuit Lines

It should be evident by now that the calculation of capacitance of a double-circuit line can be quite involved if rigorous analysis is followed. In practice, however, sufficient accuracy is obtained if we assume that the charges are uniformly distributed and that the charge q_a is divided equally between the two phase A conductors. We further assume that the line is transposed. As a result, capacitance formulae similar in nature to those for the single-circuit line emerge.

Consider a double-circuit line with phases A, B, C, A', B', and C' placed in positions 1, 2, 3, 1', 2', and 3' respectively in segment I of the transposition cycle. The situation is shown in Figure 4-29. In segment II, the phase conductors are rotated such that they occupy the positions shown in Figure 4-30. For segment III, we have Figure 4-31. Clearly, six terms will be present in each potential relation. As before we deal with phase A only.

The voltage of phase A in segment I is given by

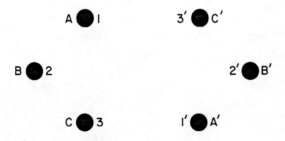

Figure 4-29. Double-Circuit Line Conductor Configuration in Cycle Segment I of Transposition.

146 The Transmission Subsystem

$$V_{A_I} = \frac{1}{2(2\pi\varepsilon_0)}\left[q_a ln\left(\frac{1}{r}\right) + q_b ln\left(\frac{1}{D_{12}}\right) + q_c ln\left(\frac{1}{D_{13}}\right)\right.$$
$$\left. + q_a ln\left(\frac{1}{D_{11'}}\right) + q_b ln\left(\frac{1}{D_{12'}}\right) + q_c ln\left(\frac{1}{D_{13'}}\right)\right] \quad (4.137)$$

In segment *II*, we have

$$V_{A_{II}} = \frac{1}{2(2\pi\varepsilon_0)}\left[q_a ln\left(\frac{1}{r}\right) + q_b ln\left(\frac{1}{D_{23}}\right) + q_c ln\left(\frac{1}{D_{12}}\right)\right.$$
$$\left. + q_a ln\left(\frac{1}{D_{22'}}\right) + q_b ln\left(\frac{1}{D_{23'}}\right) + q_c ln\left(\frac{1}{D_{21'}}\right)\right] \quad (4.138)$$

In segment III, we have

$$V_{A_{III}} = \frac{1}{2(2\pi\varepsilon_0)}\left[q_a ln\left(\frac{1}{r}\right) + q_b ln\left(\frac{1}{D_{13}}\right) + q_c ln\left(\frac{1}{D_{23}}\right)\right.$$
$$\left. + q_a ln\left(\frac{1}{D_{33'}}\right) + q_b ln\left(\frac{1}{D_{31'}}\right) + q_c ln\left(\frac{1}{D_{32'}}\right)\right] \quad (4.139)$$

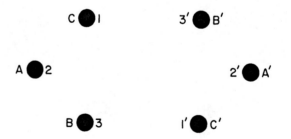

Figure 4-30. Double-Circuit Line Configuration for Segment *II*.

4.4 Line Capacitance

The voltage on phase A' in segment I is given by

$$V_{A'_I} = \frac{1}{2(2\pi\varepsilon_0)} \left[q_a ln\left(\frac{1}{r}\right) + q_b ln\left(\frac{1}{D_{1'2}}\right) + q_c ln\left(\frac{1}{D_{1'3}}\right) \right.$$

$$\left. + q_a ln\left(\frac{1}{D_{11'}}\right) + q_b ln\left(\frac{1}{D_{1'2'}}\right) + q_c ln\left(\frac{1}{D_{1'3'}}\right) \right] \quad (4.140)$$

In segment II, we have

$$V_{A'_{II}} = \frac{1}{2(2\pi\varepsilon_0)} \left[q_a ln\left(\frac{1}{r}\right) + q_b ln\left(\frac{1}{D_{2'3}}\right) q_c ln\left(\frac{1}{D_{2'1}}\right) \right.$$

$$\left. + q_a ln\left(\frac{1}{D_{22'}}\right) + q_b ln\left(\frac{1}{D_{2'3'}}\right) + q_c ln\left(\frac{1}{D_{2'1'}}\right) \right] \quad (4.141)$$

In segment III, we have

$$V_{A'_{III}} = \frac{1}{2(2\pi\varepsilon_0)} \left[q_a ln\left(\frac{1}{r}\right) + q_b ln\left(\frac{1}{D_{3'1}}\right) + q_c ln\left(\frac{1}{D_{3'2}}\right) \right.$$

$$\left. + q_a ln\left(\frac{1}{D_{33'}}\right) + q_b ln\left(\frac{1}{D_{3'1'}}\right) + q_c ln\left(\frac{1}{D_{3'2'}}\right) \right] \quad (4.142)$$

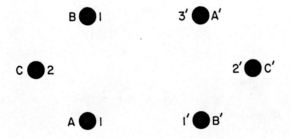

Figure 4-31. Double-Circuit Line Configuration for Segment III.

As we have done in the inductance case, we calculate the average voltage and use the balance condition

$$q_a + q_b + q_c = 0$$

to obtain

$$V_A = \frac{q_a}{12(2\pi\varepsilon_0)} \ln\left[\frac{(D_{12}D_{1'2'}D_{12'}D_{1'2})(D_{23}D_{2'3'}D_{2'3}D_{23'})(D_{13}D_{1'3'}D_{13'}D_{1'3})}{(r^6)(D_{11'}^2 D_{22'}^2 D_{33'}^2)}\right]$$

(4.143)

As a result,

$$C_{AN} = \frac{2\pi\varepsilon_0}{\ln\left(\dfrac{\text{GMD}}{\text{GMR}}\right)}$$

(4.144)

Where as before for the inductance case, we define

$$\text{GMD} = \left(D_{AB_{eq}} D_{BC_{eq}} D_{AC_{eq}}\right)^{1/3}$$

(4.145)

$$D_{AB_{eq}} = \left(D_{12} D_{1'2'} D_{12'} D_{1'2}\right)^{1/4}$$

(4.146)

$$D_{BC_{eq}} = \left(D_{23} D_{2'3'} D_{2'3} D_{23'}\right)^{1/4}$$

(4.147)

$$D_{AC_{eq}} = \left(D_{13} D_{1'3'} D_{13'} D_{1'3}\right)^{1/4}$$

(4.148)

The GMR is given by

$$\text{GMR} = (r_A r_B r_C)^{1/3}$$

(4.149)

with

$$r_A = (rD_{11'})^{1/2}$$

(4.150)

$$r_B = (rD_{22'})^{1/2}$$

(4.151)

$$r_C = (rD_{33'})^{1/2}$$

(4.152)

If we wish to include the effect of the earth in the calculation, a simple extension of the above analysis will do the job. For the three-phase, double-circuit line as shown in Figure 4-32, we have

$$V_{A_I} = \frac{1}{2(2\pi\varepsilon_0)}\left[q_a \ln\left(\frac{H_1}{r}\right) + q_b \ln\left(\frac{H_{12}}{D_{12}}\right) + q_c \ln\left(\frac{H_{13}}{D_{13}}\right)\right.$$
$$\left. + q_a \ln\left(\frac{H_{11'}}{D_{11'}}\right) + q_b \ln\left(\frac{H_{12'}}{D_{12'}}\right) + q_c \ln\left(\frac{H_{13'}}{D_{13'}}\right)\right]$$

(4.153)

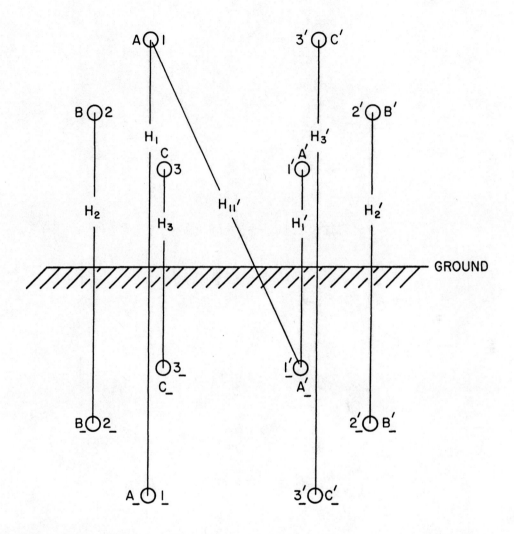

Figure 4-32. Double-Circuit Line with Ground Effect.

for phase A in segment I. In segment II, we have

$$V_{A_{II}} = \frac{1}{2(2\pi\varepsilon_0)} \left[q_a ln\left(\frac{H_2}{r}\right) + q_b ln\left(\frac{H_{23}}{D_{23}}\right) + q_c ln\left(\frac{H_{12}}{D_{12}}\right) \right.$$
$$\left. + q_a ln\left(\frac{H_{22'}}{D_{22'}}\right) + q_b ln\left(\frac{H_{23'}}{D_{23'}}\right) + q_c ln\left(\frac{H_{21'}}{D_{21'}}\right) \right] \quad (4.154)$$

In segment III, we have

$$V_{A_{III}} = \frac{1}{2(2\pi\varepsilon_0)} \left[q_a ln\left(\frac{H_3}{r}\right) + q_b ln\left(\frac{H_{13}}{D_{13}}\right) + q_c ln\left(\frac{H_{23}}{D_{23}}\right) \right.$$
$$\left. + q_a ln\left(\frac{H_{33'}}{D_{33'}}\right) + q_b ln\left(\frac{H_{31'}}{D_{31'}}\right) + q_c ln\left(\frac{H_{32'}}{D_{32'}}\right) \right] \quad (4.155)$$

For phase A' in segment I, we have

$$V_{A'_{I}} = \frac{1}{2(2\pi\varepsilon_0)} \left[q_a ln\left(\frac{H_{1'}}{r}\right) + q_b ln\left(\frac{H_{1'2}}{D_{1'2}}\right) + q_c ln\left(\frac{H_{1'3}}{D_{1'3}}\right) \right.$$
$$\left. + q_a ln\left(\frac{H_{11'}}{D_{11'}}\right) + q_b ln\left(\frac{H_{1'2'}}{D_{1'2'}}\right) + q_c ln\left(\frac{H_{1'3'}}{D_{1'3'}}\right) \right] \quad (4.156)$$

In segment II, we have

$$V_{A'_{II}} = \frac{1}{2(2\pi\varepsilon_0)} \left[q_a ln\left(\frac{H_{2'}}{r}\right) + q_b ln\left(\frac{H_{2'3}}{D_{2'3}}\right) + q_c ln\left(\frac{H_{2'1}}{D_{2'1}}\right) \right.$$
$$\left. + q_a ln\left(\frac{H_{22'}}{D_{22'}}\right) + q_b ln\left(\frac{H_{2'3'}}{D_{2'3'}}\right) + q_c ln\left(\frac{H_{2'1'}}{D_{2'1'}}\right) \right] \quad (4.157)$$

In segment III, we have

$$V_{A'_{III}} = \frac{1}{2(2\pi\varepsilon_0)} \left[q_a ln\left(\frac{H_{3'}}{r}\right) + q_b ln\left(\frac{H_{3'1}}{D_{3'1}}\right) + q_c ln\left(\frac{H_{3'2}}{D_{3'2}}\right) \right.$$
$$\left. + q_a ln\left(\frac{H_{33'}}{D_{33'}}\right) + q_b ln\left(\frac{H_{3'1'}}{D_{3'1'}}\right) + q_c ln\left(\frac{H_{3'2'}}{D_{3'2'}}\right) \right] \quad (4.158)$$

As a result, the average phase A and A' voltage is

$$V_A = \frac{q_a}{12(2\pi\varepsilon_0)} \left\{ ln\left[\frac{(D_{12}D_{1'2'}D_{12'}D_{1'2})(D_{23}D_{2'3'}D_{2'3}D_{23'})(D_{13}D_{1'3'}D_{1'3}D_{13'})}{((r^6)(D_{11'}^2 D_{22'}^2 D_{33'}^2))}\right] \right.$$
$$\left. + ln\left[\frac{(H_1 H_{1'} H_2 H_{2'} H_3 H_{3'} H_{11'}^2 H_{22'}^2 H_{33'}^2)}{(H_{12}H_{1'2'}H_{12'}H_{1'2})(H_{13}H_{1'3'}H_{1'3}H_{13'})(H_{23}H_{2'3'}H_{2'3}H_{23'})}\right] \right\}$$

$$(4.159)$$

The above expression bears striking similarities to that obtained for the single-circuit, three-phase line. We can thus write

$$C_{AN} = \frac{2\pi\varepsilon_0}{ln\left(\dfrac{GMD}{GMR}\right) + \alpha} \qquad (4.160)$$

where GMD and GMR are as given by Eqs. (4.145) and (4.149). Also we defined

$$\alpha = ln\left(\frac{H_s}{H_m}\right) \qquad (4.161)$$

$$H_s = \left(H_{s_1} H_{s_2} H_{s_3}\right)^{1/3} \qquad (4.162)$$

with

$$H_{s_1} = \left(H_1 H_{1'} H_{11'}^2\right)^{1/4} \qquad (4.163)$$

$$H_{s_2} = \left(H_2 H_{2'} H_{22'}^2\right)^{1/4} \qquad (4.164)$$

$$H_{s_3} = \left(H_3 H_{3'} H_{33'}^2\right)^{1/4} \qquad (4.165)$$

and

$$H_m = \left(H_{m_{12}} H_{m_{13}} H_{m_{23}}\right)^{1/3} \qquad (4.166)$$

$$H_{m_{12}} = \left(H_{12} H_{1'2'} H_{12'} H_{1'2}\right)^{1/4} \qquad (4.167)$$

$$H_{m_{13}} = \left(H_{13} H_{1'3'} H_{13'} H_{1'3}\right)^{1/4} \qquad (4.168)$$

$$H_{m_{23}} = \left(H_{23} H_{2'3'} H_{23'} H_{2'3}\right)^{1/4} \qquad (4.169)$$

Capacitance of Bundle-Conductor Lines

It should be evident by now that it is sufficient to consider a single-phase line to reach conclusions that can be readily extended to the three-phase case. We use this in the present discussion pertaining to bundle-conductor lines.

Consider a single-phase line with bundle conductor having N subconductors on a circle of radius A. Each subconductor has a radius of r. Phase A will have a charge q_a uniformly distributed among the N subconductors.

Although the charge on phase B will be distributed, it is practical to concentrate the charge $-q_a$ on phase B in the phase center situated a distance D from the center of phase A. The voltages on each of the subconductors of phase A are assumed equal to V_A. Thus we may take subconductor 1 to derive our desired relationship.

In accordance with the foregoing discussion, we state that

$$V_A = \frac{1}{2\pi\varepsilon_0}\left[\frac{q_a}{N}\ln\left(\frac{1}{r}\right) + \frac{q_a}{N}\ln\left(\frac{1}{D_{12}}\right) + \cdots + \frac{q_a}{N}\ln\left(\frac{1}{D_{1N}}\right) - q_a\ln\left(\frac{1}{D}\right)\right]$$

This reduces to

$$V_A = \frac{q_a}{2\pi\varepsilon_0}\ln\left[\frac{D}{(rD_{12}D_{13}\cdots D_{1N})^{1/N}}\right] \qquad (4.170)$$

From Figure 4-7 as we have seen before, and using Eq. (4.51), we conclude that

$$V_A = \frac{q_a}{2\pi\varepsilon_0}\ln\left\{\frac{D}{\left[rN(A)^{N-1}\right]^{1/N}}\right\} \qquad (4.171)$$

As a result, we obtain

$$C_{AN} = \frac{2\pi\varepsilon_0}{\ln\left\{\dfrac{D}{\left[rN(A)^{N-1}\right]^{1/N}}\right\}} \quad \text{farads per meter} \qquad (4.172)$$

The extension of the above result to the three-phase case is clearly obtained by replacing D by the GMD. Thus

$$C_{AN} = \frac{2\pi\varepsilon_0}{\ln\left\{\dfrac{\text{GMD}}{\left[rN(A)^{N-1}\right]^{1/N}}\right\}} \qquad (4.173)$$

with

$$\text{GMD} = (D_{AB}D_{BC}D_{AC})^{1/3} \qquad (4.174)$$

The capacitive reactance in megaohms calculated from Eqs. (4.171) and (4.172) for 60 Hz and 1 mile of line using the base 10 logarithm would be as follows:

$$X_c = 0.0683\log\left\{\frac{\text{GMD}}{\left[rN(A)^{N-1}\right]^{1/N}}\right\} \qquad (4.175)$$

$$X_c = X'_a + X'_d \qquad (4.176)$$

This capacitive reactance can be divided into two parts

$$X'_a = 0.0683\log\left\{\frac{1}{\left[rN(A)^{N-1}\right]^{1/N}}\right\} \qquad (4.177)$$

and

$$X'_d = 0.0683\log(\text{GMD}) \qquad (4.178)$$

If the bundle spacing S is specified rather than the radius A of the circle on which the conductors lie, then as before,

$$A = \frac{S}{2\sin\left(\frac{\pi}{N}\right)} \quad \text{for } N > 1 \qquad (4.179)$$

Example 4-13

Find the capacitance to neutral for the three-phase, 735-kV, bundle-conductor line shown in Figure 4-33 with subconductor outside diameter of 1.16 in. and subconductor spacing of 18 in.

Solution

The GMD is obtained as

$$\text{GMD} = [(50)(50)(100)]^{1/3} = 62.996 \text{ ft}$$

with

$$r = \frac{1.16}{(2)(12)} = 0.0483$$

$$N = 4$$

$$S = 2A\sin\left(\frac{\pi}{N}\right)$$

$$18 = 2A\sin 45°$$

$$A = 1.0607 \text{ ft}$$

Thus

$$\text{GMR} = \left[rN(A)^{N-1}\right]^{1/N}$$

$$= \left[(0.0483)(4)(1.0607)^3\right]^{1/4}$$

$$= 0.693 \text{ ft}$$

As a result,

$$C_{AN} = \frac{1}{(18 \times 10^9)\ln\left(\frac{62.996}{0.693}\right)}$$

$$= 12.319 \times 10^{-12} \text{ farads per meter}$$

The capacitive reactance is obtained as

$$X_{d'} = (0.06833 \times 10^6)\log(62.996) = 1.2295 \times 10^5$$

$$X_{a'} = (0.06833 \times 10^6)\log\left(\frac{1}{0.693}\right) = 1.0883 \times 10^4$$

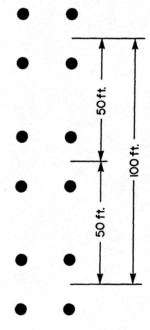

Figure 4-33. Configuration for Example 4-13.

Thus

$$X_c = X_{d'} + X_{a'}$$
$$= 1.3383 \times 10^5 \quad \text{ohms} \cdot \text{mile to neutral}$$

4.5 TWO-PORT NETWORKS

A network can have two terminals or more, but many of the networks of importance in electric energy system studies are those with four terminals arranged in two pairs. The box in Figure 4-34 indicates a two-terminal pair network, which might contain a transmission line model or a transformer model, to name a few in our power system applications. The box is sometimes called a *coupling network*, or *four-pole*, or a *two-terminal pair*. The term *two-port network* is in common use. It is a common mistake to call it a four-terminal network. In fact, the two-port network is a *restricted* four-terminal network since we require that the current at one terminal of a pair must be equal and opposite to the current at the other terminal of that pair. An obvious example of a nonrestricted four-terminal network is a three-phase load with its neutral connection brought out to a fourth terminal.

Useful examples of two-port network configurations are shown in Figure 4-35. These are: the π-network, the V-network, and the T-network. Note that these three networks are completely interchangeable. By this we mean that the T-network, for example, can be replaced by an equivalent π, using the Y-Δ conversion.

Two important problems arise in the application of two-port network theory to electric energy systems. These are:

1. *The transfer problem*: It is required to find the currents in terms of both voltages, or to find the voltages in terms of both currents.
2. *The transmission problem*: It is required to find voltage and current at one pair of terminals in terms of quantities at the other pair.

The transfer problem is easily handled as follows. The node equations for the two-port network yield

$$\begin{bmatrix} I_s \\ -I_r \end{bmatrix} = \begin{bmatrix} y_{ss} & y_{sr} \\ y_{rs} & y_{rr} \end{bmatrix} \begin{bmatrix} V_s \\ V_r \end{bmatrix} \quad (4.180)$$

Figure 4-34. A Two-Terminal Pair or Two-Port Network.

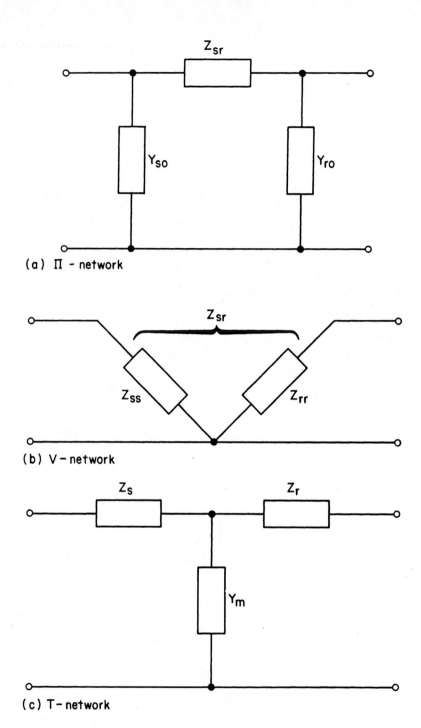

Figure 4-35. Examples of Two-Port Networks: (a) The π-Network; (b) The V-Network; (c) The T-Network.

In a bilateral network, $y_{sr} = y_{rs}$. The physical meaning of the admittance parameters is as follows:

$$y_{ss} = \frac{I_s}{V_s} \text{ for } V_r = 0$$

$$y_{sr} = \frac{I_s}{V_r} \text{ for } V_s = 0$$

$$y_{rs} = \frac{-I_r}{V_s} \text{ for } V_r = 0$$

$$y_{rr} = \frac{-I_r}{V_r} \text{ for } V_s = 0$$

Of course, $V_s = 0$ means that the s terminals are short-circuited. Thus the y's are seen to be ratios of easily measured currents and voltages. The parameters y_{ss} and y_{rr} are called *driving-point admittances*, and y_{sr} and y_{rs} are the *transfer admittances*. Sometimes the term *short-circuit* is added to these notations.

The loop equations yield

$$\begin{bmatrix} V_s \\ V_r \end{bmatrix} = \begin{bmatrix} Z_{ss} & Z_{sr} \\ Z_{rs} & Z_{rr} \end{bmatrix} \begin{bmatrix} I_s \\ -I_r \end{bmatrix} \tag{4.181}$$

Here the physical interpretation of the driving and transfer impedances follows in a manner similar to the above treatment.

The transmission problem is handled by assuming a pair of equations of the form

$$V_s = AV_r + BI_r \tag{4.182}$$
$$I_s = CV_r + DI_r \tag{4.183}$$

to represent the two-port network. In matrix form, we thus have

$$\begin{bmatrix} V_s \\ I_s \end{bmatrix} = \begin{bmatrix} A & B \\ C & D \end{bmatrix} \begin{bmatrix} V_r \\ I_r \end{bmatrix}$$

The values of the A, B, C, and D parameters are given in terms of the driving point and transfer admittances by:

$$A = \frac{-y_{rr}}{y_{rs}} \tag{4.184}$$

$$B = \frac{-1}{y_{rs}} \tag{4.185}$$

$$C = y_{sr} - \frac{y_{ss} y_{rr}}{y_{rs}} \tag{4.186}$$

$$D = \frac{-y_{ss}}{y_{rs}} \tag{4.187}$$

We also have the following relations between the driving-point and transfer impedances and the $ABCD$ parameters:

$$A = \frac{Z_{ss}}{Z_{rs}} \qquad (4.188)$$

$$B = \frac{Z_{ss}Z_{rr}}{Z_{rs}} - Z_{sr} \qquad (4.189)$$

$$C = \frac{1}{Z_{rs}} \qquad (4.190)$$

$$D = \frac{Z_{rr}}{Z_{rs}} \qquad (4.191)$$

For bilateral networks (and this includes all ordinary passive electrical networks), the transfer parameters are equal (that is, $y_{sr} = y_{rs}$ in one form, and $Z_{sr} = Z_{rs}$ in the other). Thus there are only three independent short-circuit admittances and only three independent open-circuit impedances. This suggests that A, B, C, and D may not all be independent. Indeed, a quick calculation verifies that

$$AD - BC = \frac{y_{sr}}{y_{rs}} = 1$$

if one uses the admittance values and a similar expression using the impedance values. Thus there are but three independent parameters in the $ABCD$ set also.

Symmetry of a two-port network reduces the number of independent parameters to two. The network is *symmetrical* if it can be turned end for end in a system without altering the behavior of the rest of the system. An example is the transmission line, as will be seen later on. To satisfy this definition, a symmetrical network must have

$$y_{ss} = y_{rr} \quad \text{and} \quad Z_{ss} = Z_{rr}$$

Introducing the first of these, it is seen that for a symmetrical two-port network,

$$A = D$$

It is instructive at this point to show how the A, B, C, and D parameters may be obtained for certain special two-port networks. Let us consider the symmetrical T-network shown in Figure 4-36. Our approach here will utilize Kirchhoff's voltage and current laws. Let the potential of

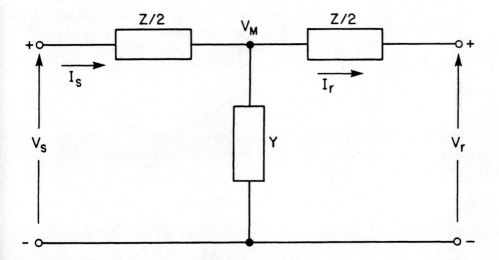

Figure 4-36. A Symmetrical T-Network.

the middle point of the T be V_M; then we obtain

$$V_s = I_s\left(\frac{Z}{2}\right) + V_M$$
$$I_s = V_M Y + I_r$$
$$V_M = V_r + I_r\left(\frac{Z}{2}\right)$$

From these equations, we can write

$$V_s = \left(1 + \frac{ZY}{2}\right)V_r + Z\left(1 + \frac{ZY}{4}\right)I_r$$
$$I_s = YV_r + \left(1 + \frac{ZY}{2}\right)I_r$$

Comparing the above expressions with Eqs. (4.182) and (4.183), we conclude that

$$A = 1 + \frac{ZY}{2} \tag{4.192}$$

$$B = Z\left(1 + \frac{ZY}{4}\right) \tag{4.193}$$

$$C = Y \tag{4.194}$$

$$D = 1 + \frac{ZY}{2} \tag{4.195}$$

We remark here that alternative approaches may be applied to determine the T-network's A, B, C, and D parameters. We will illustrate some of these in considering other networks.

We now turn our attention to another important two-port network that plays a fundamental role in power system analysis—this is the symmetrical π-network. Figure 4-37 shows a symmetrical π-network. We will use the short-circuit admittance parameters to obtain the $ABCD$ parameters. The admittance looking into either end with the other short-circuited is

$$y_{ss} = y_{rr} = \frac{1}{Z} + \frac{1}{2}Y$$

If one end, short-circuited, is carrying current I, the applied voltage between terminals at the other end must be ZI. The ratio is the short-circuit transfer admittance; thus

$$y_{sr} = y_{rs} = \frac{-1}{Z}$$

These short-circuit admittances are now substituted to find the $ABCD$ values of the symmetrical π-network as shown below:

$$A = \left(1 + \frac{ZY}{2}\right) \tag{4.196}$$

$$B = Z \tag{4.197}$$

$$C = Y\left(1 + \frac{ZY}{4}\right) \tag{4.198}$$

$$D = A \tag{4.199}$$

Our final special network is the L-network shown in Figure 4-38. Here we will use the physical interpretation of the $ABCD$ parameters to our advantage as follows. With the receiving end open-circuit, i.e., $I_r = 0$, we have

$$V_s = V_r$$

and

$$I_s = YV_r$$

from which

$$A = 1 \tag{4.200}$$

$$C = Y \tag{4.201}$$

With the receiving end short-circuited, i.e., $V_r = 0$, we have

$$\frac{I_r}{I_s} = \frac{\frac{1}{Y}}{Z + \frac{1}{Y}}$$

or

$$I_s = (1 + ZY)I_r$$

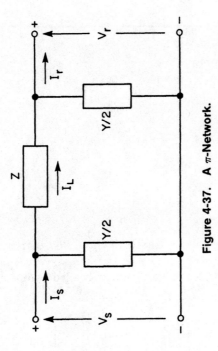

Figure 4-37. A π-Network.

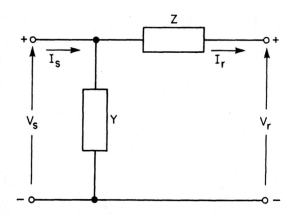

Figure 4-38. An L-Network.

and
$$V_s = ZI_r$$
Thus
$$D = 1 + ZY \tag{4.202}$$
$$B = Z \tag{4.203}$$

One of the most valued aspects of the *ABCD* parameters is that they are readily combined to find overall parameters when networks are connected in cascade. Figure 4-39 shows two cascaded two-port networks. We can write

$$\begin{bmatrix} V_s \\ I_s \end{bmatrix} = \begin{bmatrix} A_1 & B_1 \\ C_1 & D_1 \end{bmatrix} \begin{bmatrix} V_M \\ I_M \end{bmatrix},$$

$$\begin{bmatrix} V_M \\ I_M \end{bmatrix} = \begin{bmatrix} A_2 & B_2 \\ C_2 & D_2 \end{bmatrix} \begin{bmatrix} V_r \\ I_r \end{bmatrix}$$

From which, eliminating (V_M, I_M), we obtain

$$\begin{bmatrix} V_s \\ I_s \end{bmatrix} = \begin{bmatrix} A_1 & B_1 \\ C_1 & D_1 \end{bmatrix} \begin{bmatrix} A_2 & B_2 \\ C_2 & D_2 \end{bmatrix} \begin{bmatrix} V_r \\ I_r \end{bmatrix}$$

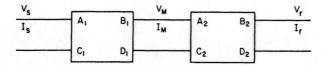

Figure 4-39. A Cascade of Two two-Port Networks.

Thus the equivalent $ABCD$ parameters of the cascade are

$$A = A_1 A_2 + B_1 C_2 \qquad (4.204)$$

$$B = A_1 B_2 + B_1 D_2 \qquad (4.205)$$

$$C = C_1 A_2 + D_1 C_2 \qquad (4.206)$$

$$D = C_1 B_2 + D_1 D_2 \qquad (4.207)$$

If three networks or more are cascaded, the equivalent $ABCD$ parameters can be obtained most easily by matrix multiplications as was done above.

The idea of cascaded networks can be used to full value in deriving the $ABCD$ parameters of networks based on manipulating elementary circuits such as a series impedance and a shunt admittance. The following are obvious relations:

$$\begin{aligned} A_Z &= 1 \\ B_Z &= Z \\ C_Z &= 0 \\ D_Z &= 1 \end{aligned} \qquad (4.208)$$

for a two-port network made up of only a series impedance Z. On the other hand,

$$\begin{aligned} A_y &= 1 \\ B_y &= 0 \\ C_y &= Y \\ D_y &= 1 \end{aligned} \qquad (4.209)$$

for a network made up of a shunt admittance. These two networks are shown in Figure 4-40.

Let us derive the $ABCD$ parameters of the symmetrical π-network of Figure 4-37 using this concept as the matrix product:

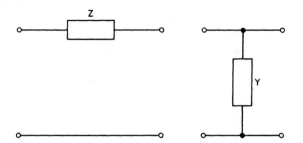

Figure 4-40. Two Elementary Networks.

$$\begin{bmatrix} A & B \\ C & D \end{bmatrix} = \begin{bmatrix} 1 & 0 \\ \frac{Y}{2} & 1 \end{bmatrix} \begin{bmatrix} 1 & Z \\ 0 & 1 \end{bmatrix} \begin{bmatrix} 1 & 0 \\ \frac{Y}{2} & 1 \end{bmatrix} = \begin{bmatrix} 1 + \frac{ZY}{2} & Z \\ Y\left(1 + \frac{ZY}{4}\right) & 1 + \frac{ZY}{2} \end{bmatrix}$$
(4.210)

You are invited to try this for the symmetrical T-network as an exercise.

4.6 TRANSMISSION LINE MODELS

The line parameters discussed in the preceding sections were obtained on a per-phase, per unit length basis. We are interested in the performance of lines with arbitrary length, say l. To be exact, one must take an infinite number of incremental lines, each with a differential length. Figure 4-41 shows the line with details of one incremental portion (dx) at a distance (x) from the receiving end.

The assumptions used in subsequent analyses are:

1. The line is operating under sinusoidal, balanced, steady-state conditions.
2. The line is transposed.

With these assumptions, we analyze the line on a per phase basis. Application of Kirchhoff's voltage and current relations yields

$$[V(x) + \Delta V] - V(x) = [I(x) + \Delta I] z \Delta x \quad (4.211)$$
$$[I(x) + \Delta I] - I(x) = V(x) y \Delta x \quad (4.212)$$

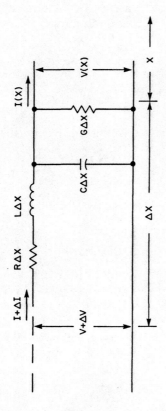

Figure 4-41. Incremental Length of the Transmission Line.

These reduce to
$$\Delta V = I(x)z\Delta x$$
$$\Delta I = V(x)y\Delta x$$
or in the limit, as $\Delta x \to 0$, we have

$$\frac{dV(x)}{dx} = zI(x) \tag{4.213}$$

$$\frac{dI(x)}{dx} = yV(x) \tag{4.214}$$

A separation of variables can be performed by differentiating the above equations and substituting to obtain

$$\frac{d^2V(x)}{dx^2} = zyV(x) \tag{4.215}$$

$$\frac{d^2I(x)}{dx^2} = zyI(x) \tag{4.216}$$

Let us introduce the propagation constant v defined as

$$v = \sqrt{zy} \tag{4.217}$$

where the series impedance per-unit length is

$$z = R + j\omega L \tag{4.218}$$

and the shunt admittance per-unit length is

$$y = G + j\omega C \tag{4.219}$$

R and L are series resistance and inductance per unit length, and G and C are shunt conductance and capacitance to neutral per unit length.

We can now write the differential Eqs. (4.215) and (4.216) as:

$$\frac{d^2V}{dx^2} = v^2 V \tag{4.220}$$

$$\frac{d^2I}{dx^2} = v^2 I \tag{4.221}$$

Equation (4.220) can be solved as an ordinary differential equation in V. The solution turns out to be

$$V(x) = A_1 \exp(vx) + A_2 \exp(-vx) \tag{4.222}$$

Now taking the derivative of V with respect to x, and using Eq. (4.213), we obtain $I(x)$ as

$$I(x) = \frac{A_1 \exp(vx) - A_2 \exp(-vx)}{Z_c} \tag{4.223}$$

Here we introduce
$$Z_c = \sqrt{\frac{z}{y}} \qquad (4.224)$$

Z_c is the characteristic (wave) impedance of the line.

The constants A_1 and A_2 may be evaluated in terms of the initial conditions at $x = 0$ (the receiving end). Thus we have
$$V(0) = A_1 + A_2$$
$$Z_c I(0) = A_1 - A_2$$

from which we can write
$$V(x) = \frac{1}{2}\{[V(0) + Z_c I(0)]\exp(vx) + [V(0) - Z_c I(0)]\exp(-vx)\} \qquad (4.225)$$
$$I(x) = \frac{1}{2}\left\{\left[I(0) + \frac{V(0)}{Z_c}\right]\exp(vx) + \left[I(0) - \frac{V(0)}{Z_c}\right]\exp(-vx)\right\} \qquad (4.226)$$

Equations (4.225) and (4.226) can be used for calculating the voltage and current at any distance x from the receiving end along the line. A more convenient form of these equations is found by using hyperbolic functions.

We recall that
$$\sinh\theta = \frac{\exp(\theta) - \exp(-\theta)}{2}$$
$$\cosh\theta = \frac{\exp(\theta) + \exp(-\theta)}{2}$$

By rearranging Eqs. (4.225) and (4.226) and substituting the hyperbolic function for the exponential terms, a new set of equations is found. These are
$$V(x) = V(0)\cosh vx + Z_c I(0)\sinh vx \qquad (4.227)$$

and
$$I(x) = I(0)\cosh vx + \frac{V(0)}{Z_c}\sinh vx \qquad (4.228)$$

We define the following *ABCD* parameters:
$$A(x) = \cosh vx \qquad (4.229)$$
$$B(x) = Z_c \sinh vx \qquad (4.230)$$
$$C(x) = \frac{1}{Z_c}\sinh vx \qquad (4.231)$$
$$D(x) = \cosh vx \qquad (4.232)$$

As a result we have
$$V(x) = A(x)V(0) + B(x)I(0)$$
$$I(x) = C(x)V(0) + D(x)I(0)$$

It is easy to verify that
$$A(x)D(x) - B(x)C(x) = 1$$

For evaluation of the voltage and current at the sending end $x = l$, it is common to write
$$V_s = V(l)$$
$$I_s = I(l)$$
$$V_r = V(0)$$
$$I_r = I(0)$$

Thus we have
$$V_s = AV_r + BI_r \tag{4.233}$$
$$I_s = CV_r + DI_r \tag{4.234}$$

The subscripts s and r stand for sending and receiving values respectively. We have from above:
$$A = A(l) = \cosh vl \tag{4.235}$$
$$B = B(l) = Z_c \sinh vl \tag{4.236}$$
$$C = C(l) = \frac{1}{Z_c} \sinh vl \tag{4.237}$$
$$D = D(l) = \cosh vl \tag{4.238}$$

It is practical to introduce the complex variable θ in the definition of the *ABCD* parameters. We define
$$\theta = vl = \sqrt{ZY} \tag{4.239}$$

As a result,
$$A = \cosh \theta \tag{4.240}$$
$$B = Z_c \sinh \theta \tag{4.241}$$
$$C = \frac{1}{Z_c} \sinh \theta \tag{4.242}$$
$$D = A \tag{4.243}$$

Observe that the total line series impedance and admittance are given by
$$Z = zl \tag{4.244}$$
$$Y = yl \tag{4.245}$$

Evaluating *ABCD* Parameters

Two methods can be employed to calculate the *ABCD* parameters of a transmission line exactly. Both assume that θ is calculated in the rectangular form
$$\theta = \theta_1 + j\theta_2$$

The first method proceeds by expanding the hyperbolic functions as follows:

$$A = \frac{e^\theta + e^{-\theta}}{2}$$
$$= \frac{1}{2}\left(e^{\theta_1}\underline{/\theta_2} + e^{-\theta_1}\underline{/-\theta_2}\right) \tag{4.246}$$

$$\sinh\theta = \frac{e^\theta - e^{-\theta}}{2}$$
$$= \frac{1}{2}\left(e^{\theta_1}\underline{/\theta_2} - e^{-\theta_1}\underline{/-\theta_2}\right).$$

$$B = \sqrt{\frac{Z}{Y}}\sinh\theta \tag{4.247}$$

$$C = \sqrt{\frac{Y}{Z}}\sinh\theta \tag{4.248}$$

Note that θ_2 is in radians to start with in the decomposition of θ.

The second method uses two well-known identities to arrive at the parameter of interest.

$$A = \cosh(\theta_1 + j\theta_2)$$
$$\cosh\theta = \cosh\theta_1 \cos\theta_2 + j\sinh\theta_1 \sin\theta_2 \tag{4.249}$$

Also we have

$$\sinh\theta = \sinh\theta_1 \cos\theta_2 + j\cosh\theta_1 \sin\theta_2 \tag{4.250}$$

Example 4-14

Find the exact $ABCD$ parameters for a 235.92-mile long, 735-kV, bundle-conductor line with four subconductors per phase with subconductor resistance of 0.1004 ohms per mile. Assume that the series inductive reactance per phase is 0.5541 ohms per mile and shunt capacitive susceptance of 7.4722×10^{-6} siemens per mile to neutral. Neglect shunt conductance.

Solution

The resistance per phase is

$$r = \frac{0.1004}{4} = 0.0251 \quad \text{ohms/mile}$$

Thus the series impedance in ohms per mile is

$$z = 0.0251 + j0.5541 \quad \text{ohms/mile}$$

The shunt admittance is

$$y = j7.4722 \times 10^{-6} \quad \text{siemens/mile}$$

For the line length,

$$Z = zl = (0.0251 + j0.5541)(235.92) = 130.86\underline{/87.41°}$$
$$Y = yl = j(7.4722 \times 10^{-6})(235.92) = 1.7628 \times 10^{-3}\underline{/90°}$$

We calculate θ as

$$\theta = \sqrt{ZY}$$
$$= \left[(130.86\underline{/87.41°})(1.7628 \times 10^{-3}\underline{/90°})\right]^{1/2}$$
$$= 0.0109 + j0.4802$$

Thus,

$$\theta_1 = 0.0109$$
$$\theta_2 = 0.4802$$

We change θ_2 to degrees. Therefore,

$$\theta_2 = (0.4802)\left(\frac{180}{\pi}\right) = 27.5117°$$

Using Eq. (4.246), we then get

$$\cosh\theta = \frac{1}{2}\left(e^{0.0109}\underline{/27.5117°} + e^{-0.0109}\underline{/-27.5117°}\right)$$
$$= 0.8870\underline{/0.3242°}$$

From the above,

$$D = A = 0.8870\underline{/0.3242°}$$

We now calculate $\sinh\theta$ as

$$\sinh\theta = \frac{1}{2}\left(e^{0.0109}\underline{/27.5117°} - e^{-0.0109}\underline{/-27.5117°}\right)$$
$$= 0.4621\underline{/88.8033°}$$

We have

$$Z_c = \sqrt{\frac{Z}{Y}} = \left(\frac{130.86\underline{/87.41°}}{1.7628 \times 10^{-3}\underline{/90°}}\right)^{1/2}$$
$$= (74234.17\underline{/-2.59})^{1/2}$$
$$= 272.46\underline{/-1.295°}$$

As a result,

$$B = Z_c \sinh\theta$$
$$= 125.904\underline{/87.508°}$$

Also
$$C = \frac{1}{Z_c}\sinh\theta$$
$$= 1.696 \times 10^{-3} \underline{/90.098°}$$

Let us employ the second method to evaluate the parameters. We find the hyperbolic functions:
$$\cosh\theta_1 = \cosh(0.0109)$$
$$= \frac{e^{0.0109} + e^{-0.0109}}{2}$$
$$= 1.000059$$
$$\sinh\theta_1 = \frac{e^{0.0109} - e^{-0.0109}}{2}$$
$$= 1.09002 \times 10^{-2}$$

(If your calculator has built-in hyperbolic functions, you can skip the intermediate steps.) We also have
$$\cos\theta_2 = \cos\left[(0.4802)\left(\frac{180}{\pi}\right)\right]$$
$$= 0.8869$$
$$\sin\theta_2 = \sin\left[(0.4802)\left(\frac{180}{\pi}\right)\right]$$
$$= 0.4619566$$

Therefore we have
$$\cosh\theta = \cosh\theta_1 \cos\theta_2 + j\sinh\theta_1 \sin\theta_2$$
$$= (1.000059)(0.8869) + j(1.09002 \times 10^{-2})(0.4619566)$$
$$= 0.8869695 \underline{/0.32527°}$$
$$\sinh\theta = \sinh\theta_1 \cos\theta_2 + j\cosh\theta_1 \sin\theta_2$$
$$= (1.09002 \times 10^{-2})(0.8869) + j(1.000059)(0.4619566)$$
$$= 0.4620851 \underline{/88.801°}$$

These results agree with the ones obtained using the first method.

Example 4-15

Find the voltage, current, and power at the sending end of the line of Example 4-14 and the transmission efficiency given that the receiving-end load is 1500 MVA at 700 kV with 0.95 PF lagging.

Solution

We have the apparent power given by

$$S_r = 1500 \times 10^6 \text{ VA}$$

The voltage to neutral is

$$V_r = \frac{700 \times 10^3}{\sqrt{3}} \text{ V}$$

Therefore,

$$I_r = \frac{1500 \times 10^6}{3\left(\dfrac{700 \times 10^3}{\sqrt{3}}\right)} \underline{/-\cos^{-1} 0.95}$$

$$= 1237.18 \underline{/-18.19°} \text{ A}$$

From Example 4-14 we have the values of the A, B, and C parameters. Thus the sending-end voltage (to neutral) is obtained as

$$V_s = AV_r + BI_r$$

$$= (0.8870 \underline{/0.3253°})\left(\frac{700 \times 10^3}{\sqrt{3}}\right)$$

$$+ (125.904 \underline{/87.508°})(1237.18 \underline{/-18.19°})$$

$$= 439.0938 \underline{/19.66°} \text{ kV}$$

The line-to-line value is obtained by multiplying the above value by $\sqrt{3}$, giving

$$V_{S_L} = 760.533 \text{ kV}$$

The sending-end current is obtained as

$$I_s = CV_r + DI_r$$

$$= (1.696 \times 10^{-3} \underline{/90.098°})\left(\frac{700 \times 10^3}{\sqrt{3}}\right)$$

$$+ (0.887 \underline{/0.3253°})(1237.18 \underline{/-18.19°})$$

$$= 1100.05 \underline{/18.49°}$$

The sending-end power factor is

$$\cos \phi_s = \cos(19.66 - 18.49)$$
$$= \cos(1.17) = 0.99979$$

As a result, the sending-end power is

$$P_s = 3(439.0938 \times 10^3)(1100.05)(0.99979)$$
$$= 1448.77 \times 10^6 \text{ MW}$$

The efficiency is

$$\eta = \frac{P_r}{P_s}$$
$$= \frac{1500 \times 10^6 \times 0.95}{1448.77 \times 10^6}$$
$$= 0.9836$$

Lumped Parameter Transmission Line Models

Lumped parameter representations of transmission lines are needed for further analysis of interconnected electric power systems. Their use enables the development of simpler algorithms for the solution of complex networks that involve transmission lines.

Here we are interested in obtaining values of the circuit elements of a π circuit, to represent accurately the terminal characteristics of the line given by

$$V_s = AV_r + BI_r$$
$$I_s = CV_r + DI_r$$

It is easy to verify that the elements of the equivalent circuit are given in terms of the $ABCD$ parameters of the line by

$$Z_\pi = B \tag{4.251}$$

and

$$Y_\pi = \frac{A-1}{B} \tag{4.252}$$

The circuit is shown in Figure 4-42.

Example 4-16

Find the equivalent π-circuit elements for the line of Example 4-14.

Solution

From Example 4-14, we have

$$A = 0.8870 \underline{/0.3242°}$$
$$B = 125.904 \underline{/87.508°}$$

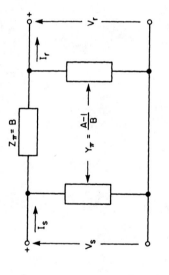

Figure 4-42. Equivalent π Model of a Transmission Line.

As a result, we have

$$Z_\pi = 125.904 \underline{/87.508°} \text{ ohms}$$

$$Y_\pi = \frac{0.8870 \underline{/0.3242°} - 1}{125.904 \underline{/87.508°}}$$

$$= 8.9851 \times 10^{-4} \underline{/89.941°} \text{ siemens}$$

Approximations to the *ABCD* Parameters of Transmission Lines

Consider the series expansion of the hyperbolic functions defining the A, B, C, and D parameters given by

$$A = 1 + \frac{\theta^2}{2} + \frac{\theta^4}{24} + \frac{\theta^6}{720} + \cdots$$

$$B = Z\left(1 + \frac{\theta^2}{6} + \frac{\theta^4}{120} + \frac{\theta^6}{5040} + \cdots\right)$$

$$C = Y\left(1 + \frac{\theta^2}{6} + \frac{\theta^4}{120} + \frac{\theta^6}{5040} + \cdots\right)$$

$$D = A$$

Using $\theta = \sqrt{ZY}$, we obtain the following:

$$A = 1 + \frac{ZY}{2} + \frac{Z^2Y^2}{24} + \frac{Z^3Y^3}{720} + \cdots \tag{4.253}$$

$$B = Z\left(1 + \frac{ZY}{6} + \frac{Z^2Y^2}{120} + \frac{Z^3Y^3}{5040} + \cdots\right) \tag{4.254}$$

$$C = Y\left(1 + \frac{ZY}{6} + \frac{Z^2Y^2}{120} + \frac{Z^3Y^3}{5040} + \cdots\right) \tag{4.255}$$

The number of terms taken into consideration when applying the above expressions will depend on the required accuracy. Usually no more than three terms are required. For overhead lines less than 500 km in length, the following approximate expressions are satisfactory:

$$A = D = 1 + \frac{ZY}{2} \tag{4.256}$$

$$B = Z\left(1 + \frac{ZY}{6}\right) \tag{4.257}$$

$$C = Y\left(1 + \frac{ZY}{6}\right) \tag{4.258}$$

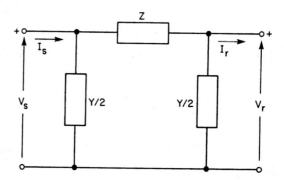

Figure 4-43. Nominal π Model of a Medium Transmission Line.

If only the first term of the expansions is used, then

$$B = Z \qquad (4.259)$$

$$\frac{A-1}{B} = \frac{Y}{2} \qquad (4.260)$$

In this case, the equivalent π circuit reduces to the nominal π, which is used generally for lines classified as medium lines (up to 250 km). Figure 4-43 shows the nominal π model of a medium transmission line. The result we obtained analytically could have been obtained easily by the intuitive assumption that the line's series impedance is lumped together and the shunt admittance Y is divided equally with each half placed at each end of the line.

A final model is the short-line (up to 80 km) model, and in this case the shunt admittance is neglected altogether. The line is thus represented only by its series impedance as shown in Figure 4-44. Let us take an example to illustrate the approximations involved.

Example 4-17

Find the nominal π and short-line representations for the line of Example 4-14. Calculate the sending-end voltage and current of the transmission line using the two representations under the conditions of Example 4-15.

Solution

For this line we have

$$Z = 130.86 \underline{/87.41°}$$

$$Y = 1.7628 \times 10^{-3} \underline{/90°}$$

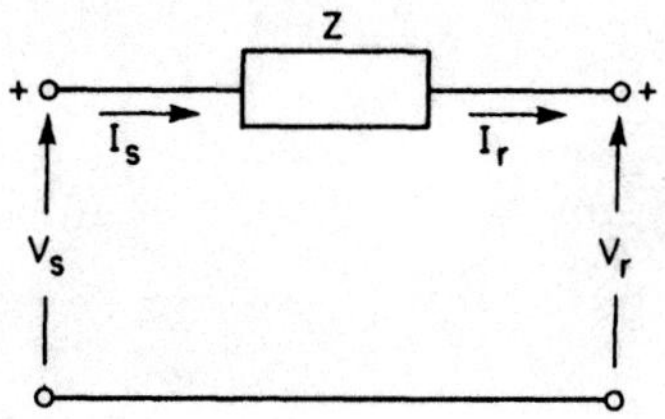

4.6 Transmission Line Models

Figure 4-44. Short-Line Model.

As a result, we have the representations shown in Figure 4-43.
From Example 4-15, we have

$$V_r = \frac{700 \times 10^3}{\sqrt{3}} \text{ V}$$

$$I_r = 1237.18 \underline{/-18.19°} \text{ A}$$

For the short-line representation we have

$$V_s = V_r + I_r Z$$

$$= \frac{700 \times 10^3}{\sqrt{3}} + (1237.18\underline{/-18.19°})(130.86\underline{/87.41°})$$

$$= 485.7682 \times 10^3 \underline{/18.16°} \text{ V}$$

For the nominal π we have

$$I_L = I_r + V_r\left(\frac{Y}{2}\right)$$

$$= 1237.18\underline{/-18.19°} + \left(\frac{700 \times 10^3}{\sqrt{3}}\right)(0.8814 \times 10^{-3}\underline{/90°})$$

$$= 1175.74\underline{/-1.4619°} \text{ A}$$

Thus,

$$V_s = V_r + I_L Z$$

$$= \frac{700 \times 10^3}{\sqrt{3}} + (1175.74 \,/\!\!-\!1.4619°)(130.86 \,/\!87.41°)$$

$$= 442.484 \times 10^3 \,/\!20.2943° \text{ V}$$

Referring back to the exact values calculated in Example 4-15, we find that the short-line approximation results in an error in the voltage magnitude of

$$\Delta V = \frac{439.0938 - 485.7682}{439.0938}$$

$$= -0.11$$

For the nominal π we have the error of

$$\Delta V = \frac{439.0938 - 442.484}{439.0938}$$

$$= -0.00772$$

which is less than 1 percent.

The sending-end current with the nominal π model is

$$I_s = I_L + V_s \left(\frac{Y}{2}\right)$$

$$= 1175.74 \,/\!\!-\!1.4619°$$

$$+ (442.484 \times 10^3 \,/\!20.2943°)(0.8814 \times 10^{-3} \,/\!90°)$$

$$= 1092.95 \,/\!17.89° \text{ A}$$

Transmission Line Model Approximation Errors

We have seen that a number of models are available for the analysis of transmission line performance. The complexity of the analysis process is least for the short-line model. A slightly more complex process is required for the nominal π model. Both models are approximations of the long-line models using the hyperbolic functions whether in the $ABCD$ form or the equivalent π model. A trade-off must be made between accuracy and model complexity.

Let us consider the error in performance calculation when using the short-line model consisting of just the line's series impedance Z in comparison with the more accurate results obtained using the nominal π model. Let us assume that V_r and I_r are given and consider the values of sending-end voltage, current, and power. We denote quantities calculated using the short-line approximation by suffix 1, whereas those for the nominal π are denoted by suffix 2. The situation is shown in Figure 4-45.

The voltages and currents calculated are

$$V_{s_1} = V_r + ZI_r$$
$$I_{s_1} = I_r$$
$$V_{s_2} = \left(1 + \frac{ZY}{2}\right)V_r + ZI_r$$
$$I_{s_2} = Y\left(1 + \frac{ZY}{4}\right)V_r + \left(1 + \frac{ZY}{2}\right)I_r$$

The errors are defined by

$$\Delta V_s = V_{s_2} - V_{s_1} \qquad (4.261)$$

$$\Delta I_s = I_{s_2} - I_{s_1} \qquad (4.262)$$

or

$$\Delta V_s = \left(\frac{ZY}{2}\right)V_r \qquad (4.263)$$

$$\Delta I_s = Y\left(1 + \frac{ZY}{4}\right)V_r + \left(\frac{ZY}{2}\right)I_r \qquad (4.264)$$

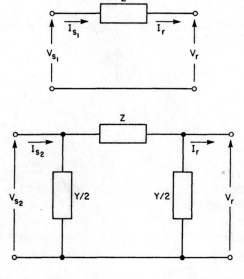

Figure 4.45. Variables Defined for Comparison of Short-Line and Nominal π Models.

For the sending-end complex power we have the error defined by
$$\Delta S_s = S_{s_2} - S_{s_1}$$
$$= V_{s_2} I^*_{s_2} - V_{s_1} I^*_{s_1}$$

This reduces to
$$\Delta S_s = (V_{s_1} + \Delta V_s)(I^*_{s_1} + \Delta I^*_s) - V_{s_1} I^*_{s_1}$$
$$= (\Delta V_s)(I^*_{s_1}) + (V_{s_1})(\Delta I^*_s) + (\Delta V_s)(\Delta I^*_s)$$

Expanding the right-hand side of the above, we obtain
$$\Delta S_s = aS_r + bS^*_r + cV^2_r + d|I_r|^2 \qquad (4.265)$$

with

$$a = -BX + \frac{B^2(R^2 + X^2)}{4} \qquad (4.266)$$

$$b = B\left[X - \frac{B}{4}(R^2 + X^2)\right] - jBR \qquad (4.267)$$

$$c = \frac{B^2 R}{4} + jB\left[\frac{3BX}{4} - \frac{B^2}{8}(R^2 + X^2) - 1\right] \qquad (4.268)$$

$$d = -\frac{jB}{2}(R^2 + X^2) \qquad (4.269)$$

where as usual,
$$Z = R + jX$$
$$Y = jB$$

Separating real and imaginary parts in the complex power error equation, we obtain

$$\Delta P_s = BR\left(\frac{BV^2_r}{4} - Q_r\right) \qquad (4.270)$$

$$\frac{\Delta Q_s}{B} = \left(\frac{B|Z|^2}{2} - 2X\right)Q_r - RP_r$$
$$+ \left(\frac{3BX}{4} - \frac{B^2}{8}|Z|^2 - 1\right)V^2_r - \left|\frac{Z}{2}\right|^2 |I_r|^2 \qquad (4.271)$$

It is clear that the errors increase with the square of the voltage and the line's shunt admittance.

The formulae obtained above are not very practical to implement. It is desirable to obtain expressions giving only upper bounds on the expected errors. These expressions turn out to be much simpler. Consider first the voltage error

$$\Delta V_s = \frac{ZY}{2}(V_r) = \frac{jB(R + jX)}{2} V_r$$

$$|\Delta V_s| = \frac{BX}{2}\left(1 + \frac{R^2}{X^2}\right)^{1/2} V_r$$

Since normally

$$\frac{R}{X} < 1$$

we have

$$1 + \frac{R^2}{X^2} < 2$$

Thus

$$|\Delta V_s| < \frac{BX}{\sqrt{2}}(V_r) \tag{4.272}$$

Define the maximum value of error (or upper bound) by

$$\Delta V_{s_u} = \frac{BX}{\sqrt{2}}(V_r)$$

Thus

$$|\Delta V_s| < \Delta V_{s_u} \tag{4.273}$$

Consider next the error in calculating the sending-end active power ΔP_s given by

$$\Delta P_s = BR\left(\frac{BV_r^2}{4} - Q_r\right)$$

For lagging Q_r, $Q_r > 0$, and with

$$Q_r < \frac{BV_r^2}{4}$$

we have

$$\Delta P_s < \frac{B^2 R V_r^2}{4} \tag{4.274}$$

Thus,

$$\Delta P_s < \Delta P_{s_u} \tag{4.275}$$

where ΔP_{s_u} is the maximum value of the error

$$\Delta P_{s_u} = \frac{B^2 R V_r^2}{4} \tag{4.276}$$

In terms of ΔV_{s_u}, we have

$$\Delta P_{s_u} = \frac{R}{2X^2}(\Delta V_{s_u})^2$$

for the condition

$$Q_r < \frac{BV_r^2}{4} \qquad (4.277)$$

For lagging Q_r and with

$$Q_r > \frac{BV_r^2}{4}$$

we have

$$-\Delta P_s = BR\left(\frac{Q_r - BV_r^2}{4}\right) \qquad (4.278)$$

Note ΔP_s is negative, indicating that the value of sending-end power calculated using the short-line approximation is higher than that using the nominal π. In this case,

$$|\Delta P_s| < BRQ_r$$

or

$$|\Delta P_{s_u}| = BRQ_r \qquad (4.279)$$

Example 4-18

For the system of Example 4-14, we have

$$B = 1.7628 \times 10^{-3}$$
$$X = 130.72$$
$$V_r = \frac{700 \times 10^3}{\sqrt{3}}$$

We calculate

$$\Delta V_{s_u} = \frac{BX}{\sqrt{2}}(V_r)$$

$$\Delta V_{s_u} = 65.851 \times 10^3 \text{ V}$$

The actual error is

$$\Delta V_s = 485.7682 \times 10^3 - 442.484 \times 10^3 = 43.28 \times 10^3 \text{ V}$$

For the power, we have

$$R = 5.92 \text{ ohms}$$
$$Q_r = 156.12 \text{ Mvar per phase}$$
$$\frac{BV_r^2}{4} = \left(\frac{1.7628 \times 10^{-3}}{4}\right)\left(\frac{700 \times 10^3}{\sqrt{3}}\right)^2$$
$$= 71.981 \times 10^6$$

Thus
$$Q_r > \frac{BV_r^2}{4}$$

We calculate
$$|\Delta P_{s_u}| = BRQ_r$$
$$= (1.7628 \times 10^{-3})(5.92)(156.12)$$
$$= 1.63 \text{ MW per phase}$$

For three phases, we have
$$|\Delta P_{s_u}| = 4.89 \text{ MW}$$

The actual values are obtained as follows for the short-line approximation:
$$\phi_s = 36.35°$$
$$\cos \phi_s = 0.805411$$
$$P_s = 3(485.7682 \times 10^3)(1237.18)(0.805411)$$
$$= 1452.11 \times 10^6 \text{ W}$$

For the medium line, we have
$$\phi_s = 20.293 - 17.89$$
$$= 2.4044°$$
$$\cos \phi_s = 0.99912$$
$$P_s = 3(442.495 \times 10^3)(1092.93)(0.99912)$$
$$= 1449.56 \times 10^6 \text{ W}$$

The actual error is
$$\Delta P_s = 1449.56 - 1452.11$$
$$= -2.55486 \text{ MW}$$

which is within the predicted range.

SOME SOLVED PROBLEMS

Problem 4-A-1

A 500-kV, double-circuit line has bundle conductors with three subconductors at 21-in. spacing. The GMR of each subconductor is 0.0485 ft. The circuit configuration is as shown in Figure 4-46. Calculate the inductive

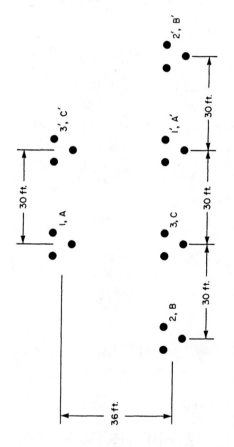

Figure 4-46. Double-Circuit Configuration for Problem 4-A-1 with Alternative 1 Phase Placement.

TABLE 4-3
Phase Placements for Line of Problem 4-A-1

	Alternative	\multicolumn{6}{c}{Physical Position}

	Alternative	1	2	3	1'	2'	3'
Phase placement	1	A	B	C	A'	B'	C'
	2	A	B	B'	C	C'	A'
	3	C	A	B	B'	A'	C'

reactance of the line in ohms per mile per phase for the alternatives of phase placement shown in the Table 4-3.

Solution

We find first the bundle's GMR as follows:

$$N = 3$$
$$S = 21 \text{ inches}$$
$$A = \frac{S}{2\sin 60°} = \frac{21}{24\frac{\sqrt{3}}{2}} = 1.0104 \text{ ft}$$

$$\text{GMR}_c = 0.0485 \text{ ft}$$
$$\text{GMR}_b = \left[3(0.0485)(1.0104)^2\right]^{1/3}$$
$$= 0.52959 \text{ ft}$$

For alternative 1, we have

$$r'_A = \left[(\text{GMR}_b)(D_{AA'})\right]^{1/2}$$
$$= \left(0.52959\sqrt{(30)^2 + (36)^2}\right)^{1/2}$$
$$= 4.9817 \text{ ft}$$
$$r'_B = \left[(\text{GMR}_b)(D_{BB'})\right]^{1/2}$$
$$= \left[0.52959(90)\right]^{1/2}$$
$$= 6.9038 \text{ ft}$$
$$r'_C = \left[(\text{GMR}_b)(D_{CC'})\right]^{1/2}$$
$$= \left[0.52959\sqrt{(30)^2 + (36)^2}\right]^{1/2}$$
$$= 4.9817 \text{ ft}$$

Thus the overall GMR is obtained as

$$\text{GMR}_p = (r'_A r'_B r'_C)^{1/3}$$
$$= 5.5541 \text{ ft}$$

We next calculate the equivalent distances:

$$D_{AB_{eq}} = \left\{ \left[\sqrt{(30)^2 + (36)^2} \right] \left[\sqrt{(60)^2 + (36)^2} \right] [(60)(30)] \right\}^{1/4}$$
$$= 49.29 \text{ ft}$$

$$D_{BC_{eq}} = \left\{ (30) \left[\sqrt{(60)^2 + (36)^2} \right] (60) \left[\sqrt{(30)^2 + (36)^2} \right] \right\}^{1/4}$$
$$= 49.29 \text{ ft}$$

$$D_{AC_{eq}} = [(36)(30)(36)(30)]^{1/4}$$
$$= 32.86 \text{ ft}$$

As a result, we obtain

$$\text{GMD} = [(49.29)(49.29)(32.86)]^{1/3}$$
$$= 43.06 \text{ ft}$$

From the above we conclude that

$$X_L = 0.2794 \log \frac{43.06}{5.5541}$$
$$= 0.2485 \text{ ohms/mile}$$

For alternative 2, shown in Figure 4-47 we have the phase GMR calculated as

$$r'_A = [0.52959(30)]^{1/2}$$
$$= 3.99 \text{ ft}$$

$$r'_B = [0.52959(30)]^{1/2}$$
$$= 3.99 \text{ ft}$$

$$r'_C = [0.52959(30)]^{1/2}$$
$$= 3.99 \text{ ft}$$

Thus,

$$\text{GMR}_p = [(3.99)(3.99)(3.99)]^{1/3}$$
$$= 3.99 \text{ ft}$$

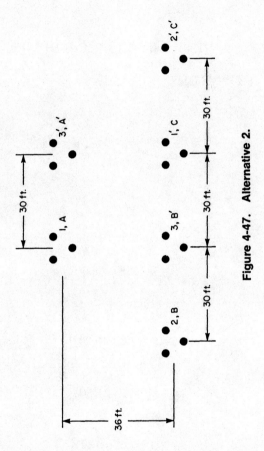

Figure 4-47. Alternative 2.

The equivalent distances are obtained as

$$D_{AB_{eq}} = \left[((36)^2 + (30)^2)(36)\left(\sqrt{(60)^2 + (36)^2}\right) \right]^{1/4}$$
$$= 48.5 \text{ ft}$$

$$D_{BC_{eq}} = [(60)(90)(30)(60)]^{1/4}$$
$$= 55.84 \text{ ft}$$

$$D_{AC_{eq}} = \left[((30)^2 + (36)^2)\left(\sqrt{(36)^2 + (60)^2}\right)(36) \right]^{1/4}$$
$$= 48.5 \text{ ft}$$

Thus

$$\text{GMD} = [(48.5)(48.5)(55.84)]^{1/3}$$
$$= 50.83 \text{ ft}$$

From which,

$$X_L = 0.2794 \log \frac{50.83}{3.99}$$
$$= 0.3088 \text{ ohms/mile}$$

For alternative 3, shown in Figure 4-48 we have

$$r'_A = [(0.52959)(90)]^{1/2}$$
$$= 6.9038 \text{ ft}$$
$$r'_B = [(0.52959)(30)]^{1/2}$$
$$= 3.99 \text{ ft}$$
$$r'_C = [(0.52959)(30)]^{1/2}$$
$$= 3.99 \text{ ft}$$

From which,

$$\text{GMR}_p = [(6.9038)(3.99)(3.99)]^{1/3}$$
$$= 4.79 \text{ ft}$$

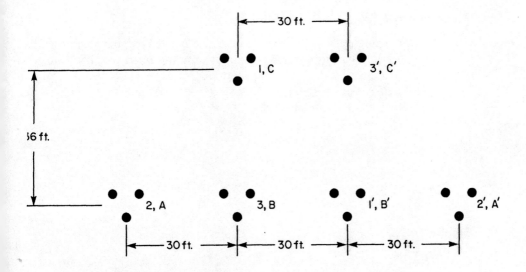

Figure 4-48. Alternative 3.

The equivalent distances are

$$D_{AB_{eq}} = [(30)(60)(30)(60)]^{1/4}$$
$$= 42.43 \text{ ft}$$
$$D_{BC_{eq}} = [(36)((30)^2 + (36)^2)(36)]^{1/4}$$
$$= 41.07 \text{ ft}$$
$$D_{AC_{eq}} = [((30)^2 + (36)^2)((60)^2 + (36)^2)]^{1/4}$$
$$= 57.26 \text{ ft}$$

Thus,

$$\text{GMD} = [(42.43)(41.07)(57.26)]^{1/3}$$
$$= 46.38 \text{ ft}$$

Consequently,

$$X_L = 0.2794 \log \frac{46.38}{4.79}$$
$$= 0.27557 \text{ ohms/mile}$$

Note that alternative 1 gives the lowest inductive reactance.

Problem 4-A-2

An important class of problems in electric power systems engineering is that of determining system parameters in terms of operating conditions rather than geometric and physical properties. One such a problem involves finding the *ABCD* parameters of a device such as a transmission line.

Show that the parameters A and B are given by

$$A = \frac{V_s I_s + V_r I_r}{V_r I_s + V_s I_r}$$

$$B = \frac{V_s^2 - V_r^2}{V_r I_s + V_s I_r}$$

Solution

We have

$$V_s = AV_r + BI_r$$

Hence

$$V_s I_s = AV_r I_s + BI_r(CV_r + AI_r)$$
$$V_s I_s + V_r I_r = AV_r I_s + (BC+1)I_r V_r + BA I_r^2$$

But

$$A^2 - BC = 1$$

Hence

$$V_s I_s + V_r I_r = AV_r I_s + A^2 I_r V_r + BA I_r^2$$
$$= A(V_r I_s + V_s I_r)$$

As a result,

$$A = \frac{V_s I_s + V_r I_r}{V_r I_s + V_s I_r}$$

And we have

$$B = \frac{V_s - AV_r}{I_r}$$

Using the relation just derived for A, we get

$$B = \frac{V_s}{I_r} - \frac{V_r}{I_r}\left(\frac{V_s I_s + V_r I_r}{V_r I_s + V_s I_r}\right)$$

$$= \frac{V_s V_r I_s + V_s^2 I_r - V_r V_s I_s - V_r^2 I_r}{I_r(V_r I_s + V_s I_r)}$$

Thus,
$$B = \frac{V_s^2 - V_r^2}{V_r I_s + V_s I_r}$$

Problem 4-A-3

Suppose we are interested in determining the transmission line circuit impedance and admittance from the A and B parameters. As a first step we substitute
$$X = e^{-\theta}$$
where as usual, $\theta = \sqrt{ZY}$. We further substitute
$$X = X_1 + jX_2$$
$$A = A_1 + jA_2$$

Show that X_1 and X_2 satisfy the following two equations
$$X_1^2 - X_2^2 - 2(A_1 X_1 - A_2 X_2) + 1 = 0$$
$$X_1 X_2 - (A_2 X_1 + A_1 X_2) = 0$$

Solution

We have
$$A = \frac{e^\theta + e^{-\theta}}{2}$$

We put $X = e^{-\theta}$. Then
$$A = \frac{\frac{1}{X} + X}{2}$$

or
$$X^2 - 2AX + 1 = 0$$

Substitute
$$X = X_1 + jX_2$$
$$A = A_1 + jA_2$$

Hence
$$X_1^2 - X_2^2 + 2jX_1 X_2 - 2[A_1 X_1 - A_2 X_2 + j(A_2 X_1 + A_1 X_2)] + 1 = 0$$

Separating real and imaginary parts,
$$X_1^2 - X_2^2 - 2(A_1 X_1 - A_2 X_2) + 1 = 0$$
$$X_1 X_2 - (A_2 X_1 + A_1 X_2) = 0$$

APPENDIX 4-A

Inductance of a Conductor Due to Internal Flux

We will consider a cylindrical conductor with a far return path for the current. With this assumption, the return path current does not affect the magnetic field of the conductor. Hence, the lines of flux are concentric with the conductor. We will assume uniform current density in the conductor. The current enclosed by a path of radius x is thus

$$I(x) = \frac{\pi x^2}{\pi r^2}(I)$$

Here I is the total current in the conductor and r is the radius of the conductor.

The magnetic field intensity $H(x)$ at any point on the circular path of radius x is obtained from

$$H(x) = \frac{1}{2\pi x} I(x)$$

This is the result of applying Ampere's circuital relationship

$$I = \int H \cdot ds.$$

With the uniform current density assumption, we have

$$H(x) = \frac{x}{2\pi r^2} I \quad \text{At/m}$$

Assuming a constant permeability μ for the conductor's material, the flux density x meters from the center is

$$B(x) = \frac{\mu x}{2\pi r^2} I$$

The flux $d\phi$ in a tubular element of thickness dx per unit length is

$$d\phi = \frac{\mu x}{2\pi r^2} I \, dx \quad \text{Wb/m}$$

We note here that the internal flux $d\phi$ links only $I(x)$, which is a fraction of the total current. Thus the flux linkages per meter of length are:

$$d\lambda = \frac{x^2}{r^2} d\phi = \frac{\mu I x^3}{2\pi r^4} dx \quad \text{Wbt/m}$$

Integrating between the limits $x = 0$ to $x = r$, we obtain the total flux linkages as

$$\lambda = \frac{\mu I}{8\pi}$$

The inductance is thus obtained using the basic relation

$$L = \frac{d\lambda}{dI}$$

For unity relative permeability, $\mu = 4\pi \times 10^{-7}$ henries per meter and hence,

$$L_i = \tfrac{1}{2} \times (10^{-7}) \quad \text{henries/m}$$

Note that the inductance due to internal flux L_i is independent of the wire size.

APPENDIX 4-B

Flux Linkages Outside a Conductor

Consider a cylindrical conductor that carries a current I. Consider a tubular element x meters from the center of the conductor with thickness dx. With the element outside the conductor (i.e., $x > r$), the magnetic field intensity at the element is

$$H(x) = \frac{I}{2\pi x}$$

The flux $d\phi$ in the element is thus

$$d\phi = \frac{\mu I}{2\pi x}(dx) \quad \text{Wb/m}$$

Now, the flux external to the conductor links all the current in the conductor. Thus the flux linkages $d\lambda$ per meter are

$$d\lambda = d\phi$$

The total flux linkages between two points P_1 and P_2 at distances D_1 and D_2 from the center are obtained as

$$\lambda_{12} = \int_{D_1}^{D_2} d\lambda = \frac{\mu I}{2\pi} \ln\left(\frac{D_2}{D_1}\right) \quad \text{Wbt/m}$$

The inductance between P_1 and P_2 is thus

$$L_{P1P2} = \frac{\mu}{2\pi} \ln\left(\frac{D_2}{D_1}\right)$$

APPENDIX 4-C

Potential Due to Single-Phase Line

We deal here with the case in which the radius of each conductor is zero. Let us consider two parallel infinitely long wires A and B situated along the lines $x = D/2$, $y = 0$, and $x = -D/2$, $y = 0$ respectively, as shown in Figure 4-49. Let us assume that a linear charge density q coulombs per meter is distributed along wire A and $-q$ coulombs per meter along wire B. The scalar potential v can be obtained from

$$v = \frac{Q}{4\pi\varepsilon_0 r}$$

$$\varepsilon_0 = \frac{1}{36\pi} \times 10^{-9}$$

Here Q is the charge and r is the distance of the observation point to the location of the charge.

The potential at point P is made up of contributions dv_P from infinitesimal charges $q\,dz'$ on A and $-q\,dz'$ on B. That is

$$dv_P = \frac{q\,dz'}{4\pi\varepsilon_0 r_1} - \frac{q\,dz'}{4\pi\varepsilon_0 r_2}$$

From the geometry of Figure 4-49, we have

$$r_1^2 = r_a^2 + z'^2$$
$$r_2^2 = r_b^2 + z'^2$$

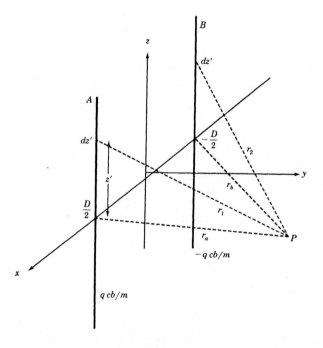

Figure 4-49. Two Infinitely long, Parallel Filaments.

Thus,
$$dv_P = \frac{q}{4\pi\varepsilon_0}\left(\frac{1}{\sqrt{r_a^2 + z'^2}} - \frac{1}{\sqrt{r_b^2 + z'^2}}\right)dz'$$

The potential at P is the integral of dv_P over the length of the wires. To evaluate the integral we will need the following for the first term:
$$z' = r_a \sinh\theta$$

Thus,
$$dz' = r_a \cosh\theta\, d\theta$$
$$\sqrt{r_a^2 + z'^2} = r_a \cosh\theta$$

For the second term we use
$$z' = r_b \sinh\phi$$

Thus,
$$dz' = r_b \cosh\phi\, d\phi$$
$$\sqrt{r_b^2 + z'^2} = r_b \cosh\phi$$

As a result,
$$dv_P = \frac{q}{4\pi\varepsilon_0}(d\theta - d\phi)$$

Assuming the wires extend from $z' = -L$ to $z' = +L$, we thus have
$$v_P = \frac{q}{4\pi\varepsilon_0}(\theta - \phi)\Big|_{z'=-L}^{z'=+L}$$

The variable θ can be expressed using the following
$$\sinh\theta = \frac{e^\theta - e^{-\theta}}{2} = \frac{z'}{r_a}$$

or
$$e^{2\theta} - 2\left(\frac{z'}{r_a}\right)e^\theta - 1 = 0$$

Solving this second-order equation in e^θ, we get
$$e^\theta = \frac{z' + \sqrt{z'^2 + r_a^2}}{r_a}$$

or
$$\theta = \ln\frac{z' + \sqrt{z'^2 + r_a^2}}{r_a}$$

Similarly,
$$\phi = \ln\frac{z' + \sqrt{z'^2 + r_b^2}}{r_b}$$

As a result,
$$v_P = \frac{q}{4\pi\varepsilon_0}\left[\ln\frac{r_b(z' + \sqrt{z'^2 + r_a^2})}{r_a(z' + \sqrt{z'^2 + r_b^2})}\right]_{z'=-L}^{z'=+L}$$

$$= \frac{q}{4\pi\varepsilon_0}\left[\ln\frac{r_b(L + \sqrt{L^2 + r_a^2})}{r_a(L + \sqrt{L^2 + r_b^2})} - \ln\frac{r_b(-L + \sqrt{L^2 + r_a^2})}{r_a(-L + \sqrt{L^2 + r_b^2})}\right]$$

The above can be expressed as

$$v_P = \frac{q}{4\pi\varepsilon_0}\left[\ln\left(\frac{1 + \sqrt{1 + \left(\frac{r_a}{L}\right)^2}}{1 + \sqrt{1 + \left(\frac{r_b}{L}\right)^2}}\right) + \ln\left(\frac{-1 + \sqrt{1 + \left(\frac{r_b}{L}\right)^2}}{-1 + \sqrt{1 + \left(\frac{r_a}{L}\right)^2}}\right)\right]$$

For very long wires $L \gg r_a, r_b$, we get

$$v_P = \frac{q}{4\pi\varepsilon_0}\ln\left(\frac{-1 + \sqrt{1 + \left(\frac{r_b}{L}\right)^2}}{-1 + \sqrt{1 + \left(\frac{r_a}{L}\right)^2}}\right) \quad L \to \infty$$

Using the expansion
$$(1 + x)^{1/2} = 1 + \left(\tfrac{1}{2}\right)x + \cdots$$

we get

$$v_P = \frac{q}{4\pi\varepsilon_0}\ln\frac{\left(\frac{1}{2}\right)\left(\frac{r_b}{L}\right)^2 + \cdots}{\left(\frac{1}{2}\right)\left(\frac{r_a}{L}\right)^2 + \cdots} \quad L \to \infty$$

$$v_P = \frac{q}{2\pi\varepsilon_0}\ln\left(\frac{r_b}{r_a}\right)$$

PROBLEMS

Problem 4-B-1

Determine the inductive reactance in ohms/mile/phase for a 345-kV, single-circuit line with ACSR 45/7 conductor for which the geometric mean radius is 0.0352 ft. Assume a horizontal phase configuration with 23.5-ft phase separation.

Problem 4-B-2

Determine the inductive reactance in ohms/mile/phase for a 345-kV, single-circuit line with ACSR 84/19 conductor for which the geometric mean radius is 0.0588 ft. Assume a horizontal phase configuration with 26-ft phase separation.

Problem 4-B-3

Calculate the inductance in henries per meter phase for the 345-kV, bundle-conductor line shown in Figure 4-50. Assume phase spacing $D_1 = 8.31$ m, bundle separation $S = 45.72$ cm, and conductor diameter is 3.038 cm.

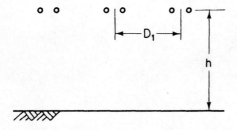

Figure 4-50. Line for Problem 4-B-3.

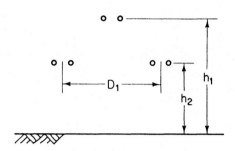

Figure 4-51. Line for Problem 4-B-4.

Problem 4-B-4

Calculate the inductance in henries per meter per phase for the 500-kV, bundle-conductor line shown in Figure 4-51. Assume phase spacing $D_1 = 12.19$ m, bundle separation $S = 45.72$ cm, and conductor diameter is 4.069 cm. Take $h_1 = 22.32$ m and $h_2 = 13.94$ m.

Problem 4-B-5

Calculate the inductive reactance in ohms/mile/phase for a 500-kV, single-circuit, two-subconductor bundle line with ACSR 84/19 subconductor for which the GMR is 0.0534 ft. Assume horizontal phase configuration with 33.5-ft phase separation. Assume bundle separation is 18 in.

Problem 4-B-6

Repeat Problem 4-B-5 for a phase separation of 36 ft.

Problem 4-B-7

Repeat Problem 4-B-5 for a phase separation of 35 ft.

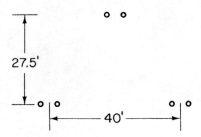

Figure 4-52. Phase Configuration for Problem 4-B-8.

Problem 4-B-8

Repeat Problem 4-B-5 with a triangular phase configuration as shown in Figure 4-52.

Problem 4-B-9

Repeat Problem 4-B-7 with an ACSR 76/19 subconductor for which the GMR is 0.0595 ft.

Problem 4-B-10

Find the inductive reactance in ohms/mile/phase for a 500-kV, single-circuit, two-subconductor bundle line with ACSR 84/19 conductor for which the GMR is 0.0588 ft. Assume horizontal phase configuration with separation of 32 ft. Bundle spacing is 18 in.

Problem 4-B-11

Calculate the inductance in henries per meter per phase for the 500-kV, bundle-conductor line shown in Figure 4-53. Assume phase spacing

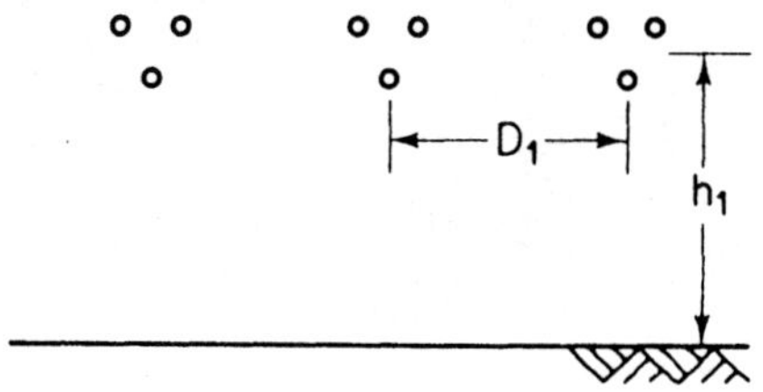

Figure 4-53. Line for Problem 4-B-11.

$D_1 = 12.19$ m, bundle separation $S = 45.72$ cm, and conductor diameter is 2.959 cm.

Problem 4-B-12

Calculate the inductance in henries per meter per phase for the 765-kV, bundle-conductor line shown in Figure 4-54. Assume phase spacing $D_1 = 13.72$ m, bundle separation $S = 45.72$ cm, and conductor diameter is 2.959 cm.

Problem 4-B-13

Find the inductive reactance in ohms/mile/phase for the 765-kV, single-circuit, bundle-conductor line with four subconductors per bundle at a spacing of 18 in., given that the subconductor GMR is 0.0385 ft. Assume horizontal phase configuration with 44.5-ft phase separation.

Problem 4-B-14

Repeat Problem 4-B-13 for bundle spacing of 24 in. and subconductor GMR of 0.0515 ft. Assume phase separation is 45 ft.

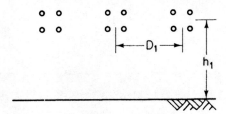

Figure 4-54. Line for Problem 4-B-12.

Problem 4-B-15

Repeat Problem 4-B-13 for a subconductor GMR of 0.0459 ft and 50-ft phase separation.

Problem 4-B-16

Calculate the inductance in henries per meter per phase for the 1100-kV, bundle-conductor line shown in Figure 4-55. Assume phase spacing $D_1 = 15.24$ m, bundle separation $S = 45.72$ cm, and conductor diameter is 3.556 cm.

Problem 4-B-17

Calculate the inductance in henries per meter per phase for the 2000-kV, bundle-conductor line shown in Figure 4-56. Assume phase spacing $D_1 = 35$ m, bundle separation $S = 45.72$ cm, and conductor diameter is 3.81 cm.

Problem 4-B-18

Calculate the inductive reactance in ohms per mile for the 500-kV, double-circuit, bundle-conductor line with three subconductors of 0.0431-ft

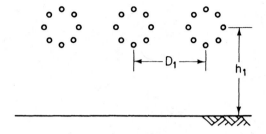

Figure 4-55. Line for Problem 4-B-16.

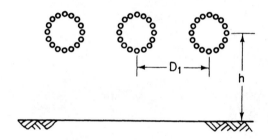

Figure 4-56. Line for Problem 4-B-17.

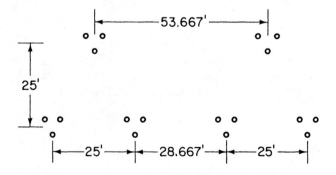

Figure 4-57. Line for Problem 4-B-18.

GMR and with 18-in. bundle separation. Assume conductor configurations as shown in Figure 4-57.

Problem 4-B-19

Calculate the inductive reactance in ohms per mile for the 345-kV, double-circuit, bundle-conductor line with two subconductors per bundle at 18-in. bundle spacing. Assume subconductor's GMR is 0.0497 ft, and conductor configuration is as shown in Figure 4-58.

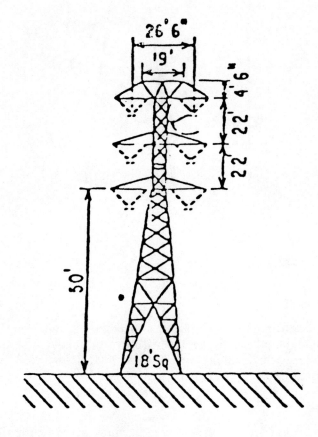

Figure 4-58. Line for Problem 4-B-19.

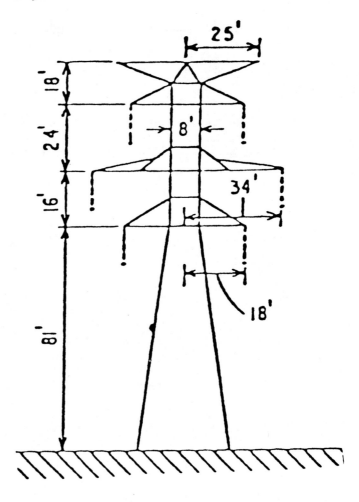

Figure 4-59. Line for Problem 4-B-20.

Problem 4-B-20

Calculate the inductive reactance in ohms per mile for the 345-kV, double-circuit, bundle-conductor line with two subconductors per bundle at 18-in. bundle spacing. Assume subconductor's GMR is 0.0373 ft, and conductor configuration is as shown in Figure 4-59.

Problem 4-B-21

Calculate the inductive reactance in ohms per mile for the 345-kV, double-circuit, bundle-conductor line with two subconductors per bundle at

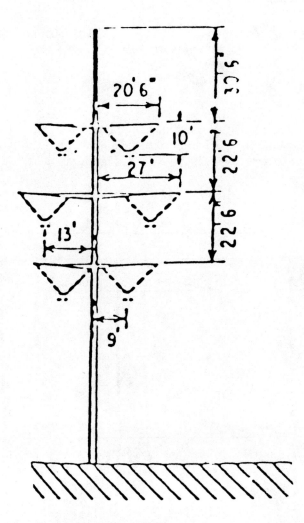

Figure 4-60. Line for Problem 4-B-21.

18-in. bundle spacing. Assume subconductor's GMR is 0.0404 ft, and conductor configuration is as shown in Figure 4-60.

Problem 4-B-22

Calculate the inductive reactance in ohms per mile for the 345-kV double-circuit, bundle-conductor line with two subconductors per bundle at

18-in. bundle spacing. Assume subconductor's GMR is 0.0497 ft, and conductor configuration is as shown in Figure 4-61.

Problem 4-B-23

Determine the capacitive reactance in ohm miles for the line of Problem 4-B-1. Assume the conductor's outside diameter is 1.063 in. Repeat by including earth effects given that the ground clearance is 51.5 ft.

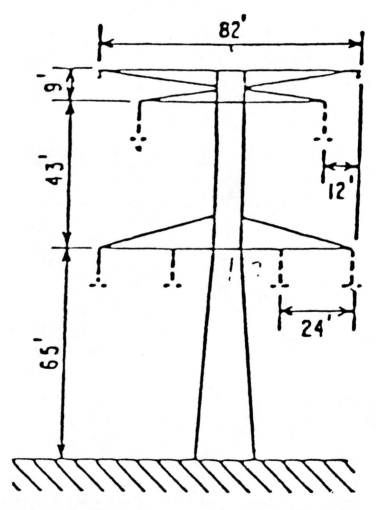

Figure 4-61. Line for Problem 4-B-22.

Problem 4-B-24

Determine the capacitive reactance in ohm miles for the line of Problem 4-B-2. Assume the conductor's outside diameter is 1.76 in. Repeat by including earth effects given that the ground clearance is 45 ft.

Problem 4-B-25

Calculate the capacitance in farads per meter per phase neglecting earth effects for the 345-kV, bundle-conductor line of Problem 4-B-3. Assume the conductor's diameter is 3.038 cm. Repeat including earth effects with $h = 13.61$ m.

Problem 4-B-26

Calculate the capacitance in farads per meter per phase neglecting earth effects for the 500-kV, bundle conductor line of Problem 4-B-4. Assume the conductor's diameter is 4.069 cm. Repeat including earth effects with $h_1 = 22.32$ m and $h_2 = 13.94$ m.

Problem 4-B-27

Determine the capacitive reactance in ohm miles for the line of Problem 4-B-5. Assume the conductor's outside diameter is 1.602 in. Repeat by including earth effects given that the ground clearance is 82 ft.

Problem 4-B-28

Determine the capacitive reactance in ohm miles for the line of Problem 4-B-6. Assume the conductor's outside diameter is 1.823 in. Repeat by including earth effects given that the ground clearance is 80 ft.

Problem 4-B-29

Determine the capacitive reactance in ohm miles for the line of Problem 4-B-7. Assume the conductor's outside diameter is 1.602 in. Repeat by including earth effects given that the ground clearance is 136 ft.

Problem 4-B-30

Determine the capacitive reactance in ohm miles for the line of Problem 4-B-8. Assume the conductor's outside diameter is 1.602 in. Neglect earth effects.

Problem 4-B-31

Determine the capacitive reactance in ohm miles for the line of Problem 4-B-9. Assume the conductor's outside diameter is 1.7 in. Neglect earth effects.

Problem 4-B-32

Determine the capacitive reactance in ohm miles for the line of Problem 4-B-10. Assume the conductor's outside diameter is 1.762 in. Repeat by including earth effects given that the ground clearance is 63 ft.

Problem 4-B-33

Calculate the capacitance in farads per meter per phase neglecting earth effects for the 500-kV, bundle-conductor line of Problem 4-B-11. Assume the conductor's diameter is 2.959 cm. Repeat including earth effects with $h_1 = 14.43$ m.

Problem 4-B-34

Calculate the capacitance in farads per meter per phase neglecting earth effects for the 765-kV, bundle-conductor line of Problem 4-B-12. Assume the conductor's diameter is 2.959 cm. Repeat including earth effects with $h_1 = 20.83$ m.

Problem 4-B-35

Determine the capacitive reactance in ohm miles for the line of Problem 4-B-13. Assume the conductor's outside diameter is 1.165 in.

Problem 4-B-36

Determine the capacitive reactance in ohm·miles for the line of Problem 4-B-14. Assume the conductor's outside diameter is 1.6 in. Repeat by including earth effects given that the ground clearance is 90 ft.

Problem 4-B-37

Calculate the capacitance in farads per meter per phase neglecting earth effect for the 1100-kV, bundle-conductor line of Problem 4-B-16.

Assume the conductor diameter is 3.556 cm. Repeat including earth effects with $h_1 = 21.34$ m.

Problem 4-B-38

Calculate the capacitance in farads per meter per phase neglecting earth effects for the 2000-kV, bundle-conductor line of Problem 4-B-17. Assume the conductor diameter is 3.81 cm. Repeat including earth effects with $h_1 = 45.00$ m.

Problem 4-B-39

Determine the capacitive reactance in ohm · mile for the line of Problem 4-B-18. Assume the conductor's outside diameter is 1.302 in. Neglect earth effect.

Problem 4-B-40

Determine the capacitive reactance in ohm miles for the line of Example 4.8. Assume the conductor's outside diameter is 1.76 in. Repeat by including earth effects given that the ground clearance is 90 ft.

Problem 4-B-41

Determine the capacitive reactance in ohm · mile for the line of Problem 4-B-19. Assume the conductor's outside diameter is 1.502 in. Neglect earth effects.

Problem 4-B-42

Determine the capacitive reactance in ohm · mile for the line of Problem 4-B-20. Assume the conductor's outside diameter is 1.165 in.

Problem 4-B-43

Determine the capacitive reactance in ohm · mile for the line of Problem 4-B-21. Assume the conductor's outside diameter is 1.196 in. Neglect earth effects.

Problem 4-B-44

Determine the capacitive reactance in ohm · mile for the line of Problem 4-B-22. Assume the conductor's outside diameter is 1.302 in.

Problem 4-B-45

Calculate the inductance per phase in henries per meter and the capacitance to neutral in farads per meter with and without earth effect for the 345-kV line shown in Figure 4-62. Assume the following:

$h_1 = 26.31$ m $\quad D_1 = 12.24$ m
$h_2 = 18.85$ m $\quad D_2 = 16.81$ m
$h_3 = 12.29$ m $\quad D_3 = 12.85$ m
$h_4 = 33.93$ m $\quad D_4 = 7.32$ m
Conductor diameter = 3.165 cm
Bundle separation = 35.72 cm

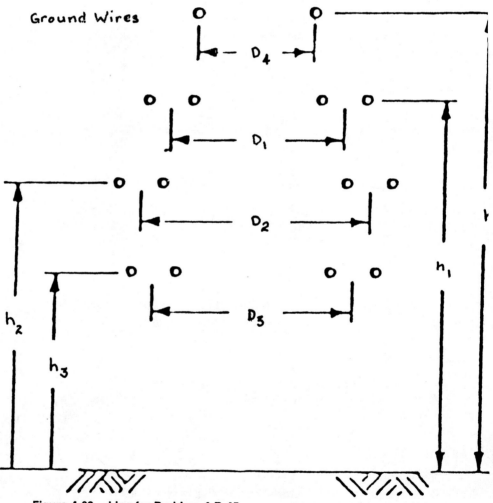

Figure 4-62. Line for Problem 4-B-45.

Problem 4-B-46

Assume that the 345-kV line of Problems 4-B-1 and 4-B-23 is 230 miles long and that the subconductor's resistance is 0.1197 ohms/mile.

A. Calculate the exact $ABCD$ parameters for the line.
B. Find the circuit elements of the equivalent π model for the line. Neglect earth effects.

Problem 4-B-47

Assume that the 345-kV line of Problems 4-B-2 and 4-B-24 is 14 miles long and that the conductor's resistance is 0.0466 ohms/mile.

A. Calculate the exact $ABCD$ parameters for the line.
B. Find the circuit elements of the equivalent π model for the line. Neglect earth effects.

Problem 4-B-48

Assume that the 345-kV line of Problems 4-B-3 and 4-B-25 is 200 km long and that the subconductor's resistance is 0.0620 ohms/km.

A. Calculate the exact $ABCD$ parameters for the line.
B. Find the circuit elements of the equivalent π model for the line. Neglect earth effects.

Problem 4-B-49

Assume that the 500-kV line of Problems 4-B-4 and 4-B-26 is 300 km long and that the subconductor's resistance is 0.0341 ohms/km.

A. Calculate the exact $ABCD$ parameters of the line.
B. Find the circuit elements of the equivalent π model for the line. Neglect earth effects.

Problem 4-B-50

Assume that the 500-kV line of Problems 4-B-11 and 4-B-33 is 250 km long and that the subconductor's resistance is 0.0656 ohms/km.

A. Calculate the exact $ABCD$ parameters for the line.
B. Find the circuit elements of the equivalent π model for the line. Neglect earth effects.

Problem 4-B-51

Assume that the 765-kV line of Problems 4-B-12 and 4-B-34 is 300 km long and that the subconductor's resistance is 0.0656 ohms/km.

A. Calculate the exact $ABCD$ parameters of the line.

B. Find the circuit elements for the equivalent π model for the line. Neglect earth effects.

Problem 4-B-52

Assume that the 1100-kV line of Problems 4-B-16 and 4-B-37 is 400 km long and that the subconductor's resistance is 0.0435 ohms/km.

A. Calculate the exact $ABCD$ parameters of the line.

B. Find the circuit elements of the equivalent π model for the line. Neglect earth effects.

Problem 4-B-53

Assume that the 2000-kV line of Problems 4-B-17 and 4-B-38 is 500 km long and that the subconductor's resistance is 0.0386 ohms/km.

A. Calculate the exact $ABCD$ parameters of the line.

B. Find the circuit elements of the equivalent π model for the line. Neglect earth effects.

Problem 4-B-54

Assume that the 345-kV line of Problem 4-B-45 is 200 km long and that the subconductor's resistance is 0.0574 ohms/km.

A. Calculate the exact $ABCD$ parameters of the line.

B. Find the circuit elements of the equivalent π model for the line. Neglect earth effects.

Problem 4-B-55

The following information is available for a single-circuit, three-phase, 345-kV, 360 mega volt amperes (MVA) transmission line:

Line length = 413 miles.

Number of conductors per phase = 2.

Bundle spacing = 18 in.

Outside conductor diameter = 1.165 in.

Conductor's GMR = 0.0374 ft.

Conductor's resistance = 0.1062 ohms/mile.

Phase separation = 30 ft.

Phase configuration is equilateral triangle.

Minimum ground clearance = 80 ft.

A. Calculate the line's inductive reactance in ohms per mile per phase.
B. Calculate the capacitive reactance including earth effects in ohm miles per phase.
C. Calculate the exact A and B parameters of the line.
D. Find the voltage at the sending end of the line if normal rating power at 0.9 PF is delivered at 345-kV at the receiving end. Use the exact formulation.
E. Repeat (d) using the short-line approximation. Find the error involved in computing the magnitude of the sending-end voltage between this method and the exact one.

Problem 4-B-56

For the transmission line of Problem 4-B-48, calculate the sending-end voltage, sending-end current, power, and power factor when the line is delivering 350 MVA at 0.9 PF lagging at rated voltage, using the following:

A. Exact formulation.
B. Nominal π approximation.
C. Short-line approximation.

Problem 4-B-57

For the transmission line of Problem 4-B-49, calculate the sending-end voltage, sending-end current, power, and power factor when the line is delivering 750 MVA at 0.9 PF lagging at rated voltage, using the following:

A. Exact formulation.
B. Nominal π approximation.
C. Short-line approximation.

Problem 4-B-58

For the transmission line of Problem 4-B-50, calculate the sending-end voltage, sending-end current, power, and power factor when the line is delivering 750 MVA at 0.9 PF lagging at rated voltage, using the following:

A. Exact formulation.
B. Nominal π approximation.
C. Short-line approximation.

Problem 4-B-59

For the transmission line of Problem 4-B-51, calculate the sending-end voltage, sending-end current, power, and power factor when the line is delivering 1800 MVA at 0.9 PF lagging at rated voltage, using the following:

A. Exact formulation.
B. Nominal π approximation.
C. Short-line approximation.

Problem 4-B-60

For the transmission line of Problem 4-B-52, calculate the sending-end voltage, sending-end current, power, and power factor when the line is delivering 4500 MVA at 0.9 PF lagging at rated voltage, using the following:

A. Exact formulation.
B. Nominal π approximation.
C. Short-line approximation.

Problem 4-B-61

For the transmission line of Problem 4-B-53, calculate the sending-end voltage, sending-end current, power, and power factor when the line is delivering 15,200 MVA at 0.9 PF lagging at rated voltage, using the following:

A. Exact formulation.
B. Nominal π approximation.
C. Short-line approximation.

Problem 4-B-62

For the transmission line of Problem 4-B-54, calculate the sending-end voltage, sending-end current, power, and power factor when the line is delivering 750 MVA at 0.9 PF lagging at rated voltage, using the following:

A. Exact formulation.
B. Nominal π approximation.
C. Short-line approximation.

Problem 4-B-63

For the conditions of Problem 4-B-58, evaluate the upper bounds on the errors in evaluating the sending-end voltage and power using the short-line approximation with the nominal π model.

Problem 4-B-64

Repeat Problem 4-B-63 for the conditions of Problem 4-B-59.

Problem 4-B-65

Repeat Problem 4-B-63 for the conditions of Problem 4-B-56.

Problem 4-B-66

Repeat Problem 4-B-63 for the conditions of Problem 4-B-57.

Problem 4-B-67

Repeat Problem 4-B-63 for the conditions of Problem 4-B-60.

Problem 4-B-68

Repeat Problem 4-B-63 for the conditions of Problem 4-B-61.

Problem 4-B-69

Repeat Problem 4-B-63 for the conditions of Problem 4-B-62.

CHAPTER V

The Load Subsystem

5.1 INTRODUCTION

The previous two chapters treated the synchronous machine, which is the major generating source in present-day electric energy systems, and the transmission lines that are used to transport the generated energy to major load centers and utilization points. The present chapter is intended to cover two major components of the system. The first is the *power transformer*, which is used in many parts of the system on the generating and distribution sides. The second is the *induction motor*, which is the workhorse in industrial and commercial electric energy utilization. Due to the similarities in the models for both components, it seems appropriate to study them under one heading.

5.2 GENERAL THEORY OF TRANSFORMER OPERATION

One of the most valuable apparatus in electric power systems is the transformer, for it enables us to utilize different voltage levels across the system for the most economical value. Generation of power at the synchronous machine level is normally at a relatively low voltage, which is most desirable economically. Stepping up of this generated voltage to high voltage, extra-high voltage, or even to ultra-high voltage is done through power transformers to suit the power transmission requirement to minimize losses and increase the transmission capacity of the lines. This transmission voltage level is then stepped down in many stages for distribution and utilization purposes.

A transformer contains two or more windings that are linked by a mutual field. The primary winding is connected to an alternating voltage source, which results in an alternating flux whose magnitude depends on the voltage and number of turns of the primary winding. The alternating flux links the secondary winding and induces a voltage in it with a value that depends on the number of turns of the secondary winding. If the primary voltage is v_1, the core flux ϕ is established such that the counter EMF e equals the impressed voltage (neglecting winding resistance). Thus,

$$v_1 = e_1 = N_1 \left(\frac{d\phi}{dt} \right) \tag{5.1}$$

Here N_1 denotes the number of turns of the primary winding. The EMF e_2 is induced in the secondary by the alternating core flux ϕ:

$$v_2 = e_2 = N_2 \left(\frac{d\phi}{dt} \right) \tag{5.2}$$

Taking the ratio of Eqs. (5.1) to (5.2), we see that

$$\frac{v_1}{v_2} = \frac{N_1}{N_2} \tag{5.3}$$

Neglecting losses, the instantaneous power is equal on both sides of the transformer, as shown below:

$$v_1 i_1 = v_2 i_2 \tag{5.4}$$

Combining Eqs. (5.3) and (5.4), we get

$$\frac{i_1}{i_2} = \frac{N_2}{N_1} \tag{5.5}$$

Thus the current ratio is the inverse of the voltage ratio. We can conclude that almost any desired voltage ratio, or ratio of transformation, can be obtained by adjusting the number of turns.

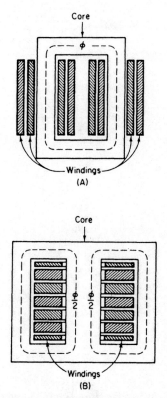

Figure 5-1. (a) Core-Type and (b) Shell-Type Transformer Construction.

Transformer action requires the existence of the flux that links the two windings. This will be obtained more effectively if an iron core is used because an iron core confines the flux to a definite path linking both windings. A magnetic material such as iron undergoes a loss of energy due to the application of alternating voltage to its B-H loop. The losses are composed of two parts. The first is called the *eddy-current loss*, and the second is the *hysteresis loss*. Eddy-current loss is basically an I^2R loss due to the induced currents in the magnetic material. To reduce these losses, the magnetic circuit is usually made of a stack of thin laminations. Hysteresis loss is caused by the energy used in orienting the magnetic domains of the material along the field. The loss depends on the material used.

Two types of construction are used, as shown in Figure 5-1. The first is denoted the *core type*, which is a single ring encircled by one or more groups of windings. The mean length of the magnetic circuit for this type is long, whereas the mean length of windings is short. The reverse is true for the *shell type*, where the magnetic circuit encloses the windings.

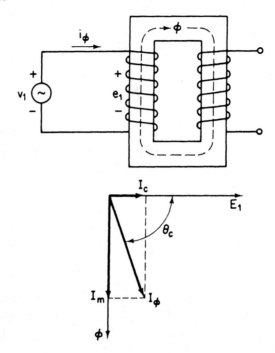

Figure 5-2. Transformer on No-Load.

Due to the nonlinearity of the *B-H* curve of the magnetic material, the primary current on no-load (for illustration purposes) will not be a sinusoid but rather a certain distorted version, which is still periodic. For analysis purposes, a Fourier analysis shows that the fundamental component is out of phase with the applied voltage. This fundamental primary current is basically made of two components. The first is in phase with the voltage and is attributed to the power taken by eddy-current and hysteresis losses and is called the *core-loss component* I_c of the exciting current I_ϕ. The component that lags e by 90° is called the magnetizing current I_m. Higher harmonics are neglected. Figure 5-2 shows the no-load phasor diagram for a single-phase transformer.

Consider an ideal transformer (with negligible winding resistances and reactances and no exciting losses) connected to a load as shown in Figure 5-3. Clearly Eqs. (5.1)–(5.5) apply. The dot markings indicate terminals of corresponding polarity in the sense that both windings encircle the core in the same direction if we begin at the dots. Thus comparing the voltages of the two windings shows that the voltages from a dot-marked terminal to an unmarked terminal will be of the same polarity for the primary and secondary windings (i.e., v_1 and v_2 are in phase). From Eqs. (5.3) and (5.5), we can write for sinusoidal steady state operation

$$\frac{V_1}{I_1} = \left(\frac{N_1}{N_2}\right)^2 \frac{V_2}{I_2}$$

But the load impedance Z_2 is

$$\frac{V_2}{I_2} = Z_2$$

Thus,

$$\frac{V_1}{I_1} = \left(\frac{N_1}{N_2}\right)^2 Z_2$$

The result is that as far as its effect is concerned, Z_2 can be replaced by an equivalent impedance Z_2' in the primary circuit. Thus,

$$Z_2' = \left(\frac{N_1}{N_2}\right)^2 Z_2 \tag{5.6}$$

The equivalence is shown in Figure 5-3.

More realistic representations of the transformer must account for winding parameters as well as the exciting current. The equivalent circuit of the transformer can be visualized by following the chain of events as we proceed from the primary winding to the secondary winding in Figure 5-4. First the impressed voltage V_1 will be reduced by a drop $I_1 R_1$ due to the primary winding resistance as well as a drop $jI_1 X_1$ due to the primary leakage represented by the inductive reactance X_1. The resulting voltage is denoted E_1. The current I_1 will supply the exciting current I_ϕ as well as the current I_2', which will be transformed through to the secondary winding. Thus,

$$I_1 = I_\phi + I_2'$$

Since I_ϕ has two components (I_c in phase with E_1 and I_m lagging E_1 by 90°), we can model its effect by the parallel combination G_c and B_m as shown in the circuit. Next E_1 and I_1 are transformed by an ideal transformer with turns ratio N_1/N_2. As a result, E_2 and I_2 emerge on the secondary side. E_2 undergoes drops $I_2 R_2$ and $jI_2 X_2$ in the secondary winding to result in the terminal voltage V_2.

Figure 5-4(b) shows the transformer's equivalent circuit in terms of primary variables. This circuit is called "circuit referred to the primary side." Note that

$$V_2' = \frac{N_1}{N_2}(V_2) \tag{5.7}$$

$$I_2' = \frac{N_2}{N_1}(I_2) \tag{5.8}$$

$$R_2' = R_2\left(\frac{N_1}{N_2}\right)^2 \tag{5.9}$$

$$X_2' = X_2\left(\frac{N_1}{N_2}\right)^2 \tag{5.10}$$

222   The Load Subsystem

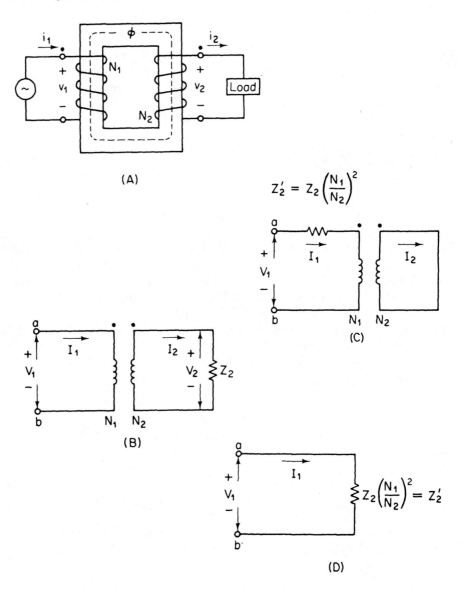

Figure 5-3. Ideal Transformer and Load and Three Equivalent Representations.

Although the equivalent circuit illustrated above is simply a *T*-network, it is customary to use approximate circuits such as shown in Figure 5-5. In the first two circuits we move the shunt branch either to the secondary or primary sides to form inverted *L*-circuits. Further simplifications are shown where the shunt branch is neglected in Figure 5-5(c) and finally with the resistances neglected in Figure 5-5(d). These last two

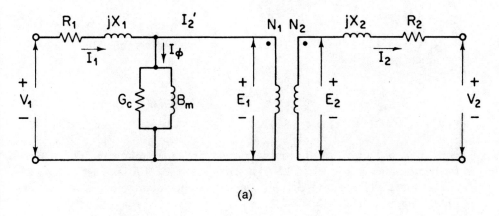

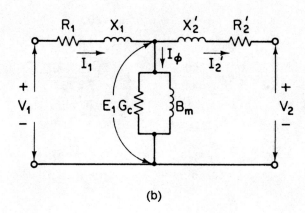

Figure 5-4. Equivalent Circuits of Transformer.

circuits are of sufficient accuracy in most power system applications. In Figure 5-5 note that

$$R_{eq} = R_1 + R'_2$$

$$X_{eq} = X_1 + X'_2$$

An example will illustrate the principles and orders of approximations involved.

Example 5-1

A 50-kVA, 400/2000 V, single-phase transformer has the following parameters:

$$R_1 = 0.02 \text{ ohm} \qquad R_2 = 0.5 \text{ ohm}$$
$$X_1 = 0.06 \text{ ohm} \qquad X_2 = 1.5 \text{ ohm}$$
$$G_c = 2 \text{ mS} \qquad B_m = -6 \text{ mS}$$

Note that G_c and B_m are given in terms of primary reference. The transformer supplies a load of 40 kVA at 2000 V and 0.8 PF lagging. Calculate the primary voltage and current using the equivalent circuits shown in Figure 5-5 and that of Figure 5-4.

Solution

Let us refer all the data to the primary (400 V) side:

$$R_1 = 0.02 \text{ ohm} \qquad X_1 = 0.06 \text{ ohm}$$
$$R_2' = 0.5\left(\frac{400}{2000}\right)^2 \qquad X_2' = 1.5\left(\frac{400}{2000}\right)^2$$
$$= 0.02 \text{ ohm} \qquad = 0.06 \text{ ohm}$$

Thus,

$$R_{eq} = R_1 + R_2' \qquad X_{eq} = X_1 + X_2'$$
$$= 0.04 \text{ ohm} \qquad = 0.12 \text{ ohm}$$

The voltage $V_2 = 2000$ V; thus

$$V_2' = 2000\left(\frac{400}{2000}\right) = 400 \text{ V}$$

The current I_2' is thus

$$|I_2'| = \frac{40 \times 10^3}{400} = 100 \text{ A}$$

The power factor of 0.8 lagging implies that

$$I_2' = 100\underline{/-36.87°} \text{ A}$$

For ease of computation, we start with the simplest circuit of Figure 5-5(d). Let us denote the primary voltage calculated through this circuit by V_{1_d}. It is clear then that

$$V_{1_d} = V_2' + jI_2'(X_{eq})$$
$$= 400\underline{/0} + j\left(100\underline{/-36.87°}\right)(0.12)$$

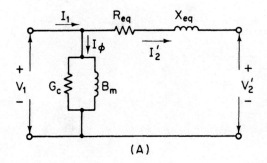

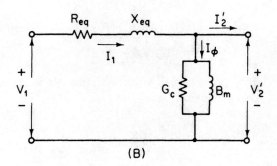

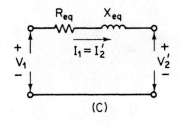

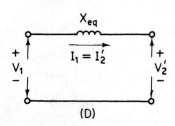

Figure 5-5. Approximate Equivalent Circuits for the Transformer.

Thus,

$$V_{1_d} = 407.31 \underline{/1.35°} \text{ V}$$

$$I_{1_d} = 100 \underline{/-36.87°} \text{ A}$$

Comparing circuits (c) and (d) in Figure 5-5, we deduce that

$$V_{1_c} = V_2' + I_2'(R_{eq} + jX_{eq}) = V_{1_d} + I_2'(R_{eq})$$

Thus,

$$V_{1_c} = 407.31\underline{/1.35°} + \left(100\underline{/-36.87}\right)(0.04)$$
$$= 410.46\underline{/1.00°} \text{ V}$$
$$I_{1_c} = I_2' = 100\underline{/-36.87°} \text{ A}$$

Let us consider circuit (a) in Figure 5-5. We can see that

$$V_{1_a} = V_{1_c} = 410.46\underline{/1.00°} \text{ V}$$

But

$$I_{1_a} = I_2' + (G_c + jB_m)V_{1_a}$$
$$= 100\underline{/-36.87°} + (2 \times 10^{-3} - j6 \times 10^{-3})\left(410.46\underline{/1.00}\right)$$
$$= 102.17\underline{/-37.68°} \text{ A}$$

Circuit (b) is a bit different since we start with V_2'' impressed on the shunt branch. Thus

$$I_{1_b} = I_2' + (G_c + jB_m)V_2''$$
$$= 100\underline{/-36.87°} + (2 \times 10^{-3} - j6 \times 10^{-3})\left(400\underline{/0}\right)$$
$$= 102.09\underline{/-37.68°} \text{ A}$$

Now

$$V_{1_b} = V_2'' + I_{1_b}(R_{eq} + jX_{eq})$$
$$= 400\underline{/0} + \left(102.09\underline{/-37.68°}\right)(0.04 + j0.12)$$
$$= 410.78\underline{/1.00°}$$

The exact equivalent circuit is now considered as shown in Figure 5-4(b). We first calculate E_1:

$$E_1 = V_2'' + I_2'(R_2' + jX_2')$$
$$= 400\underline{/0} + \left(100\underline{/-36.87°}\right)(0.02 + j0.06)$$
$$= 405.22\underline{/0.51°}$$

Now

$$I_1 = I_2' + E_1(G_c + jB_m)$$
$$= 100\underline{/-36.87°} + \left(405.22\underline{/0.51}\right)(2 \times 10^{-3} - j6 \times 10^{-3})$$
$$= 102.13\underline{/-37.68°} \text{ A}$$

Thus,

$$V_1 = E_1 + I_1(R_1 + jx_1)$$
$$= 405.22\underline{/0.51} + (102.13\underline{/-37.68})(0.02 + j0.06)$$
$$= 410.63\underline{/1.01°}\ V$$

The values of V_1 and I_1 calculated using each of the five circuits are tabulated in Table 5-1 with the angle of V_1 denoted by θ_1 and the angle of I_1 denoted by ψ_1. The largest error in percent is 0.8085 percent in calculating $|V_1|$ using circuit (d) as opposed to the exact circuit. This confirms our earlier statements about common practice in taking equivalent circuits for power transformers.

Transformer Performance Measures

Two important performance measures are of interest when choosing transformers. These are the voltage regulation and efficiency of the transformer. The voltage regulation is a measure of the variation in the secondary voltage when the load is varied from zero to rated value at a constant power factor. The percentage voltage regulation (P.V.R.) is thus given by

$$\text{P.V.R.} = 100 \frac{|V_{2(\text{no load})}| - |V_{2\,\text{rated}}|}{|V_{2\,\text{rated}}|} \qquad (5.13)$$

If we neglect the exciting current and refer the equivalent circuit to the secondary side, we have by inspection of Figure 5-6,

$$\text{P.V.R.} = 100 \frac{\left|\frac{V_1}{a}\right| - |V_2|}{|V_2|}$$

TABLE 5-1

Values of V_1 and I_1 as Calculated Using Different Approximate Circuits for Example 5-1

	Exact	(a)	(b)	(c)	(d)
V_1	410.63	410.46	410.78	410.46	407.31
θ_1	1.01	1.00	1.00	1.00	1.35
I_1	102.13	102.17	102.09	100	100
ψ_1	-37.68	-37.68	-37.68	-37.87	-36.87

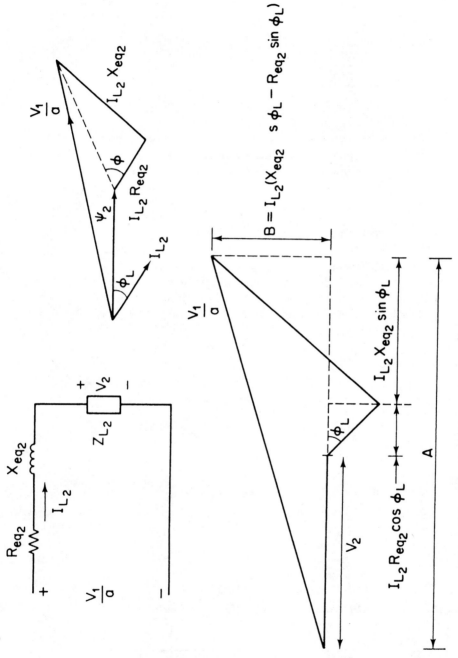

Figure 5-6. Transformer Approximate Equivalent Circuit and Associated Phasor Diagrams for Voltage Regulation Derivation.

5.2 General Theory of Transformer Operation

where a is the transformer ratio. Thus

$$a = \frac{N_1}{N_2}$$

From the phasor diagram we have

$$\left|\frac{V_1}{a}\right| = \sqrt{A^2 + B^2} = A\left[1 + \left(\frac{B^2}{A^2}\right)\right]^{1/2}$$

We use the following approximation:

$$\left|\frac{V_1}{a}\right| = A\left(1 + \frac{B^2}{2A^2} + \cdots\right) \cong A + \frac{B^2}{2A}$$

Hence the percentage voltage regulation is

$$\text{P.V.R.} = \left[\frac{\left(A - V_2 + \frac{B^2}{2A}\right)}{V_2}\right] 100 \cong \left[\frac{\left(A - V_2 + \frac{B^2}{2V_2}\right)}{V_2}\right] 100$$

In terms of transformer constants, we get

$$\text{P.V.R.} \cong 100 \left[\frac{I_{L_2}(R_{\text{eq}_2} \cos \phi_L + X_{\text{eq}_2} \sin \phi_L)}{V_2}\right.$$

$$\left. + \frac{1}{2}\left\{\frac{I_{L_2}(X_{\text{eq}_2} \cos \phi_L - R_{\text{eq}_2} \sin \phi_L)}{V_2}\right\}^2\right] \quad (5.14)$$

The efficiency of the transformer is the ratio of output (secondary) power to the input (primary) power. Formally the efficiency is η:

$$\eta = \frac{P_2}{P_1} \quad (5.15)$$

If we deal with the transformer as referred to the secondary side, we have

$$P_2 = |V_2||I_L|\cos \phi_L$$

where I_L is the load current. The input power P_1 is the sum of the output power and power loss in the transformer. Thus

$$P_1 = P_2 + P_l$$

The power loss in the transformer is made of two parts: the I^2R loss and the core loss P_c. Thus

$$P_l = P_c + |I_L|^2(R_{\text{eq}})$$

As a result, the efficiency is given by

$$\eta = \frac{|V_2||I_L|\cos\phi_L}{|V_2||I_L|\cos\phi_L + P_c + |I_L|^2(R_{eq})} \quad (5.16)$$

The following example utilizes results of Example 5-1 to illustrate the computations involved.

Example 5-2

Find the P.V.R. and efficiency for the transformer of Example 5-1.

Solution

Let us apply the basic formula of Eq. (5.14). We have from Example 5-1:

$$V_2 = 2000 \text{ V}$$
$$I_{L_2} = 20 \text{ A}$$
$$R_{eq_2} = 0.04\left(\frac{2000}{400}\right)^2 = 1 \text{ ohm}$$
$$X_{eq_2} = 0.12\left(\frac{2000}{400}\right)^2 = 3 \text{ ohms}$$

Thus substituting in Eq. (5.14), we get

$$\text{P.V.R.} = 100\left\{\frac{20[1(0.8) + 3(0.6)]}{2000} + \frac{1}{2}\left[\frac{20[3(0.8) - 1(0.6)]}{2000}\right]^2\right\}$$

$$= 2.600 \text{ percent}$$

Let us compare this with the result of applying Eq. (5.13) with no approximations. Using the results of circuit (c) for Example 5-1, we have for load conditions,

$$V_1 = 410.46 \text{ V}$$
$$V_2' = 400 \text{ V}$$

Referred to secondary, we have

$$V_1' = 410.46\left(\frac{2000}{400}\right) = 2052.30 \text{ V}$$

This is V_2 on no-load. Thus,

$$\text{P.V.R.} = 100\left(\frac{2052.30 - 2000}{2000}\right)$$

$$= 2.62 \text{ percent}$$

5.2 General Theory of Transformer Operation

To calculate the efficiency we need only to apply the basic definition. Take the results of the exact circuit. The input power is

$$P_1 = V_1 I_1 \cos \phi_1$$
$$= (410.63)(102.13)(\cos 38.69)$$
$$= 32733.99 \text{ W}$$
$$P_2 = V_2 I_2 \cos \phi_2$$
$$= (400)(100)(0.8)$$
$$= 32,000 \text{ W}$$

Thus,

$$\eta = \frac{32,000}{32,733.99} = 0.97758$$

The efficiency of a transformer varies with the load current I_L. It attains a maximum when

$$\frac{\partial \eta}{\partial I_L} = 0$$

Using Eq. (5.15) the derivative is

$$\frac{\partial \eta}{\partial I_L} = \frac{P_1 \left(\frac{\partial P_2}{\partial |I_L|} \right) - P_2 \left(\frac{\partial P_1}{\partial |I_L|} \right)}{P_1^2}$$

Thus the condition for maximum power is

$$\frac{P_1}{P_2} = \frac{\frac{\partial P_1}{\partial |I_L|}}{\frac{\partial P_2}{\partial |I_L|}}$$

Using Eq. (5.16) we get

$$\frac{P_1}{P_2} = \frac{|V_2| \cos \phi_L + 2|I_L| R_{eq}}{|V_2| \cos \phi_L}$$

This reduces to

$$P_1 = P_2 + 2|I_L|^2 (R_{eq})$$

Thus for maximum efficiency we have

$$P_l = P_c + |I_L|^2 (R_{eq})$$

As a result, the maximum efficiency occurs for

$$P_c = |I_L|^2 (R_{eq}) \tag{5.17}$$

That is, when the $I^2 R$ losses equal the core losses, maximum efficiency is attained.

Example 5-3

Find the maximum efficiency of the transformer of Example 5-1 under the same power factor and voltage conditions.

Solution

We need first the core losses. These are obtained from the exact equivalent circuit as

$$P_c = |E_1|^2 (G_c)$$
$$= (405.22)^2 (2 \times 10^{-3})$$
$$= 328.41 \text{ W}$$

For maximum efficiency,

$$P_c = I_L^2 (R_{eq})$$

Referred to the primary, we thus have

$$328.41 = I_L^2 (0.04)$$

Thus for maximum efficiency,

$$I_L = 90.61 \text{ A}$$

$$\eta_{max} = \frac{V_2'|I_L|\cos\phi_L}{V_2'|I_L|\cos\phi_L + 2P_c}$$

$$= \frac{(400)(90.61)(0.8)}{(400)(90.61)(0.8) + 2(328.41)}$$

$$= 0.97785$$

5.3 TRANSFORMER CONNECTIONS

Single-phase transformers can be connected in a variety of ways. To start with, consider two single-phase transformers A and B. They can be connected in four different combinations provided that the polarities are observed. Figure 5-7 illustrates a series-series connection where the primaries of the two transformers are connected in series whereas the secondaries are connected in series. Figure 5-8 illustrates the series-parallel connection and the parallel-series connection. Note that when windings are connected in parallel, those having the same voltage and polarity are paralleled. When connected in series, windings of opposite polarity are joined in one junction. Coils of unequal voltage ratings may be series-connected either aiding or opposing.

Figure 5-9 shows two transformers A and B connected in parallel with their approximate equivalent circuit indicated as well. Assume that Z_A is

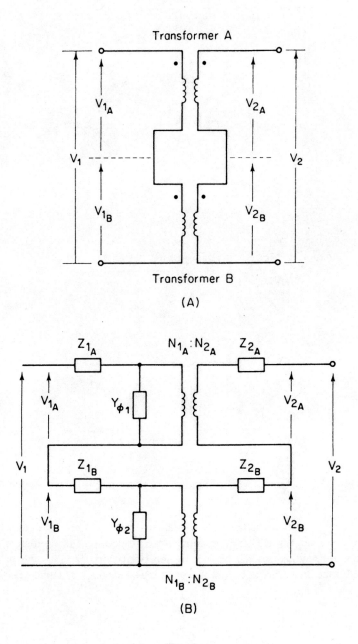

Figure 5-7. Two Transformers with Primaries in Series and Secondaries in Series. (a) Connection Diagram. (b) Exact Equivalent Circuit.

234 The Load Subsystem

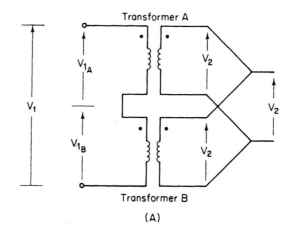

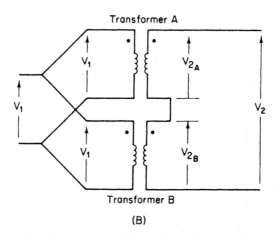

Figure 5-8. Series-Parallel and Parallel-Series Connections for Single-Phase Transformers.

the ohmic equivalent impedance of transformer A referred to its secondary side. Similarly, Z_B is the ohmic equivalent impedance of transformer B referred to its secondary side. Z_L is the load impedance. V_1 is the primary voltage on both transformers. Let the primary to secondary turns ratios be

$$a_A = \frac{N_{1A}}{N_{2A}}$$

$$a_B = \frac{N_{1B}}{N_{2B}}$$

The two ratios should be identical for the parallel connection to make sense.

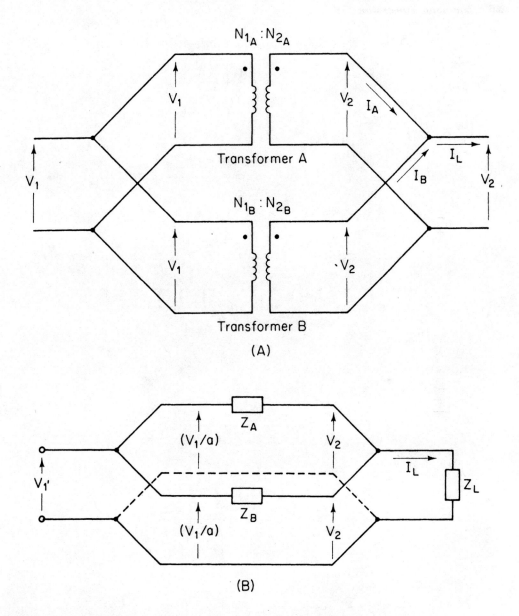

Figure 5-9. Parallel-Connected Single-Phase Transformers. (a) Connection Diagram. (b) Equivalent Circuit.

The current delivered by transformer A is I_A. Thus

$$I_A = \frac{\left(\dfrac{V_1}{a}\right) - V_2}{Z_A}$$

Similarly,

$$I_B = \frac{\left(\dfrac{V_1}{a}\right) - V_2}{Z_B}$$

Thus the load current I_L is

$$I_L = I_A + I_B$$

or

$$I_L = V_1\left(\frac{1}{aZ_A} + \frac{1}{aZ_B}\right) - V_2\left(\frac{1}{Z_A} + \frac{1}{Z_B}\right)$$

But

$$V_2 = I_L Z_L$$

Thus,

$$I_L = V_1\left(\frac{1}{aZ_A} + \frac{1}{aZ_B}\right) - I_L Z_L\left(\frac{1}{Z_A} + \frac{1}{Z_B}\right)$$

or

$$I_L = \frac{\dfrac{V_1}{a}}{Z_L + \left(\dfrac{1}{\dfrac{1}{Z_A} + \dfrac{1}{Z_B}}\right)} \tag{5.18}$$

Three-Winding Transformers

The three-winding transformer is used in many parts of the power system for the economy achieved when using three windings on the one core. Figure 5-10 shows a three-winding transformer with a practical equivalent circuit. The impedances Z_1, Z_2, and Z_3 are calculated from the three

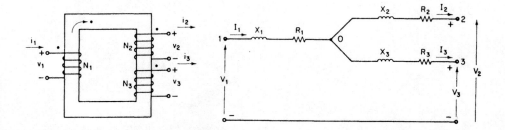

Figure 5-10. Three-Winding Transformer and Its Practical Equivalent Circuit.

impedances obtained by considering each pair of windings separately with

$$Z_1 = \frac{Z_{12} + Z_{13} - Z_{23}}{2} \quad (5.19)$$

$$Z_2 = \frac{Z_{12} + Z_{23} - Z_{13}}{2} \quad (5.20)$$

$$Z_3 = \frac{Z_{13} + Z_{23} - Z_{12}}{2} \quad (5.21)$$

The I^2R or load loss for a three-winding transformer can be obtained from analysis of the equivalent circuit shown.

Example 5-4

Consider a three-winding transformer with the particulars shown in the equivalent circuit referred to the primary side given in Figure 5-11. Assuming V_1 is the reference, calculate the following:

A. The secondary and tertiary voltages referred to the primary side.
B. The apparent powers and power factors at the primary, secondary, and tertiary terminals.
C. The transformer efficiency.

Assume that

$$I_2 = 50\underline{/-30°}$$
$$I_3 = 50\underline{/-35°}$$

Solution

The primary current is

$$I_1 = I_2 + I_3$$
$$= 99.9048\underline{/-32.5°}$$

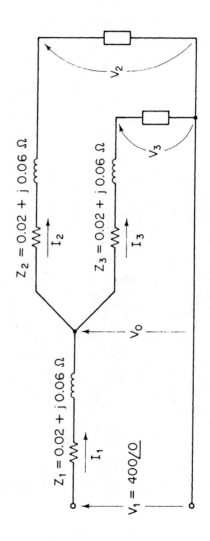

Figure 5-11. Circuit for Example 5-4.

5.3 Transformer Connections

Now the voltage at point 0 is

$$V_0 = V_1 - I_1 Z_1$$
$$= 400 - (99.9048)(\underline{/-32.5°})(0.02 + j0.06)$$
$$= 395.114\underline{/-0.577°} \text{ V}$$

The secondary voltage is obtained referred to the primary as

$$V_2 = V_0 - I_2 Z_2$$
$$= 395.114\underline{/-0.577°} - (50\underline{/-30°})(0.02 + j0.06)$$
$$= 392.775\underline{/-0.887°}$$

The tertiary voltage is obtained referred to the primary as

$$V_3 = V_0 - I_3 Z_3$$
$$= 395.114\underline{/-0.577°} - (50\underline{/-35°})(0.02 + j0.06)$$
$$= 392.598\underline{/-0.856°}$$

The apparent power into the load connected to the secondary winding is thus

$$S_2 = V_2 I_2^*$$
$$= 19638.75\underline{/29.113°}$$

As a result,

$$PF_2 = \cos(29.113°) = 0.87366$$

Similarly for the tertiary winding, we get

$$S_3 = V_3 I_3^*$$
$$= 19629.9\underline{/34.144°}$$

As a result

$$PF_3 = \cos(34.144°) = 0.82763$$

The apparent power at the primary side is

$$S_1 = V_1 I_1^*$$
$$= 39961.92\underline{/32.5°}$$

As a result,

$$PF_1 = \cos(32.5°) = 0.84339$$

The active powers are

$$P_2 = 19638.75 \cos 29.113°$$
$$= 17157.627 \text{ W}$$

$$P_3 = 19629.9 \cos 34.144°$$
$$= 16246.285 \text{ W}$$
$$P_1 = 39961.92 \cos 32.5°$$
$$= 33703.5415 \text{ W}$$

The efficiency is therefore

$$\eta = \frac{P_2 + P_3}{P_1}$$
$$= 0.99111$$

The Autotransformer

The basic idea of the autotransformer is permitting the interconnection of the windings electrically. Figure 5-12 shows a two-winding transformer connected in an autotransformer step-up configuration. We will assume the same voltage per turn; i.e.,

$$\frac{V_1}{N_1} = \frac{V_2}{N_2}$$

The rating of the transformer when connected in a two-winding configuration is

$$S_{\text{rated}} = V_1 I_1 = V_2 I_2 \tag{5.22}$$

In the configuration chosen, the apparent power into the load is

$$S_0 = (V_1 + V_2) I_2$$
$$= V_2 I_2 \left(1 + \frac{N_1}{N_2}\right) \tag{5.23}$$

The input apparent power is

$$S_i = V_1 (I_1 + I_2)$$
$$= V_1 I_1 \left(1 + \frac{N_1}{N_2}\right)$$

Thus the rating of the autotransformer is higher than the original rating of the two-winding configuration. Note that each winding passes the same current in both configurations, and as a result the losses remain the same. Due to the increased power rating, the efficiency is thus improved.

Autotransformers are generally used when the ratio is 3:1 or less. Two disadvantages are the lack of electric isolation between primary and secondary and the increased short-circuit current over that for the corresponding two-winding configuration.

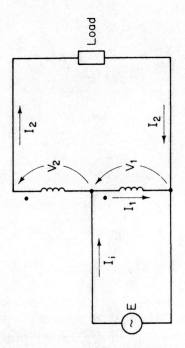

Figure 5-12. Step-Up Autotransformer.

241

Example 5-5

A 30-kVA, 2.4/0.6-kV transformer is connected as a step-up autotransformer from a 2.4-kV supply. Calculate the currents in each part of the transformer and the load rating. Neglect losses.

Solution

With reference to Figure 5.12, the primary winding rated current is

$$I_1 = \frac{30}{2.4} = 12.5 \text{ A}$$

The secondary rated current is

$$I_2 = \frac{30}{0.6} = 50 \text{ A}$$

Thus the load current is

$$I_L = 50 \text{ A}$$

The load voltage is

$$V_L = V_1 + V_2 = 3 \text{ kV}$$

As a result, the load rating is

$$S_L = V_L I_L = 150 \text{ kVA}$$

Note that

$$I_i = I_1 + I_2$$
$$= 62.5 \text{ A}$$
$$V_i = V_1 = 2.4 \text{ kV}$$

Thus,

$$S_i = (2.4)(62.5) = 150 \text{ kVA}$$

Three-Phase Transformer Connections

For three-phase system applications it is possible to install three-phase transformer units or banks made of three single-phase transformers connected in the desired three-phase configurations. The latter arrangement is advantageous from a reliability standpoint since it is then possible to install a single standby single-phase transformer instead of a three-phase unit. This provides a considerable cost saving. We have seen that there are two possible three-phase connections; the Y-connection and the Δ-connection. We thus see that three-phase transformers can be connected in four different ways. In the Y/Y connection, both primary and secondary windings are connected in Y. In addition, we have Δ/Δ, Y/Δ, or Δ/Y connections. The Y-connected windings may or may not be grounded.

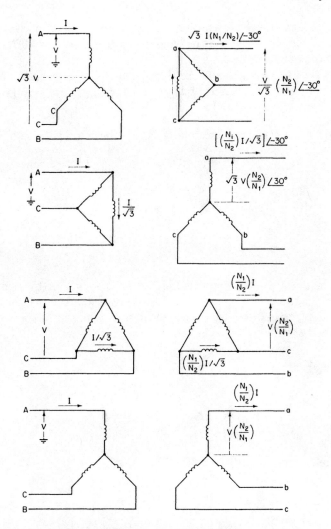

Figure 5-13. Three-Phase Transformer Connections.

The Y/Δ configuration is used for stepping down from a high voltage to a medium or low voltage. This provides a grounding neutral on the high-voltage side. Conversely, the Δ/Y configuration is used in stepping up to a high voltage. The Δ/Δ connection enables one to remove one transformer for maintenance while the other two continue to function as a three-phase bank (with reduced rating) in an open-delta or V-connection. The difficulties arising from the harmonic contents of the exciting current associated with the Y/Y connection make it seldom used.

In Figure 5-13, the four common three-phase transformer connections are shown along with the voltage and current relations associated with the transformation. It is important to realize that the line-to-ground voltages on the Δ side lead the corresponding Y-side values by 30° and that the line currents on the Δ side also lead the currents on the Y side by 30°. The proof of this statement is given now.

Consider the Y/Δ three-phase transformer shown in Figure 5-14. The secondary voltage E_s is given in terms of the line-to-ground voltages by

$$E_s = V_{an} - V_{cn}$$

Assuming phase sequence a-b-c, taking V_{an} as the reference, we have

$$V_{cn} = V_{an}\underline{/120°}$$

As a result,

$$E_s = \left[1 - 1\underline{/120°}\right] V_{an}$$

or

$$E_s = \sqrt{3}\, V_{an}\underline{/-30°}$$

This last result can be verified either analytically or by reference to the phasor diagram in Figure 5-14. Assuming that each winding of the primary has N_1 turns and that each secondary winding has a number of turns N_2, we have

$$E_p = \frac{N_1}{N_2}(E_s)$$

or

$$E_p = \frac{N_1}{N_2}\sqrt{3}\, V_{an}\underline{/-30°}$$

But the line-to-ground voltage on the Y side is

$$V_{An} = E_p$$

Thus we have

$$V_{An} = \frac{N_1}{N_2}\sqrt{3}\, V_{an}\underline{/-30°}$$

We can conclude that the Δ-side line-to-ground secondary voltage V_{an} leads the Y-side line-to-ground primary voltage V_{An} by 30°.

Turning our attention now to the current relations, we start by

$$I_x = \frac{N_1}{N_2}(I_A)$$

$$I_y = \frac{N_1}{N_2}(I_B) = \frac{N_1}{N_2}(I_A)\underline{/-120°}$$

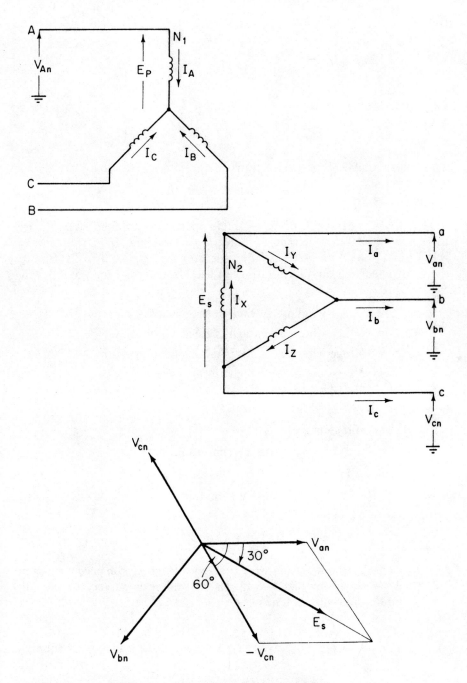

Figure 5-14. A Y-Δ Transformer and a Phasor Diagram.

But
$$I_a = I_x - I_y$$
$$= \frac{N_1}{N_2}(I_A)(1 - 1\underline{/-120°})$$

This reduces to
$$I_a = \frac{N_1}{N_2}\sqrt{3}\, I_A\underline{/30°}$$

Thus the secondary line current leads the primary current by 30°.

Example 5-6

A three-phase bank of three single-phase transformers steps up the three-phase generator voltage of 13.8 kV (line-to-line) to a transmission voltage of 138 kV (line-to-line). The generator rating is 41.5 MVA. Specify the voltage, current, and kVA ratings of each transformer for the following connections:

    A. Low-voltage windings Δ, high-voltage windings Y.
    B. Low-voltage windings Y, high-voltage windings Δ.
    C. Low-voltage windings Y, high-voltage windings Y.
    D. Low-voltage windings Δ, high-voltage windings Δ.

Solution

The low voltage is given by
$$V_1 = 13.8 \text{ kV (line-to-line)}$$
The high voltage is given by
$$V_2 = 138 \text{ kV (line-to-line)}$$
The apparent power is
$$|S| = 41.5 \text{ MVA}$$

A. Consider the situation with the low-voltage windings connected in Δ, as shown in Figure 5-15. Each winding is subject to the full line-to-line voltage. Thus
$$E_p = 13.8 \text{ kV}$$

The power per winding is $|S|/3$; thus the current in each winding is
$$I_p = \frac{41.5 \times 10^6}{(3)(13.8 \times 10^3)} = 1002.42 \text{ A}$$

5.3 Transformer Connections

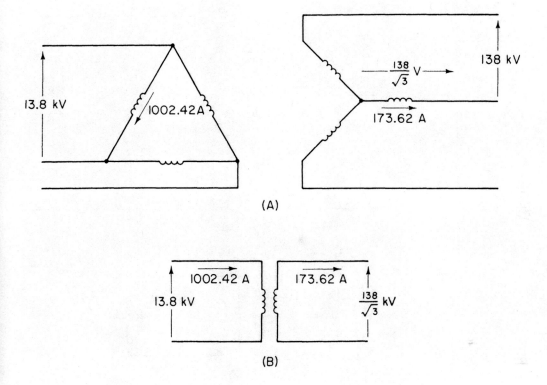

Figure 5-15. (a) Δ-Y Transformer with Variables Indicated. (b) Single Transformer Loading.

With the secondary connected in Y, the voltage on each winding is the line-to-ground value

$$E_s = \frac{138}{\sqrt{3}} = 79.67 \text{ kV}$$

The current in each winding is obtained as

$$I_s = \frac{41.5 \times 10^6}{(3)(79.67 \times 10^3)} = 173.62 \text{ A}$$

The kVA rating of each transformer is thus

$$|S_1| = E_p I_p = E_s I_s = 13.83 \text{ MVA}$$

248 The Load Subsystem

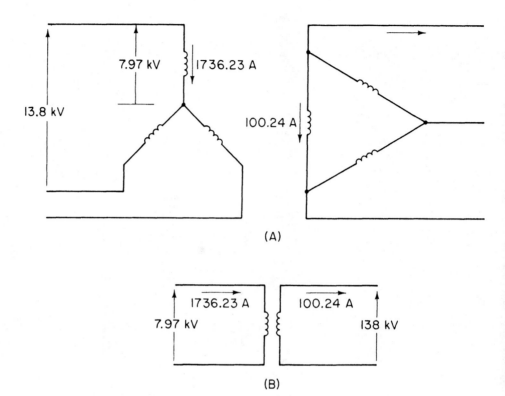

Figure 5-16. (a) Y-Δ Transformer with Variables Indicated for Example 5-6. (b) Single Transformer Loading.

B. When the low-voltage windings are connected in Y, the voltage on each winding is the line-to-ground value

$$E_p = \frac{13.8}{\sqrt{3}} = 7.97 \text{ kV}$$

The current is

$$I_p = \frac{41.5 \times 10^6}{(3)(7.97)(10^3)} = 1736.23 \text{ A}$$

With the secondary windings connected in Δ, the voltage on each winding is

$$E_s = 138 \text{ kV}$$

The current is calculated as

$$I_s = \frac{41.5 \times 10^6}{3(138 \times 10^3)} = 100.24 \text{ A}$$

The kVA rating of each transformer is therefore

$$|S_1| = E_p I_p = E_s I_s = 13.83 \text{ MVA}$$

The arrangement is shown in Figure 5-16.

C. With low-voltage windings connected in Y, from the solution to part (b) we have

$$E_p = 7.97 \text{ kV}$$
$$I_p = 1736.23 \text{ A}$$

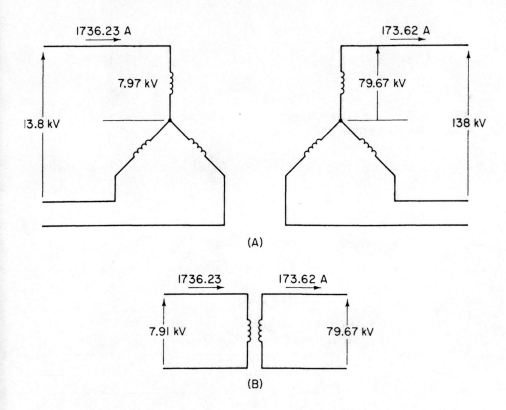

Figure 5-17. (a) Y-Y Transformer with Variables Indicated for Example 5-6. (b) Single Transformer Loading.

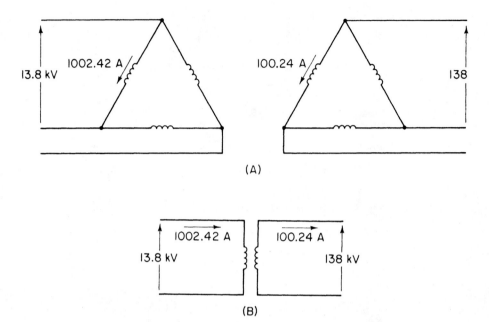

Figure 5-18. (a) Δ-Δ Transformer with Variables Indicated for Example 5-6. (b) Single Transformer Loading.

With high-voltage windings connected in Y, from the solution to part (a) we have

$$E_s = 79.67 \text{ kV}$$
$$I_s = 173.62 \text{ A}$$

This arrangement is shown in Figure 5-17.

TABLE 5-2

Comparison of Single Transformer Ratings for Different Three-Phase Connections

	Δ-Y	Y-Δ	Y-Y	Δ-Δ
E_p (kV)	13.8	7.97	7.97	13.8
I_p (A)	1002.42	1736.23	1736.23	1002.42
E_s (kV)	79.67	138	79.67	138
I_s (A)	173.62	100.24	173.62	100.24

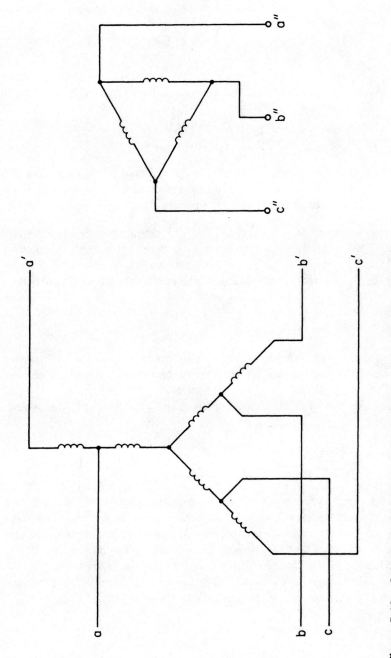

Figure 5-19. Schematic Diagram of a Three-Winding Autotransformer.

251

D. With low-voltage windings connected in Δ, from the solution to part (a) we get

$$E_p = 13.8 \text{ kV}$$
$$I_p = 1002.42 \text{ A}$$

With high-voltage windings connected in Δ, from the solution to part (b) we get

$$E_s = 138 \text{ kV}$$
$$I_s = 100.24 \text{ A}$$

The situation is shown in Figure 5-18. Table 5-2 summarizes the voltage and current ratings for the single-phase transformers associated with each transformer connection.

Three-phase autotransformers are usually Y-Y connected with the neutral grounded. A third (tertiary) Δ-connected set of windings is included to carry the third harmonic component of the exciting current. A schematic diagram of a three-phase autotransformer with a Δ-tertiary is shown in Figure 5-19.

Control Transformers

Transformers are used not only to step up or step down bulk power voltages but also as a means for controlling the operations of the power system. Two examples of control transformer applications involve (1) tap changing under load (TCUL) transformers, and (2) the regulating transformer.

Load Tap Changing

The intended use of the TCUL transformer is to maintain a constant voltage at a point in the system by changing the transformation ratio by increasing or decreasing the number of active turns in one winding with respect to another winding. This is performed while not interfering with the load. In practice, a voltage measuring device actuates the motor that drives the tap changer. If the actual voltage is higher than a desired upper limit, the motor will change to the next lower tap voltage; similarly, a voltage lower than the desired will cause a change to the next higher up.

The Regulating Transformer

The main purpose of the regulating transformer is to change (by a small amount) the voltage magnitude and phase angle at a certain point in the system. Figure 5-20 shows the arrangement of a regulating transformer.

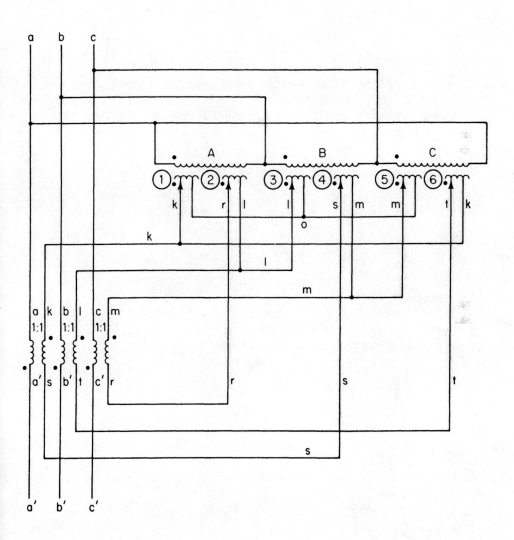

Figure 5-20. Schematic of Regulating Transformer.

For analysis purposes let us assume that

$$V_{an} = V\underline{/0}$$
$$V_{bn} = V\underline{/-120°}$$
$$V_{cn} = V\underline{/+120°}$$

The primary windings of the transformers A, B, and C are connected in Δ. The secondary windings 1, 3, and 5 are connected in Y with their voltages adjustable. Recalling the phase-shift property in Δ-Y transformers, we have

$$V_{ko} = \frac{V_m}{\sqrt{3}}\underline{/30°}$$

$$V_{lo} = \frac{V_m}{\sqrt{3}}\underline{/-90°}$$

$$V_{mo} = \frac{V_m}{\sqrt{3}}\underline{/150°}$$

The magnitude of V_m can be controlled in a small range and is utilized for adjusting the magnitude of the three-phase voltage set $V_{a'}$, $V_{b'}$, and $V_{c'}$. The tertiary windings 2, 4, and 6 have voltages

$$V_{rl} = V_\phi\underline{/30°} \tag{5.24}$$
$$V_{sm} = V_\phi\underline{/-90°} \tag{5.25}$$
$$V_{tk} = V_\phi\underline{/150°} \tag{5.26}$$

The magnitude V_ϕ is adjustable and is used for control of the phase angle of the voltages $V_{a'}$, $V_{b'}$, and $V_{c'}$.

We can derive the voltages V_{km}, V_{lk}, V_{ml} from V_{ko}, V_{lo}, V_{mo} as

$$V_{km} = V_m\underline{/0}$$
$$V_{lk} = V_m\underline{/-120°}$$
$$V_{ml} = V_m\underline{/+120°}$$

Note that V_{km}, V_{lk}, and V_{ml} are in phase with the system voltages V_{an}, V_{bn}, and V_{cn}. The voltages V_{rl}, V_{sm}, and V_{tk} are 90° out of phase with the same voltages. The incremental voltages ΔV_a, ΔV_b, and ΔV_c are given by

$$\Delta V_a = V_{ks}$$
$$\Delta V_b = V_{lt}$$
$$\Delta V_c = V_{mr}$$

or

$$\Delta V_a = V_{km} - V_{sm} = V_m \underline{/0} - V_\phi \underline{/-90°} \qquad (5.27)$$
$$\Delta V_b = V_{lk} - V_{tk} = V_m \underline{/-120°} - V_\phi \underline{/150°} \qquad (5.28)$$
$$\Delta V_c = V_{ml} - V_{rl} = V_m \underline{/+120°} - V_\phi \underline{/30°} \qquad (5.29)$$

The ΔV values are added in series in each phase to give

$$V_{a'n} = V_{an} + \Delta V_a \qquad (5.30)$$
$$V_{b'n} = V_{bn} + \Delta V_b \qquad (5.31)$$
$$V_{c'n} = V_{cn} + \Delta V_c \qquad (5.32)$$

A phasor diagram of the voltages in the system is shown in Figure 5-21.

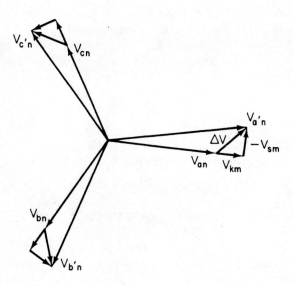

Figure 5-21. Output Voltages of Regulating Transformer.

5.4 THREE-PHASE INDUCTION MOTORS

A major proportion of industrial and commercial motor requirements is served by the induction motor due to its simplicity, reliability, and low cost, combined with reasonable overload capacity, minimal service requirements, and good efficiency. An induction motor utilizes alternating current supplied to the stator directly. The rotor receives power by induction effects. The stator windings of an induction motor are similar to those of the synchronous machine. The rotor may be one of two types. In the *wound rotor motor*, windings similar to those of the stator are employed with terminals connected to insulated slip rings mounted on the shaft. The rotor terminals are made available through carbon brushes bearing on the slip rings. The second type is called the *squirrel-cage rotor*, where the windings are simply conducting bars embedded in the rotor and short-circuited at each end by conducting end rings.

When the stator is supplied by a balanced three-phase source, it will produce a magnetic field that rotates at synchronous speed as determined by the number of poles and applied frequency f_s. Thus

$$n_s = \frac{120 f_s}{P} \text{ r/min} \tag{5.33}$$

The rotor runs at a steady speed n_r r/min in the same direction as the rotating stator field. The speed n_r is very close to n_s when the motor is running light, and is lower as the mechanical load is increased. The difference $(n_s - n_r)$ is termed the *slip* and is commonly defined as a per unit value s. Thus

$$s = \frac{n_s - n_r}{n_s} \tag{5.34}$$

As a result of the relative motion between stator and rotor, induced voltages will appear in the rotor with a frequency f_r called the *slip frequency*. Thus

$$f_r = s f_s \tag{5.35}$$

From the above we can conclude that the induction motor is simply a transformer with a secondary frequency f_r.

An equivalent circuit of the three-phase induction motor can be developed on the basis of the above considerations and transformer models treated in Section 5.2. Looking into the stator terminals, we find that the applied voltage V_s will supply the resistive drop $I_s R_1$ as well as the inductive voltage $jI_s X_1$ and the counter EMF E_1 where I_s is the stator current and R_1 and X_1 are the stator effective resistance and inductive reactance respectively. In a manner similar to that employed for the analysis of the transformer, we model the magnetizing circuit by the shunt conductance G_c and inductive susceptance $-jB_m$.

The rotor's induced voltage E_{2s} is related to the stator EMF E_1 by

$$E_{2s} = sE_1 \tag{5.36}$$

This is due simply to the relative motion between stator and rotor. The rotor current I_{rs} is equal to the current I_r in the stator circuit. The induced EMF E_{2s} supplies the resistive voltage component $I_r R_2$ and inductive component $jI_r(sX_2)$. R_2 is the rotor resistance, and X_2 is the rotor inductive reactance on the basis of the stator frequency. Thus

$$E_{2s} = I_r R_2 + jI_r(sX_2)$$

or

$$sE_1 = I_r R_2 + jI_r(sX_2) \tag{5.37}$$

From the above we conclude that

$$\frac{E_1}{I_r} = \frac{R_2}{s} + jX_2$$

The complete equivalent circuit of the induction motor is shown in Figure 5-22.

If we consider the active power flow into the induction machine, we find that the input power P_s supplies the stator I^2R losses as well as the core losses. The remaining power denoted by the air-gap power P_g is that transferred to the rotor circuit. Part of the air-gap power is expended as rotor I^2R losses with the remainder being the mechanical power delivered to the motor shaft. We can express the air-gap power as

$$P_g = 3I_r^2 \left(\frac{R_2}{s} \right) \tag{5.38}$$

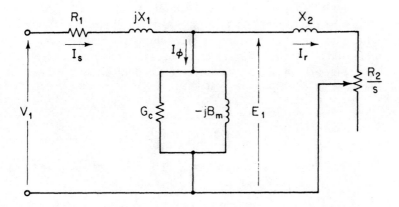

Figure 5-22. Equivalent Circuit for a Three-Phase Induction Motor.

The rotor I^2R losses are given by

$$P_{lr} = 3I_r^2 R_2 \tag{5.39}$$

As a result, the mechanical power output (neglecting mechanical losses) is

$$P_r = P_g - P_{lr}$$
$$= 3I_r^2 \frac{(1-s)}{s} R_2 \tag{5.40}$$

The last formula suggests a splitting of R_2/s into the sum of R_2 representing the rotor resistance and a resistance

$$\frac{1-s}{s}(R_2)$$

which is the equivalent resistance of the mechanical load. As a result, it is customary to modify the equivalent circuit to the form shown in Figure 5-23.

The torque T developed by the motor is related to P_r by

$$T = \frac{P_r}{\omega_r} \tag{5.41}$$

with ω_r being the angular speed of the rotor. Thus

$$\omega_r = \omega_s(1-s) \tag{5.42}$$

The angular synchronous speed ω_s is given by

$$\omega_s = \frac{2\pi n_s}{60} \tag{5.43}$$

As a result, the torque is given by

$$T = \frac{3I_r^2(R_2)}{s\omega_s} \tag{5.44}$$

The torque is slip-dependent. It is customary to utilize a simplified equivalent circuit for the induction motor in which the shunt branch is moved to the voltage source side. This situation is shown in Figure 5-24. The stator resistance and shunt branch can be neglected in many instances.

On the basis of the approximate equivalent circuit, we can find the rotor current as

$$I_r = \frac{V_1}{R_1 + \dfrac{R_2}{s} + jX_T} \tag{5.45}$$

At starting, we have $\omega_r = 0$; thus $s = 1$. The rotor starting current is hence given by

$$I_{r_{st}} = \frac{V_1}{(R_1 + R_2) + jX_T}$$

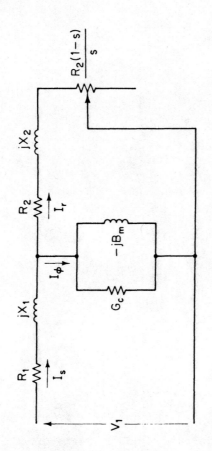

Figure 5-23. Modified Equivalent Circuit of the Induction Motor.

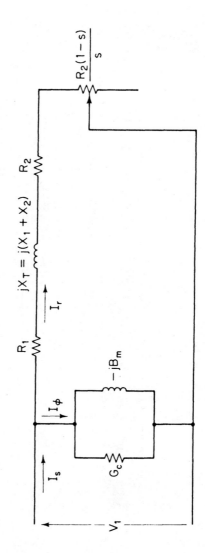

Figure 5-24. Approximate Equivalent Circuit of the Induction Motor.

It is clear that the motor starting current is much higher than the normal (or full-load) current. Depending on the motor type, the starting current can be as high as six to seven times the normal current. We consider now an example.

Example 5-7

A 15-hp, 220-V, three-phase, 60-Hz, six-pole, Y-connected induction motor has the following parameters per phase:

$R_1 = 0.128$ ohm
$R_2 = 0.0935$ ohm
$X_T = 0.496$ ohm
$G_c = 5.4645 \times 10^{-3}$
$B_m = 0.125$ S

The rotational losses are equal to the stator hysteresis and eddy-current losses. For a slip of 3 percent, find the following:

A. The line current and power factor.
B. The horsepower output.
C. The starting torque.

Solution

A. The voltage specified is line-to-line value as usual. Utilizing the approximate equivalent circuit of Figure 5-24, the rotor current can be seen to be given by

$$I_r = \frac{\frac{220}{\sqrt{3}}}{\left(0.128 + \frac{0.0935}{0.03}\right) + j0.496}$$

$$= 38.7 \underline{/-8.69°} \text{ A}$$

The no-load current I_ϕ is obtained as

$$I_\phi = \frac{220}{\sqrt{3}}(5.4645 \times 10^{-3} - j0.125)$$

$$= 0.69 - j15.88 \text{ A}$$

As a result, the line current (stator current) is

$$I_s = I_r + I_\phi$$

$$= 44.6 \underline{/-29.15°}$$

Since V_1 is taken as reference, we conclude that

$$\phi_s = 29.15°$$
$$\cos \phi_s = 0.873$$

B. The air-gap power is given by

$$P_g = 3I_r^2\left(\frac{R_2}{s}\right) = 3(38.7)^2\left(\frac{0.0935}{0.03}\right) = 14000 \text{ W}$$

The mechanical power to the shaft is

$$P_m = (1-s)P_g = 13580 \text{ W}$$

The core losses are

$$P_c = 3E_1^2(G_c) = 264 \text{ W}$$

The rotational losses are thus

$$P_{rl} = 264 \text{ W}$$

As a result, the net output mechanical power is

$$P_{out} = P_m - P_{rl}$$
$$= 13316 \text{ W}$$

Therefore, in terms of horsepower, we get

$$\text{hp}_{out} = \frac{13316}{746} = 17.85 \text{ hp}$$

C. At starting $s = 1$:

$$|I_r| = \frac{\frac{220}{\sqrt{3}}}{(0.128 + 0.0935) + j0.496} = 234 \text{ A}$$

$$P_g = 3(234)^2(0.0935) = 15{,}336 \text{ W}$$

$$\omega_s = \frac{2\pi(60)}{3} = 40\pi$$

$$T = \frac{P_g}{\omega_s} = \frac{15336}{40\pi} = 122.0 \text{ N.m.}$$

The torque developed by the motor can be derived in terms of the motor parameters and slip using the expressions given before.

$$T = \frac{3|V_1|^2}{\omega_s} \frac{\frac{R_2}{s}}{\left(R_1 + \frac{R_2}{s}\right)^2 + X_T^2}$$

Neglecting stator resistance, we have

$$T = \frac{3|V_1|^2}{\omega_s} \frac{\frac{R_2}{s}}{\left(\frac{R_2}{s}\right)^2 + X_T^2}$$

The maximum torque occurs for

$$\frac{\partial T}{\partial \left(\frac{R_2}{s}\right)} = 0$$

The result is

$$\left[\left(\frac{R_2}{s}\right)^2 + X_T^2\right] - \left(\frac{R_2}{s}\right)\left(\frac{2R_2}{s}\right) = 0$$

This gives the slip at which maximum torque occurs as

$$s_{\max_T} = \frac{R_2}{X_T}$$

The value of maximum torque is

$$T_{\max} = \frac{3|V_1|^2}{2\omega_s X_T}$$

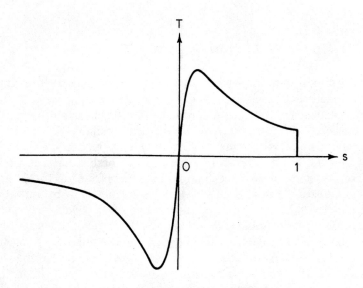

Figure 5-25. Torque-Slip Characteristics for Induction Motor.

The torque-slip variations are shown in Figure 5-25. We now take our next example.

Example 5-8

The rotor resistance and reactance of a squirrel-cage induction motor rotor at standstill are 0.1 ohm per phase and 0.8 ohm per phase respectively. Assuming a transformer ratio of unity, from the eight-pole stator having a phase voltage of 120 V at 60 Hz to the rotor secondary, calculate the following:

A. Rotor starting current per phase.

B. The value of slip producing maximum torque.

Solution

A. At starting $s = 1$:

$$I_r = \frac{120}{0.1 + j0.8}$$

$$= 148.84 \underline{/-82.87} \text{ A}$$

B.

$$s_{\max_T} = \frac{R_r}{X_T} = \frac{0.1}{0.8} = 0.125$$

Classification of Induction Motors

Integral-horsepower, three-phase squirrel-cage motors are available from manufacturers' stock in a range of standard ratings up to 200 hp at standard frequencies, voltages, and speeds. (Larger motors are regarded as special-purpose.) Several standard designs are available to meet various starting and running requirements. Representative torque-speed characteristics of four designs are shown in Figure 5-26. These curves are typical of 1,800 r/min (synchronous-speed) motors in ratings from 7.5 to 200 hp.

The induction motor meets the requirements of substantially constant-speed drives. Many motor applications, however, require several speeds or a continuously adjustable range of speeds. The synchronous speed of an induction motor can be changed by (1) changing the number of poles or (2) varying the line frequency. The slip can be changed by (1) varying the line voltage, (2) varying the rotor resistance, or (3) inserting voltages of the appropriate frequency in the rotor circuits. A discussion of the details of speed control mechanisms is beyond the scope of this work. A common classification of induction motors is as follows.

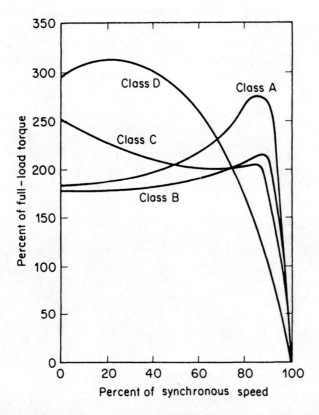

Figure 5-26. Typical Torque-Speed Curves for 1,800 r/min General-Purpose Induction Motors.

Class A

Normal starting torque, normal starting current, low slip. This design has a low-resistance, single-cage rotor. It provides good running performance at the expense of starting. The full-load slip is low and the full-load efficiency is high. The maximum torque usually is over 200 percent of full-load torque and occurs at a small slip (less than 20 percent). The starting torque at full voltage varies from about 200 percent of full-load torque in small motors to about 100 percent in large motors. The high starting current (500 to 800 percent of full-load current when started at rated voltage) is the disadvantage of this design.

Class B

Normal starting torque, low starting current, low slip. This design has approximately the same starting torque as the Class A with only 75 percent of the starting current. The full-load slip and efficiency are good (about the same as for the Class A). However, it has a slightly decreased power factor and a lower maximum torque (usually only slightly over 200 percent of full-load torque being obtainable). This is the commonest design in the 7.5 to 200-hp range of sizes used for constant-speed drives where starting-torque requirements are not severe.

Class C

High starting torque, low starting current. This design has a higher starting torque with low starting current but somewhat lower running efficiency and higher slip than the Class A and Class B designs.

Class D

High starting torque, high slip. This design produces very high starting torque at low starting current and high maximum torque at 50 to 100-percent slip, but runs at a high slip at full load (7 to 11 percent) and consequently has low running efficiency.

SOME SOLVED PROBLEMS

Problem 5-A-1

The equivalent impedance referred to the primary of a 2300/230-V, 500-kVA, single-phase transformer is

$$Z = 0.2 + j0.6 \text{ ohm}$$

Calculate the percentage voltage regulation (P.V.R.) when the transformer delivers rated capacity at 0.8 power factor lagging at rated secondary voltage. Find the efficiency of the transformer at this condition given that core losses at rated voltage are 2 kW.

Solution

The secondary current referred to the primary side is

$$I_2' = \frac{500 \times 10^3}{2300} = 217.39 \underline{/-36.87°}$$

Thus the primary voltage at rated load is

$$\begin{aligned} V_1 &= V_2' + I_2'Z \\ &= 2300\underline{/0} + \left(217.39\underline{/-36.87}\right)(0.2 + j0.6) \\ &= 2414.31\underline{/1.86°} \text{ V} \end{aligned}$$

As a result, we calculate
$$\text{P.V.R.} = 100\left(\frac{V_1 - V_2'}{V_2'}\right) = 100\left(\frac{2414.31 - 2300}{2300}\right) = 5\%$$

The efficiency is calculated as
$$\eta = \frac{(500 \times 10^3)(0.8)}{(500 \times 10^3 \times 0.8) + (217.39)^2(0.2) + 2 \times 10^3}$$
$$= 0.9722$$

Problem 5-A-2

A 500/100 V, two-winding transformer is rated at 5 kVA. The following information is available:

A. The maximum efficiency of the transformer occurs when the output of the transformer is 3 kVA.

B. The transformer draws a current of 3 A, and the power is 100 W when a 100-V supply is impressed on the low-voltage winding with the high-voltage winding open-circuit.

Find the rated efficiency of the transformer at 0.8 PF lagging.

Solution

The core losses are 100 W from the specifications of part (b). From part (a), the I^2R loss at 3-kVA load is thus 100 W. For a 5-kVA load, the I^2R loss is
$$I^2R_{eq} = 100\left(\tfrac{5}{3}\right)^2 = 277.78 \text{ W}$$

The efficiency is
$$\eta = \frac{P_{out}}{P_{out} + I^2R_{eq} + P_{core}}$$
$$= \frac{5 \times 10^3 \times 0.8}{5 \times 10^3 \times 0.8 + 277.78 + 100}$$
$$= 0.9137$$

Problem 5-A-3

The no-load input power to a 50-kVA, 2300/230-V, single-phase transformer is 200 VA at 0.15 PF at rated voltage. The voltage drops due to resistance and leakage reactance are 0.012 and 0.018 times rated voltage when the transformer operates at rated load. Calculate the input power and power factor when the load is 30 kW at 0.8 PF lagging at rated voltage.

Solution

The rated load current is

$$I_r = \frac{50 \times 10^3}{230} = 217.39 \text{ A}$$

The no-load current from the specifications of the problem is

$$I_0 = \frac{200}{230} \underline{/-\cos^{-1} 0.15}$$

$$= 0.87 \underline{/-81.37°}$$

We will assume that the equivalent circuit of Figure 5-5(a) applies and that variables are referred to the secondary side. The resistive voltage drop is

$$\Delta V_r = I_r R_{eq} = (0.012)(230)$$

Thus,

$$(217.39) R_{eq} = 2.76$$

As a result

$$R_{eq} = 0.0127 \text{ ohm}$$

Similarly, we obtain

$$X_{eq} = 0.0190 \text{ ohm}$$

The primary voltage referred to the secondary is

$$V_1' = V_2 + I Z_{eq}$$

For 30 kW at a 0.8 PF, we have

$$I = \frac{30 \times 10^3}{230 \times 0.8} \underline{/-\cos^{-1} 0.8}$$

Thus we calculate

$$V_1' = 233.52 \underline{/0.3°}$$

The primary current referred to the secondary is

$$I_1' = I + I_0$$

$$= 163.67 \underline{/-37.08°}$$

Consequently, the phase angle at the primary side is

$$\phi_1 = 0.3 + 37.08 = 37.38°$$
$$\cos \phi_1 = 0.7946$$

The input power is

$$P_1 = V_1' I_1' \cos \phi_1$$
$$= 30.3708 \text{ kW}$$

Problem 5-A-4

To identify the equivalent circuit parameters of a 100-kVA, 4-kV/1-kV transformer, a short-circuit test is performed with the power input of 2.5 kW at

$$V_1 = 224 \text{ V} \quad \text{and} \quad I_1 = 25 \text{ A}$$

Determine the parameters R_{eq} and X_{eq} of the transformer referred to the primary.

Solution

With a short-circuit on the secondary winding, we have with reference to Figure 5-27

$$P_{sc} = I_{sc}^2 (R_{eq})$$

$$2500 = (25)^2 (R_{eq})$$

This yields

$$R_{eq} = 4 \text{ ohms}$$

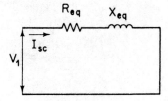

Figure 5-27. Equivalent Circuit for Problem 5-A-4.

We also have
$$P_{sc} = V_1 I_1 \cos \phi_1$$
$$2500 = (224)(25)\cos \phi_1$$
$$\phi_1 = -63.49°$$

But
$$\tan -\phi_1 = \frac{X_{eq}}{R_{eq}}$$
$$X_{eq} = 4 \tan 63.49°$$
$$= 8 \text{ ohms}$$

Problem 5-A-5

Determine the number of poles, the slip, and the frequency of the rotor currents at rated load for three-phase, 5-hp induction motors rated at:

A. 220 V, 50 Hz, 1440 r/min.
B. 120 V, 400 Hz, 3800 r/min.

Solution

We use $P = 120 f/n$, to obtain P, using n_r, the rotor speed given.

A.
$$P = \frac{120 \times 50}{1440} = 4.17$$

But P should be an even number. Therefore, take $P = 4$. Hence
$$n_s = \frac{120 f}{P} = \frac{120 \times 50}{4} = 1500 \text{ r/min}$$

The slip is thus given by
$$s = \frac{n_s - n_r}{n_s} = \frac{1500 - 1440}{1500} = 0.04$$

The rotor frequency is
$$f_r = s f_s = 0.04 \times 50 = 2 \text{ Hz}$$

B.
$$P = \frac{120 \times 400}{3800} = 12.63$$

Take $P = 12$.
$$n_s = \frac{120 \times 400}{12} = 4000 \text{ r/min}$$
$$s = \frac{4000 - 3800}{4000} = 0.05$$
$$f_r = 0.05 \times 400 = 20 \text{ Hz}$$

Problem 5-A-6

The full-load slip of a squirrel-cage induction motor is 0.05, and the starting current is five times the full-load current. Neglecting the stator core and copper losses as well as the rotational losses, obtain:

A. The ratio of starting torque (st) to the full-load torque (fld).
B. The ratio of maximum (max) to full-load torque and the corresponding slip.

Solution

$$s_{fld} = 0.05 \quad \text{and} \quad I_{st} = 5I_{fld}$$

$$\left(\frac{I_{st}}{I_{fld}}\right)^2 = \frac{\left(\frac{R_2}{0.05}\right)^2 + X_T^2}{R_2^2 + X_T^2} = (5)^2$$

This gives

$$\frac{R_2}{X_T} = \sqrt{\frac{24}{375}} \cong 0.25$$

A.

$$T = \frac{3I_r^2(R_2)}{s\omega_s}$$

$$\frac{T_{st}}{T_{fld}} = \frac{I_{st}^2}{I_{fld}^2}\left(\frac{s_{fld}}{s_{st}}\right) = (5)^2 \frac{0.05}{1} = 1.25$$

B.

$$s_{max_T} = \frac{R_2}{X_T} = 0.25$$

$$\frac{T_{max}}{T_{fld}} = \frac{I_{max}^2}{I_{fld}^2}\left(\frac{s_{fld}}{s_{max_T}}\right)$$

$$= \left(\frac{s_{fld}}{s_{max_T}}\right)\frac{\left(\frac{R_2}{s_{fld}}\right)^2 + X_T^2}{(2X_T^2)}$$

$$= \frac{s_{fld}}{s_{max_T}} \frac{\left(\frac{s_{max_T}}{s_{fld}}\right)^2 + 1}{2}$$

$$= \frac{0.05}{0.25}\left[\frac{(5)^2 + 1}{2}\right]$$

Thus

$$\frac{T_{max}}{T_{fld}} = 2.6$$

Problem 5-A-7

Speed control of induction motors of the wound-rotor type can be achieved by inserting additional rotor resistance (R_a). In addition to this, torque control at a given speed can be achieved using this method. Let

$$K = \frac{3V^2}{\omega_s}$$

$$a_1 = \frac{R_2}{s_1}$$

$$a_2 = \frac{R_2 + R_a}{s_2}$$

$$\alpha = \frac{T_2}{T_1}$$

Where suffix 1 refers to operating conditions without additional rotor resistance, whereas suffix 2 refers to conditions with the additional rotor resistance. Neglect stator resistance.

A. Show that the torque ratio is given by

$$\alpha = \frac{a_2}{a_1} \frac{(a_1^2 + x_T^2)}{(a_2^2 + x_T^2)}$$

B. For equal torque $T_1 = T_2$ or $\alpha = 1$, show that the additional rotor resistance needed is

$$R_a = R_2 \left(\frac{s_2}{s_1} - 1 \right)$$

C. Show that the rotor currents are the same for conditions of part (b).

Solution

We have

$$T = \frac{3V^2 \frac{R_2}{s}}{\omega_s \left[\left(\frac{R_2}{s} \right)^2 + X_T^2 \right]}$$

or
$$T = K\frac{a}{a^2 + X_T}$$

Introducing a resistance in the rotor winding will change a. Let the original torque be T_1. Thus

$$T_1 = K\left(\frac{a_1}{a_1^2 + X_T^2}\right)$$

Let the new torque with additional resistance be T_2. Thus

$$T_2 = K\left(\frac{a_2}{a_2^2 + X_T^2}\right)$$

A. For
$$\frac{T_2}{T_1} = \alpha$$

we have
$$\alpha = \frac{(a_1^2 + X_T^2)}{(a_2^2 + X_T^2)} \frac{a_2}{a_1}$$

B. For $\alpha = 1$, i.e., equal torque, we have
$$a_2(a_1^2 + X_T^2) = a_1(a_2^2 + X_T^2)$$
or
$$(a_2 - a_1)(X_T^2 - a_1 a_2) = 0$$

Thus for $T_1 = T_2$, we get $a_1 = a_2$ or
$$\frac{R_2}{s_1} = \frac{R_2 + R_a}{s_2}$$
or
$$R_a = R_2\left(\frac{s_2}{s_1} - 1\right)$$

C. The current in the rotor circuit is
$$I_r = \frac{V}{\sqrt{\left(\frac{R_2}{s}\right)^2 + X_T^2}} = \frac{V}{\sqrt{a^2 + X_T^2}}$$

$$\left(\frac{I_{r_1}}{I_{r_2}}\right)^2 = \frac{a_2^2 + X_T^2}{a_1^2 + X_T^2} = 1 \quad \text{for } T_1 = T_2$$

Thus for equal torque, equal current flows into the rotor circuit.

Problem 5-A-8

The rotor resistance and reactance of a wound-rotor induction motor at standstill are 0.1 ohm per phase and 0.8 ohm per phase, respectively. Assuming a transformer ratio of unity, from the eight-pole stator having a phase voltage of 120 V at 60 Hz to the rotor secondary, find the additional rotor resistance required to produce maximum torque at:

A. Starting $s = 1$.

B. A speed of 450 r/min.

Use results of Problem 5-A-7 and neglect stator parameters.

Solution

Given that
$$R_2 = 0.1 \text{ ohms}$$
$$X_2 = 0.8 \text{ ohms}$$

We get for maximum torque operation
$$s_1 = s_{\max_T} = \frac{R_2}{X_2} = 0.125$$

A.
$$s_2 = s_{st} = 1$$

Thus using
$$R_a = R_2 \left(\frac{s_2}{s_1} - 1 \right)$$

we get
$$R_a = (0.1)\left[(0.125)^{-1} - 1 \right]$$
$$= 0.7 \text{ ohms}$$

B.
$$n_s = \frac{120 f}{P}$$
$$= \frac{(120)(60)}{8} = 900 \text{ r/min}$$

For $n_r = 450$ r/min,
$$s_2 = \frac{900 - 450}{900} = 0.5$$
$$R_a = (0.1)\left(\frac{0.5}{0.125} - 1 \right)$$
$$= 0.3 \text{ ohms}$$

Problem 5-A-9

A. Derive the approximate equivalent circuit for a three-phase wound-rotor induction motor with balanced three-phase capacitors inserted in the rotor circuit.

B. Show that for the motor described in part (a) the rotor current is given by

$$I_r = \frac{V}{\left[R_1 + \dfrac{R_2}{s}\right] + j\left(X_T - \dfrac{x_c}{s^2}\right)}$$

where x_c is the capacitive reactance inserted per phase in the rotor circuit.

C. Show that the internal torque developed is

$$T = \frac{3V^2(R_2)}{\omega_s} \cdot \frac{1}{s\left[\left(R_1 + \dfrac{R_2}{s}\right)^2 + \left(X_T - \dfrac{x_c}{s^2}\right)^2\right]}$$

D. Show that for maximum power factor, the rotor current is given by

$$I_r = \frac{V}{R_1 + R_2\sqrt{\dfrac{X_T}{x_c}}}$$

Solution

Inserting a capacitor bank in the rotor circuit of a wound-rotor induction motor leads to an equivalent circuit representation as shown in Figure 5-28. Evidently all expressions for the performance of the motor can be obtained by replacing x_T by

$$x_T - \frac{X_c}{s^2}$$

Thus

$$I_r = \frac{V}{\left(R_1 + \dfrac{R_2}{s}\right) + j\left(X_T - \dfrac{x_c}{s^2}\right)}$$

The mechanical power is thus

$$P_m = \frac{(3|I_r|^2)R_2(1-s)}{s}$$

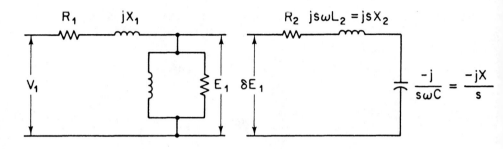

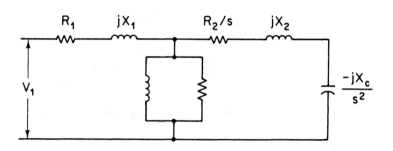

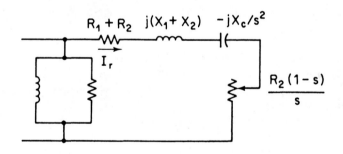

Figure 5-28. Development of Equivalent Circuit for Problem 5-A-9.

and the torque is

$$T = \left(\frac{3V^2 R_2}{\omega_s}\right) \frac{1}{s\left[\left(R_1 + \frac{R_2}{s}\right)^2 + \left(X_T - \frac{x_c}{s^2}\right)^2\right]}$$

The maximum power factor (neglecting the magnetizing circuit) occurs

when
$$X_T = \frac{x_c}{s^2}$$

In this case,
$$I_{r_{max_{P_F}}} = \frac{V}{R_1 + \frac{R_2}{s}}$$
$$= \frac{V}{R_1 + R_2\sqrt{\frac{X_T}{x_c}}}$$

Obviously for a given speed, there is an x_c that yields the maximum power factor.

PROBLEMS

Problem 5-B-1

A 100-kVA, 400/2000 V, single-phase transformer has the following parameters

$R_1 = 0.01$ ohm $\qquad R_2 = 0.25$ ohm
$X_1 = 0.03$ ohm $\qquad X_2 = 0.75$ ohms
$G_c = 2.2$ mS $\qquad B_m = 6.7$ mS

Note that G_c and B_m are given in terms of primary reference. The transformer supplies a load of 90 kVA at 2000 V and 0.8 PF lagging. Calculate the primary voltage and current using the equivalent circuits shown in Figure 5-5.

Problem 5-B-2

Find the P.V.R. and efficiency for the transformer of Problem 5-B-1.

Problem 5-B-3

Find the maximum efficiency of the transformer of Problem 5-B-1, under the same conditions.

Problem 5-B-4

Repeat Example (5-4) for
$$I_2 = 40\underline{/-30°}$$
$$I_3 = 40\underline{/-35°}$$

Problem 5-B-5

A 50-kVA, 2.4/0.6 kV transformer is connected as a step-up autotransformer from a 2.4-kV supply. Calculate the currents in each part of the transformer and the load rating. Neglect losses.

Problem 5-B-6

A three-phase bank of three single-phase transformers steps up the three-phase generator voltage of 13.8 kV (line-to-line) to a transmission voltage of 138 kV (line-to-line). The generator rating is 83 MVA. Specify the voltage, current, and kVA ratings of each transformer for the following connections:

- A. Low-voltage windings Δ, high-voltage windings Y
- B. Low-voltage windings Y, high-voltage windings Δ
- C. Low-voltage windings Y, high-voltage windings Y
- D. Low-voltage windings Δ, high-voltage windings Δ

Problem 5-B-7

The equivalent impedance referred to the secondary of a 13.8/138 kV, 83 MVA, three-phase, Δ/Y connected transformer is

$$Z = 2 + j13.86 \text{ ohms}$$

Calculate the percentage voltage regulation when the transformer delivers rated capacity at 0.8 PF lagging at rated secondary voltage. Find the efficiency of the transformer at this condition given that core losses at rated voltage are 76.5 kW.

Problem 5-B-8

A two winding transformer is rated at 50 kVA. The maximum efficiency of the transformer occurs when the output of the transformer is 35 kVA. Find the rated efficiency of the transformer at 0.8 PF lagging given that the no load losses are 200 W.

Problem 5-B-9

The no-load input to a 5 kVA, 500/100-V, single-phase transformer is 100 W at 0.15 PF at rated voltage. The voltage drops due to resistance and leakage reactance are 0.01 and 0.02 times the rated voltage when the transformer operates at rated load. Calculate the input power and power factor when the load is 3 kW at 0.8 PF lagging at rated voltage.

Problem 5-B-10

To identify the equivalent circuit parameters of a 30 kVA, 2400/240-V transformer, a short-circuit test is performed with the power input of 1050 W at $V_1 = 70$ V and $I_1 = 18.8$ A. Determine the parameters R_{eq} and X_{eq} of the transformer referred to the primary.

Problem 5-B-11

Consider the three-winding transformer of Example (5-4). This time assume that the loads on the secondary and tertiary are specified by

$$|S_2| = |S_3| = 20 \text{ kVA}$$

The secondary winding's load has a power factor of 0.9 lagging. The tertiary windings load has a power factor of 0.8 lagging at a voltage of 400 V referred to the primary. Calculate the primary voltage and current for this loading condition as well as the voltage at the secondary terminals referred to the primary side.

Problem 5-B-12

A multiple-loaded high-voltage line is shown in Figure 5-29. The transformers are modeled by nominal T-networks, whereas the transmission lines are modeled by their nominal π-networks. The voltage at E is 110 kV and the load is 12 MVA at 0.9 PF lagging. The load at C is 18 MVA at 0.95 PF lagging. The circuit parameters are:

$R_{T2} = 6.8 \; \Omega$ $R_{T1} = 1.45 \; \Omega$
$X_{T2} = j106 \; \Omega$ $X_{T_1} = j26.6 \; \Omega$

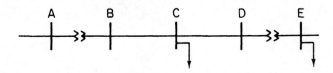

Figure 5-29. System Configuration for Problem 5-B-12.

$R_{mT2} = 504 \text{ k}\Omega$ $R_{mT1} = 186 \text{ k}\Omega$
$X_{mT2} = j46.1 \text{ k}\Omega$ $X_{mT1} = 17.3 \text{ k}\Omega$
$R_{CD} = 6.55 \text{ }\Omega$ $R_{BC} = 3.57 \text{ }\Omega$
$X_{CD} = j22 \text{ }\Omega$ $X_{BC} = j12 \text{ }\Omega$
$Y_{CD} = j0.15896 \times 10^{-3}$ siemens $Y_{BC} = j0.0862069 \times 10^{-3}$ siemens

A. Draw the equivalent circuit of the line.

B. Use KVL and KCL to obtain the voltages, currents, and power factors at points D, C, B, and A. What is the efficiency of transmission?

Problem 5-B-13

Repeat Problem (5-B-12) using the $ABCD$ parameters approach.

Problem 5-B-14

For the system of Problems (5-B-12) and (5-B-13) obtain the voltage regulation at E with load at C on, at C with load E on, and at C and E with no loading at either of the points.

Problem 5-B-15

Determine the number of poles, the slip, and the frequency of the rotor currents at rated load for three-phase, induction motors rated at:

A. 2200V, 60 Hz, 588 r/min.
B. 120 V, 600 Hz, 873 r/min.

Problem 5-B-16

A 50-HP, 440-V, three-phase, 60-Hz, six-pole, Y-connected induction motor has the following parameters per phase:

$R_2 = 0.12$ ohm
$R_1 = 0.1$ ohm
$G_c = 6.2 \times 10^{-3}$ siemens
$X_T = 0.75$ ohm
$B_m = 0.07$ siemens

The rotational losses are equal to the stator hysteresis and eddy-current losses. For a slip of 3 percent, find the following

    A. The line current and power factor.

    B. The horsepower output.

    C. The starting torque.

Problem 5-B-17

The rotor resistance and reactance of a squirrel-cage induction motor rotor at standstill are 0.12 ohm per phase and 0.7 ohm per phase respectively. Assuming a transformer ratio of unity, from the eight-pole stator having a phase voltage of 254 at 60 Hz to the rotor secondary, calculate the following

    A. Rotor starting current per phase

    B. The value of slip producing maximum torque.

Problem 5-B-18

The full-load slip of a squirrel-cage induction motor is 0.05, and the starting current is four times the full-load current. Neglecting the stator core and copper losses as well as the rotational losses, obtain:

    A. The ratio of starting torque to the full-load torque.

    B. The ratio of maximum to full-load torque and the corresponding slip.

Problem 5-B-19

The rotor resistance and reactance of a wound-rotor induction motor at standstill are 0.12 ohm per phase and 0.7 ohm per phase, respectively. Assuming a transformer ratio of unity, from the eight-pole stator having a phase voltage of 254 V at 60 Hz to the rotor secondary, find the additional rotor resistance required to produce maximum torque at:

    A. Starting $s=1$

    B. A speed of 450 r/min.

Use results of problem (5-A-7) and neglect stator parameters.

Problem 5-B-20

A 400-V, four-pole, three-phase, 50-Hz, wound-rotor induction motor has the following parameters:

    $R_1 = 10$ ohms/phase $X_1 = 24$ ohms/phase
    $R_2 = 10$ ohms/phase $X_2 = 24$ ohms phase

A three-phase Y-connected capacitor bank with 20 ohms capacitive reactance per phase is inserted in the rotor circuit. It is required to:

A. Find the starting current and torque

B. Find the current and torque at a slip of 0.05.

Neglect magnetizing and core effects.

CHAPTER VI

Analysis of Interconnected Systems

6.1 INTRODUCTION

The previous three chapters treated aspects of modeling the major components of an electric power system for analysis and design purposes. It is the intent of the present chapter to discuss a number of aspects when these components form parts of an interconnected power system. The goal here is to obtain an overall model for an interconnected system.

We start by considering the problem of reducing parts of the interconnected systems to produce equivalent representation of smaller size but maintaining the electrical performance characteristics unchanged. The reader will find a certain degree of overlap between this part and the treatment of Chapter 4. It is common practice in utility systems to use the per unit system for specifying system parameters and variables, and this is discussed in Section 6.3.

The formulation of the network equations in the nodal admittance form is the basis for proven techniques used in the analysis of interconnected power systems and is detailed in Section 6.4. The nature of the system dictates that the solution to the network equations cannot be obtained in a closed form. This is the load-flow problem formulated in Sections 6.5 and 6.6. Here we emphasize the nonlinear nature of the problem and motivate the use of iterative techniques for obtaining a solution. As will become evident, good initial estimates of the solution are important, and a technique for getting started is treated in Section 6.7. There are many excellent numerical solution methods for solving the load-flow problem. We choose here to introduce the Newton-Raphson method in Section 6.8 and its application in Section 6.9. The presentation is given in a format suitable for small-computer (and even programmable calculator) implementation.

6.2 REDUCTION OF INTERCONNECTED SYSTEMS

In the analysis of interconnected systems, certain tools can prove useful in reducing parts of the system to single equivalents. These tools are quite simply based on network reduction techniques. To illustrate the point, let us consider the system shown in Figure 6-1. This is part of a much larger system; however, it serves our present purposes.

It is clear that lines 1, 2, and 3 are in parallel between buses C and MO. Similarly, lines 4, 5, and 6 are in parallel between buses MO and A. A reduction in subsequent computational effort would result if we were able to represent the network between C and A by a single equivalent line. It is also evident that lines 8 and 9 are in parallel and can be replaced by a single equivalent line. We illustrate the procedure using the following data:

$$Z_1 = Z_2 = 3.339 + j77.314 \text{ ohms}$$
$$Y_1 = Y_2 = j1.106095 \times 10^{-3} \text{ siemens}$$
$$Z_3 = 3.346 + j77.299 \text{ ohms}$$
$$Y_3 = j1.106065 \times 10^{-3} \text{ siemens}$$
$$Z_4 = 3.202 + j73.964 \text{ ohms}$$
$$Y_4 = j1.058342 \times 10^{-3} \text{ siemens}$$
$$Z_5 = Z_6 = 3.194 + j73.962 \text{ ohms}$$
$$Y_5 = Y_6 = j1.058139 \times 10^{-3} \text{ siemens}$$
$$Z_7 = 2.655 + j61.384 \text{ ohms}$$
$$Y_7 = j0.878286 \times 10^{-3} \text{ siemens}$$
$$Z_8 = 2.452 + j56.738 \text{ ohms}$$
$$Y_8 = j0.811774 \times 10^{-3} \text{ siemens}$$
$$Z_9 = 2.451 + j56.742 \text{ ohms}$$
$$Y_9 = j0.811781 \times 10^{-3} \text{ siemens}$$
$$Z_{10} = 0.676 + j20.483 \text{ ohms}$$
$$Y_{10} = j0.286090 \times 10^{-3} \text{ siemens}$$

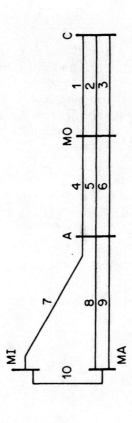

Figure 6-1. Part of a Network to Illustrate Concepts in Network Reduction.

286   Analysis of Interconnected Systems

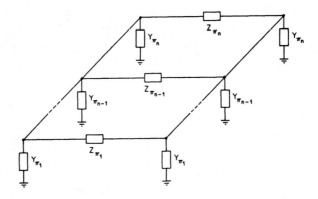

Figure 6-2. Diagram Showing *n* Parallel Lines.

Given the series impedance Z and shunt admittance Y for a line, we can easily calculate the A, B, C, and D parameters of each of the lines. Having done so, we proceed to obtain the parallel combinations for the lines by first representing the lines using the equivalent π circuit and then combining the circuits in parallel. Figure 6-2 shows n parallel lines each with series impedance Z_{Π_i} and shunt admittance Y_{Π_i}. The equivalent single line has elements $Z_{\Pi_{eq}}$ and $Y_{\Pi_{eq}}$, given by

$$\frac{1}{Z_{\Pi_{eq}}} = \frac{1}{Z_{\Pi_1}} + \frac{1}{Z_{\Pi_2}} + \cdots + \frac{1}{Z_{\Pi_n}} \tag{6.1}$$

$$Y_{\Pi_{eq}} = Y_{\Pi_1} + Y_{\Pi_2} + \cdots + Y_{\Pi_n} \tag{6.2}$$

The above formulae provide the basis for the expressions of the equivalent A and B parameters of the parallel lines. We have shown in Chapter 4 that

$$Z_\Pi = B \tag{6.3}$$

Thus Eq. (6.1) tells us that

$$\frac{1}{B_{eq}} = \frac{1}{B_1} + \frac{1}{B_2} + \cdots + \frac{1}{B_n} \tag{6.4}$$

Moreover we have

$$Y_\Pi = \frac{A-1}{B} \tag{6.5}$$

Thus, using Eq. (6.5) in Eq. (6.2) gives

$$\frac{A_{eq}-1}{B_{eq}} = \frac{A_1-1}{B_1} + \frac{A_2-1}{B_2} + \cdots + \frac{A_n-1}{B_n}$$

As a result of Eq. (6.4), we get

$$\frac{A_{eq}}{B_{eq}} = \frac{A_1}{B_1} + \frac{A_2}{B_2} + \cdots + \frac{A_n}{B_n} \qquad (6.6)$$

Equations (6.4) and (6.6) are the desired expressions for the equivalent representation of parallel lines.

Referring back to our example system, we see that by applying the results of the above discussion we can reduce the network between C and MO to one equivalent line, that between MO and A to one equivalent line, and that between A and MA to one equivalent line. It can further be seen that the network between C and A can further be reduced to a single equivalent representation. To do this, we need to discuss the equivalents of cascaded lines or in general cascaded two-port networks.

Probably the most intuitively obvious means is again network reduction. Here we obtain the equivalent π representations of the two cascaded lines, and then with the help of series and parallel combinations as well as the Y-Δ transformations, we obtain the desired single equivalent. We can also use a simple formula that can be derived by considering the cascade shown in Figure 6-3. We have

$$V_{s_1} = A_1 V_{r_1} + B_1 I_{r_1} \qquad (6.7)$$

But the receiving end for line 1 is the sending end of line 2. Thus

$$V_{r_1} = V_{s_2} = A_2 V_{r_2} + B_2 I_{r_2} \qquad (6.8)$$

and

$$I_{r_1} = I_{s_2} = C_2 V_{r_2} + D_2 I_{r_2} \qquad (6.9)$$

As a result, using Eqs. (6.8) and (6.9) in Eq. (6.7), we have

$$V_{s_1} = (A_1 A_2 + B_1 C_2) V_{r_2} + (A_1 B_2 + B_1 D_2) I_{r_2}$$

For the equivalent line, we have

$$V_{s_1} = A_{eq} V_{r_2} + B_{eq} I_{r_2}$$

Comparing the above two equations, we conclude that

$$A_{eq} = A_1 A_2 + B_1 C_2 \qquad (6.10)$$
$$B_{eq} = A_1 B_2 + B_1 D_2 \qquad (6.11)$$

We will now proceed with a numerical example using network reduction techniques. Application of the A, B equivalent formulae is easy and should give the same results.

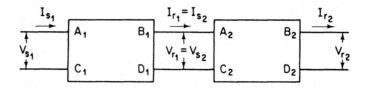

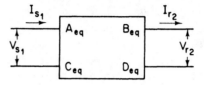

Figure 6-3. Two Cascaded Lines and Their Equivalent.

Example 6-1

Let us consider lines 1 and 2 of Figure 6-1. The series impedance and shunt admittance for each of the lines are given below:

$$Z_1 = Z_2 = 3.339 + j77.314 \text{ ohms}$$
$$Y_1 = Y_2 = j1.106095 \times 10^{-3} \text{ siemens}$$

The A and B parameters for each of the two lines are calculated as

$$A_1 = A_2 = 0.957243 \underline{/0.110529°}$$
$$B_1 = B_2 = 77.3861 \underline{/87.5271°}$$

As a result, the equivalent π representation is calculated as

$$Z_{\Pi_1} = Z_{\Pi_2} = 77.3861 \underline{/87.5271°}$$
$$Y_{\Pi_1} = Y_{\Pi_2} = 5.53048 \times 10^{-4} \underline{/90°}$$

For line 3, we have

$$Z_3 = 3.346 + j77.299 \text{ ohms}$$
$$Y_3 = j1.106065 \times 10^{-3} \text{ siemens}$$

The A and B parameters of the line are calculated as

$$A_3 = 0.957253 \underline{/0.110757°}$$
$$B_3 = 77.3714 \underline{/87.5214°}$$

As a result, the equivalent π representation is

$$Z_{\Pi_3} = 77.3713 \underline{/87.5214°} \text{ ohms}$$
$$Y_{\Pi_3} = 5.53033 \times 10^{-4} \underline{/90°} \text{ siemens}$$

6.2 Reduction of Interconnected Systems

The equivalent of the three parallel lines from C to MO is therefore obtained as the equivalent Π representation with series elements as shown below:

$$\frac{1}{Z_{\Pi_{CMO}}} = \frac{1}{Z_{\Pi_1}} + \frac{1}{Z_{\Pi_2}} + \frac{1}{Z_{\Pi_3}}$$

The result is

$$Z_{\Pi_{CMO}} = 25.7937 \underline{/87.5252°} \text{ ohms}$$

For the shunt admittance we have

$$Y_{\Pi_{CMO}} = Y_{\Pi_1} + Y_{\Pi_2} + Y_{\Pi_3}$$

The result is

$$Y_{\Pi_{CMO}} = 1.65913 \underline{/90°} \text{ siemens}$$

Line 4 has circuit parameters of

$$Z_4 = 3.202 + j73.964 \text{ ohms}$$
$$Y_4 = j1.058342 \times 10^{-3} \text{ siemens}$$

The A and B parameters for the line are calculated as

$$A_4 = 0.960861 \underline{/0.101037°}$$
$$B_4 = 74.0333 \underline{/87.5211°}$$

As a result, the equivalent π representation is

$$Z_{\Pi_4} = 73.0712 \underline{/87.5537°} \text{ ohms}$$
$$Y_{\Pi_4} = 5.2917 \times 10^{-4} \underline{/90°} \text{ siemens}$$

Lines 5 and 6 have the following series impedance and shunt admittance each:

$$Z_5 = Z_6 = 3.194 + j73.962 \text{ ohms}$$
$$Y_5 = Y_6 = j1.058139 \times 10^{-3} \text{ siemens}$$

The A and B parameters are calculated as

$$A_5 = A_6 = 0.960870 \underline{/0.100764°}$$
$$B_5 = B_6 = 74.0309 \underline{/87.5273°}$$

Thus, the equivalent π representation is

$$Z_{\Pi_5} = Z_{\Pi_6} = 74.0309 \underline{/87.5272°} \text{ ohms}$$
$$Y_{\Pi_5} = Y_{\Pi_6} = 5.29069 \times 10^{-4} \underline{/90°} \text{ siemens}$$

The equivalent of the three parallel lines from MO to A is therefore obtained using

$$\frac{1}{Z_{\Pi_{MOA}}} = \frac{1}{Z_{\Pi_4}} + \frac{1}{Z_{\Pi_5}} + \frac{1}{Z_{\Pi_6}}$$

The result is

$$Z_{\Pi_{MOA}} = 24.6772 \underline{/87.5252°} \text{ ohms}$$

For the shunt admittance, we have

$$Y_{\Pi_{MOA}} = Y_{\Pi_4} + Y_{\Pi_5} + Y_{\Pi_6}$$

The result is

$$Y_{\Pi_{MOA}} = 1.5873 \times 10^{-3} \underline{/90°} \text{ siemens}$$

We are now in a position to consider the reduction of the circuit to one single equivalent from C to A. The equivalent circuit is shown in Figure 6-4. The simplest means is to employ a Y-Δ transformation. We first replace the two parallel admittances $Y_{\Pi_{MOA}}$ and $Y_{\Pi_{CMO}}$ by the equivalent

$$Y_{MO} = Y_{\Pi_{MOA}} + Y_{\Pi_{CMO}}$$

This turns out to be

$$Y_{\Pi_{MO}} = 3.24644 \underline{/90°} \text{ siemens}$$

The resulting circuit is shown in Figure 6-5.

The elements of the Δ are obtained now. For the element between C and A, we obtain the impedance value using the formula

$$Z_{CA} = Z_{\Pi_{MOA}} + Z_{\Pi_{CMO}} + Y_{MO} Z_{\Pi_{CMO}} Z_{\Pi_{MOA}}$$

The result is

$$Z_{CA} = 48.40656 \underline{/87.6308°} \text{ ohms}$$

The element from C to neutral has the impedance

$$Z_{CO} = Z_{\Pi_{CMO}} + \frac{1}{Y_{MO}} + \frac{Z_{\Pi_{CMO}}}{Y_{MO} Z_{\Pi_{MOA}}}$$

The result is

$$Z_{CO} = 604.227 \underline{/-89.8944°} \text{ ohms}$$

Its admittance is

$$Y_{CO} = 1.6550 \times 10^{-3} \underline{/-89.8944°} \text{ siemens}$$

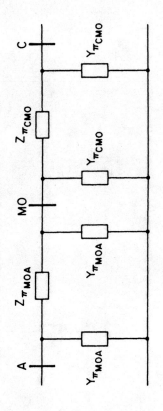

Figure 6-4. Equivalent Circuit with Buses *A*, *MO*, and *C*.

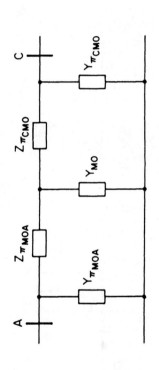

Figure 6-5. Same Network as in Figure 6-4 but with Parallel Reduction.

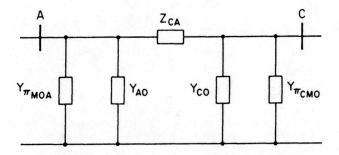

Figure 6-6. Result of Y-Δ Transformation for Network between C and A.

The element from A to neutral has the impedance

$$Z_{AO} = Z_{\Pi_{MOA}} + \frac{1}{Y_{MO}} + \frac{Z_{\Pi_{MOA}}}{Y_{MO} Z_{\Pi_{CMO}}}$$

The result is

$$Z_{AO} = 578.07 \, \underline{/-89.894°} \text{ ohms}$$

Its admittance is

$$Y_{AO} = 1.7299 \times 10^{-3} \, \underline{/89.894°} \text{ siemens}$$

The resulting circuit is shown in Figure 6-6.

The final step is to combine Y_{CO} and $Y_{\Pi_{CMO}}$ in the parallel combination denoted Y_C, and Y_{AO} and $Y_{\Pi_{MOA}}$ in the parallel combination denoted Y_A. Thus,

$$Y_C = Y_{CO} + Y_{\Pi_{CMO}}$$

The numerical value is

$$Y_C = 3.31413 \times 10^{-3} \, \underline{/89.9473°} \text{ siemens}$$

Also

$$Y_A = Y_{AO} + Y_{\Pi_{MOA}}$$

with numerical value

$$Y_A = 3.31719 \times 10^{-3} \, \underline{/89.9449°} \text{ siemens}$$

The resulting circuit is shown in Figure 6-7.

The following example uses line 8 and 9 to illustrate the use of the A and B formulae to obtain parallel combination equivalents.

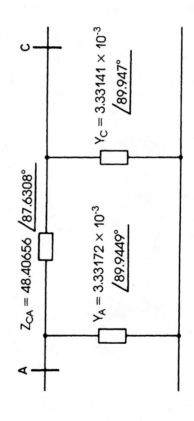

Figure 6-7. Final Reduced Circuit between Buses C and A.

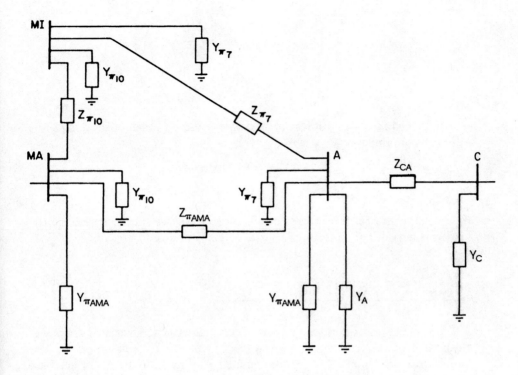

Figure 6-8. Final Reduced Network for System of Example 6.2.

Example 6-2

The series impedance and shunt admittances for lines 8 and 9 result in the following A and B parameters:

$$A_8 = 0.97697 \, \underline{/0.58367°}$$
$$B_8 = 56.7909 \, \underline{/87.5254°}$$
$$A_9 = 0.97697 \, \underline{/0.58343°}$$
$$B_9 = 56.7949 \, \underline{/87.5266°}$$

The B_{eq} is obtained as

$$B_{AMA}^{-1} = B_8^{-1} + B_9^{-1}$$

The result is

$$B_{AMA} = 28.39647 \, \underline{/87.52603°}$$

Next we calculate A_{AMA}:

$$A_{AMA} = B_{AMA}\left(\frac{A_8}{B_8} + \frac{A_9}{B_9}\right)$$

The result is

$$A_{AMA} = .9769704 \,\underline{/0.058355°}$$

The equivalent π representation of the lines between buses A and MA are thus obtained as

$$Z_{\Pi_{AMA}} = 28.396467\,\underline{/87.52603°}\text{ ohms}$$

$$Y_{\Pi_{AMA}} = 8.11778 \times 10^{-4}\,\underline{/90°}\text{ siemens}$$

As a result of the process carried out above, we can now produce the reduced network of our example system. This is shown in Figure 6-8.

6.3 THE PER UNIT SYSTEM

The per unit (p.u.) value representation of electrical variables in power system problems is favored by the electric power systems engineer. The numerical per unit value of any quantity is its ratio to a chosen base quantity of the same dimension. Thus a per unit quantity is a *normalized* quantity with respect to the chosen base value. The per unit value of a quantity is thus defined as

$$\text{p.u. value} = \frac{\text{Actual value}}{\text{Reference or base value of the same dimension}} \quad (6.12)$$

In an electrical network, five quantities are usually involved in the calculations. These are the current I, the voltage V, the complex power S, the impedance Z, and the phase angles. The angles are dimensionless; the other four quantities are completely described by knowledge of only two of them. It is thus clear that an arbitrary choice of two base quantities will fix the other base quantities. Let $|I_b|$ and $|V_b|$ represent the base current and base voltage expressed in kiloamperes and kilovolts, respectively. The product of the two gives the base complex power in megavoltamperes (MVA)

$$|S_b| = |V_b||I_b|\text{ MVA} \quad (6.13)$$

The base impedance will also be given by

$$|Z_b| = \frac{|V_b|}{|I_b|} = \frac{|V_b|^2}{|S_b|}\text{ ohms} \quad (6.14)$$

The base admittance will naturally be the inverse of the base impedance.

Thus

$$|Y_b| = \frac{1}{|Z_b|} = \frac{|I_b|}{|V_b|}$$

$$= \frac{|S_b|}{|V_b|^2} \text{ siemens} \qquad (6.15)$$

The nominal voltage of lines and equipment is almost always known as well as the apparent (complex) power in megavoltamperes, so these two quantities are usually chosen for base value calculation. The same megavoltampere base is used in all parts of a given system. One base voltage is chosen; all other base voltages must then be related to the one chosen by the turns ratios of the connecting transformers.

From the definition of per unit impedance, we can express the ohmic impedance Z_Ω in the per unit value $Z_{\text{p.u.}}$ as

$$Z_{\text{p.u.}} \triangleq \frac{Z_\Omega}{|Z_b|}$$

$$= \frac{Z_\Omega |I_b|}{|V_b|}$$

Thus

$$Z_{\text{p.u.}} = \frac{Z_\Omega |S_b|}{|V_b|^2} \text{ p.u.} \qquad (6.16)$$

As for admittances, we have

$$Y_{\text{p.u.}} \triangleq \frac{1}{Z_{\text{p.u.}}} = \frac{|V_b|^2}{Z_\Omega |S_b|} = Y_S \frac{|V_b|^2}{|S_b|} \text{ p.u.} \qquad (6.17)$$

It is interesting to note that $Z_{\text{p.u.}}$ can be interpreted as the ratio of the voltage drop across Z with base current injected to the base voltage. This can be verified by inspection of the expression

$$Z_{\text{p.u.}} = \frac{Z_\Omega I_b}{V_b}$$

An example will illustrate the procedure.

Example 6-3

Consider line 3 of Example 6-1 and assume

$$S_b = 100 \text{ MVA}$$
$$V_b = 735 \text{ kV}$$

We thus have

$$Z_{p.u.} = Z_\Omega \cdot \frac{S_b}{|V_b|^2} = Z_\Omega \cdot \frac{100}{(735)^2}$$

$$= 1.85108 \times 10^{-4}(Z_\Omega)$$

For $R = 3.346$ ohms, we obtain

$$R_{p.u.} = (3.346)(1.85108 \times 10^{-4}) = 6.19372 \times 10^{-4}$$

For $X = 77.299$ ohms, we obtain

$$X_{p.u.} = (77.299)(1.85108 \times 10^{-4}) = 1.430867 \times 10^{-2}$$

For the admittance we have

$$Y_{p.u.} = Y_S \cdot \frac{|V_b|^2}{S_b}$$

$$= Y_S \frac{(735)^2}{100}$$

$$= 5.40225 \times 10^3 (Y_S)$$

For $Y = 1.106065 \times 10^{-3}$ siemens, we obtain

$$Y_{p.u.} = (5.40225 \times 10^3)(1.106065 \times 10^{-3})$$

$$= 5.97524$$

Given an impedance in per unit on a given base S_{b_o} and V_{b_o}, it is sometimes required to obtain the per unit value referred to a new base set S_{b_n} and V_{b_n}. The conversion expression is obtained as follows:

$$Z_\Omega = Z_{p.u._o} \left(\frac{|V_{b_o}|^2}{|S_{b_o}|} \right)$$

Also the same impedance Z_Ω in ohms is given referred to the new base by

$$Z_\Omega = Z_{p.u._n} \left(\frac{|V_{b_n}|^2}{|S_{b_n}|} \right)$$

Thus equating the above two expressions, we get

$$Z_{p.u._n} = Z_{p.u._o} \frac{|S_{b_n}|}{|S_{b_o}|} \cdot \frac{|V_{b_o}|^2}{|V_{b_n}|^2} \tag{6.18}$$

which is our required conversion formula. The admittance case simply follows the inverse rule. Thus,

$$Y_{p.u._n} = Y_{p.u._o} \frac{|S_{b_o}|}{|S_{b_n}|} \cdot \frac{|V_{b_n}|^2}{|V_{b_o}|^2} \tag{6.19}$$

Example 6-4

Convert the impedance and admittance values of Example 6-3 to the new base of 200 MVA and 345 kV.

Solution

We have

$$Z_{p.u._o} = 6.19372 \times 10^{-4} + j1.430867 \times 10^{-2}$$

for a 100-MVA, 735-kV base. With a new base of 200 MVA and 345 kV, we have, using the impedance conversion formula,

$$Z_{p.u._n} = Z_{p.u._o} \left(\frac{200}{100}\right) \cdot \left(\frac{735}{345}\right)^2$$
$$= 9.0775 \, Z_{p.u._o}$$

Thus

$$Z_{p.u._n} = 5.6224 \times 10^{-3} + j1.2989 \times 10^{-1} \text{ p.u.}$$

As a check, we use the fundamental formula for conversion from ohmic values to per unit for the new base. We do this only for the resistance. Thus,

$$R_{p.u.} = (3.346)\left[\frac{200}{(345)^2}\right]$$
$$= 5.6224 \times 10^{-3} \text{ p.u.}$$

which agrees with our result.

For the admittance we have

$$Y_{p.u._n} = Y_{p.u._o}\left(\frac{100}{200}\right) \cdot \left(\frac{345}{735}\right)^2$$
$$= 0.11016 Y_{p.u._o}$$

Thus

$$Y_{p.u._n} = (5.97524)(0.11016)$$
$$= 0.65825 \text{ p.u.}$$

6.4 NETWORK NODAL ADMITTANCE FORMULATION

Consider the reduced power system network of Section 6.2 with bus *MI* relabeled as 1, *MA* as 2, *A* as 3, and *C* as 4. The system is reproduced in Figure 6-9 with generating capabilities as well as loads indicated. Buses 1, 2, and 3 are buses having generation capabilities as well as loads. Bus 3 is a load bus with no real generation. Bus 4 is a net generation bus.

300 Analysis of Interconnected Systems

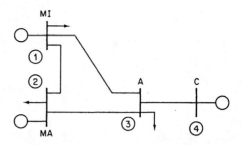

Figure 6-9. Single-Line Diagram to Illustrate Nodal Matrix Formulation.

Using the π equivalent representation for each of the lines, we obtain the network shown in Figure 6-10. Note that this is the same network as that of Figure 6-8 except for the labeling of buses. Let us examine this network in which we exclude the generator and load branches. We can write the current equations as

$$I_1 = V_1 Y_{10} + (V_1 - V_2) Y_{L_{12}} + (V_1 - V_3) Y_{L_{13}}$$
$$I_2 = V_2 Y_{20} + (V_2 - V_1) Y_{L_{12}} + (V_2 - V_3) Y_{L_{23}}$$
$$I_3 = V_3 Y_{30} + (V_3 - V_1) Y_{L_{13}} + (V_3 - V_4) Y_{L_{34}} + (V_3 - V_2) Y_{L_{23}}$$
$$I_4 = V_4 Y_{40} + (V_4 - V_3) Y_{L_{34}}$$

We introduce the following admittances:

$$Y_{11} = Y_{10} + Y_{L_{12}} + Y_{L_{13}}$$
$$Y_{22} = Y_{20} + Y_{L_{12}} + Y_{L_{23}}$$
$$Y_{33} = Y_{30} + Y_{L_{13}} + Y_{L_{23}} + Y_{L_{34}}$$
$$Y_{44} = Y_{40} + Y_{L_{34}}$$
$$Y_{12} = Y_{21} = -Y_{L_{12}}$$
$$Y_{13} = Y_{31} = -Y_{L_{13}}$$
$$Y_{23} = Y_{32} = -Y_{L_{23}}$$
$$Y_{34} = Y_{43} = -Y_{L_{34}}$$

Thus the current equations reduce to

$$I_1 = Y_{11} V_1 + Y_{12} V_2 + Y_{13} V_3 + 0 V_4$$
$$I_2 = Y_{21} V_1 + Y_{22} V_2 + Y_{23} V_3 + 0 V_4$$
$$I_3 = Y_{13} V_1 + Y_{23} V_2 + Y_{33} V_3 + Y_{34} V_4$$
$$I_4 = 0 V_1 + 0 V_2 + Y_{43} V_3 + Y_{44} V_4$$

Note that $Y_{14} = Y_{41} = 0$, since buses 1 and 4 are not connected; also $Y_{24} = Y_{42} = 0$ since buses 2 and 4 are not connected.

The above set of equations can be written in the nodal-matrix current equation form:

$$\mathbf{I}_{bus} = \mathbf{Y}_{bus} \mathbf{V}_{bus} \qquad (6.20)$$

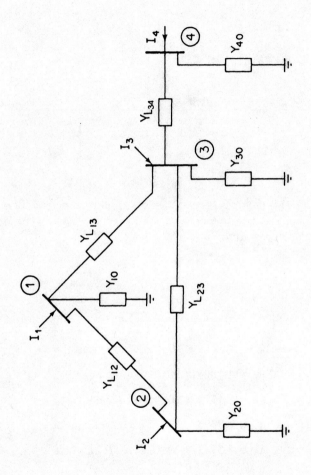

Figure 6-10. Equivalent Circuit for System of Figure 6-9.

where the current vector is defined as

$$\mathbf{I}_{bus} = \begin{bmatrix} I_1 \\ I_2 \\ I_3 \\ I_4 \end{bmatrix}$$

The voltage vector is defined as

$$\mathbf{V}_{bus} = \begin{bmatrix} V_1 \\ V_2 \\ V_3 \\ V_4 \end{bmatrix}$$

The admittance matrix is defined as

$$\mathbf{Y}_{bus} = \begin{bmatrix} Y_{11} & Y_{12} & Y_{13} & Y_{14} \\ Y_{12} & Y_{22} & Y_{23} & Y_{24} \\ Y_{13} & Y_{23} & Y_{33} & Y_{34} \\ Y_{14} & Y_{24} & Y_{34} & Y_{44} \end{bmatrix}$$

We note that the bus admittance matrix $\mathbf{Y}_{bus}$ is symmetric. The following example illustrates the numerical procedure for calculating the bus admittance matrix for our system.

Example 6-5

From Examples 6-1 and 6-2, we have the shunt admittances:

$$Y_{\Pi_7} = 4.3914 \times 10^{-3} \, \underline{/90°} \text{ siemens}$$

$$Y_{\Pi_{10}} = 1.4304 \times 10^{-4} \, \underline{/90°} \text{ siemens}$$

$$Y_{\Pi_{AMA}} = 8.1177 \times 10^{-4} \, \underline{/90°} \text{ siemens}$$

$$Y_A = 3.3171 \times 10^{-3} \, \underline{/89.945°} \text{ siemens}$$

$$Y_C = 3.3141 \times 10^{-3} \, \underline{/89.947°} \text{ siemens}$$

We also have the series impedances:

$$Z_{\Pi_7} = 61.4414 \, \underline{/87.534°} \text{ ohms}$$

$$Z_{\Pi_{10}} = 20.4941 \, \underline{/87.109°} \text{ ohms}$$

$$Z_{\Pi_{AMA}} = 28.3965 \, \underline{/87.526°} \text{ ohms}$$

$$Z_{C_A} = 48.4066 \, \underline{/87.631°} \text{ ohms}$$

6.4 Network Nodal Admittance Formulation

With our bus renumbering, we thus have

$$Y_{10} = Y_{\Pi_{10}} + Y_{\Pi_7}$$
$$= 1.4304 \times 10^{-4} \underline{/90°} + 4.3914 \times 10^{-3} \underline{/90°}$$
$$= 5.8218 \times 10^{-4} \underline{/90°} \text{ siemens}$$

$$Y_{20} = Y_{\Pi_{10}} + Y_{\Pi_{AMA}}$$
$$= 1.4304 \times 10^{-4} \underline{/90°} + 8.1177 \times 10^{-4} \underline{/90°}$$
$$= 9.5482 \times 10^{-4} \underline{/90°} \text{ siemens}$$

$$Y_{30} = Y_{\Pi_{AMA}} + Y_A + Y_{\Pi_7}$$
$$= 8.1177 \times 10^{-4} \underline{/90°} + 3.3171 \times 10^{-3} \underline{/89.945°}$$
$$+ 4.3914 \times 10^{-3} \underline{/90°}$$
$$= 4.5681 \times 10^{-3} \underline{/89.96°} \text{ siemens}$$

$$Y_{40} = Y_C$$
$$= 3.3141 \times 10^{-3} \underline{/89.947°} \text{ siemens}$$

$$Y_{L_{12}} = \frac{1}{Z_{\Pi_{10}}} = 4.8794 \times 10^{-2} \underline{/-88.109°} \text{ siemens}$$

$$Y_{L_{13}} = \frac{1}{Z_{\Pi_7}} = 1.6275 \times 10^{-2} \underline{/-87.523°} \text{ siemens}$$

$$Y_{L_{23}} = \frac{1}{Z_{\Pi_{AMA}}} = 3.5216 \times 10^{-2} \underline{/-87.526°} \text{ siemens}$$

$$Y_{L_{34}} = \frac{1}{Z_{CA}} = 2.0658 \times 10^{-2} \underline{/-87.631°} \text{ siemens}$$

With the above data available, we proceed to calculate the elements of the bus admittance matrix. The self-admittance elements are calculated first as the sum of all admittances connected to the bus considered. As a result,

$$Y_{11} = Y_{10} + Y_{L_{12}} + Y_{L_{13}}$$
$$= 5.8218 \times 10^{-4} \underline{/90°} + 4.8794 \times 10^{-2} \underline{/-88.109°}$$
$$+ 1.6276 \times 10^{-2} \underline{/-87.523°}$$
$$= 6.4487 \times 10^{-2} \underline{/-87.9447°} \text{ siemens}$$

$$Y_{22} = Y_{20} + Y_{L_{12}} + Y_{L_{23}}$$
$$= 9.5482 \times 10^{-4} \underline{/90°} + 4.8794 \times 10^{-2} \underline{/-88.109°}$$
$$+ 3.5215 \times 10^{-2} \underline{/-87.526°}$$
$$= 8.3055 \times 10^{-2} \underline{/-87.841°} \text{ siemens}$$

$$Y_{33} = Y_{30} + Y_{L_{13}} + Y_{L_{23}} + Y_{L_{34}}$$
$$= 4.5681 \times 10^{-3}\,\underline{/89.96°} + 1.6276 \times 10^{-2}\,\underline{/-87.523°}$$
$$+ 3.5216 \times 10^{-2}\,\underline{/-87.526°} + 2.0658 \times 10^{-2}\,\underline{/-87.631°}$$
$$= 6.7586 \times 10^{-2}\,\underline{/-87.3876°}$$
$$Y_{44} = Y_{40} + Y_{L_{34}}$$
$$= 3.3143 \times 10^{-3}\,\underline{/89.947°} + 2.0658 \times 10^{-2}\,\underline{/-87.631°}$$
$$= 1.7347 \times 10^{-2}\,\underline{/-87.1683°}$$

The mutual admittance elements are simply the admittances of the lines connecting the two buses considered with a sign change. Thus,

$$Y_{12} = -Y_{L_{12}} = 4.8794 \times 10^{-2}\,\underline{/91.890°}$$
$$Y_{13} = -Y_{L_{13}} = 1.6276 \times 10^{-2}\,\underline{/92.477°}$$
$$Y_{14} = 0$$
$$Y_{23} = -Y_{L_{23}} = 3.5216 \times 10^{-2}\,\underline{/92.474°}$$
$$Y_{24} = 0$$
$$Y_{34} = -Y_{L_{34}} = 2.0658 \times 10^{-2}\,\underline{/92.369°}$$

This completely defines our bus admittance matrix elements.

6.5 THE GENERAL FORM OF THE LOAD-FLOW EQUATIONS

The above example can be generalized to the case of n buses. In this case, each of vectors $\mathbf{I}_{bus}$ and $\mathbf{V}_{bus}$ are $n \times 1$ vectors. The bus admittance matrix becomes an $n \times n$ matrix with elements

$$Y_{ij} = Y_{ji} = -Y_{L_{ij}} \tag{6.21}$$

$$Y_{ii} = \sum_{j=0}^{n} Y_{L_{ij}} \tag{6.22}$$

where the summation is over the set of all buses connected to bus i including the ground (node 0).

We recall that bus powers S_i rather than the bus currents I_i are, in practice, specified. We thus use

$$I_i^* = \frac{S_i}{V_i}$$

As a result, we have

$$\frac{P_i - jQ_i}{V_i^*} = \sum_{j=1}^{n} (Y_{ij} V_j) \qquad (i=1,\ldots,n) \qquad (6.23)$$

These are the static load-flow equations. Each equation is complex, and therefore we have $2n$ real equations.

The nodal admittance matrix current equation can be written in the power form:

$$P_i - jQ_i = (V_i^*) \sum_{j=1}^{n} (Y_{ij} V_j) \qquad (6.24)$$

The bus voltages on the right-hand side can be substituted for using either the rectangular form:

$$V_i = e_i + jf_i$$

or the polar form:

$$V_i = |V_i| e^{j\theta_i}$$
$$= |V_i| \underline{/\theta_i}$$

Rectangular Form

If we choose the rectangular form, then we have by substitution,

$$P_i = e_i \left(\sum_{j=1}^{n} (G_{ij} e_j - B_{ij} f_j) \right) + f_i \left(\sum_{j=1}^{n} (G_{ij} f_j + B_{ij} e_j) \right) \qquad (6.25)$$

$$Q_i = f_i \left(\sum_{j=1}^{n} (G_{ij} e_j - B_{ij} f_j) \right) - e_i \left(\sum_{j=1}^{n} (G_{ij} f_j + B_{ij} e_j) \right) \qquad (6.26)$$

where the admittance is expressed in the rectangular form:

$$Y_{ij} = G_{ij} + jB_{ij} \qquad (6.27)$$

Polar Form

On the other hand, if we choose the polar form, then we have

$$P_i = |V_i| \sum_{j=1}^{n} |Y_{ij}| |V_j| \cos(\theta_i - \theta_j - \psi_{ij}) \qquad (6.28)$$

$$Q_i = |V_i| \sum_{j=1}^{n} |Y_{ij}| |V_j| \sin(\theta_i - \theta_j - \psi_{ij}) \qquad (6.29)$$

where the admittance is expressed in the polar form:

$$Y_{ij} = |Y_{ij}| \underline{/\psi_{ij}} \qquad (6.30)$$

Hybrid Form

An alternative form of the load-flow equations is the hybrid form, which is essentially the polar form with the admittances expressed in rectangular form. Expanding the trigonometric functions, we have

$$P_i = |V_i| \sum_{j=1}^{n} |Y_{ij}||V_j|\left[\cos(\theta_i - \theta_j)\cos\psi_{ij} + \sin(\theta_i - \theta_j)\sin\psi_{ij}\right] \quad (6.31)$$

$$Q_i = |V_i| \sum_{j=1}^{n} |Y_{ij}||V_j|\left[\sin(\theta_i - \theta_j)\cos\psi_{ij} - \cos(\theta_i - \theta_j)\sin\psi_{ij}\right] \quad (6.32)$$

Now we use

$$Y_{ij} = |Y_{ij}|(\cos\psi_{ij} + j\sin\psi_{ij})$$
$$= G_{ij} + jB_{ij} \quad (6.33)$$

Separating the real and imaginary parts, we obtain

$$G_{ij} = |Y_{ij}|\cos\psi_{ij} \quad (6.34)$$
$$B_{ij} = |Y_{ij}|\sin\psi_{ij} \quad (6.35)$$

so that the power-flow equations reduce to

$$P_i = |V_i| \sum_{j=1}^{n} |V_j|\left[G_{ij}\cos(\theta_i - \theta_j) + B_{ij}\sin(\theta_i - \theta_j)\right] \quad (6.36)$$

$$Q_i = |V_i| \sum_{j=1}^{n} |V_j|\left[G_{ij}\sin(\theta_i - \theta_j) - B_{ij}\cos(\theta_i - \theta_j)\right] \quad (6.37)$$

A simple example will illustrate the procedure.

Example 6-6

For the network shown in Figure 6-11, it is required to write down the load-flow equations:

A. In polar form.
B. In hybrid form.
C. In rectangular form.

Solution

The nodal admittances are given below:

$$Y_{11} = 4 - j5 \qquad Y_{22} = 4 - j10 \qquad Y_{33} = 8 - j15$$
$$Y_{12} = 0 \qquad Y_{13} = -4 + j5 \qquad Y_{23} = -4 + j10$$

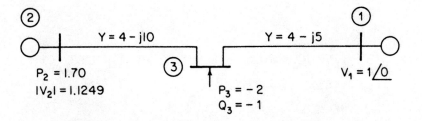

Figure 6-11. Single-Line Diagram for System of Example 6-6.

We also have

$$|V_1| = 1 \qquad \theta_1 = 0$$
$$P_2 = 1.7 \qquad |V_2| = 1.1249$$
$$P_3 = -2 \qquad Q_3 = -1$$

The load-flow equations are obtained utilizing the formula

$$P_i - jQ_i = V_i^* \Sigma Y_{ij} V_j$$

A. We have in polar form,

$$Y_{11} = 6.4031 \underline{/-51.34°}$$
$$Y_{12} = 0$$
$$Y_{13} = 6.4031 \underline{/128.66°}$$
$$Y_{22} = 10.77 \underline{/-68.199°}$$
$$Y_{23} = 10.77 \underline{/111.80°}$$
$$Y_{33} = 17.00 \underline{/-61.928°}$$
$$V_1 = 1\underline{/0}$$
$$V_2 = 1.1249 \underline{/\theta_2}$$
$$V_3 = |V_3| \underline{/\theta_3}$$

For bus 1, we have
$$P_1 - jQ_1 = V_1^*(Y_{11}V_1 + Y_{12}V_2 + Y_{13}V_3)$$
Thus,
$$P_1 - jQ_1 = 6.4031\underline{/-51.34°} + 6.4031|V_3|\underline{/128.66 + \theta_3}$$
For bus 2, we have
$$P_2 - jQ_2 = V_2^*(Y_{12}V_1 + Y_{22}V_2 + Y_{23}V_3)$$
Thus we have
$$1.7 - jQ_2 = 1.1249\underline{/-\theta_2}\left[\left(10.77\underline{/-68.199°}\right)\left(1.1249\underline{/\theta_2}\right)\right.$$
$$\left. + \left(10.77\underline{/111.80°}\right)\left(|V_3|\underline{/\theta_3}\right)\right]$$
This reduces to
$$1.7 - jQ_2 = 13.628\underline{/-68.199°} + 12.115|V_3|\underline{/111.80 + \theta_3 - \theta_2}$$
For bus 3, we have
$$P_3 - jQ_3 = V_3^*(Y_{13}V_1 + Y_{23}V_2 + Y_{33}V_3)$$
Thus,
$$-2 + j1 = |V_3|\underline{/-\theta_3}\left[\left(6.4031\underline{/128.66°}\right) + \left(10.77\underline{/111.80}\right)\left(1.1249\underline{/\theta_2}\right)\right.$$
$$\left. + \left(17\underline{/-61.928}\right)\left(|V_3|\underline{/\theta_3}\right)\right]$$
This reduces to
$$-2 + j1 = 6.4031|V_3|\underline{/128.66 - \theta_3} + 12.115|V_3|\underline{/111.80 + \theta_2 - \theta_3}$$
$$+ 17.00|V_3|^2\underline{/-61.928°}$$

Separating the real and imaginary parts, we obtain for bus 1,
$$P_1 = 4 + 6.4031|V_3|\cos(128.66 + \theta_3) \quad (6.38)$$
$$-Q_1 = -5 + 6.4031|V_3|\sin(128.66 + \theta_3) \quad (6.39)$$
For bus 2 we get
$$1.7 = 5.0612 + 12.115|V_3|\cos(111.8 + \theta_3 - \theta_2) \quad (6.40)$$
$$-Q_2 = -12.653 + 12.115|V_3|\sin(111.8 + \theta_3 - \theta_2) \quad (6.41)$$
For bus 3 we have
$$-2 = 6.4031|V_3|\cos(128.66 - \theta_3)$$
$$+ 12.115|V_3|\cos(111.8 + \theta_2 - \theta_3) + 8|V_3|^2 \quad (6.42)$$
$$1 = 6.4031|V_3|\sin(128.66 - \theta_3)$$
$$+ 12.115|V_3|\sin(111.8 + \theta_2 - \theta_3) - 15|V_3|^2 \quad (6.43)$$

6.5 The General Form of the Load-Flow Equations

The above six equations define the load-flow problem in polar form.

B. The hybrid form is obtained by simply expanding the trigonometric functions in the above six equations. Thus for bus 1 we write

$$P_1 = 4 - 4|V_3|\cos\theta_3 - 5|V_3|\sin\theta_3 \quad (6.44)$$

$$-Q_1 = -5 + 5|V_3|\cos\theta_3 - 4|V_3|\sin\theta_3 \quad (6.45)$$

For bus 2 we write

$$1.7 = 5.0612 - 4.5|V_3|\cos(\theta_3 - \theta_2)$$
$$-11.25|V_3|\sin(\theta_3 - \theta_2) \quad (6.46)$$

$$-Q_2 = -12.654 + 11.25|V_3|\cos(\theta_3 - \theta_2)$$
$$-4.5|V_3|\sin(\theta_3 - \theta_2) \quad (6.47)$$

For bus 3 we write

$$-2 = -4|V_3|\cos\theta_3 + 5|V_3|\sin\theta_3 - 4.5|V_3|\cos(\theta_2 - \theta_3)$$
$$-11.25|V_3|\sin(\theta_2 - \theta_3) + 8|V_3|^2 \quad (6.48)$$

$$1 = 5|V_3|\cos\theta_3 + 4|V_3|\sin\theta_3 + 11.25|V_3|\cos(\theta_2 - \theta_3)$$
$$-4.5|V_3|\sin(\theta_2 - \theta_3) - 15|V_3|^2 \quad (6.49)$$

C. To obtain the rectangular form, we have for bus 1,

$$P_1 - jQ_1 = (1 + j0)[(4 - j5)(1 + j0) + (-4 + j5)(e_3 + jf_3)]$$

This yields

$$P_1 = 4 - 4e_3 - 5f_3 \quad (6.50)$$
$$-Q_1 = -5 + 5e_3 - 4f_3 \quad (6.51)$$

For bus 2 we have

$$1.7 - jQ_2 = (e_2 - jf_2)[(4 - j10)(e_2 + jf_2) + (-4 + j10)(e_3 + jf_3)]$$

This reduces to

$$1.7 - jQ_2 = (4 - j10)(e_2^2 + f_2^2) + (e_2 - jf_2)(-4 + j10)(e_3 + jf_3)$$

Using

$$|V_2|^2 = e_2^2 + f_2^2 = (1.1249)^2$$

we obtain

$$1.7 - jQ_2 = 5.0616 - j12.654 + (-4 + j10)$$
$$[(e_2 e_3 + f_2 f_3) + j(f_3 e_2 - f_2 e_3)]$$

Separating real and imaginary parts, we get

$$1.7 - 5.0616 - 4(e_2 e_3 + f_2 f_3) - 10(f_3 e_2 - f_2 e_3) \quad (6.52)$$
$$-Q_2 = -12.654 + 10(e_2 e_3 + f_2 f_3) - 4(f_3 e_2 - f_2 e_3) \quad (6.53)$$

We should include

$$(1.1249)^2 = e_2^2 + f_2^2 \quad (6.54)$$

For bus 3 we have

$$-2 + j1 = (e_3 - jf_3)[(-4 + j5) + (-4 + j10)(e_2 + jf_2)$$
$$+ (8 - j15)(e_3 + jf_3)]$$

Separating real and imaginary parts, we get

$$-2 = -4e_3 + 5f_3 - e_3(4e_2 + 10f_2) + f_3(10e_2 - 4f_2) + 8(e_3^2 + f_3^2) \quad (6.55)$$

$$+1 = 5e_3 + 4f_3 + e_3(10e_2 - 4f_2) + f_3(4e_2 + 10f_2) - 15(e_3^2 + f_3^2) \quad (6.56)$$

The seven equations (6.50) through (6.56) form the required load flow set.

6.6 THE LOAD-FLOW PROBLEM

The load-flow (or power-flow) problem is concerned with the solution for the static operating condition of an electric power transmission system. Load-flow calculations are performed in power system planning, operational planning, and operation control. The static operating state of the system is defined by the constraints on power and/or voltage at the network buses.

Normally buses are categorized as follows:

1. A *load bus* (*P-Q* bus) is one at which $S_i = P_i + jQ_i$ is specified. In Example 6-6, bus 3 is a load bus.

2. A *generator bus* (*P-V* bus) is a bus with specified injected active power and a fixed voltage magnitude. In our example, bus 2 is a generator bus.

3. A *system reference* or *slack* (*swing*) *bus* is one at which both the magnitude and phase angle of the voltage are specified. It is customary to choose one of the available *P-V* buses as slack and to regard its active power as the unknown. In the previous example, bus 1 is the reference bus.

As we have seen before, each bus is modeled by two equations. In all we have $2N$ equations in $2N$ unknowns. These are $|V|$ and θ at the load buses, Q and θ at the generator buses, and the P and Q at the slack bus.

Let us emphasize here that due to the bus classifications, it is not necessary for us to solve the $2N$ equations simultaneously. A reduction in the required number of equations can be effected. Returning to our example system, we note that although the number of unknowns is six, namely P_1, Q_1, Q_2, $|V_3|$, θ_2, and θ_3, the first three can be computed as a result of the last three. Indeed we need to solve the following equations in $|V_3|$, θ_2, and θ_3:

$$1.7 = 5.0612 - 4.5|V_3|\cos(\theta_3 - \theta_2) - 11.25|V_3|\sin(\theta_3 - \theta_2)$$
$$-2 = -4|V_3|\cos\theta_3 + 5|V_3|\sin\theta_3 - 4.5|V_3|\cos(\theta_2 - \theta_3)$$
$$-11.25|V_3|\sin(\theta_2 - \theta_3) + 8|V_3|^2$$
$$1 = 5|V_3|\cos\theta_3 + 4|V_3|\sin\theta_3 + 11.25|V_3|\cos(\theta_2 - \theta_3)$$
$$-4.5|V_3|\sin(\theta_2 - \theta_3) - 15|V_3|^2$$

With the solution at hand it is only a matter of substitution to get P_1, Q_1, and Q_2 using

$$P_1 = 4 - 4|V_3|\cos\theta_3 - 5|V_3|\sin\theta_3$$
$$-Q_1 = -5 + 5|V_3|\cos\theta_3 - 4|V_3|\sin\theta_3$$
$$-Q_2 = -12.653 + 11.25|V_3|\cos(\theta_3 - \theta_2) - 4.5|V_3|\sin(\theta_3 - \theta_2)$$

The foregoing discussion leads us to specifying the necessary equations for a full solution:

1. At load buses, two equations for active and reactive powers are needed.
2. At generator buses, with $|V_j|$ specified, only the active power equation is needed.

Solution for the primary unknowns $|V_i|$ and θ_i at load buses and θ_i at generator buses is thus possible. This is followed by evaluating the secondary unknowns P_i and Q_i at the slack bus and Q_i at the generator buses using the active and reactive equations for the slack bus and the reactive power equations for the generator buses.

Before we get into the discussion of the solution to the load-flow equations, let us first consider some of their characteristics using some simple examples.

Nonlinearity of the Load-Flow Problem

Consider the two bus system shown in Figure 6-12. Assume that the load at bus 2 is specified and bus 1 is the reference bus with unity voltage magnitude and zero phase angle. We can show that the load-flow problem reduces to solving

$$\alpha x^4 + \beta x^2 + \gamma = 0 \tag{6.57}$$

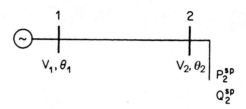

Figure 6-12. Example Network to Illustrate the Nonlinearity of the Load-Flow Problem.

This follows since we have for bus 2 the following two equations on the basis of Eq. (6.28) and (6.29):

$$P_2^{sp} = |V_2||Y_{12}|\cos(\theta_2 - \theta_1 - \psi_{12}) + |V_2|^2|Y_{22}|\cos(\psi_{22})$$
$$Q_2^{sp} = |V_2||Y_{12}|\sin(\theta_2 - \theta_1 - \psi_{12}) - |V_2|^2|Y_{22}|\sin(\psi_{22})$$

Since this is a load bus, we have

$$P_2 = P_2^{sp}$$
$$Q_2 = Q_2^{sp}$$

where superscript (sp) signifies a specified quantity.

The two power equations yield:

$$\cos(\theta_2 - \theta_1 - \psi_{12}) = \frac{P_2^{sp} - G_{22}|V_2|^2}{|V_2||Y_{12}|}$$

$$\sin(\theta_2 - \theta_1 - \psi_{12}) = \frac{Q_2^{sp} + B_{22}|V_2|^2}{|V_2||Y_{12}|}$$

Note that we use

$$G_{22} = |Y_{22}|\cos\psi_{22}$$
$$B_{22} = |Y_{22}|\sin\psi_{22}$$

Squaring and adding, we obtain

$$|Y_{12}|^2|V_2|^2 = (S_2^{sp})^2 + |Y_{22}|^2|V_2|^4 + 2|V_2|^2(B_{22}Q_2^{sp} - G_{22}P_2^{sp})$$

where

$$(S_2^{sp})^2 = (P_2^{sp})^2 + (Q_2^{sp})^2 \qquad (6.58)$$

Let us define

$$\alpha = |Y_{22}|^2 \tag{6.59}$$
$$\beta = 2(B_{22}Q_2^{sp} - G_{22}P_2^{sp}) - |Y_{12}|^2 \tag{6.60}$$
$$\gamma = (S_2^{sp})^2 \tag{6.61}$$

Our unknown is $|V_2|$, which is replaced by x. Thus we have

$$\alpha x^4 + \beta x^2 + \gamma = 0$$

where

$$x = |V_2|$$

The solution to the quartic equation is straightforward since we can solve first for x^2 as

$$x^2 = \frac{-\beta \pm \sqrt{\beta^2 - 4\alpha\gamma}}{2\alpha} \tag{6.62}$$

Since x^2 cannot be imaginary, we have a first condition requiring that

$$\beta^2 - 4\alpha\gamma \geq 0$$

From the definitions of α, β, and γ, we can show that

$$\beta^2 - 4\alpha\gamma = |Y_{12}|^4 - 4|Y_{12}|^2(B_{22}Q_2^{sp} - G_{22}P_2^{sp}) - 4(B_{22}P_2^{sp} + G_{22}Q_2^{sp})^2$$

Thus for a meaningful solution to exist, we need to satisfy the condition

$$|Y_{12}|^4 \geq 4\left[(B_{22}P_2^{sp} + G_{22}Q_2^{sp})^2 + |Y_{12}|^2(B_{22}Q_2^{sp} - G_{22}P_2^{sp})\right] \tag{6.63}$$

A second condition can be obtained if we observe that x cannot be imaginary, requiring that x^2 be positive. Observing that α and γ are positive by their definition leads us to conclude that

$$\left|\sqrt{\beta^2 - 4\alpha\gamma}\right| \leq |\beta|$$

For x^2 to be positive, we need

$$\beta \leq 0$$

or

$$2(B_{22}Q_2^{sp} - G_{22}P_2^{sp}) - |Y_{12}|^2 \leq 0 \tag{6.64}$$

Let us consider the following numerical example.

Example 6-7

The bus admittance matrix elements of interest are

$$Y_{12} = -1.1757 + j10.516$$
$$Y_{22} = 1.1757 - j10.272$$

314 Analysis of Interconnected Systems

We calculate

$$\alpha = |Y_{22}|^2 = 106.9$$
$$\beta = 2(-10.272 Q_2^{sp} - 1.1757 P_2^{sp}) - 111.97$$
$$\gamma = P_2^{sp^2} + Q_2^{sp^2}$$

Let us consider first the case with

$$P_2^{sp} = -0.8$$
$$Q_2^{sp} = -0.6$$

Here we have

$$\gamma = 1$$
$$\beta = -97.76$$

Thus,

$$x^2 = \frac{97.76 \pm \sqrt{(97.76)^2 - 4 \times 106.9}}{(2)(106.9)}$$
$$= 0.90415$$

As a result,

$$x = 0.95087$$

or

$$|V_2| = 0.95087 \text{ p.u.}$$

Next, consider the case with

$$P_2^{sp} = -4$$
$$Q_2^{sp} = -3$$

As a result, we have

$$\gamma = 25$$
$$\beta = -40.93$$

The solution is

$$x^2 = \frac{40.93 \pm \sqrt{(40.93)^2 - (4)(25)(106.9)}}{2 \times 106.9}$$
$$= 0.19144 \pm j0.44409$$

Clearly no practical solution exists. From the value of β we see that the second condition is satisfied. However, the first condition is violated.

Let us pause here and underline the conclusions to be drawn from consideration of the foregoing system. First it is evidently clear that there may be some specified operating conditions for which no solution exists. The second point is that more than one solution can exist. The choice can be narrowed down to a practical answer using further considerations.

It is now evident that except for very simple networks, the load-flow problem results in a set of simultaneous algebraic equations that cannot be solved in closed form. It is thus necessary to employ numerical iterative techniques that start by assuming a set of values for the unknowns and then repeatedly improve on their values in an organized fashion until (hopefully) a solution satisfying the load flow equations is reached. The next section considers the question of getting estimates (initial guess) for the unknowns.

6.7 GETTING STARTED

It is important to have a good approximation to the load-flow solution, which is then used as a starting estimate (or initial guess) in the iterative procedure. A fairly simple process can be used to evaluate a good approximation to the unknown voltages and phase angles. The process is implemented in two stages: the first calculates the approximate angles, and the second calculates the approximate voltage magnitudes.

Busbar Voltage Angles Approximation

In this stage we make the following assumptions:

1. All angles are small, so that $\sin\theta \cong \theta, \cos\theta \cong 1$.
2. All voltage magnitudes are 1 p.u.

Applying these assumptions to the active power equations for the generator buses and load buses in hybrid form, we obtain

$$P_i = \sum_{j=1}^{N} (G_{ij}) + B_{ij}(\theta_i - \theta_j)$$

This is a system of $N-1$ simultaneous linear equations in θ_i, which is then solved to obtain the busbar voltage angle approximations.

Example 6-8

For the system of Example 6-6, bus 2 is a generator bus and bus 3 is a load bus. The equations for evaluating the angles approximately are obtained as

$$P_2 = 1.7 = 4 - 4 + 10(\theta_2 - \theta_3)$$
$$P_3 = -2 = -4 + 5(\theta_3 - 0) - 4 + 10(\theta_3 - \theta_2) + 8$$

As a result, we have
$$1.7 = 10(\theta_2 - \theta_3)$$
$$-2 = 10(\theta_3 - \theta_2) + 5\theta_3$$

The approximate values of θ_2 and θ_3 are thus obtained as
$$\theta_2 = 0.11 \text{ radians}$$
$$\theta_3 = -0.06 \text{ radians}$$

Expressed in degrees, we have
$$\theta_2 = 6.3°$$
$$\theta_3 = -3.44°$$

Busbar Voltage Magnitude Approximation

The calculation of voltage magnitudes employs the angles provided by the above procedure. The calculation is needed only for load buses. We use a current formula derived from the original power form:

$$\frac{P_i + jQ_i}{V_i} = \sum Y_{ij}^* V_j^*$$

The left-hand-side can be expanded to the following:

$$\frac{1}{|V_i|}(P_i + jQ_i)(\cos\theta_i - j\sin\theta_i) = \sum_j |V_j|(G_{ij} - jB_{ij})$$
$$\times (\cos\theta_j - j\sin\theta_j)$$

The imaginary part of this equation is

$$\frac{Q_i \cos\theta_i - P_i \sin\theta_i}{|V_i|} = -\sum_j |V_j| A_{ij}$$

where
$$A_{ij} = B_{ij} \cos\theta_j + G_{ij} \sin\theta_j$$

Let us represent each unknown voltage magnitude as
$$|V_i| = 1 + \Delta V_i$$

We also assume that
$$\frac{1}{1 + \Delta V_i} \simeq 1 - \Delta V_i$$

Thus we have
$$(Q_i \cos\theta_i - P_i \sin\theta_i)(1 - \Delta V_i)$$
$$= -\left[A_{ii}(1 + \Delta V_i) + \underset{\substack{\text{(other} \\ \text{load} \\ \text{buses)}}}{\sum} A_{ij}(1 + \Delta V_j) + \underset{\substack{\text{(all} \\ \text{generator} \\ \text{buses)}}}{\sum} A_{ij}|V_j| \right]$$

The first summation in the above equation is over all other load buses and the second summation is over all generator buses. Grouping all knowns onto the left-hand-side and the unknowns onto the right-hand-side, we get

$$Q_i \cos \theta_i - P_i \sin \theta_i + A_{ii} + \underbrace{\sum A_{ij}|V_j|}_{\substack{\text{(slack \&} \\ \text{generator} \\ \text{buses)}}} + \underbrace{\sum A_{ij}}_{\substack{\text{(other} \\ \text{load} \\ \text{buses)}}}$$

$$= (Q_i \cos \theta_i - P_i \sin \theta_i - A_{ii})(\Delta V_i) - \underbrace{\sum A_{ij}(\Delta V_j)}_{\substack{\text{(other} \\ \text{load} \\ \text{buses)}}}$$

Writing the above equation for all load bus bars gives a linear system of simultaneous equations in the unknowns ΔV_i.

Example 6-9

For the system of Example 6-6, only bus 3 needs to be treated. The above formula is applied using the following numerical values:

$$\theta_1 = 0°$$
$$\theta_2 = 6.3°$$
$$\theta_3 = -3.44°$$

$B_{13} = 5$ $G_{13} = -4$
$B_{23} = 10$ $G_{23} = -4$
$B_{33} = -15$ $G_{33} = 8$
$Q_3 = -1$ $P_3 = -2$

Thus,

$$A_{31} = 5\cos(0) - 4\sin(0) = 5$$
$$A_{32} = 10\cos(6.3°) - 4\sin(6.3°) = 9.5007$$
$$A_{33} = -15\cos(-3.44°) + 8\sin(-3.44°) = -15.453$$

Our formula is thus

$$(-1)\cos(-3.44°) + 2\sin(-3.44°) - 15.453 + 5 + (9.5007)(1.1249)$$
$$= [(-1)\cos(-3.44°) + 2\sin(-3.44°) + 15.453]\,\Delta V_3$$

This provides us with

$$\Delta V_3 = -0.061659$$

As a result, the approximate value of the voltage magnitude at bus bar 3 is

$$|V_3| = 0.93834$$

Let us examine how close are the derived approximate values to satisfying the load flow equations. These are given by

$$P_2 = 5.0612 - 4.5|V_3|\cos(\theta_3 - \theta_2) - 11.25|V_3|\sin(\theta_3 - \theta_2)$$
$$P_3 = -4|V_3|\cos\theta_3 + 5|V_3|\sin\theta_3 - 4.5|V_3|\cos(\theta_2 - \theta_3)$$
$$- 11.25|V_3|\sin(\theta_2 - \theta_3) + 8|V_3|^2$$
$$-Q_3 = 5|V_3|\cos\theta_3 + 4|V_3|\sin\theta_3 + 11.25|V_3|\cos(\theta_2 - \theta_3)$$
$$- 4.5|V_3|\sin(\theta_2 - \theta_3) - 15|V_3|^2$$

Substituting we obtain

$$P_2 = 2.6854$$
$$P_3 = -2.9318$$
$$Q_3 = -0.94059$$

Note that the specified values are

$$P_2 = 1.7$$
$$P_3 = -2$$
$$Q_3 = -1$$

The accuracies of the calculated angles and voltage magnitudes compared with a full ac load-flow solution vary from problem to problem. It is noted, however, that the results are much more reliable than the commonly used flat-start process where all voltages are assumed to be $1\underline{/0}$.

6.8 NEWTON-RAPHSON METHOD

The Newton-Raphson (NR) method is widely used for solving nonlinear equations. It transforms the original nonlinear problem into a sequence of linear problems whose solutions approach the solution of the original problem. The method can be applied to one equation in one unknown or to a system of simultaneous equations with as many unknowns as equations.

One-Dimensional Case

Let $F(x)$ be a nonlinear equation. Any value of x that satisfies $F(x) = 0$ is a root of $F(x)$. To find a particular root, an initial guess for x in the vicinity of the root is needed. Let this initial guess be x_0. Thus

$$F(x_0) = \Delta F_0$$

where ΔF_0 is the error since x_0 is not a root. The situation can be shown graphically as in Figure 6-13. A tangent is drawn at the point on the curve

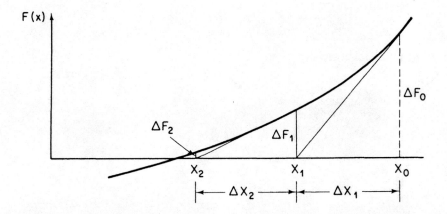

Figure 6-13. Illustrating the Newton-Raphson Method.

corresponding to x_0, and is projected until it intercepts the x-axis to determine a second estimate of the root. Again the derivative is evaluated, and a tangent line is formed to proceed to the third estimate of x. The line generated in this process is given by

$$y(x) = F(x_n) + F'(x_n)(x - x_n) \tag{6.65}$$

which, when $y(x) = 0$, gives the recursion formula for iterative estimates of the root:

$$x_{n+1} = x_n - \frac{F(x_n)}{F'(x_n)} \tag{6.66}$$

A numerical example will illustrate the process.

Example 6-10

Compute the cube root of 64, which is 4, using the Newton-Raphson method.

$$F(x) = x^3 - 64$$

Solution

We have
$$F'(x) = 3x^2$$
Let the initial guess be
$$x_0 = 5$$
Our iterations proceed according to
$$x_{n+1} = x_n - \Delta x_n$$
$$\Delta x_n = \frac{F(x_n)}{F'(x_n)}$$
$$= \frac{x_n^3 - 64}{3x_n^2}$$

For the first iteration we have
$$\Delta x_1 = \frac{125 - 64}{75} = 0.8133$$
Thus
$$x_1 = 5 - 0.8133 = 4.1867$$
For the second iteration we have
$$\Delta x_2 = \frac{(4.1867)^3 - 64}{3(4.1867)^2}$$
$$= 0.1785$$
$$x_2 = 4.1867 - 0.1785$$
$$= 4.0082$$

The process is continued until the desired accuracy has been achieved.

N-Dimensional Case

The single dimensional concept of the Newton-Raphson method can be extended to N dimensions. All that is needed is an N-dimensional analog of the first derivative. This is provided by the Jacobian matrix. Each of the N rows of the Jacobian matrix is composed of the partial derivatives of one of the equations of the system with respect to each of the N variables.

An understanding of the general case can be gained from the specific example $N = 2$. Assume that we are given the two nonlinear equations F_1, F_2. Thus,

$$F_1(x_1, x_2) = 0 \qquad F_2(x_1, x_2) = 0 \qquad (6.67)$$

6.8 Newton-Raphson Method

The Jacobian matrix for this 2×2 system is

$$\begin{bmatrix} \dfrac{\partial F_1}{\partial x_1} & \dfrac{\partial F_1}{\partial x_2} \\ \dfrac{\partial F_2}{\partial x_1} & \dfrac{\partial F_2}{\partial x_2} \end{bmatrix} \tag{6.68}$$

If the Jacobian matrix is numerically evaluated at some point $(x_1^{(k)}, x_2^{(k)})$, the following linear relationship is established for small displacements $(\Delta x_1, \Delta x_2)$:

$$\begin{bmatrix} \dfrac{\partial F_1^{(k)}}{\partial x_1} & \dfrac{\partial F_1^{(k)}}{\partial x_2} \\ \dfrac{\partial F_2^{(k)}}{\partial x_1} & \dfrac{\partial F_2^{(k)}}{\partial x_2} \end{bmatrix} \begin{bmatrix} \Delta x_1^{(k+1)} \\ \Delta x_2^{(k+1)} \end{bmatrix} = \begin{bmatrix} \Delta F_1^{(k)} \\ \Delta F_2^{(k)} \end{bmatrix} \tag{6.69}$$

A recursive algorithm can be developed for computing the vector displacements $(\Delta x_1, \Delta x_2)$. Each displacement is a solution to the related linear problem. With a good initial guess and other favorable conditions, the algorithm will converge to a solution of the nonlinear problem. We let $(x_1^{(0)}, x_2^{(0)})$ be the initial guess. Then the errors are

$$\Delta F_1^{(0)} = -F_1\big[x_1^{(0)}, x_2^{(0)}\big], \qquad \Delta F_2^{(0)} = -F_2\big[x_1^{(0)}, x_2^{(0)}\big] \tag{6.70}$$

The Jacobian matrix is then evaluated at the trial solution point $[x_1^{(0)}, x_2^{(0)}]$. Each element of the Jacobian matrix is computed from an algebraic formula for the appropriate partial derivative using $x_1^{(0)}, x_2^{(0)}$. Thus,

$$\begin{bmatrix} \dfrac{\partial F_1^{(0)}}{\partial x_1} & \dfrac{\partial F_1^{(0)}}{\partial x_2} \\ \dfrac{\partial F_2^{(0)}}{\partial x_1} & \dfrac{\partial F_2^{(0)}}{\partial x_2} \end{bmatrix} \begin{bmatrix} \Delta x_1^{(1)} \\ \Delta x_2^{(1)} \end{bmatrix} = \begin{bmatrix} \Delta F_1^{(0)} \\ \Delta F_2^{(0)} \end{bmatrix} \tag{6.71}$$

This system of linear equations is then solved directly for the first correction. The correction is then added to the initial guess to complete the first iteration:

$$\begin{bmatrix} x_1^{(1)} \\ x_2^{(1)} \end{bmatrix} = \begin{bmatrix} x_1^{(0)} \\ x_2^{(0)} \end{bmatrix} + \begin{bmatrix} \Delta x_1^{(1)} \\ \Delta x_2^{(1)} \end{bmatrix} \tag{6.72}$$

Equations (6.71) and (6.72) are rewritten using matrix symbols and a general superscript h for the iteration count:

$$[J^{h-1}][\Delta x^h] = [\Delta F^{h-1}] \tag{6.73}$$

$$x^h = x^{h-1} + \Delta x^h \tag{6.74}$$

Analysis of Interconnected Systems

The algorithm is repeated until ΔF^h satisfies some tolerance. In most solvable problems, it can be made practically zero. A numerical example will illustrate the procedure.

Example 6-11

Use the Newton-Raphson method to solve the following equations:

$$F_1 = x_1^2 + x_2^2 - 5x_1 = 0$$
$$F_2 = x_1^2 - x_2^2 + 1.5x_2 = 0$$

Solution

The partial derivatives are

$$\frac{\partial F_1}{\partial x_1} = 2x_1 - 5 \qquad \frac{\partial F_1}{\partial x_2} = 2x_2$$

$$\frac{\partial F_2}{\partial x_1} = 2x_1 \qquad \frac{\partial F_2}{\partial x_2} = -2x_2 + 1.5$$

Let us take as initial guess

$$x_1^{(0)} = 3 \qquad x_2^{(0)} = 3$$

As a result,

$$F_1^0 = (3)^2 + (3)^2 - 5 \times 3 = 3$$
$$F_2^0 = (3)^2 - (3)^2 + 1.5 \times 3 = 4.5$$

The Jacobian elements are

$$\frac{\partial F_1^{(0)}}{\partial x_1} = (2)(3) - 5 = 1 \qquad \frac{\partial F_1^{(0)}}{\partial x_2} = (2)(3) = 6$$

$$\frac{\partial F_2^{(0)}}{\partial x_1} = (2)(3) = 6 \qquad \frac{\partial F_2^{(0)}}{\partial x_2} = (-2)(3) + 1.5 = -4.5$$

Thus,

$$\begin{bmatrix} 1 & 6 \\ 6 & -4.5 \end{bmatrix} \begin{bmatrix} \Delta x_1^{(1)} \\ \Delta x_2^{(1)} \end{bmatrix} = \begin{bmatrix} -3 \\ -4.5 \end{bmatrix}$$

The solution is

$$\Delta x_1^{(1)} = -1$$
$$\Delta x_2^{(0)} = -0.333$$

6.8 Newton-Raphson Method

Thus the first estimate is

$$x_1^{(1)} = x_1^{(0)} + \Delta x_1^{(1)} = 3 - 1 = 2$$
$$x_2^{(1)} = x_2^{(0)} + \Delta x_2^{(1)} = 3 - 0.333 = 2.667$$

We now repeat the above procedure:

$$F_1^{(1)} = (2)^2 + (2.667)^2 - 5(2) = 1.1129$$
$$F_2^{(1)} = (2)^2 - (2.667)^2 + 1.5(2.667) = 0.8876$$

The Jacobian elements are as follows:

$$\frac{\partial F_1^{(1)}}{\partial x_1} = (2)(2) - 5 = -1 \qquad \frac{\partial F_1^{(1)}}{\partial x_2} = (2)(2.667) = 5.334$$

$$\frac{\partial F_2^{(1)}}{\partial x_1} = (2)(2) = 4 \qquad \frac{\partial F_2^{(1)}}{\partial x_2} = (-2)(2.667) + 1.5 = -3.834$$

Thus,

$$\begin{bmatrix} -1 & 5.334 \\ 4 & -3.834 \end{bmatrix} \begin{bmatrix} \Delta x_1^{(2)} \\ \Delta x_2^{(2)} \end{bmatrix} = \begin{bmatrix} -1.1129 \\ -0.8876 \end{bmatrix}$$

The solution is

$$\Delta x_1^{(2)} = -0.5143$$
$$\Delta x_2^{(2)} = -0.3051$$

Thus the second estimate is

$$x_1^{(2)} = x_1^{(1)} + \Delta x_1^{(2)} = 2 - 0.5143 = 1.4857$$
$$x_2^{(2)} = x_2^{(1)} + \Delta x_2^{(2)} = 2.667 - 0.3051 = 2.3619$$

Repeating the above procedure, the following sequence of estimates is obtained:

Iteration	x_1	x_2
3	1.2239	2.1738
4	1.0935	2.0733
5	1.0316	2.0248
6	1.0065	2.0051
7	1.0004	2.0003
8	1.00000189	2.00000149

It is clear that the sequence is converging in on the solution point:

$$x_1 = 1.00 \quad \text{and} \quad x_2 = 2.00$$

To emphasize the importance of the initial guess to the process, we observe that the above problem has four solutions that can be obtained using different initial guesses. The first at $x_1 = 1.00$ and $x_2 = 2.00$ was obtained with $x_1^{(0)} = 3$ and $x_2^{(0)} = 3$ above. If we start with $x_1^{(0)} = 10$ and $x_2^{(0)} = 10$, the following sequence is obtained:

Iteration	x_1	x_2
0	10	10
1	5.46	5.9
2	3.1961	3.8748
3	2.0688	2.8839
4	1.5099	2.4097
5	1.2336	2.1848
6	1.0979	2.077
7	1.0336	2.0264
8	1.0072	2.0056
9	1.0005	2.0004
10	1.0000027	2.00000212

This is the same solution point.

If we start at $x_1^{(0)} = -10$ and $x_2^{(0)} = -10$, we obtain the second solution $x_1 = 0$ and $x_2 = 0$. The sequence of iterates then is as follows:

Iteration	x_1	x_2
0	-10	-10
1	-4.58	-4.266
2	-1.895	-1.451
3	-0.591	-0.171
4	-0.069	-0.130
5	-0.005	-0.01
6	-2.56×10^{-5}	-5.5973×10^{-5}
7	-7.5790×10^{-10}	-1.6507×10^{-9}
8	-8×10^{-19}	-1×10^{-18}
9	-2×10^{-28}	-2×10^{-28}
10	0	0

The third solution at $x_1 = 0.91028$ and $x_2 = 1.9294$ can be obtained starting at $x_1^{(0)} = -10$ and $x_2^{(0)} = 10$. The sequence of iterates is as follows:

Iteration	x_1	x_2
0	-10	10
1	-4.2898	4.637
2	-1.5204	2.077
3	-0.27275	1.067
4	0.3209	1.4023
5	0.65022	1.7483
6	0.79677	1.8381
7	0.86984	1.8978
8	0.90068	1.9219
9	0.90943	1.9288
10	0.91027	1.9294
11	0.91028	1.9294

The fourth solution is at $x_1 = 3.08972$ and $x_2 = -2.42945$, which is obtained starting with $x_1^{(0)} = 10$ and $x_2^{(0)} = -20$ with the attendant sequence of iterates given below:

Iteration	x_1	x_2
0	10	-20
1	6.1511	-10.1933
2	4.2748	-5.4214
3	3.4266	-3.2743
4	3.1353	-2.5429
5	3.0908	-2.4321
6	3.0897	-2.4295
7	3.0897	-2.42945
8	3.08972	-2.42945

Figure 6-14 shows the four solution points as the intersection of the two locii F_1 and F_2 in the two-dimensional representation. To illustrate how the converged solution depends on the initial guess, Table 6-1 lists iteration values for a few initial guesses.

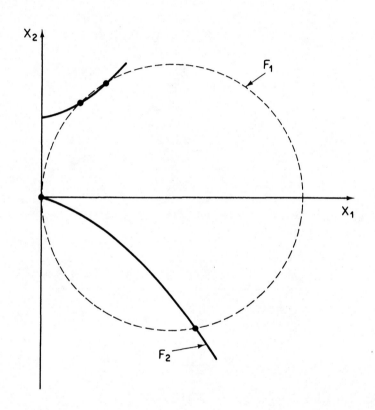

Figure 6-14. Solution Points for Example 6-11.

TABLE 6-1

Guess	x_1	x_2
1	0.5 −0.08 −0.01 −0.00	0.5 0.17 −0.02 −0.00
2	0.5 0.729 0.83716 0.8876 0.9065 0.91013 0.91028	2 1.7917 1.8716 1.9117 1.9265 1.9293 1.9294
3	1.5 0.784 0.98768 0.99057 1.0013 1.000	4 2.4773 2.0469 1.9936 2.0010 2.000
4	2.5 3.5875 3.1757 3.0933 3.0897 3.0897	−1 −3.625 −2.635 −2.438 −2.4295 −2.4295
5	−1 −0.245 −0.0155 −1.1669 × 10^{-4} −3.7784 × 10^{-9} −8 × 10^{-18} −2 × 10^{-27}	−1 −0.140 0.0185 −7.2626 × 10^{-5} 5.5609 × 10^{-9} −1.1 × 10^{-17} 0
6	−0.5 0.175 0.648 0.79379 0.86865 0.90024 0.90936 0.91027 0.91028	1 1.15 1.8983 1.8342 1.8969 1.9216 1.9287 1.9294 1.9294
7	2.5 3.4375 3.1318 3.0905 3.0897 3.0897	−5 −3.125 −2.5136 −2.4310 −2.4295 −2.4294

6.9 THE NEWTON-RAPHSON METHOD FOR LOAD-FLOW SOLUTION

There are several distinctly different ways of applying the Newton-Raphson method to solving the load-flow equations. We illustrate a popular version employing the polar form. As we have seen for each generator bus (except for the slack bus), we have the active power equation and the corresponding unknown phase angle θ_i. We write this equation in the form

$$\Delta P_i = P_i^{\text{sch}} - P_i = 0$$

For each load bus we have the active and reactive equations and the unknowns $|V_i|$ and θ_i. We write the two equations in the form

$$\Delta P_i = P_i^{\text{sch}} - P_i = 0$$
$$\Delta Q_i = Q_i^{\text{sch}} - Q_i = 0$$

In the above equations, the superscript "sch" denotes the scheduled or specified bus active or reactive powers. Using the polar form, we have

$$P_i = |V_i| \sum_{j=1}^{n} |Y_{ij}||V_j|\cos(\theta_i - \theta_j - \psi_{ij})$$

$$Q_i = |V_i| \sum_{j=1}^{n} |Y_{ij}||V_j|\sin(\theta_i - \theta_j - \psi_{ij})$$

Before we proceed with the formulation, let us illustrate the above concepts for the system of Example 6-6.

Example 6-12

Recalling that in Example 6-6, bus 2 is a generator bus and bus 3 is a load bus, we can write three equations in the unknowns θ_2, θ_3, and $|V_3|$. Thus,

$$\Delta P_2 = 1.7 - \left[5.0612 - 4.5|V_3|\cos(\theta_3 - \theta_2)\right.$$
$$\left. - 11.25|V_3|\sin(\theta_3 - \theta_2)\right] = 0$$
$$\Delta P_3 = -2 - \left[-4|V_3|\cos\theta_3 + 5|V_3|\sin\theta_3 - 4.5|V_3|\cos(\theta_2 - \theta_3)\right.$$
$$\left. - 11.25|V_3|\sin(\theta_2 - \theta_3) + 8|V_3|^2\right] = 0$$
$$\Delta Q_3 = 1 - \left[5|V_3|\cos\theta_3 + 4|V_3|\sin\theta_3 + 11.25|V_3|\cos(\theta_2 - \theta_3)\right.$$
$$\left. - 4.5|V_3|\sin(\theta_2 - \theta_3) - 15|V_3|^2\right] = 0$$

We now proceed to show the application of the Newton-Raphson method to solve the load-flow problem. The incremental corrections to estimates of the unknowns are obtained as the solution to the linear system

6.9 The Newton-Raphson Method for Load-Flow Solution

of equations. Thus,

$$\frac{\partial P_2}{\partial \theta_2}(\Delta\theta_2) + \frac{\partial P_2}{\partial \theta_3}(\Delta\theta_3) + \frac{\partial P_2}{\partial |V_3|}\Delta|V_3| = \Delta P_2$$

$$\frac{\partial P_3}{\partial \theta_2}(\Delta\theta_2) + \frac{\partial P_3}{\partial \theta_3}(\Delta\theta_3) + \frac{\partial P_3}{\partial |V_3|}\Delta|V_3| = \Delta P_3$$

$$\frac{\partial Q_3}{\partial \theta_2}(\Delta\theta_2) + \frac{\partial Q_3}{\partial \theta_3}(\Delta\theta_3) + \frac{\partial Q_3}{\partial |V_3|}\Delta|V_3| = \Delta Q_3$$

To simplify the calculation as will be seen later, the third term in each of the equations is modified so that we solve for $(\Delta|V_3|/|V_3|)$. We thus write

$$\frac{\partial P_2}{\partial \theta_2}(\Delta\theta_2) + \frac{\partial P_2}{\partial \theta_3}(\Delta\theta_3) + \left(|V_3|\frac{\partial P_2}{\partial |V_3|}\right)\frac{\Delta|V_3|}{|V_3|} = \Delta P_2$$

$$\frac{\partial P_3}{\partial \theta_2}(\Delta\theta_2) + \frac{\partial P_3}{\partial \theta_3}(\Delta\theta_3) + \left(|V_3|\frac{\partial P_3}{\partial |V_3|}\right)\frac{\Delta|V_3|}{|V_3|} = \Delta P_3$$

$$\frac{\partial Q_3}{\partial \theta_2}(\Delta\theta_2) + \frac{\partial Q_3}{\partial \theta_3}(\Delta\theta_3) + \left(|V_3|\frac{\partial Q_3}{\partial |V_3|}\right)\frac{\Delta|V_3|}{|V_3|} = \Delta Q_3$$

In matrix notation we have

$$\begin{bmatrix} \frac{\partial P_2}{\partial \theta_2} & \frac{\partial P_2}{\partial \theta_3} & \left(|V_3|\frac{\partial P_2}{\partial |V_3|}\right) \\ \frac{\partial P_3}{\partial \theta_2} & \frac{\partial P_3}{\partial \theta_3} & \left(|V_3|\frac{\partial P_3}{\partial |V_3|}\right) \\ \frac{\partial Q_3}{\partial \theta_2} & \frac{\partial Q_3}{\partial \theta_3} & \left(|V_3|\frac{\partial Q_3}{\partial |V_3|}\right) \end{bmatrix} \begin{bmatrix} \Delta\theta_2 \\ \Delta\theta_3 \\ \frac{\Delta|V_3|}{|V_3|} \end{bmatrix} = \begin{bmatrix} \Delta P_2 \\ \Delta P_3 \\ \Delta Q_3 \end{bmatrix}$$

Solving for $\Delta\theta_2$, $\Delta\theta_3$, and $(\Delta|V_3|/|V_3|)$, we thus obtain the new estimates at the $(n+1)$th iteration:

$$\theta_2^{(n+1)} = \theta_2^{(n)} + \Delta\theta_2$$
$$\theta_3^{(n+1)} = \theta_3^{(n)} + \Delta\theta_3$$
$$|V_3|^{(n+1)} = |V_3|^{(n)} + \Delta|V_3|$$

In order to simplify notation in describing the programming,

$$H_{ij} = \frac{\partial P_i}{\partial \theta_j} \qquad N_{ij} = \frac{\partial P_i}{\partial |V_j|}|V_j|$$

$$J_{ij} = \frac{\partial Q_i}{\partial \theta_j} \qquad L_{ij} = \frac{\partial Q_i}{\partial |V_j|}|V_j|$$

TABLE 6-2

Generalized Linear Equation Set-Up for Newton Raphson Method for Load Flow Solution

$$
\begin{bmatrix}
H_{22} & H_{23} & \cdots & H_{2m} & H_{2(m+1)} & \cdots & H_{2n} & N_{2(m+1)} & \cdots & N_{2n} \\
H_{32} & H_{33} & \cdots & H_{3m} & H_{3(m+1)} & \cdots & H_{3n} & N_{3(m+1)} & \cdots & N_{3n} \\
\vdots & \vdots & & \vdots & \vdots & & \vdots & \vdots & & \vdots \\
H_{m2} & H_{m3} & \cdots & H_{mm} & H_{m(m+1)} & \cdots & H_{mn} & N_{m(m+1)} & \cdots & N_{mn} \\
H_{(m+1)2} & H_{(m+1)3} & \cdots & H_{(m+1)m} & H_{(m+1)(m+1)} & \cdots & H_{(m+1)n} & N_{(m+1)(m+1)} & \cdots & N_{(m+1)n} \\
\vdots & \vdots & & \vdots & \vdots & & \vdots & \vdots & & \vdots \\
H_{n2} & H_{n3} & \cdots & H_{nm} & H_{n(m+1)} & \cdots & H_{nn} & N_{n(m+1)} & \cdots & N_{nn} \\
J_{(m+1)2} & \cdots & & J_{(m+1)m} & J_{(m+1)(m+1)} & \cdots & J_{(m+1)n} & L_{(m+1)(m+1)} & \cdots & L_{(m+1)n} \\
\vdots & & & \vdots & \vdots & & \vdots & \vdots & & \vdots \\
J_{n2} & \cdots & & J_{nm} & J_{n(m+1)} & \cdots & J_{nn} & L_{n(m+1)} & \cdots & L_{nn}
\end{bmatrix}
\begin{bmatrix}
\Delta\theta_2 \\
\Delta\theta_3 \\
\vdots \\
\Delta\theta_m \\
\Delta\theta_{m+1} \\
\Delta\theta_{m+2} \\
\vdots \\
\Delta\theta_n \\
\dfrac{\Delta|V_{m+1}|}{|V_{m+1}|} \\
\vdots \\
\dfrac{\Delta|V_n|}{|V_n|}
\end{bmatrix}
=
\begin{bmatrix}
\Delta P_2 \\
\Delta P_3 \\
\vdots \\
\Delta P_m \\
\Delta P_{m+1} \\
\Delta P_{m+2} \\
\vdots \\
\Delta P_n \\
\Delta Q_{m+1} \\
\vdots \\
\Delta Q_n
\end{bmatrix}
$$

6.9 The Newton-Raphson Method for Load-Flow Solution

Assuming that bus 1 is the slack bus, that buses 2,..., m are generator buses, and that buses $m+1, m+2,..., n$ are load buses, the general form for a system of arbitrary size is as shown in Table 6-2.

In condensed form, we have

$$\left[\begin{array}{c|c} H & N \\ \hline J & L \end{array}\right] \left[\begin{array}{c} \Delta\theta \\ \left(\dfrac{\Delta V}{V}\right) \end{array}\right] = \left[\begin{array}{c} \Delta P \\ \Delta Q \end{array}\right]$$

Evaluation of the Derivatives

We have seen that the partial derivatives are needed. The power equations in polar form are given by Eqs. (6.28) and (6.29) as

$$P_i = |V_i| \sum_{j=1}^{n} |Y_{i,j}||V_j|\cos(\theta_i - \theta_j - \psi_{ij})$$

$$Q_i = |V_i| \sum_{j=1}^{n} |Y_{i,j}||V_j|\sin(\theta_i - \theta_j - \psi_{ij})$$

We find the derivatives of P_i with respect to θ_i as

$$\frac{\partial P_i}{\partial \theta_i} = -|V_i| \sum_{\substack{j=1 \\ j \neq i}}^{n} |Y_{i,j}||V_j|\sin(\theta_i - \theta_j - \psi_{ij})$$

Using Eq. (6.29), we can see that

$$\frac{\partial P_i}{\partial \theta_i} = -\left[Q_i - |V_i|^2|Y_{ii}|\sin(-\psi_{ii})\right]$$

Thus we have

$$\frac{\partial P_i}{\partial \theta_i} = -Q_i - |V_i|^2 B_{ii}$$

The derivatives of P_i with respect to $\theta_j, j \neq i$ are given by

$$\frac{\partial P_i}{\partial \theta_j} = |V_i||V_j||Y_{i,j}|\sin(\theta_i - \theta_j - \psi_{ij}) \quad (j \neq i)$$

To avoid trigonometric functions we expand and use the rectangular form to obtain, after some algebra,

$$\frac{\partial P_i}{\partial \theta_j} = a_{ij}f_i - b_{ij}e_i \quad (j \neq i)$$

with
$$a_{ij} = G_{ij}e_j - B_{ij}f_j$$
$$b_{ij} = G_{ij}f_j + B_{ij}e_j$$

The partial derivatives of Q_i with respect to θ_i are
$$\frac{\partial Q_i}{\partial \theta_i} = |V_i| \sum_{\substack{j=1 \\ j \neq i}}^{n} |V_j||Y_{ij}|\cos(\theta_i - \theta_j - \psi_{ij})$$

Using Eq. (6.28), we can see that
$$\frac{\partial Q_i}{\partial \theta_i} = \left(P_i - |V_i|^2|Y_{ii}|\cos\psi_{ii}\right)$$

Thus,
$$\frac{\partial Q_i}{\partial \theta_i} = P_i - |V_i|^2(G_{ii})$$

The partial derivatives of Q_i with respect to θ_j, $j \neq i$ are
$$\frac{\partial Q_i}{\partial \theta_j} = -|V_i||V_j||Y_{ij}|\cos(\theta_i - \theta_j - \psi_{ij}) \qquad (j \neq i)$$

Again to avoid trigonometric expansions, we use the rectangular form to obtain
$$\frac{\partial Q_i}{\partial \theta_j} = -a_{ij}e_i - b_{ij}f_i$$

The derivatives with respect to $|V_i|$ are obtained as follows:
$$\frac{\partial P_i}{\partial |V_i|} = 2|V_i||Y_{ii}|\cos\psi_{ii} + \sum_{\substack{j=1 \\ j \neq i}}^{n} |V_j||Y_{ij}|\cos(\theta_i - \theta_j - \psi_{ij})$$

To simplify the calculation, we multiply both sides by $|V_i|$ to obtain
$$|V_i|\frac{\partial P_i}{\partial |V_i|} = |V_i|^2|Y_{ii}|\cos\psi_{ii} + \sum_{j=1}^{n} |V_i||V_j|\cos(\theta_i - \theta_j - \psi_{ij})$$

This reduces to
$$|V_i|\frac{\partial P_i}{\partial |V_i|} = P_i + |V_i|^2(G_{ii})$$

We next have
$$\frac{\partial P_i}{\partial |V_j|} = |V_i||Y_{ij}|\cos(\theta_i - \theta_j - \psi_{ij}) \qquad (j \neq i)$$

6.9 The Newton-Raphson Method for Load-Flow Solution

Again we multiply by $|V_j|$ to obtain

$$|V_j|\frac{\partial P_i}{\partial |V_j|} = |V_i||V_j||Y_{i,j}|\cos(\theta_i - \theta_j - \psi_{ij})$$

This reduces to

$$|V_j|\frac{\partial P_i}{\partial |V_j|} = a_{ij}e_i + b_{ij}f_i$$

For the reactive power we have

$$\frac{\partial Q_i}{\partial |V_i|} = -2|V_i||Y_{ii}|\sin\psi_{ii} + \sum_{\substack{j=1 \\ j\neq i}}^{n} |V_j||Y_{ij}|\sin(\theta_i - \theta_j - \psi_{ij})$$

Again multiplying by $|V_i|$, we obtain

$$|V_i|\frac{\partial Q_i}{\partial |V_i|} = Q_i - |V_i|^2 B_{ii}$$

Finally we have

$$\frac{\partial Q_i}{\partial |V_j|} = |V_i||Y_{ij}|\sin(\theta_i - \theta_j - \psi_{ij}) \qquad (j \neq i)$$

This reduces to

$$|V_j|\frac{\partial Q_i}{\partial |V_j|} = a_{ij}f_i - b_{ij}e_i$$

To summarize, in terms of the Van Ness variables, we have

$$H_{ij} = L_{ij} = a_{ij}f_i - b_{ij}e_i$$
$$N_{ij} = -J_{ij} = a_{ij}e_i + b_{ij}f_i \qquad (i \neq j \text{ off diagonals})$$
$$H_{ii} = -Q_i - B_{ii}V_i^2$$
$$L_{ii} = Q_i - B_{ii}V_i^2$$
$$N_{ii} = P_i + G_{ii}V_i^2$$
$$J_{ii} = P_i - G_{ii}V_i^2$$

Here we use

$$a_{ij} = G_{ij}e_j - B_{ij}f_j$$
$$b_{ij} = G_{ij}f_j + B_{ij}e_j$$

From the standpoint of computation, we use the rectangular form of the power equations:

$$P_i = (e_i) \sum_{j=1}^{n} (G_{ij}e_j - B_{ij}f_j) + (f_i) \sum_{j=1}^{n} (G_{ij}f_j + B_{ij}e_j)$$

$$Q_i = (f_i) \sum_{j=1}^{n} (G_{ij}e_j - B_{ij}f_j) - e_i \sum_{j=1}^{n} (G_{ij}f_j + B_{ij}e_j)$$

In terms of the a_{ij} and b_{ij} variables, we have

$$P_i = (e_i) \sum_{j=1}^{n} a_{ij} + (f_i) \sum_{j=1}^{n} b_{ij}$$

$$Q_i = (f_i) \sum_{j=1}^{n} a_{ij} - (e_i) \sum_{j=1}^{n} b_{ij}$$

Example 6-13

Recall the following data for Example 6-6:

$$G_{11} = 4 \qquad B_{11} = -5$$
$$G_{12} = 0 \qquad B_{12} = 0$$
$$G_{13} = -4 \qquad B_{13} = 5$$
$$G_{22} = 4 \qquad B_{22} = -10$$
$$G_{23} = -4 \qquad B_{23} = 10$$
$$G_{33} = 8 \qquad B_{33} = -15$$

The Specified Voltages are

$$|V_1| = 1.00$$
$$|V_2| = 1.1249$$
$$\theta_1 = 0$$

The Specified Powers are

$$P_2^{\text{sch}} = 1.7$$
$$P_3^{\text{sch}} = -2$$
$$Q_3^{\text{sch}} = -1$$

From Examples 6-8 and 6-9, the following starting estimates were obtained:

$$\theta_2^{(0)} = 6.3°$$
$$\theta_3^{(0)} = -3.44°$$
$$|V_3|^{(0)} = 0.93834$$

6.9 The Newton-Raphson Method for Load-Flow Solution

We will now illustrate the calculation of the Newton-Raphson correction terms. The a and b variables are first calculated. The relevant formulae are given by

$$a_{23} = G_{23}e_3 - B_{23}f_3$$
$$b_{23} = G_{23}f_3 + B_{23}e_3$$
$$a_{32} = G_{32}e_2 - B_{32}f_2$$
$$b_{32} = G_{32}f_2 + B_{32}e_2$$

With the given admittances, we have

$$a_{23} = -4e_3 - 10f_3$$
$$b_{23} = -4f_3 + 10e_3$$
$$a_{32} = -4e_2 - 10f_2$$
$$b_{32} = -4f_2 + 10e_2$$

The calculation of the Jacobian's off-diagonal terms is performed first, using the following:

$$H_{23} = a_{23}f_2 - b_{23}e_2$$
$$H_{32} = a_{32}f_3 - b_{32}e_3$$
$$N_{23} = a_{23}e_2 + b_{23}f_2$$
$$J_{32} = -(a_{32}e_3 + b_{32}f_3)$$

The active and reactive powers are then calculated using the specified values in rectangular form:

$$Q_2 = 12.654 + f_2 a_{23} - e_2 b_{23}$$
$$Q_3 = -5e_3 - 4f_3 + f_3 a_{32} - e_3 b_{32} + 15|V_3|^2$$
$$P_3 = e_3(a_{32} - 4) + f_3(b_{32} + 5) + 8|V_3|^2$$

Using H_{23}, H_{32}, and J_{32}, we can write

$$Q_2 = 12.654 + H_{23}$$
$$Q_3 = -5e_3 - 4f_3 + H_{32} + 15|V_3|^2$$
$$P_3 = -4e_3 + 5f_3 - J_{32} + 8|V_3|^2$$

The remaining Jacobian elements can be calculated as follows:

$$H_{22} = -Q_2 - B_{22}|V_2|^2$$
$$H_{33} = -Q_3 - B_{33}|V_3|^2$$
$$N_{33} = P_3 + G_{33}|V_3|^2$$
$$L_{33} = Q_3 - B_{33}|V_3|^2$$
$$J_{33} = P_3 - G_{33}|V_3|^2$$

Using the specified values, we have
$$H_{22} = -Q_2 + 10(1.1249)^2 = -Q_2 + 12.654$$
$$H_{33} = -Q_3 + 15|V_3|^2$$
$$N_{33} = P_3 + 8|V_3|^2$$
$$L_{33} = Q_3 + 15|V_3|^2$$
$$J_{33} = P_3 - 8|V_3|^2$$

We now proceed to calculate the elements of the first iteration. With the initial guess voltage magnitude and phase angles, we have
$$e_2^{(0)} = 1.1181$$
$$f_2^{(0)} = 0.12344$$
$$e_3^{(0)} = 0.93665$$
$$f_3^{(0)} = -0.0563$$

The a's and b's are thus
$$a_{23} = -4(0.93665) - 10(-0.0563) = -3.1836$$
$$b_{23} = -4(-0.0563) + 10(0.93665) = 9.5917$$
$$a_{32} = -4(1.1181) - 10(0.12344) = -5.7068$$
$$b_{32} = -4(0.12344) + 10(1.1181) = 10.6873$$

The off-diagonal Jacobian elements are
$$H_{23} = (-3.1836)(0.12344) - (9.5917)(1.1181) = -11.1175$$
$$H_{32} = (-5.7068)(-0.0563) - (10.6873)(0.93665) = -9.6889$$
$$N_{23} = (-3.1836)(1.1181) + (9.5917)(0.12344) = -2.3756$$
$$J_{32} = -[(-5.7068)(0.93665) + (10.6873)(-0.0563)] = 5.9470$$

The powers are now computed as
$$Q_2 = 12.654 + (-11.1175) = 1.5365$$
$$Q_3 = -5(0.93665) - 4(-0.0563) + (-9.6889) + 15(0.93834)^2$$
$$= -0.9397$$
$$P_3 = -4(0.93665) + 5(-0.0563) - (5.9470) + 8(0.93834)^2$$
$$= -2.9313$$

The remaining Jacobian elements are
$$H_{22} = -1.5365 + 12.654 = 11.1175$$
$$H_{33} = +0.9397 + 15(0.93834)^2 = 14.1470$$
$$N_{33} = -2.9313 + 8(0.93834)^2 = 4.1126$$

6.9 The Newton-Raphson Method for Load-Flow Solution

$$L_{33} = -0.9397 + 15(0.93834)^2 = 12.2675$$
$$J_{33} = -2.9313 - 8(0.93834)^2 = -9.9751$$

The error increments are

$$\Delta Q_3 = Q_3^{\text{sch}} - Q_3 = -1 + 0.9397 = -0.0603$$
$$\Delta P_3 = P_3^{\text{sch}} - P_3 = -2 + 2.9313 = 0.9313$$

To find ΔP_2 we still have to find P_2. This is given by

$$P_2 = 5.0616 + N_{23} = 2.6860$$

Thus,

$$\Delta P_2 = P_2^{\text{sch}} - P_2 = 1.7 - 2.6860 = -0.9860$$

As a result of the above calculations, we have

$$\begin{bmatrix} H_{22} & H_{23} & N_{23} \\ H_{32} & H_{33} & N_{33} \\ J_{32} & J_{33} & L_{33} \end{bmatrix} \begin{bmatrix} \Delta \theta_2 \\ \Delta \theta_3 \\ \dfrac{\Delta |V_3|}{|V_3|} \end{bmatrix} = \begin{bmatrix} \Delta P_2 \\ \Delta P_3 \\ \Delta Q_3 \end{bmatrix}$$

Numerically we have to solve the following equation in the increments:

$$\begin{bmatrix} 11.1175 & -11.1175 & -2.3756 \\ -9.6889 & 14.147 & 4.1126 \\ 5.9470 & -9.9751 & 12.2675 \end{bmatrix} \begin{bmatrix} \Delta \theta_2 \\ \Delta \theta_3 \\ \dfrac{\Delta |V_3|}{|V_3|} \end{bmatrix} = \begin{bmatrix} -0.9860 \\ 0.9313 \\ -0.0603 \end{bmatrix}$$

The solution is

$$\Delta \theta_2 = -0.08100864$$
$$\Delta \theta_3 = 0.00029175$$
$$\dfrac{\Delta |V_3|}{|V_3|} = 0.03459649$$

Note that the angles obtained are in radians.

The new estimates of the unknowns are

$$\theta_2^{(1)} = 6.3 + (-0.081)\left(\dfrac{180}{\pi}\right) = 1.6858546°$$

$$|V_3|^{(1)} = 0.93834(1 + 0.0345965) = 0.97080327$$

$$\theta_3^{(1)} = -3.44 + (0.00029175)\left(\dfrac{180}{\pi}\right) = -3.4233189°$$

The procedure is repeated to obtain further new estimates.

The following table gives the results of a few further iterations.

Iteration	θ_2^0	$|V_3|$	θ_3^0
1	1.658546	0.970803	-3.4233189
2	1.399516	0.96655	-3.719
3	1.396215	0.96652	-3.7224
4	1.3962147	0.96652	-3.722415

In order to appreciate the effect of starting estimates on the progress of the iterations, we try the flat start:

$$\theta_2^{(0)} = 0$$
$$|V_3|^{(0)} = 1.00$$
$$\theta_3^{(0)} = 0$$

The following table gives the results of the iterative process:

Iteration	θ_2	$|V_3|$	θ_3
1	1.96	0.9724	-3.2025
2	1.404	0.9666	-3.7158
3	1.3962	0.96652	-3.722414
4	1.3962148	0.96652	-3.7224

Comparing the two processes we note that the first process is better on the second iteration. However, both can be declared successful in reaching the solution after the fourth iteration.

The question of which is the most efficient iterative technique to solve the load-flow problem has resulted in a tremendous number of proposed techniques. It is beyond the scope of this text to outline many of the proposed variations. The Newton-Raphson method has gained a wide acceptability in industry circles, and as a result there are a number of available computer packages that are based on this powerful method and sparsity-directed programming.

SOME SOLVED PROBLEMS

Problem 6-A-1

Consider the simple electric power system shown in Figure 6-15. The load-flow solution for this system can be obtained in a systematic manner without resorting to iterative techniques. It is required to carry out the

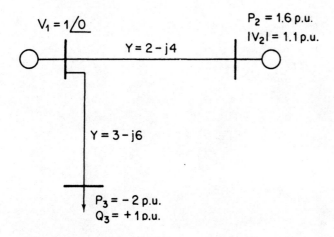

Figure 6-15. System for Problem 6-A-1.

following calculations:

A. Write down the elements of the bus admittance matrix **Y**.
B. Using the active power equation at bus 2 (generator bus), calculate the phase angel θ_2. (Using polar form leads to a simple calculation; note that θ_2 will be in the first quadrant.)
C. Using both the active and reactive power equations at bus 3 (load bus), calculate $|V_3|$ and hence θ_3. (θ_3 should be in the fourth quadrant.)
D. Calculate the active real power generated at bus 1.
E. Find the total active power losses in this system.

Solution

We have

$Y_{11} = 5 - j10$ $Y_{22} = 2 - j4$ $Y_{33} = 3 - j6$
$\quad = 11.18\underline{/-63.43°}$ $\quad = 4.47\underline{/-63.43°}$ $\quad = 6.71\underline{/-63.43°}$

$Y_{12} = -2 + j4$ $Y_{13} = -3 + j6$ $Y_{23} = 0$
$\quad = 4.47\underline{/116.57°}$ $\quad = 6.71\underline{/116.57}$

340 *Analysis of Interconnected Systems*

At bus 2,
$$P_2 = |V_2|[|Y_{12}||V_1|\cos(\theta_2 - \theta_1 - \Psi_{12}) + |Y_{22}||V_2|\cos(\theta_2 - \theta_2 - \Psi_{22})$$
$$+ |Y_{23}||V_3|\cos(\theta_2 - \theta_3 - \Psi_{23})]$$

Thus,
$$1.6 = 1.1[(4.47)(1)\cos(\theta_2 - 116.57) + (4.47)(1.1)\cos(-63.43)]$$

As a result,
$$\cos(\theta_2 - 116.57) = -0.16669$$
$$\theta_2 - 116.57 = \pm 99.59535$$
$$\theta_2 = 216.16 \quad \text{or} \quad 16.97465°$$

Take
$$\theta_2 = 16.97465°$$

For bus 3, we have
$$P_3 = |V_3|[|Y_{31}||V_1|\cos(\theta_3 - \theta_1 - \Psi_{31}) + |Y_{33}||V_3|\cos(-\Psi_{33})]$$

Substituting, we get
$$-2 = |V_3|[(6.71)(1)\cos(\theta_3 - 116.57) + (6.71)|V_3|\cos(63.43)]$$

Thus
$$\frac{-2}{6.71} = |V_3|^2 \cos(63.43) + |V_3|\cos(\theta_3 - 116.57) \qquad (A)$$

Also we have
$$Q_3 = |V_3|[|Y_{31}||V_1|\sin(\theta_3 - \theta_1 - \Psi_{31}) + |Y_{33}||V_3|\sin(-\Psi_{33})]$$
$$1 = |V_3|[6.71\sin(\theta_3 - 116.57) + (6.71)|V_3|\sin(63.43°)]$$
$$\frac{1}{6.71} = |V_3|^2\sin(63.43) + |V_3|\sin(\theta_3 - 116.57) \qquad (B)$$

Equations (A) and (B) are combined to give
$$\left[\frac{2}{6.71} + |V_3|^2\cos(63.43)\right]^2 + \left[\frac{1}{6.71} - |V_3|^2\sin(63.43)\right]^2 = |V_3|^2$$

or
$$|V_3|^4 + \frac{4}{6.71}|V_3|^2[\cos(63.43) - 0.5\sin(63.43)] + \frac{5}{(6.71)^2} = |V_3|^2$$

This gives
$$|V_3|^4 - |V_3|^2 + \frac{1}{9} = 0$$

The solution is
$$|V_3|^2 = \frac{1 \pm \sqrt{5}/3}{2}$$

Take the positive sign:
$$|V_3|^2 = 0.8727$$
The solution for $|V_3|$ is
$$|V_3| = 0.9342$$
As a result, substituting in Eq. (A),
$$\frac{-2}{6.71} = 0.8727\cos(63.43) + 0.9342\cos(\theta_3 - 116.57)$$
Thus,
$$\cos(\theta_3 - 116.57) = -0.7369014307$$
or
$$\theta_3 - 116.57 = \pm 137.468$$
As a result
$$\theta_3 = -20.898$$
We now obtain P_1 as
$$P_1 = |V_1|[|Y_{11}||V_1|\cos(-\Psi_{11}) + |Y_{12}||V_2|\cos(\theta_1 - \theta_2 - \Psi_{12})$$
$$+ |Y_{13}||V_3|\cos(\theta_1 - \theta_3 - \Psi_{13})]$$
$$= 0.9937$$

Problem 6-A-2

Use the Newton-Raphson method to find the roots of the equation
$$F(x) = x^3 - 6x^2 + 11x - 6$$
Assume the following initial guesses:

(1) $x = 0$
(2) $x = 0.5$
(3) $x = 1.5$
(4) $x = 2.5$
(5) $x = 4$

Solution

$$F = x^3 - 6x^2 + 11x - 6$$

Thus
$$F' = 3x^2 - 12x + 11$$

The increments are
$$\Delta = \frac{-F}{F'} = \frac{-x^3 + 6x^2 - 11x + 6}{3x^2 - 12x + 11}$$

The new estimates are given by
$$x^{n+1} = x^n + \Delta$$

(1) $x^{(0)} = 0$
$x^{(1)} = 0.545$
$x^{(2)} = 0.84895$
$x^{(3)} = 0.97467$
$x^{(4)} = 0.99909$
$x^{(5)} = 1.000$

(2) $x^{(0)} = 0.5$
$x^{(1)} = 0.826$
$x^{(2)} = 0.96769$
$x^{(3)} = 0.99854$
$x^{(4)} = 1.000$

(3) $x^{(0)} = 1.5$
$x^{(1)} = 3.00$

(4) $x^{(0)} = 2.5$
$x^{(1)} = 1.00$

(5) $x^{(0)} = 4$
$x^{(1)} = 3.45$
$x^{(2)} = 3.15$
$x^{(3)} = 3.03$
$x^{(4)} = 3.0009$
$x^{(5)} = 3.0000$

Note that the equation has roots at $x = 1$, 2, and 3. We have converged in on the first and third roots, depending on the starting estimate.

Problem 6-A-3

It is required to solve the following two equations in x_1 and x_2 using the Newton-Raphson method:

$$F_1 = x_1^2 + x_2^2 - 4x_1 = 0$$
$$F_2 = x_1^2 + x_2^2 - 8x_1 + 12 = 0$$

A. Find the expressions for the elements of the Jacobian matrix and find the correction increments Δx_1 and Δx_2.

B. Calculate the first five iterations to find estimates of the solution using the following initial guesses.

(i) $x_1 = 2$, $x_2 = 4$
(ii) $x_1 = -5$, $x_2 = -5$
(iii) $x_1 = -0.1$, $x_2 = 1$

Solution

A. The Jacobian elements are as follows:

$$\frac{\partial F_1}{\partial x_1} = (2x_1 - 4) \quad \frac{\partial F_1}{\partial x_2} = (2x_2)$$

$$\frac{\partial F_2}{\partial x_1} = (2x_1 - 8) \quad \frac{\partial F_2}{\partial x_2} = (2x_2)$$

As a result, we have

$$\begin{bmatrix} 2x_1 - 4 & 2x_2 \\ 2x_1 - 8 & 2x_2 \end{bmatrix} \begin{bmatrix} \Delta x_1 \\ \Delta x_2 \end{bmatrix} = \begin{bmatrix} -(x_1^2 + x_2^2 - 4x_1) \\ -(x_1^2 + x_2^2 - 8x_1 + 12) \end{bmatrix}$$

The solution for the increments is

$$\Delta x_1 = 3 - x_1$$

$$\Delta x_2 = \frac{x_1^2 - x_2^2 - 6x_1 + 12}{2x_2}$$

As a result, the new estimates of the solution are given by

$$x_1^{(n+1)} = x_1^{(n)} + (3 - x_1^{(n)}) = 3$$

Thus no matter what the starting point is, the estimate of x_1 is 3.

$$x_2^{(n+1)} = x_2^{(n)} + \frac{x_1^{2(n)} - x_2^{2(n)} - 6x_1^{(n)} + 12}{2x_2^{(n)}}$$

$$= \frac{x_1^{2(n)} + x_2^{2(n)} - 6x_1^{(n)} + 12}{2x_2^{(n)}}$$

B. With $x_1^{(n)} = 3$, we have

$$x_2^{(n+1)} = \frac{x_2^{2(n)} + 3}{2x_2^{(n)}}$$

Iteration	x_1	x_2
0	2	4
1	3	2.5
2	3	1.85
3	3	1.74
4	3	1.7321
5	3	1.73205

Iteration	x_1	x_2
0	-5	-5
1	3.00	-9.2
2	3.00	-4.763
3	3.00	-2.6964
4	3.00	-1.9045
5	3.00	-1.7399
6	3.00	-1.7321
7	3.00	-1.73205

Iteration	x_1	x_2
0	-0.1	$+1$
1	3.00	6.805
2	3.00	3.623
3	3.00	2.225
4	3.00	1.787
5	3.00	1.7329
6	3.00	1.73205

Problem 6-A-4

Apply the Newton-Raphson method to solve the load-flow equations for the system of Example 6-6 with initial guess as shown below:

$$\theta_2^{(0)} = 0$$
$$|V_3|^{(0)} = 0.9$$
$$\theta_3^{(0)} = 0$$

Solution

The following are the iteration results:

Iteration	θ_2	$|V_3|$	θ_3
0	0	0.9	0
1	2.6873	0.9794	-3.2276
2	1.4376	0.9668	-3.698
3	1.3962	0.96652	-3.7224

Note that this guess results in less iterations (one to be precise) than with the flat start.

PROBLEMS

Problem 6-B-1

For the network shown in Figure 6-16 find the equivalent load impedance Z_E in ohms, given that

$$P_E = 88.47 \text{ MW}$$
$$Q_E = 29.22 \text{ MVAR}$$

The voltage magnitude at bus E is

$$|V_E| = 1.031 \times 230 \text{ kV}$$

Problem 6-B-2

Assume that a per unit system is used for the network of problem (6-B-1) such that

$$V_b = 230 \text{ kV}$$
$$P_b = 100 \text{ MW}$$

Convert the load impedance Z_E of problem (6-B-1) to this system and hence reduce the network by eliminating bus E. The line data in per unit are

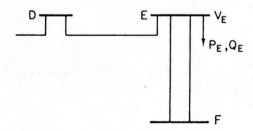

Figure 6-16. Network for Problem (6-B-1).

given by

$$Z_{DE} = (0.725 + j4.045)10^{-2} \quad \text{p.u.}$$
$$Y_{DE} = j0.28233 \quad \text{p.u.}$$

Each of the lines connecting buses E and F have parameters

$$Z = 0.0096 + j0.0953 \quad \text{p.u.}$$
$$Y = j0.18409 \quad \text{p.u.}$$

Assume that lines are represented by a nominal-π circuit.

Problem 6-B-3

For the two bus system shown in Fig. 6-17 with the operating conditions indicated, it is required to find the equivalent load impedance Z_A. Assume the voltage is given in per unit on a 230 kV basis. Power base is 100 MW.

Problem 6-B-4

Find the Thevenin's equivalent at bus B of the two bus system shown in Fig. 6-18. Assume that the transmission line is represented by its nominal-π model and that impedance and admittance values are given in per unit on 230 kV and 100 MVA base. Assume Z_A as obtained in Problem (6-B-3).

Problem 6-B-5

For the three bus system shown in Fig. 6-19 it is required to perform a network reduction such that bus C is eliminated while retaining the neutral

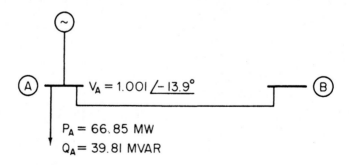

Figure 6-17. Network for Problem (6-B-3).

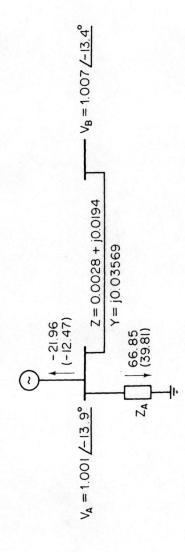

Figure 6-18. Network for Problem (6-B-4).

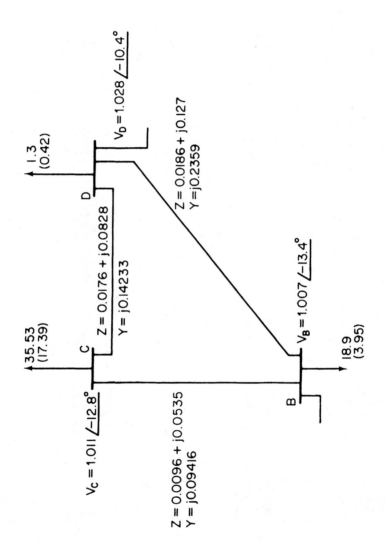

Figure 6-19. Network for Problem (6-B-5).

bus. All loads shown are in MW (MVAR), while voltages and impedances are given in per unit to 230 kV and 100 MVA base.

Problem 6-B-6

For the network shown in Figure 6-20, eliminate buses G and H while retaining ground. Assume same voltage and power bases as in Problem (6-B-5).

Problem 6-B-7

Eliminate buses I, L, and M from the network of Figure 6-21, assuming the same voltage and power base as in Problem (6-B-5)

Problem 6-B-8

Consider the problem of finding the roots of the following function

$$f(x) = x^4 - 81 = 0$$

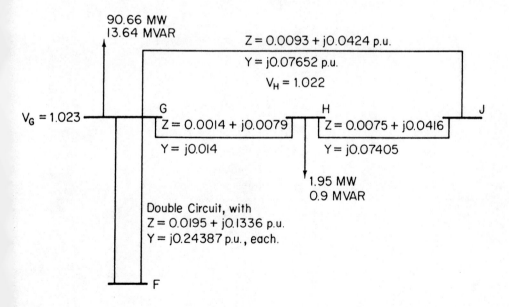

Figure 6-20. Network for Problem (6-B-6).

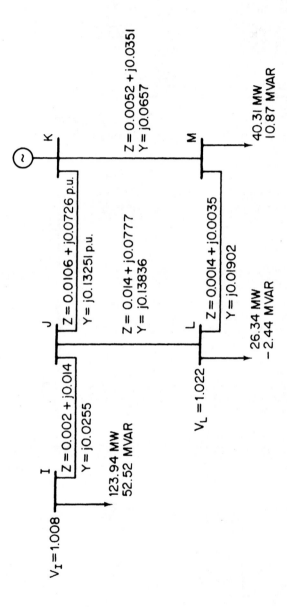

Figure 6-21. Network for Problem (6-B-7).

Apply Newton-Raphson method to find the roots using $x^{(0)} = -10$ and $x^{(0)} = +10$.

Problem 6-B-9

Repeat problem (6-B-8) for the function
$$f(x) = 2x^3 - 9x^2 + 27 = 0$$
with initial guess of $x^{(0)} = -4$ and $x^{(0)} = 4$.

Problem 6-B-10

Consider the function $f(x)$ given by
$$f(x) = x^4 + 5x^3 - 12x^2 - 36x = 0$$
There are 4 roots to this function, namely
$$x^* = -6, -2, 0, \text{ and } 3.$$
The purpose of this problem is to investigate the behavior of Newton-Raphson algorithm as the initial guess of the solution is varied. It may be helpful here to start by an approximate plot of the function. With the aid of a program for Newton-Raphson method verify that
a- with an initial guess of $x^{(0)} = -8$, the iterates converge to $x^* = -6$
b- with an initial guess of $x^{(0)} = -4$, the iterates converge to $x^* = -2$
c- with an initial guess of $x^{(0)} = -1$, the iterates converge to $x^* = 3$
d- with an initial guess of $x^{(0)} = +1$, the iterates converge to $x^* = 0$
e- with an initial guess of $x^{(0)} = +6$, the iterates converge to $x^* = 3$

Problem 6-B-11

Consider the same function of problem (6-B-10), to illustrate the sensitivity of the converged solution to initial guess, verify that with an initial guess of $x^{(0)} = -4.63$, the iterates converge to $x^* = -6$, while for $x^{(0)} = -4.62$, the iterates converge to $x^* = 3$.

Problem 6-B-12

Consider the system of nonlinear equations given by
$$f_1(\mathbf{x}) = x_1^2 + 9x_2^2 - 54x_2 + 45 = 0$$
$$f_2(\mathbf{x}) = 9x_1^2 + x_2^2 - 36 = 0$$

With the aid of a Newton-Raphson iterative algorithm find solutions to the system using the following initial guess values:

a- $x_1(0) = -10$, $\quad x_2(0) = -10$
b- $x_1(0) = -10$, $\quad x_2(0) = 10$
c- $x_1(0) = 10$, $\quad x_2(0) = 10$
d- $x_1(0) = 10$, $\quad x_2(0) = -10$

Problem 6-B-13

Consider the system of nonlinear equations

$$f_1(x_1, x_2) = x_1^2 + 9x_2^2 - 54x_2 + 45 = 0$$
$$f_2(x_1, x_2) = 9x_1^2 + x_2^2 - 1 = 0$$

with the aid of a Newton-Raphson algorithm find solutions to the system using the following initial guess values:

a- $x_1^{(0)} = 3$ $x_2^{(0)} = 3$
b- $x_1^{(0)} = 4$ $x_2^{(0)} = 4$
c- $x_1^{(0)} = -4$ $x_2^{(0)} = 4$
d- $x_1^{(0)} = -3$ $x_2^{(0)} = 3$

Problem 6-B-14

Consider the system of nonlinear equations given by

$$f_1(x_1, x_2) = 9x_1^2 + x_2^2 - 36 = 0$$
$$f_2(x_1, x_2) = x_1^2 + 9x_2^2 + 36x_1 x_2 - 36 = 0$$

with the aid of a Newton-Raphson algorithm find solutions to the system using the following initial guess values:

a- $x_1^{(0)} = 3$ $x_2^{(0)} = -2$
b- $x_1^{(0)} = -1$ $x_2^{(0)} = 3$
c- $x_1^{(0)} = -6$ $x_2^{(0)} = 3$
d- $x_1^{(0)} = 1$ $x_2^{(0)} = -2$

Problem 6-B-15

For the system of example (6-6) with $|V_2| = 1.1$, verify that the load flow solution is

$$\theta_2 = 2.6436° \qquad \theta_3 = -2.96°$$
$$|V_3| = 0.950134$$

Problem 6-B-16

For the system of example (6-6) with $|V_2| = 1.5$, verify that the load flow solution is

$$\theta_2 = -18.28° \qquad \theta_3 = -17.66°$$
$$|V_3| = 1.18471$$

Comment on the value of the calculated Q_2.

Problem 6-B-17

Apply Newton-Raphson method to find the operating conditions for the system of example (6-6) with all data unchanged except for the voltage at bus 2 which is now $|V_2|=1$.

Problem 6-B-18

Repeat problem (6-B-17) for $|V_2|=1.2$

Problem 6-B-19

Repeat problem (6-B-17) for $|V_2|=1.3$

Problem 6-B-20

Repeat problem (6-B-17) for $|V_2|=1.4$

Problem 6-B-21

For the system of example (6-6) with $P_2=1$, verify that the load flow solution is

$$\theta_2 = -11.3346° \qquad \theta_3 = -12.3082°$$
$$|V_3| = 0.942846$$

Problem 6-B-22

For the system of example (6-6), with $P_2=2$, verify that the load flow solution is

$$\theta_2 = 6.20277° \qquad \theta_3 = -0.527812°$$
$$|V_3| = 0.971874$$

Problem 6-B-23

For the system of example (6-6), with $P_2=2.4$ verify that the load flow solution is

$$\theta_2 = 12.2594° \qquad \theta_3 = 3.46851°$$
$$|V_3| = 0.975816$$

Problem 6-B-24

Apply Newton-Raphson method to find the operating conditions for the system of example (6-6) with all data unchanged except for the power generated at bus 2 which is now $P_2=1.2$.

Problem 6-B-25

Repeat problem (6-B-24) for $P_2 = 1.4$

Problem 6-B-26

Repeat problem (6-B-24) for $P_2 = 1.6$

Problem 6-B-27

Repeat problem (6-B-24) for $P_2 = 1.8$

Problem 6-B-28

Repeat problem (6-B-24) for $P_2 = 2.2$

Problem 6-B-29

Apply Newton-Raphson method to find the operating conditions for the system of example (6-6) with all data unchanged except that for the load at bus 3 which is now $P_3 = -3$

Problem 6-B-30

Repeat problem (6-B-29) for $P_3 = -1$

Problem 6-B-31

Apply Newton-Raphson method to find the operating conditions for the system of example (6-6) with $P_3 = -3$ and $Q_3 = 1$

Problem 6-B-32

For the system of example (6-6), assume that
$$|V_3| = 0.894994$$
$$P_3 = -3$$
$$Q_3 = -1$$

Calculate θ_2, θ_3 and Q_2

CHAPTER VII

High-Voltage Direct-Current Transmission

7.1 INTRODUCTION

In 1954, transmission of electric power using high-voltage direct current (HVDC) became a commercial reality with the commissioning of a 20-MW, 90-km link between the Swedish mainland and Gotland Island. Over twenty additional HVDC schemes have become operational throughout the world since that time. In less than 30 years, a typical system's power rating had increased to 1800 MW. Transmission voltages and currents had increased over the same time span from 100 kV to ±533 kV and from 200 to 2000 amperes respectively.

The use of ac for the Swedish underwater link was not possible because the intermediate reactive compensation required for cable transmission was not feasible. The availability of a type of mercury-arc valve invented by U. Lamm in Sweden during World War II made this first major dc underwater link possible. Its reliable and economic operation justified

TABLE 7-1
Major HVDC Systems in Operation (1980)*

System and Year Operational	Capacity, MW	DC Voltage, kV	Length of Route, km	Value Type	Six-Pulse Bridge Rating kV	kA
Moscow-Kashira U.S.S.R. (experimental), 1950	30	200	113 (overhead)	Mercury arc	200	0.15
Gotland, Sweden, 1954	20 (1954) 30 (1970)	100 150	98 (cable)	Mercury arc and thyristor	50	0.2
Cross-Channel, England-France, 1961	160	±100	65 (cable)	Mercury arc	100	0.8
Volgograd-Donbass, U.S.S.R., 1962	750	±400	472 (overhead)	Mercury arc	100	0.9
Konti-skan Denmark-Sweden, 1965	250	±250	95 (overhead) 85 (cable)	Mercury arc	125	1.0
Sakuma I, Japan, 1965	300	2 × 125	0	Mercury arc	125	1.2
New Zealand, 1965	600	±250	567 (overhead) 38 (cable)	Mercury arc	125	1.2
Sardinia-Italy, 1967	200	200	290 (overhead) 120 (cable)	Mercury arc	100	1.0
Vancouver Stage III, Canada, 1968/69	312	+260	41 (overhead) 32 (cable)	Mercury arc	130	1.2
Pacific Intertie Stage I, U.S., 1970	1440	±400	1354 (overhead)	Mercury arc	133	1.8
Eel River, Canada, 1972	320	2 × 80	0	Thyristor, air-cooled and insulated	40	2.0
Nelson River Bipole I, Canada, 1973/75	810 (1973) 1080 (1975)	+150 −300 ±300	890 (overhead)	Mercury arc	150	1.8
Kingsnorth, England, 1975	640	±266	82 (cable, 3 substations)	Mercury arc	133	1.2
Cabora-Bassa, Mozambique-South Africa, 1975	960	±266	1410 (overhead)	Thyristor, oil-cooled and insulated	133	1.8
Nelson River Bipole I,	1620	±450	890 (overhead)	Mercury arc	150	1.8

Project	Capacity (MW)	Voltage (kV)	Length (km)	Type		
Canada, 1975/76 Vancouver Stages IV and V, Canada, 1976/78	552 (1976)	+260 −140 +260	41 (overhead) 32 (cable)	Thyristor, air-cooled and insulated	140	1.72 (winter)
Tri-States, U.S. 1976	792 (1978) (winter) 100	−280 50	0	Thyristor, air-cooled and insulated	25	2.0
Cabora-Bassa	1440 (1977)	+266 −533	1410 (overhead)	Thyristor, oil-cooled and insulated	133	1.8
Mozambique–South Africa, 1977/79 Square Butte, U.S., 1977	1920 (1979) 500	±533 ±250	745 (overhead)	Thyristor, air-cooled and insulated	125	1.0
Skagerrak, Norway-Denmark, 1976/77	500	±250	100 (overhead) 130 (cable)	Thyristor, air-cooled and insulated	125	1.0
EPRI Compact Substation, U.S., 1978	100	100 (400 kV to ground)	0.6 (cable)	Thyristor, freon-cooled and SF$_6$ insulated	50	1.0
CU, U.S., 1978	1000	±400	656 (overhead)	Thyristor, air-cooled and insulated	200	1.25
Inga-Shaba, Zaire, Stage I, 1976	560	±500	1700 (overhead)	Thyristor, air-cooled and insulated	250	0.56
Nelson River Bipole 2, Canada, 1978/81	900 (1978) 1800 (1981)	±250 ±500	920 (overhead)	Thyristor, water-cooled and air-insulated	125	1.8
Shin-Shinano, Japan, 1978	150	125	0	Thyristor, oil-cooled and insulated		
Hokkaido-Honshu, Japan, 1979	150	±250	124 (overhead) 44 (cable)	Thyristor, air-cooled and insulated		

*Source: EPRI Journal

later connections between Sweden and Denmark, between England and France, between the main islands of New Zealand, and between the island of Sardinia and Italy. HVDC has since been used in overland connections in the United States, Canada, England, Japan, U.S.S.R., Zaire, and between Mozambique and South Africa. Table 7-1 shows a listing of major dc projects in operation today. Our purpose in this chapter is to offer an introduction to elements of HVDC transmission.

7.2 MAIN APPLICATIONS OF HVDC

HVDC transmission is advantageous in the following areas of application:

1. For long underwater cable crossings (wider than 32 km). In six of the first seven commercial schemes, submarine cables are the medium of power transfer. The success of the Gotland scheme justified later underwater connections as mentioned earlier. A 25-km submarine cable between New Brunswick and Prince Edward Island was completed in 1977. The initial operation was at 138 kV ac. The design is such that the forecast increase in transmission capacity will be met by HVDC operation at 1200 kV.

2. For long-distance, bulk-power transmission by overhead lines, when the savings in cost of a dc line would more than compensate for the cost of converter stations. For the same power capability, the cost per unit length of a dc line is lower than that of an ac line. In Figure 7-1, we show the comparative costs of ac and dc overhead lines versus distance of transmission. The break-even distance is the abscissa of the intersection of the dc transmission cost with the ac transmission cost. If the transmission distance is longer than the break-even distance, then dc is cheaper than ac. The break-even distance varies with the power transmitted, the transmission voltage, the type of terrain, the cost of equipment, and other factors. This particular aspect will be treated later on in the chapter. Thus dc transmission plays a significant role in situations where it is more economical to generate power at the minemouth, hydrosite, or gaswell and to transmit it electrically to the load center.

3. The dc systems have an inherent short-time overload capacity that can be used for damping system oscillations. Two systems when interconnected by ac lines sustain instability. A dc link interconnecting the two would overcome this difficulty. The Eel River tie, Canada, has operated in this mode for the past several years. The Stegall project in Nebraska was constructed to connect east-west

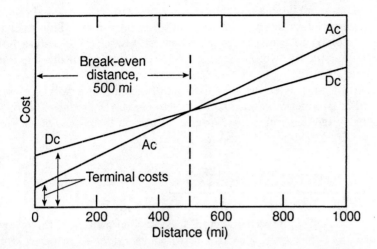

Figure 7-1. Comparative Costs of ac and dc Overhead Lines for Various Distances.

systems in the United States at a point along what might be termed the "electric continental divide."
4. A requirement to provide an intertie between two systems without raising the short-circuit level appreciably can be met by using a HVDC link.
5. Two systems having different frequencies may be tied together through a dc interconnection.
6. For transmission in underground metropolitan cable systems where long distances are involved.

Some Limitations of HVDC Transmission

The lack of HVDC circuit breakers is regarded as a limitation to HVDC transmission. In ac circuits, circuit breakers take advantage of the current zeros occurring twice per cycle. The arc does not restrike between contacts because the design is such that the breakdown strength of the arc path between contacts is increased so rapidly as to enable extinction. Grid control in converter valves on radial lines is used to block the dc temporarily. The realization of multiterminal systems requires the use of HVDC circuit breakers. A number of breaker concepts have been described, and several laboratory prototypes have been developed. With the availability of these commercially, utility planners can proceed with serious consideration of multiterminal HVDC systems.

The reliability and maintenance of converters have been a major problem for dc systems with mercury-arc converters. This difficulty has been resolved in projects using thyristor valves. These valves have little overload capacity, which can present a problem when one bipolar line is involved. In this case it is not possible to meet the requirement of 100 percent half-hour overload capability to take care of a pole outage.

The production of harmonics due to converter operation leads to audio frequency telephone line interference. Filters on both sides of the dc system are required to suppress these harmonics.

7.3 HVDC CONVERTERS

A HVDC converter is simply a controlled switch suitable for HVDC transmission purposes. The converter can conduct in either direction depending on the controlled times of closing and opening as well as on the

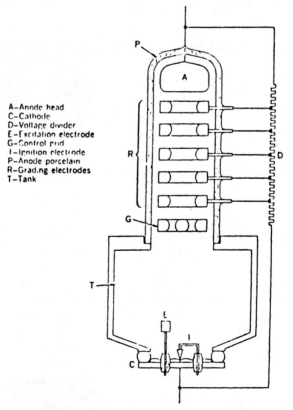

Figure 7-2. A Single-Anode Mercury-Arc Valve.
(Source: IEEE *Spectrum*, August 1966, p. 77.)

circuit's EMF. In general, unidirectional conduction is desired, and devices with this property are called *valves*. The vacuum and vapor or gas-filled tubes with thermionic cathodes, mercury-arc tubes with mercury-pool cathode, and solid-state devices (thyristors) give typical valve characteristics.

The mercury-arc rectifier is a steel vessel that is evacuated (see Figure 7-2). A mercury pool at the bottom serves as the cathode. One or more anodes are also present in the vessel. A small area on the mercury surface is raised to incandescence when an arc is drawn from the surface by some auxiliary means. The spot will emit electrons that will be attracted to the positive anode. The action maintains itself provided that the current flowing to the cathode is maintained above a critical value. The action of the mercury-arc rectifier requires the cathode spot to be produced first by some auxiliary device before the arc can be established. This is done by using a small auxiliary anode. The process is called *ignition*. The use of mercury is advantageous since it is a liquid metal that can easily vaporize and condense, returning by gravity to the cathode without loss of material. Mercury is a good electrical conductor and is easily ionized with a low arc drop.

Voltage drops at the anode and cathode surfaces and in the arc contribute to the loss in the rectifier chamber. The cathode drop produces a loss of 8 watts per ampere of arc current. At the anode, the drop is less than 10 volts. In the arc there is a drop of roughly 10 V/m of arc path. This depends on the conditions of load, temperature, and vacuum. The total anode-cathode voltage is usually called the *arc drop*.

To avoid the concentration of voltage close to the anode during the inverse period, a succession of intermediate or "grading" electrodes is used in the path between the anode and cathode. These are connected to a voltage divider between the anode and cathode. The symbol for a controlled mercury-arc valve is shown in Figure 7-3. Symbols for a controlled valve of any type (mercury-arc or solid-state) are shown in Figure 7-4.

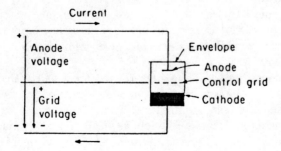

Figure 7-3. Symbol for Controlled Mercury-Arc Valve.

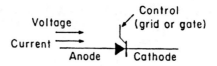

Figure 7-4. Symbol for Controlled Valve.

The top part of the ac half-wave is cut out and transposed without grid control. The dc voltage thus attains a maximum, which, in zero-load condition, is only 5 percent lower than the peak of the phase-to-phase ac voltage wave. Delaying the commutation through grid control displaces the transposed one-third part of the ac voltage wave to yield a section with a lower mean value; thus the dc voltage will be lower. At 90 (electrical phase) degrees delay, the dc voltage will be zero, and at greater delays, the dc voltage polarity will reverse. Since the current must retain the same direction, the power flow also reverses, and the convertor functions as an inverter. This is the principle of inverter operation conversion of direct current into alternating current.

The use of solid-state devices known as *silicon-controlled rectifiers*, or *thyristors*, showed great potential for the production of valves with the required high-current, high-voltage ratings. The thyristor is a special type of diode that in order to start conduction must have not only a positive anode-cathode voltage but also a high enough positive pulse applied to a third electrode called the *gate*. Once conduction starts, the voltage across the thyristor drops to a low value, and the current rises to a value limited by the external circuit only.

The anode-to-cathode resistance of a thyristor in the conducting state is very low; nevertheless, it is not zero and passage of the large anode current produces heat. Cooling is therefore necessary for normal operation. Air, oil, or water-cooling systems are in current use in conjunction with thyristor valves.

To achieve the voltage ratings desired for a valve, thyristors may be strung together in series; increases in the ratings of thyristors reduce the number needed and result in higher efficiency and lower cost. Thyristors 77 mm in diameter and rated as high as 3800 V are available for HVDC converters, and larger thyristors rated at 5000 V or higher are being developed. The refinement of thyristors and thyristor valve design has increased the reliability of converter valves dramatically.

Advantages of Thyristor Valves

The problem of arc-back has plagued mercury-arc rectifier applications. This is essentially the failure of rectifying action, which results in the flow of the electron stream in the reverse direction due to the formation of a cathode spot on an anode. A bypass valve around the bridge circuit was introduced to short-circuit the bridge and permit the current to flow around the problem section. This valve is not necessary in the case of solid-state valves.

With thyristor valves it is possible to tap into any point of the thyristor chain. Consequently it is possible to adopt the 12-pulse bridge configuration. This has a higher rms value of the rectified waveform and a higher harmonic range. In the 6-pulse bridge circuit used with mercury-arc converters, the significant harmonic components are the 5th, 7th, 11th, and 13th, in contrast with only 11th and 13th in the case of the 12-pulse bridge. Thus there is significant saving in cost of filters to reduce these harmonics when thyristor valves are used.

The availability of high-current thyristor devices makes it possible to replace the series-parallel configurations that were necessary to obtain the required current ratings in earlier converter designs. A standard single-bridge converter with several high-voltage devices coupled in parallel is used. This eliminates the need to control series valves. Moreover, there is no need to install the parallel converters at the same site. With a reliable system of communications between sites, it is possible to site each branch of the circuit on its own. Thus an increased system reliability is achieved since the loss of one site does not result in the loss of the whole.

Fiber-optic devices are used for simultaneous triggering of the devices as well as for monitoring purposes. A fiber-optic guide bundle carries the image of a light source at ground potential up to photo detectors that operate the triggering circuit of each thyristor.

7.4 CLASSIFICATION OF DIRECT-CURRENT LINKS

Direct-current links are classified according to the number of conductors used. The *monopolar* link has one conductor (usually of negative polarity) and uses ground or sea return. A monopolar link is shown in Figure 7-5. Two conductors, one positive, the other negative, are used in a *bipolar* link. The neutral points given by the junctions between the converters are grounded at one or both ends. Each terminal of a bipolar link

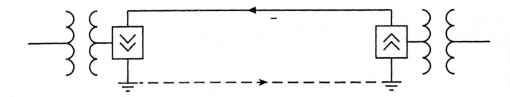

Figure 7-5. A Monopolar Link.

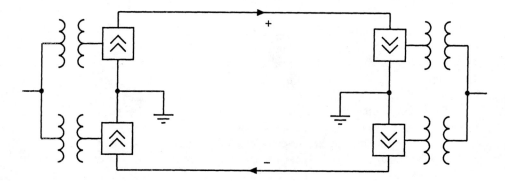

Figure 7-6. A Bipolar Link.

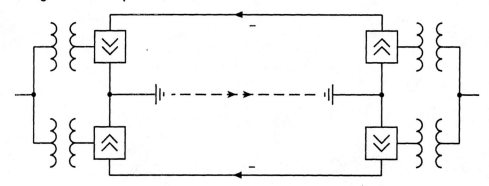

Figure 7-7. A Homopolar Link.

has two converters of equal-rated voltage in series on the dc side. Figure 7-6 shows a bipolar link. The *homopolar* link has two or more conductors, all having the same polarity (usually negative), and always operates with ground return. This is advantageous since on fault the entire converter can be connected to the remaining nonfaulted conductor or conductors. The homopolar link is shown in Figure 7-7.

7.5 SOME ADVANTAGES OF HVDC TRANSMISSION

The primary economy in dc transmission is that only two conductors per circuit are needed rather than the three required for ac. Consequently, dc transmission towers carry less conductor dead weight, and they can be smaller, less costly to fabricate, and easier to erect.

For the same amount of power transmitted over the same size conductors, line losses are smaller with dc than with ac. Neglecting ac skin effect, ac line losses are 33 percent greater than dc line losses. The basis for this comparison is given in the following example.

Example 7-1

Show that the ratio of ac line loss to the corresponding dc loss is 1.33 assuming equal power transfer and equal peak voltages for both options.

Solution

Assume three-phase ac is compared with bipolar dc. The ac power is

$$P_{ac} = 3V_{ph}I_L$$

and the dc power is

$$P_{dc} = 2V_d I_d$$

Now the rms voltage to neutral V_{ph} in ac is related to the maximum or peak value through

$$V_{ph} = \frac{1}{\sqrt{2}} V_{max_{n_{ac}}}$$

For dc, we have

$$V_d = V_{max_{n_{dc}}}$$

Therefore, for equal power transfer,

$$P_{ac} = P_{dc}$$

we have

$$3V_{ph}I_L = 2V_d I_d$$

$$\frac{3}{\sqrt{2}} V_{max_{n_{ac}}} I_L = 2V_{max_{n_{dc}}} I_d$$

Thus assuming

$$V_{max_{n_{ac}}} = V_{max_{n_{dc}}}$$

we conclude that
$$I_d = \frac{3}{2\sqrt{2}} I_L$$
$$= 1.06 I_L$$

The power loss in each case is given by
$$P_{L_{ac}} = 3 I_L^2 R_L$$
$$P_{L_{dc}} = 2 I_d^2 R_L$$

Thus,
$$\frac{P_{L_{ac}}}{P_{L_{dc}}} = \frac{3}{2} \left(\frac{1}{1.06} \right)^2$$
$$= 1.33$$

Note that even though the power transfer is the same, the dc line is simpler and cheaper, with two conductors as opposed to three. This leads to requiring two-thirds of the insulators that would be required for the ac line.

The next example illustrates that if the lines are designed such that the line losses are equal, the voltage level in the dc case will be less than the corresponding ac voltage.

Example 7-2

A new dc transmission system is compared with a three-phase ac system transmitting the same power and having the same losses and size of conductor. Assume that the direct voltage for breakdown of an insulator string is equal to the peak value of the alternating voltage to cause breakdown. Show that the dc line will not only have two conductors instead of three for the ac line, but in addition the insulation level will only be 87 percent of that of the ac line.

Solution

The ac power transmitted is given by
$$P_{ac} = 3 V_{ph} I_L$$
where we assumed unity power factor. The dc power transmitted is
$$P_{dc} = 2 V_d I_d$$
The corresponding power losses are
$$P_{L_{ac}} = 3 I_L^2 R$$
$$P_{L_{dc}} = 2 I_d^2 R$$

7.5 Some Advantages of HVDC Transmission

For equal line losses, we get

$$I_d = \sqrt{\frac{3}{2}}\, I_L$$

For equal transmitted power, we get

$$3V_{ph}I_L = 2V_d I_d$$

Combining the above two results, we conclude that

$$V_d = \sqrt{\frac{3}{2}}\, V_{ph}$$

Assuming that the direct voltage for breakdown of an insulator is equal to the peak value of the alternating voltage to cause breakdown, the insulating level of ac line $= \sqrt{2}\, V_{ph}(k_1)$, and insulation level of dc line $= V_d k_2$ where k_1 and k_2 are multiplying factors. We can assume for simplicity:

$$k_1 = k_2$$

Hence

$$\frac{\text{Direct-current insulation level}}{\text{Alternating-current insulation level}} = \frac{V_d}{\sqrt{2}\, V_{ph}}$$

$$= \frac{\sqrt{3}}{2} = 0.87$$

Thus the dc line will not only have two conductors instead of three (of the same size) for the ac line, but in addition the insulation level will only be 87 percent of that of the ac line.

The previous two examples clearly deal with two extreme cases. In the first, savings in the line loss are achieved while maintaining equal insulation level. The reverse is the case treated in the second example. The following example gives a basis for weighing loss reduction versus insulation level reduction.

Example 7-3

Assume that a design choice calls for a ratio y of the losses in the dc case to losses in the ac alternative. Thus,

$$\frac{P_{L_{dc}}}{P_{L_{ac}}} = y \qquad (7.1)$$

This choice leads to a specific ratio of dc voltage to peak ac voltage. Show that in this case

$$\frac{\text{Insulation level dc}}{\text{Insulation level ac}} = 0.5\sqrt{\frac{3}{y}}$$

Assume equal power transfer in both cases.

Solution

The loss ratio gives

$$\frac{2I_d^2 R_{dc}}{3I_L^2 R_{ac}} = y \qquad (7.2)$$

Assuming equal resistance, we have from Eq. (7.2):

$$\frac{I_d}{I_L} = \sqrt{\frac{3y}{2}} \qquad (7.3)$$

However, since the power transmitted using ac is the same as for the dc, we have for unity power factor

$$3V_{ph}I_{ac} = 2I_d V_d \qquad (7.4)$$

Combining Eqs. (7.3) and (7.4), we thus have

$$\frac{V_d}{V_{ph}} = \sqrt{\frac{3}{2y}}$$

From which,

$$\frac{\text{Insulation level dc}}{\text{Insulation level ac}} = \frac{V_d}{\sqrt{2}\, V_{ph}} = \frac{0.87}{\sqrt{y}} \qquad (7.5)$$

This result is shown graphically in Figure 7-8. The reader is reminded that we are discussing the advantages of HVDC, which is the focus of our attention.

In the event of a single-line fault on a dc line, the remaining conductors will still be functional through the use of ground return. This enables the repair of faulty sections without considerable reduction in service level. The fact that each conductor can act as an independent circuit is a contributing factor to the better reliability of dc transmission lines.

Switching surges on dc lines are lower than those on ac lines. In ac overhead lines, attempts are made to limit them to peak values of two or three times the normal maximum voltage value as opposed to 1.7 times in the dc case. Radio interference and corona losses during foul weather are lower in the dc case than in the ac case.

The ac resistance of a conductor is commonly known to be higher than its dc resistance, due to skin effect.

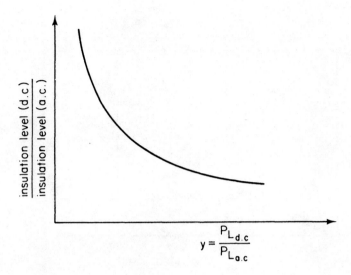

Figure 7-8. Variation of Insulation Ratio with Loss Ratio for Equal Power Transfer.

A dc transmission link has no stability problem. This is contrasted to the ac case where a steady stability limit exists. Operating beyond this limit will mean loss of synchronism as discussed earlier.

Interconnecting systems by a dc link will not increase short-circuit currents of the ac systems nearly as much as interconnection via ac links. This can save on synchronous condenser requirements in the system.

The transient reactance of some hydro plants is abnormally low (to raise the stability limit), necessitating a higher generator cost. This would not be required if dc transmission is used.

One of the most important economies achieved can be appreciated if we observe that using dc transmission, the prime-mover speed need not be confined to correspond to 50 or 60 Hz, but could rather be chosen for best economy.

The reactive power produced by the cable's shunt capacitance greatly exceeds that consumed by the series inductance. This is due to loading below surge impedance level to avoid overheating. In a 60-Hz cable 40–80 km long, the charging current alone equals the rated current. Shunt compensation could theoretically rectify this problem. However, this is difficult to implement in submarine cable applications. Direct-current cables have no such limitations.

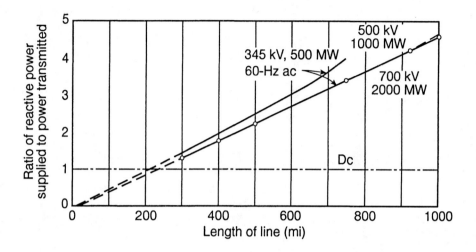

Figure 7-9. Reactive Power Requirements of Long Overhead ac and dc Lines at Full Load as a Function of the Length of Line.

A dc line itself does not require reactive power. The converters at both ends of the line draw reactive power from the ac system connected. This is independent of the line length, in contrast with ac where reactive power consumption varies almost linearly with distance, as shown in Figure 7-9.

The voltage at an open end of a long line is considerably higher than rated. This phenomenon is called the *Ferranti effect* and is a limiting factor for ac lines. The voltage at an open end of a long line presents a special problem when the line is put in service by first connecting it to the main ac system. It is not feasible to close both ends at exactly the same instant. Clearly this difficulty is avoided in dc lines.

7.6 SOME ECONOMIC CONSIDERATIONS

The result of Example 7-3 indicates that a saving (reduction) in line losses results in an increase in the required voltage (insulation level). This is clear by inspection of Figure 7-8 showing the variation of dc insulation level with the losses, assuming $P_{L_{ac}}$ and the ac insulation levels as the reference. A compromise choice can be made based on the economic trade-offs.

7.6 Some Economic Considerations

The annual variable cost of operating a dc line can be assumed to be the sum of two parts. The first is cost of power lost in transmission, and the second is the annual amortized capital cost of line and terminal stations (converters). The cost of dc losses can be expressed in terms of y (the loss ratio) as

$$\text{Cost of dc losses} = By \tag{7.6}$$

where we define B as the annual cost of losses in the ac case. The line and terminal costs vary directly with the required insulation level, which in turn is voltage-dependent. In terms of the loss ratio y, we thus write in the light of Eq. (7.5),

$$\begin{array}{l}\text{Voltage dependent}\\ \text{dc capital costs}\\ \text{amortized annually}\end{array} = \frac{0.87\tilde{A}}{\sqrt{y}} \tag{7.7}$$

The constant $\tilde{A}$ is expressed as

$$\tilde{A} = qA$$

Here we take the voltage-dependent ac capital cost as the basis for comparison:

$$A = \text{Voltage-dependent ac equipment capital cost amortized annually}$$

The factor q is introduced to account for the cost of terminal stations in relation to line costs. As a result of the foregoing assumptions, the total annual costs of operating the dc line are

$$C_{dc} = \frac{0.87\tilde{A}}{\sqrt{y}} + By \tag{7.8}$$

The variation of the costs C_{dc} with loss ratio y is shown in Figure 7-10. The corresponding ac costs are

$$C_{ac} = A + B \tag{7.9}$$

The following example gives the basis for economic choice of y, the loss ratio.

Example 7-4

Show that for minimum total annual cost of the dc line,

Cost of losses = 0.5(cost of voltage-dependent equipment)

Solution

For minimum cost,

$$\frac{\partial C_{dc}}{\partial y} = 0$$

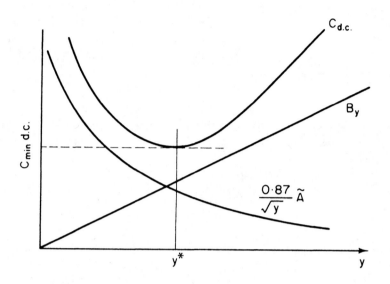

Figure 7-10. Variation of dc Costs with Loss Ratio *y*.

This requires, in view of Eq. (7.8),

$$B - \frac{0.87\tilde{A}}{2(y)^{3/2}} = 0 \qquad (7.10)$$

or

$$By = \frac{1}{2}\left(\frac{0.87}{\sqrt{y}}\right)\tilde{A}$$

which implies that at the optimum

    Cost of losses = 0.5(cost of voltage-dependent equipment) (7.11)

To get a feeling for the numerical implications of the above, we have the following example.

Example 7-5

The annual cost of losses for a dc line is 4×10^6. Assuming that the design is optimal, find the corresponding capital cost of voltage-dependent equipment amortized annually. Calculate the ratio of losses in the dc case to that of the equivalent ac option. Assume that annual cost of the latter is 5×10^6.

Solution

From the previous example, we have by Eq. (7.11),

Annual capital cost of voltage-dependent equipment $= 2By$

But
$$By = 4 \times 10^6$$

Thus the required cost $= 2 \times 4 \times 10^6 = 8 \times 10^6$ dollars per year.

We are also given
$$B = 5 \times 10^6$$
$$y(5 \times 10^6) = 4 \times 10^6$$

Thus the required ratio is
$$y = 0.8$$

Breakeven Between ac and dc

From Example 7-4, at the optimum we have by Eq. (7.10):

$$By^* = \frac{1}{2}\left(0.5\sqrt{\frac{3}{y^*}}\right)\tilde{A}$$

Thus the minimum cost of the dc line in operation is according to Eq. (7.8),

$$C_{\min} = By^* + 0.5\tilde{A}\sqrt{\frac{3}{y^*}}$$

$$= 3By^* \qquad (7.12)$$

But we have from Eq. (7.10),

$$(y^*)^3 = \frac{3\tilde{A}^2}{16B^2}$$

Thus Eq. (7.12) is rewritten as

$$C_{\min} = 3B\left(\frac{3\tilde{A}^2}{16B^2}\right)^{1/3}$$

$$= 1.72(\tilde{A}^2 B)^{1/3} \qquad (7.13)$$

The capital cost of the ac transmission option is given according to Example 7-4's definition of Eq. (7.9) as

$$C_{ac} = A + B \tag{7.14}$$

In order to compare the two options available, we take the difference in costs as the measure. We thus define the savings S as

$$S = C_{ac} - C_{dc} \tag{7.15}$$

This reduces on using Eqs. (7.13) and (7.14) to

$$S = A + B - 1.72 B^{1/3} \tilde{A}^{2/3} \tag{7.16}$$

Defining the constants,

$$p = 1.72 \left(\frac{\tilde{A}}{A}\right)^{2/3} \tag{7.17}$$

$$z = \left(\frac{A}{B}\right)^{1/3} \tag{7.18}$$

Then Eq. (7.16) can be written as

$$S = B(1 - pz^2 + z^3) \tag{7.19}$$

Clearly S can take on negative values, indicating a cutoff value where the dc option is no longer more economic than the corresponding ac alternative.

For illustration purposes it is more convenient to deal with the ratio (S/B), denoted here by $\tilde{S}$:

$$\tilde{S} = \frac{S}{B} \tag{7.20}$$

This has a minimum with respect to the variable z, obtained by setting the derivative equal to zero. Thus

$$\frac{\partial \tilde{S}}{\partial z} = -2pz + 3z^2$$
$$= 0$$

This occurs at a value of z_m:

$$z_m = \tfrac{2}{3} p \tag{7.21}$$

Here the minimum is given by substituting Eq. (7.21) into Eq. (7.20). Thus,

$$\tilde{S}_{min} = 1 - \tfrac{4}{27} p^3 \tag{7.22}$$

The variation of $\tilde{S}_{min}$ with the ratio $(\tilde{A}/A)$ is obtained by substituting in Eq. (7.22) from Eq. (7.17):

$$\tilde{S}_{min} = 1 - \frac{3}{4}\left(\frac{\tilde{A}}{A}\right)^2 \tag{7.23}$$

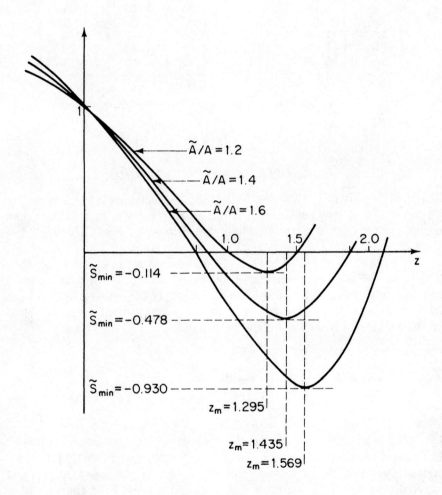

Figure 7-11. Variation of $\tilde{S}$ with z for Different $\tilde{A}/A$ Ratios.

TABLE 7-2
Relevant Points on the $\tilde{S}$ versus z Characteristics

$\tilde{A}/A$	p	z_1	z_2	z_m	$\tilde{S}_m$
1.2	1.942	1.072	1.495	1.295	-0.114
1.4	2.153	0.8810	1.865	1.435	-0.478
1.6	2.353	0.8033	2.133	1.569	-0.930
1.8	2.545	0.7454	2.367	1.697	-1.442
2.0	2.730	0.7021	2.580	1.820	-2.015

Inspection of Eqs. (7.22) and (7.23) shows that for positive values of p:

$$\tilde{S}_{\min} \geq 0, \text{ for } p \leq 1.8899 \text{ or } \frac{\tilde{A}}{A} \leq 1.1547 \quad (7.24)$$

$$\tilde{S}_{\min} < 0 \text{ for } p > 1.8899 \text{ or } \frac{\tilde{A}}{A} > 1.1547 \quad (7.25)$$

The implication of Eq. (7.24) is that $\tilde{S}$ (and consequently S) is positive for all values of z as long as $(\tilde{A}/A)$ is less than 1.1547. From Eq. (7.15), we can conclude that $C_{ac} > C_{dc}$ for all values of z defined by Eq. (7.18) if $\tilde{A}/A < 1.1547$.

For values of $\tilde{A}/A > 1.1547$, $S_{\min}$ and hence a range of S are negative. The value of $\tilde{S}$ will be zero for two values of z denoted by z_1 and z_2 and will be a minimum at z_m. Table 7-2 lists a few values of the ratio $(\tilde{A}/A)$ and the corresponding values of z_1, z_2, z_m, and $\tilde{S}_{\min}$. The general shape of the $\tilde{S}$ versus z variation is shown in Figure 7-11.

7.7 CONVERTER CIRCUITS: CONFIGURATIONS AND PROPERTIES

Groups of valves can be connected in various ways to form a converter. In this section we examine some possible configurations of converter circuits and study the basic properties that are useful in the design of HVDC converter circuits. It is instructive to begin with a study of the case of a single-phase ac power supply. The full-wave rectifier circuit is studied first.

Single-Phase Full-Wave Rectifier

A single-phase, full-wave rectifier circuit is shown in Figure 7-12. A transformer with a center-tapped secondary winding and two valves 1 and 2 are used. The cathodes of the valves are connected through a large smoothing reactor to the dc load.

7.7 Converter Circuits: Configurations and Properties

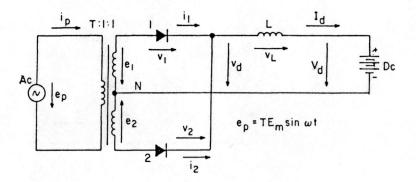

Figure 7-12. Single-Phase, Full-Wave Rectifier Circuit.

The anode voltages to neutral e_1 and e_2 have 180°-phase displacement, and can be expressed as

$$e_1(t) = E_m \sin \omega t$$

$$e_2(t) = E_m \sin(\omega t - 180°)$$

The waveforms are shown in Figure 7-13, starting at $\omega t = -\pi$. The cathodes of the two valves have a common voltage $v_d(t)$. When a valve is conducting, the voltage difference between its anode and cathode (denoted *valve voltage*) is zero. Thus $v_d(t)$ is always equal to the anode voltage of the conducting valve that has a higher anode voltage than the other valve. Thus $v_d(t)$ is made of the positive half-waves of e_1 and e_2. For the first period we have

$$v_d(t) = e_1(t) \qquad (0 \leq \omega t \leq \pi)$$
$$= e_2(t) \qquad (\pi \leq \omega t \leq 2\pi)$$

The reactor L is assumed to filter the ripple in $v_d(t)$; i.e., the voltage V_d on the load side is assumed to be maintained at a constant value due to the reactor action. This is essentially an averaging operation, and thus

$$V_d = v_{d_{av}}$$
$$= \frac{1}{\pi} \int_0^\pi v_d(t) d(\omega t)$$

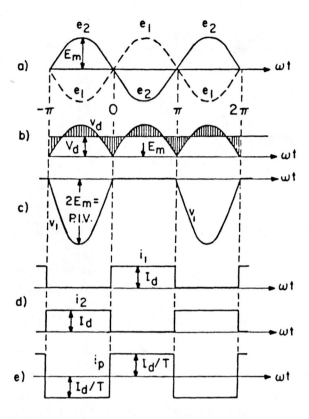

Figure 7-13. Waveforms for Single-Phase, Full-Wave Rectifier.

The direct voltage is thus given by

$$V_d = \frac{1}{\pi} \int_0^\pi E_m \sin \omega t \, d(\omega t)$$

$$= \frac{E_m}{\pi}(-\cos \omega t)_0^\pi$$

$$V_d = \frac{2E_m}{\pi} = 0.6366 E_m \tag{7.26}$$

The inverse relation is

$$E_m = 1.5708 V_d \tag{7.27}$$

Peak Inverse Voltage (PIV)

Kirchhoff's voltage law applied to the loop, including the two valves and the transformer secondary, leads to the following relation:

$$v_1 - v_2 = e_1 - e_2$$

Ideal valve characteristics are assumed so that when valve 1 is conducting, $v_1 = 0$. Thus $e_2 - e_1$ appears across valve 2. Similarly, $e_1 - e_2$ appears across

7.7 Converter Circuits: Configurations and Properties

valve 1 when valve 2 is conducting. The voltage v_1 is shown in Figure 7-13(c); its peak is $2E_m$ as evident from the figure. This value is important in identifying the voltampere rating of the valve and is commonly referred to as the *peak inverse voltage* (PIV). Thus for a single-phase full-wave rectifier, we have

$$\text{PIV} = 2E_m = \pi V_d$$
$$= 3.1416 V_d \qquad (7.28)$$

Peak-to-Peak Ripple (PPR)

The voltage $v_1(t)$ varies from 0 to E_m as evident from inspection of the voltage waveform. We thus have the peak-to-peak ripple (PPR) defined as the difference between the highest and lowest values of the output waveform given for this circuit by

$$\text{PPR} = E_m$$

In terms of the dc voltage, this is then

$$\text{PPR} = 1.5708 V_d \qquad (7.29)$$

The PPR is a useful measure of the quality of the dc voltage.

Value Current Relationships

Each valve conducts during the one-half cycle when the associated anode voltage is the higher of the two anode voltages. The valve current wave is a rectangular pulse of height I_d and duration π. Thus

$$i_1(t) = I_d \qquad (0 \leq \omega t \leq \pi)$$
$$= 0 \qquad (\pi \leq \omega t \leq 2\pi)$$

The average valve current is

$$I_{av} = \frac{\int_0^{2\pi} i_1(t) d(\omega t)}{2\pi}$$
$$= \frac{\int_0^{\pi} I_d d(\omega t)}{2\pi}$$
$$I_{av} = \frac{I_d}{2} \qquad (7.30)$$

The effective valve current is

$$I_{eff}^2 = \frac{\int_0^{2\pi} i_1^2(t) d(\omega t)}{2\pi}$$
$$= \frac{\int_0^{\pi} I_d^2 d(\omega t)}{2\pi}$$
$$= \frac{I_d^2}{2}$$

Thus
$$I_{eff} = \frac{I_d}{\sqrt{2}} \qquad (7.31)$$

Transformer Current Relationships

The transformer secondary currents are the same as the valve currents. The MMF of the secondary winding is proportional to $i_1 - i_2$ and has an average of zero; thus no dc component of the MMF exists. Hence there is no tendency to saturate the core. If the turns ratio of the transformer is T, then the primary MMF is $Ti_p = i_1 - i_2$. The primary current i_p is thus

$$i_p(t) = \frac{I_d}{T} \qquad (0 \leqslant \omega t \leqslant \pi)$$

$$= \frac{-I_d}{T} \qquad (\pi \leqslant \omega t \leqslant 2\pi)$$

The effective value of this wave is

$$I_{p_{eff}} = \frac{I_d}{T} \qquad (7.32)$$

VA Rating of Valve

The voltampere rating of a valve is defined as the product of its average current and its peak inverse voltage (PIV). In our case, the average current is ($I_d/2$), and the PIV is given by Eq. (7.28). Thus,

$$VA_v = \pi V_d \left(\frac{I_d}{2} \right)$$

$$= \frac{\pi}{2} (P_d) \qquad (7.33)$$

where the dc power delivered is

$$P_d = V_d I_d \qquad (7.34)$$

VA Rating of Transformer

The voltampere rating of a transformer winding is the product of its rms voltage and rms current. For each half of the secondary winding this is obtained from

$$i_{1_{rms}} = \sqrt{\frac{(I_d)^2}{2}} = 0.707 I_d \qquad (7.35)$$

and

$$\begin{aligned} e_{1_{rms}} &= 0.707 E_m \\ &= (0.707)(1.5708 V_d) \\ &= 1.1107 V_d \end{aligned} \qquad (7.36)$$

Observe that the rms value is the same as the effective value. The voltampere rating of the whole winding is thus given by

$$VA_{t_s} = (2)(0.707I_d)(1.1107V_d)$$

$$VA_{t_s} = 1.5708P_d \tag{7.37}$$

For the primary winding we have

$$e_{p_{rms}} = 0.707TE_m$$
$$= 1.1107TV_d \tag{7.38}$$

and

$$i_{p_{rms}} = \frac{I_d}{T} \tag{7.39}$$

Thus the voltampere rating for the primary winding is obtained as

$$VA_{t_p} = (1.1107TV_d)\left(\frac{I_d}{T}\right)$$

or

$$VA_{t_p} = 1.1107P_d \tag{7.40}$$

Pulse Number p

The number of cycles of ripple (pulsations) of direct voltage for every cycle of alternating voltage is called the *pulse number*. For the single-phase, full-wave rectifier we have

$$p = 2$$

as can be verified from inspection of Figure 7-13.

Three-Phase Converters

In major bulk-power applications, three-phase circuits are preferred to single-phase ones. From the HVDC application point of view, we prefer three-phase arrangements since the ripple of the direct voltage is smaller in magnitude and higher in frequency than the corresponding values for the single-phase case, as will be seen in the following analysis.

Three-Phase One-Way Circuit

The one-way circuit is the simplest three-phase converter circuit and serves as a step in illustrating other connections. The arrangement is shown in Figure 7-14.

The voltages e_a, e_b, and e_c are balanced, and the anode voltages with respect to N are equal to the corresponding transformer secondary voltages.

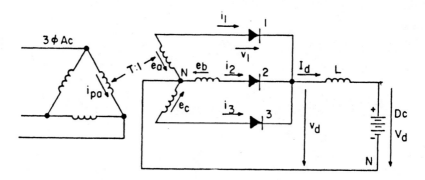

Figure 7-14. Three-Phase One-Way Circuit Configuration.

We thus have

$$e_a(t) = E_m \cos(\omega t)$$
$$e_b(t) = E_m \cos(\omega t - 120°)$$
$$e_c(t) = E_m \cos(\omega t + 120°)$$

The cathodes of the three valves have a common voltage to neutral $v_d(t)$. As in the previous rectifier circuit, $v_d(t)$ will take on the value of the highest of the anode voltages e_a, e_b, and e_c. Inspection of the waveforms in Figure 7-14 reveals that $e_a(t)$ is higher in value than $e_b(t)$ and $e_c(t)$ for the range

$$\left(-\frac{\pi}{3} \leqslant \omega t \leqslant \frac{\pi}{3}\right)$$

and therefore $v_d(t) = e_a(t)$ over that range. Similarly, we conclude that $v_d(t) = e_b(t)$ over the range

$$\left(\frac{\pi}{3} \leqslant \omega t \leqslant \pi\right)$$

and $v_d(t) = e_c(t)$ over the range

$$\left(\pi \leqslant \omega t \leqslant \frac{5\pi}{3}\right)$$

7.7 Converter Circuits: Configurations and Properties

Thus $v_d(t)$ is the upper envelope of the voltages e_a, e_b, and $e_c(t)$.
The direct voltage is the average of $v_d(t)$, and we thus have

$$V_d = \left(\frac{1}{\frac{2\pi}{3}}\right) \int_{-\pi/3}^{\pi/3} (E_m)\cos(\omega t)\, d(\omega t)$$

$$= \frac{3E_m}{2\pi}\left(2\sin\frac{\pi}{3}\right)$$

$$= \frac{3\sqrt{3}}{2\pi} E_m$$

This reduces to

$$V_d = 0.827 E_m \tag{7.41}$$

with the inverse relation

$$E_m = 1.2092 V_d \tag{7.42}$$

Peak Inverse Voltage

As an example, consider the voltage across valve 1. In the range

$$\left(-\frac{\pi}{3} \leq \omega t \leq \frac{\pi}{3}\right)$$

the valve is conducting and hence $v_1(t) = 0$ in this range. Beyond that for the range

$$\left(\frac{\pi}{3} \leq \omega t \leq \pi\right)$$

valve 2 conducts and the voltage across valve 1 is

$$v_1(t) = e_a(t) - e_b(t)$$

The situation in the range

$$\left(\pi \leq \omega t \leq \frac{5\pi}{3}\right)$$

is such that valve 3 conducts and

$$v_1(t) = e_a(t) - e_c(t)$$

The voltage waveform $v_1(t)$ is shown in Figure 7-15(b).
For the range

$$\left(\frac{\pi}{3} \leq \omega t \leq \pi\right)$$

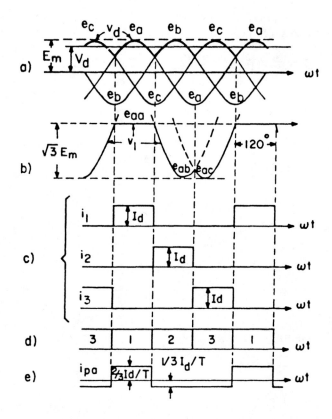

Figure 7-15. Waveforms for Three-Phase One-Way Circuit.

we have

$$v_1(t) = E_m[\cos(\omega t) - \cos(\omega t - 120°)]$$
$$= \sqrt{3}\, E_m \cos(\omega t + 30°)$$

Clearly the absolute value of the minimum of $v_1(t)$ is $\sqrt{3}\, E_m$. Thus we conclude that the peak inverse voltage is given by

$$\text{PIV} = \sqrt{3}\, E_m$$

In terms of the dc voltage,

$$\text{PIV} = (\sqrt{3})(1.2092) V_d$$

or

$$\text{PIV} = 2.0944 V_d \tag{7.43}$$

7.7 Converter Circuits: Configurations and Properties

Peak-to-Peak Ripple

With reference to Figure 7-15(b), the PPR is the difference between the voltages at $t=0$ and $t=\pi/3\omega$. Thus,

$$\text{PPR} = e_a(0) - e_a\left(\frac{\pi}{3}\right)$$
$$= E_m - E_m \cos\left(\frac{\pi}{3}\right)$$
$$= 0.5 E_m$$

In terms of the dc voltage, this is given by

$$\text{PPR} = 0.5(1.2092) V_d$$

or

$$\text{PPR} = 0.6046 V_d \qquad (7.44)$$

This is clearly lower in value than the PPR for a single-phase circuit as is evident by inspection of Eq. (7.29).

Valve Current Relationships

Each valve conducts during the one-third cycle when the associated anode voltage is the highest. The wave of valve current is a rectangular pulse of height I_d and duration $2\pi/3$. For example,

$$i_1(t) = I_d \qquad \left(-\frac{\pi}{3} \leq \omega t \leq \frac{\pi}{3}\right)$$
$$= 0 \qquad \left(\frac{\pi}{3} \leq \omega t \leq \frac{5\pi}{3}\right)$$

The average valve current is thus

$$I_{av} = \frac{\int_{-\pi/3}^{5\pi/3} i_1(t) d(\omega t)}{2\pi}$$
$$= \frac{\int_{-\pi/3}^{\pi/3} I_d d(\omega t)}{2\pi}$$
$$= \frac{\frac{2\pi I_d}{3}}{2\pi}$$

Thus

$$I_{av} = \frac{I_d}{3} \qquad (7.45)$$

The effective valve current is obtained as follows:

$$I_{eff}^2 = \frac{\int_{-\pi/3}^{5\pi/3} i_1^2(t)d(\omega t)}{2\pi}$$

$$= \frac{\int_{-\pi/3}^{\pi/3} I_d^2 d(\omega t)}{2\pi}$$

$$= \frac{I_d^2}{3}$$

Thus,

$$I_{eff} = \frac{I_d}{\sqrt{3}} \qquad (7.46)$$

Transformer Current Relationships

The transformer secondary winding current is the same as the valve's. The primary side does not contain a dc component. Therefore, the average of the primary current is zero. The expression for the primary current to satisfy this requirement is

$$i_{pa}(t) = \frac{i_1(t) - I_{av}}{T}$$

$$= \frac{i_1(t) - \left(\frac{I_d}{3}\right)}{T}$$

This is further given by

$$i_{pa}(t) = \frac{I_d - \left(\frac{I_d}{3}\right)}{T}$$

$$= 2\left(\frac{I_d}{3T}\right) \qquad \left(-\frac{\pi}{3} \leq \omega t \leq \frac{\pi}{3}\right)$$

$$i_{pa}(t) = \frac{0 - \left(\frac{I_d}{3}\right)}{T}$$

$$= \frac{-I_d}{3T} \qquad \left(\frac{\pi}{3} < \omega t \leq \frac{5\pi}{3}\right)$$

7.7 Converter Circuits: Configurations and Properties

The effective value is given by

$$I_{pa_{eff}}^2 = \frac{\int_{-\pi/3}^{\pi/3}\left(\frac{2I_d}{3T}\right)^2 d(\omega t) + \int_{\pi/3}^{5\pi/3}\left(\frac{I_d}{3T}\right)^2 d(\omega t)}{2\pi}$$

$$= \frac{\frac{4I_d^2}{9T^2}\left(\frac{2\pi}{3}\right) + \frac{I_d^2}{9T^2}\left(\frac{4\pi}{3}\right)}{2\pi}$$

$$= \frac{2I_d^2}{9T^2}$$

Thus the effective value of the transformer primary current is

$$I_{pa_{eff}} = \frac{\sqrt{2}\,I_d}{3T}$$

$$= \frac{0.4714 I_d}{T} \tag{7.47}$$

VA Rating of Valve

The voltampere rating of the valve is obtained using the definition

$$VA_v = I_{av}(\text{PIV})$$

Thus substituting for I_{av} and the peak inverse voltage in terms of the direct-current quantities, we get

$$VA_v = \left(\frac{I_d}{3}\right)(2.0944)V_d$$

$$= \left(\frac{2.0944}{3}\right)P_d \tag{7.48}$$

where as before

$$P_d = V_d I_d$$

VA Rating of Transformer

The transformer secondary voltage has an effective value of

$$E_{eff_s} = \frac{E_m}{\sqrt{2}}$$

$$= \frac{1.2092 V_d}{\sqrt{2}}$$

$$= 0.855 V_d \tag{7.49}$$

The effective value of the transformer's secondary winding's current is

$$I_{eff_s} = \frac{I_d}{\sqrt{3}}$$

Thus the voltampere rating of the secondary is

$$VA_{t_s} = (3)(0.855V_d)\left(\frac{I_d}{\sqrt{3}}\right)$$

$$= 1.481 V_d I_d$$

$$= 1.481 P_d \qquad (7.50)$$

The primary voltage has an effective value of

$$E_{eff_p} = 0.855 T V_d \qquad (7.51)$$

The effective value of the primary current is

$$I_{pa_{eff}} = \frac{0.4714 I_d}{T}$$

Therefore we have for the VA rating of the transformer's primary,

$$VA_{tp} = (3)(0.855 T V_d)\left(\frac{0.4714 I_d}{T}\right)$$

As a result,

$$VA_{tp} = 1.2091 P_d \qquad (7.52)$$

Pulse Number

The pulse number is the number of cycles of ripple of v_d per cycle of the alternating voltage. This is given by 3 in the three-phase one-way circuit.

Three-Phase, Two-Way (Graetz) Circuit

The three-phase, two-way circuit arrangement is shown in Figure 7-16. The transformer's secondary windings feed groups of three valves each. Each group may feed a separate dc load. If the two loads are balanced, the neutral (shown dashed) may be omitted.

The cathodes of the upper group of valves (1, 3, 5) are connected to the anodes of the lower group. The common potential of the cathodes of these valves, $v_{du}(t)$, is equal to the most positive anode voltage.

The common potential of the anodes of the lower group of valves (2, 4, 6), $v_{dl}(t)$, is equal to the most negative cathode voltage.

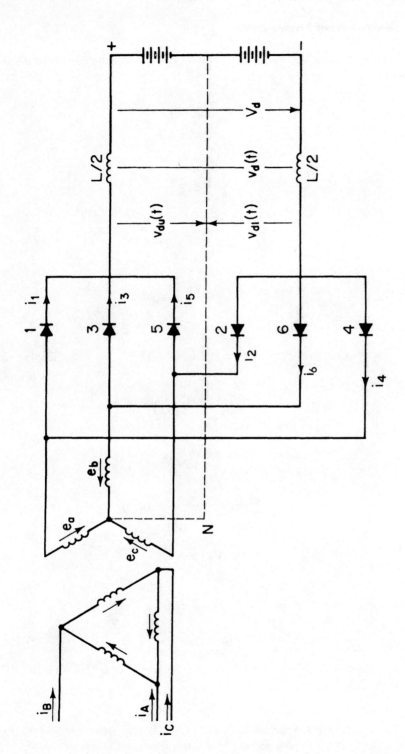

Figure 7-16. Three-Phase, Two-Way (Graetz) Circuit Configuration.

Assuming that the secondary transformer voltages are given by a balanced set, then

$$e_a(t) = E_m \cos \omega t$$
$$e_b(t) = E_m \cos(\omega t - 120°)$$
$$e_c(t) = E_m \cos(\omega t + 120°)$$

For the range

$$\left(-\frac{\pi}{3} \leqslant \omega t \leqslant \frac{\pi}{3}\right)$$

e_a is the most positive voltage, and hence valve 1 conducts. Therefore,

$$v_{du}(t) = e_a(t) = E_m \cos \omega t \qquad \left(-\frac{\pi}{3} \leqslant \omega t \leqslant \frac{\pi}{3}\right)$$

For the range

$$\left(0 \leqslant \omega t \leqslant \frac{2\pi}{3}\right)$$

e_c is the most negative, and hence valve 2, conducts. Therefore,

$$v_{dl}(t) = e_c(t) = E_m \cos(\omega t + 120°) \qquad \left(0 \leqslant \omega t \leqslant \frac{2\pi}{3}\right)$$

The output voltage is

$$v_d(t) = v_{du}(t) - v_{dl}(t)$$

For the range

$$0 \leqslant \omega t \leqslant \frac{\pi}{3}$$

we have

$$v_d(t) = E_m [\cos \omega t - \cos(\omega t + 120°)]$$
$$= \sqrt{3} \, E_m \cos(\omega t - 30°)$$

The direct voltage V_d is therefore given by

$$V_d = \frac{1}{\frac{\pi}{3}} \int_0^{\pi/3} \sqrt{3} \, E_m \cos(\theta - 30°) \, d\theta$$

$$= \frac{3\sqrt{3}}{\pi} (E_m) \qquad (7.53)$$

Note that the direct voltage here is double that for the single-way circuit. The inverse relation is

$$E_m = 0.6046 V_d \qquad (7.54)$$

The waveforms are shown in Figure 7-17. The transformer secondary line-to-neutral voltages are shown in Figure 7-17(a). These are also the

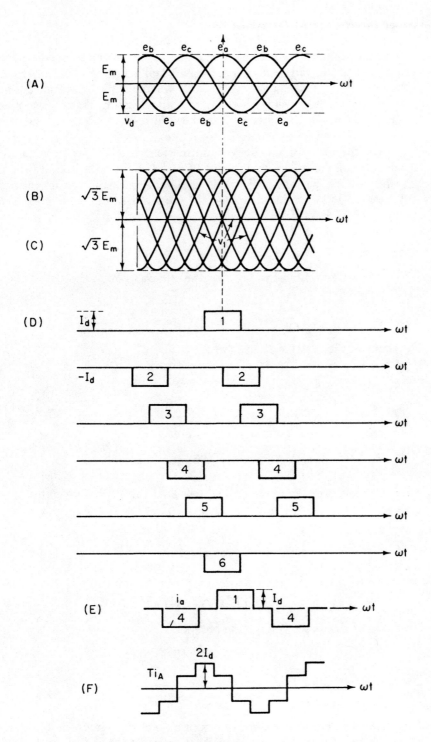

Figure 7-17. Waveforms for the Graetz Circuit.

voltages of the anodes of the lower group of valves and of the cathodes of the upper group, all with respect to neutral point N. The difference in ordinates between the upper and lower envelopes is the instantaneous direct voltage on the valve side of the smoothing reactor. This is replotted in Figure 7-17(b) as the envelope of the line-to-line voltages. The voltage across valve 1 is also shown in Figure 7-17(c).

On the basis of the above information we can now proceed to the evaluation of the design parameters of interest. We will evaluate the peak inverse voltage, peak-to-peak ripple, and valve and transformer ratings. In all cases it will become evident that the Graetz circuit is more efficient than the two previous circuits.

Peak Inverse Voltage

An analysis similar to that for the one-way circuit can be carried out. This gives

$$\text{PIV} = \sqrt{3}\, E_m$$

In terms of the direct voltage, this is

$$\text{PIV} = \frac{\pi}{3} V_d$$
$$= 1.047 V_d \tag{7.55}$$

This is the same as for the one-way circuit.

Peak-to-Peak Ripple

Inspection of Figure 7-17(b) reveals that the PPR is given by

$$\text{PPR} = v_d\left(\frac{\pi}{6}\right) - v_d(0)$$
$$= \sqrt{3}\, E_m(\cos 0 - \cos 30°)$$

As a result,

$$\text{PPR} = 0.2321 E_m$$

In terms of the direct voltage, we obtain

$$\text{PPR} = (0.2321)(0.6046) V_d$$

or

$$\text{PPR} = 0.1403 V_d \tag{7.56}$$

We observe here that the peak-to-peak ripple of this circuit is considerably less than that for the one-way circuit.

7.7 Converter Circuits: Configurations and Properties

Valve Current Relationships

The load current is always carried by two valves in series, one from the upper half bridge and one from the lower. Each valve conducts for one-third cycle, as in the one-way circuit. Commutation in one group, however, is staggered with respect to commutation in the other group. Considering both groups, commutation occurs every one-sixth cycle (60°). In Figure 7-16, as well as in the diagrams of other converter circuits, the valves are numbered in the order in which they fire (begin to conduct). Commutation occurs from valve 1 to valve 3, then from 2 to 4, 3 to 5, 4 to 6, 5 to 1, and 6 to 2. The current waveforms are shown in Figure 7-17(d).

As before for the one-way circuit, we conclude that the average valve current is

$$I_{av} = \frac{I_d}{3} \tag{7.57}$$

The effective value is

$$I_{eff} = \frac{I_d}{\sqrt{3}} \tag{7.58}$$

Transformer Current Relationships

The current in each phase of the Y-connected secondary windings is the difference of the currents of two valves, the numbers of which differ by 3; for example, $i_a = i_1 - i_4$—see Figure 7-17(e).

$$i_a(t) = I_d \qquad \left(-\frac{\pi}{3} \leq \omega t \leq \frac{\pi}{3}\right)$$

$$= 0 \qquad \left(\frac{\pi}{3} \leq \omega t \leq \frac{2\pi}{3}\right)$$

$$= -I_d \qquad \left(\frac{2\pi}{3} \leq \omega t \leq \frac{4\pi}{3}\right)$$

The effective value is given by

$$I_{a_{eff}}^2 = \frac{\int_{-\pi/3}^{\pi/3} I_d^2 \, d(\omega t) + \int_{2\pi/3}^{4\pi/3} I_d^2 \, d(\omega t)}{2\pi}$$

$$= \frac{2 I_d^2}{3}$$

As a result,

$$I_{a_{eff}} = \sqrt{\frac{2}{3}} \, I_d$$

$$= 0.8165 I_d \tag{7.59}$$

VA Rating of Valve

The voltampere rating of the valve is

$$VA_v = I_{av}(\text{PIV})$$
$$= \left(\frac{I_d}{3}\right)(1.047 V_d)$$

Thus,

$$VA_v = 0.349 P_d \qquad (7.60)$$

VA Rating of Transformer

The effective line-to-neutral secondary voltage is

$$E_{eff_s} = \frac{E_m}{\sqrt{2}}$$
$$= 0.4275 V_d \qquad (7.61)$$

The effective value of the phase secondary current from Eq. (7.59) is

$$I_{eff} = 0.8165 I_d$$

Thus the voltampere rating of the secondary is

$$VA_{t_s} = (3)(0.4275 V_d)(0.8165 I_d)$$

This reduces to

$$VA_{t_s} = 1.0472 P_d \qquad (7.62)$$

The rating of the primary winding is the same as that for the secondary.

Pulse Number

The number of pulsations of v_d ripple for one alternating voltage cycle is clearly six, which is the pulse number for this circuit. This is double that for the one-way, three-phase circuit.

Other Converter Circuits

Additional converter circuit configurations result from rearranging valve groups. The Graetz circuit can be considered as a series combination of two one-way circuits where one group of valves has a common anode connection and the other group has a common cathode connection. Another circuit referred to as the *cascade circuit* is obtained by having both groups with a common cathode connection each. The transformer secondary windings are connected in a double Y with 180° phase shift between the two Y connections. This circuit is analyzed in Problem 7-A-6.

A third alternative is to connect the two groups in parallel on the dc side. Since the pulse ripple of the two groups is staggered, an indirect parallel connection is used. Here the two similar poles of the groups are connected in common to the corresponding pole of the dc line. The opposite poles are connected through an autotransformer (interphase transformer), whose center tap is connected to the other pole of the dc line. This circuit is analyzed in Problem 7-A-7. The interphase transformer may be omitted to result in the six-phase diametrical circuit treated in Problem 7-A-8. Instead of using two three-phase circuits, a cascade of three single-phase rectifiers may be considered. This alternative is discussed in connection with Problem 7-A-9.

All of the circuits discussed above utilize a three-phase supply with six valves and have a pulse number of six. Table 7-3 lists the pertinent properties of these circuits. The table shows that the most advantageous circuit is the Graetz configuration. It is characterized by a low peak inverse voltage, a low transformer voltampere rating for both the primary and secondary sides, and a low valve voltampere rating. It should also be noted that the transformer connection is the simplest in the case of the Graetz circuit. Due to all of the above advantages, this circuit is commonly referred to as the *three-phase bridge converter circuit* and is the most commonly used one for HVDC applications.

TABLE 7-3

Comparison of Converter Circuits

	Graetz Circuit	Cascade of 2-3 Phases	Y-Y Interphase	Six-Phase Diametrical	Cascade of 3 Single-Phases
DC ripple voltage	$0.140V_d$	$0.140V_d$	$0.140V_d$	$0.140V_d$	$0.140V_d$
Peak inverse voltage	$1.047V_d$	$1.047V_d$	$2.094V_d$	$2.094V_d$	$1.047V_d$
Transformer secondary rms voltage	$0.428V_d$	$0.428V_d$	$0.855V_d$	$0.740V_d$	$0.370V_d$
Peak valve current	$1.000I_d$	$1.000I_d$	$0.500I_d$	$1.000I_d$	$1.000I_d$
Average valve current	$0.333I_d$	$0.333I_d$	$0.167I_d$	$0.167I_d$	$0.500I_d$
Transformer secondary rms current	$0.816I_d$	$0.577I_d$	$0.289I_d$	$0.4081I_d$	$0.707I_d$
All valves VA	$2.094P_d$	$2.094P_d$	$2.094P_d$	$2.094P_d$	$3.142P_d$
Tranformer primary VA	$1.047P_d$	$1.047P_d$	$1.047P_d$	$1.283P_d$	$1.111P_d$
Transformer secondary VA	$1.047P_d$	$1.481P_d$	$1.481P_d$	$1.814P_d$	$1.571P_d$

7.8 ANALYSIS OF THE THREE-PHASE BRIDGE CONVERTER

Having established that the Graetz circuit is the most useful for HVDC application, we now analyze the performance of the circuit that we will refer to as the three-phase bridge converter. A number of assumptions are made here. The first is the familiar balanced steady-state sinusoidal operation of the three-phase ac side. The direct current is assumed constant and ripple-free. Ideal valves are assumed so that the forward resistance is zero and the backward (inverse) resistance is infinite. The last assumption stipulates that the valves ignite at equal intervals of one-sixth of the ac cycle.

We assume the following variation of the source voltages with time:

$$e_a(t) = E_m \cos(\omega t + 60°)$$
$$e_b(t) = E_m \cos(\omega t - 60°)$$
$$e_c(t) = E_m \cos(\omega t - 180°)$$

Consequently, the line-to-line voltages are

$$e_{ac}(t) = \sqrt{3}\, E_m \cos(\omega t + 30°)$$
$$e_{ba}(t) = \sqrt{3}\, E_m \cos(\omega t - 90°)$$
$$e_{cb}(t) = \sqrt{3}\, E_m \cos(\omega t - 210°)$$

These six voltage waves are shown in Figure 7-18.

Let us consider the case of two valves conducting, as shown in Figure 7-19 with the nonconducting valves omitted. The instantaneous direct voltage V_d across the bridge on the valve side of the dc reactor consists of 60° arcs of the alternating line-to-line voltages, shown by the shaded area A_0 in Figure 7-18 for the period in which valves 1 and 3 conduct. The average direct voltage V_{do} is found by integrating the instantaneous voltages over any 60° period. For ωt replaced by θ, we have

$$V_{do} = \frac{3}{\pi} \int_{-\pi/3}^{0} e_{ac}\, d\theta = \frac{3}{\pi} \int_{-60°}^{0} \sqrt{3}\, E_m \cos(\theta + 30°)\, d\theta$$

Thus performing the integration required, we get

$$V_{do} = \frac{3\sqrt{3}\, E_m}{\pi} = 1.65 E_m \tag{7.63}$$

where E_m is the maximum value of the ac phase voltage. Since we normally

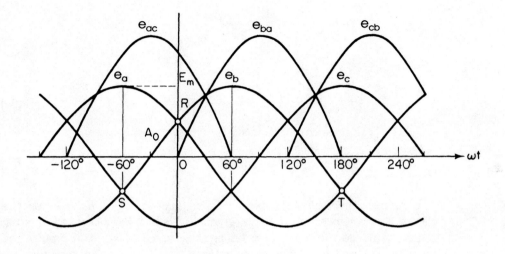

Figure 7-18. Voltage Waveforms Associated with Three-Phase Bridge Converter.

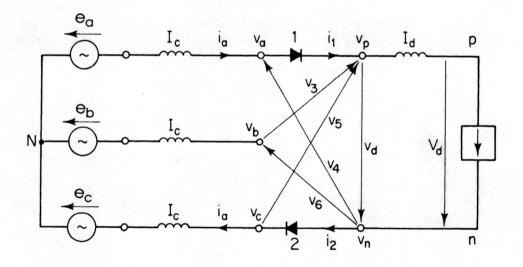

Figure 7-19. Bridge Converter with Valves 1 and 2 Conducting.

work with rms line voltages on the ac side, we substitute

$$E_m = \frac{\sqrt{2}\,(V_L)}{\sqrt{3}}$$

As a result,

$$V_{do} = \frac{3\sqrt{2}}{\pi} V_L \tag{7.64}$$

where V_L is the transformer's secondary line voltage (rms).

Delay Angle

An uncontrolled valve begins to conduct (ignites) as soon as the voltage across it becomes positive. If valves 1 and 2 have been conducting, valve 3 ignites as soon as e_b becomes greater than e_a, that is, at point R, as shown in Figure 7-18. This instant is taken as $\omega t = 0$. At the same instant, under the present assumptions, valve 1 is extinguished (ceases to conduct). Valves having control grids can be made to delay ignition but not to advance it. The delay angle is denoted by α and corresponds to a time delay of α/ω sec. If delayed this long, valve 3 ignites when $\omega t = \alpha$; valve 4 when $\omega t = \alpha + 60°$; valve 5 when $\omega t = \alpha + 120°$; and so on. The delay α cannot exceed 180°. The ignition delay affects the direct voltage as shown in the following analysis.

Instantaneous direct voltages for various values of α are shown in Figure 7-20. The direct voltage is obtained as before except for an increase in the two integration limits by α. Thus,

$$V_d = \frac{3}{\pi} \int_{\alpha-\pi/3}^{\alpha} e_{ac} \, d\theta = \frac{3}{\pi} \int_{\alpha-\pi/3}^{\alpha} \sqrt{3}\, E_m \cos(\theta + 30°) \, d\theta$$

$$= \frac{3\sqrt{3}}{\pi} E_m \cos \alpha$$

In terms of V_{do} (average direct voltage with no delay angle), we have

$$V_d = V_{do} \cos \alpha \tag{7.65}$$

We thus see that the average direct voltage is reduced by $\cos \alpha$, which is referred to as the *delay factor*.

Inversion

The delay angle α can range from 0 to almost 180°; thus $\cos \alpha$ can range from 1 to -1. It follows that V_d can range from V_{do} to $-V_{do}$. Now I_d cannot reverse; hence a negative V_d implies a reversed power flow. This is

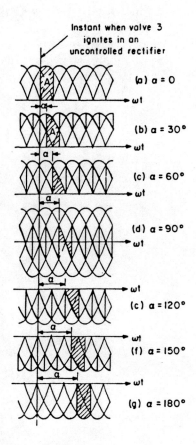

Figure 7-20. Instantaneous Direct Voltage (shown by heavy line) of Bridge Converter with Ignition Delay Angle α but No Overlap.

essentially a process of conversion of dc into ac, and is called inversion in contrast with rectification. It is obvious that grid control is needed for inversion.

Current and Phase Relations with Delay Only

The alternating line current wave consists of rectangular pulses of amplitude I_d and width $(2\pi/3)$ rad, as shown in Figure 7-21. The root-mean-square value of the secondary current (the full wave) is given by

$$I_{rms}^2 = \frac{1}{\pi} \int_{-\pi/2}^{\pi/2} i^2 \, d\Psi$$

$$= \frac{1}{\pi} \int_{-\pi/3}^{\pi/3} I_d^2 \, d\Psi$$

$$= 2 \frac{I_d^2}{3}$$

As a result

$$I = \sqrt{\frac{2}{3}} \, I_d \qquad (7.66)$$

Using Fourier analysis, this periodic wave is resolved into the sum of sinusoidal waves of different harmonics. The peak value of the fundamental frequency component is obtained as

$$\sqrt{2} \, I_{L1} = \frac{2}{\pi} \int_{-\pi/3}^{\pi/3} I_d \cos\Psi \, d\Psi$$

$$= \frac{2\sqrt{3}}{\pi} I_d$$

The rms value of the fundamental component of the current is thus given by

$$I_{L1} = \frac{\sqrt{6}}{\pi} (I_d) = 0.780 \, I_d \qquad (7.67)$$

The power on the ac side is given by

$$P_{ac} = \sqrt{3} \, V_L I_{L1} \cos\phi \qquad (7.68)$$

where $\cos\phi$ is the fundamental frequency's power factor. We recall from Eq. (7.64) that

$$V_{do} = \frac{3\sqrt{2}}{\pi} (V_L)$$

As a result, we can express the ac power of Eq. (7.68) in terms of the dc

7.8 Analysis of the Three-Phase Bridge Converter

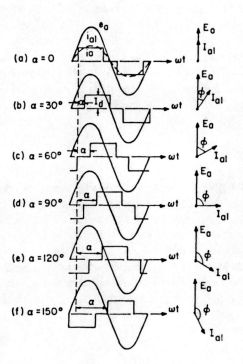

Figure 7-21. Relation between Ignition Delay and Phase Displacement.

voltage and current using Eqs. (7.64) and (7.67). Thus,

$$P_{ac} = V_{do} I_d \cos\phi \tag{7.69}$$

The dc power is given by

$$P_{dc} = V_d I_d \tag{7.70}$$

or using Eq. (7.65),

$$P_{dc} = (V_{do} \cos\alpha) I_d \tag{7.71}$$

For negligible converter losses, the ac power must equal the dc power. We thus conclude by equating Eqs. (7.69) and (7.71) that

$$\cos \phi = \cos \alpha \qquad (7.72)$$

This means that the power factor angle ϕ (angle by which the fundamental line current lags the source phase voltage) is equal to the delay factor $\cos \alpha$.

We conclude from the above discussion that with no ignition delay, the fundamental component of the line current is in phase with the source voltage. Ignition delay α shifts the current wave and its fundamental component by angle $\phi = \alpha$ as shown in Figure 7-21. Thus the converter draws reactive power from the ac system in the presence of ignition delay.

Overlap Angle

Due to the presence of inductance in the ac source transformers, the transfer of currents between phases can only occur at a finite rate. The time required is called the *overlap* or *commutation time*. The overlap angle is denoted μ, and consequently the overlap time is μ/ω seconds. In normal operation, μ is less than 60°.

Figure 7-22 shows the effects of overlap angle on the number of valves conducting simultaneously. During commutation, three valves conduct simultaneously, but between commutations only two valves conduct. A new commutation begins every 60° and lasts for an angle μ. Thus the angular interval when two valves conduct is $(60° - \mu)$.

Analysis of Operation with Overlap

The interval in which valves 1 and 2 conduct ends at $\omega t = \alpha$; at this time valve 3 ignites. In the next interval, the effective circuit is that shown in Figure 7-23 with valves 1, 2, and 3 conducting. During this interval the direct current is transferred from valve 1 to valve 3. The end of the interval is at $\omega t = \delta$, where δ is called the extinction angle and is given by

$$\delta = \alpha + \mu \qquad (7.73)$$

The boundary conditions on the currents for the beginning of commutation at $t = \alpha/\omega$ is

$$i_1\left(\frac{\alpha}{\omega}\right) = I_d, \quad \text{and} \quad i_3\left(\frac{\alpha}{\omega}\right) = 0 \qquad (7.74)$$

At the end of commutation $t = \delta/\omega$,

$$i_1\left(\frac{\delta}{\omega}\right) = 0, \quad \text{and} \quad i_3\left(\frac{\delta}{\omega}\right) = I_d \qquad (7.75)$$

7.8 Analysis of the Three-Phase Bridge Converter

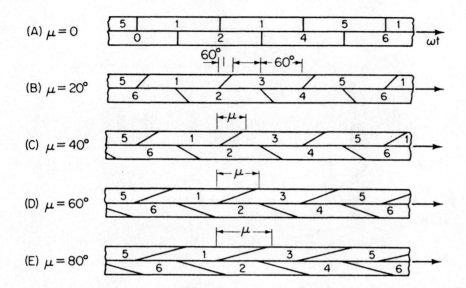

Figure 7-22. Effect of Overlap Angle μ on the Number of Valves Conducting Simultaneously.

For the loop containing phase a, valve 1, valve 3, and phase b, we write KVL as

$$e_b - e_a = L_c \left(\frac{di_3}{dt} - \frac{di_1}{dt} \right) \qquad (7.76)$$

where L_c is the transformer secondary inductance. Now we have

$$I_d = i_3 + i_1 \qquad (7.77)$$

Recall that I_d is assumed ripple-free. Thus differentiating both sides of Eq. (7.77), we obtain

$$0 = \frac{di_3}{dt} + \frac{di_1}{dt} \qquad (7.78)$$

Also we have from our basic assumptions for the secondary voltages:

$$e_b - e_a = \sqrt{3} \, E_m \sin \omega t \qquad (7.79)$$

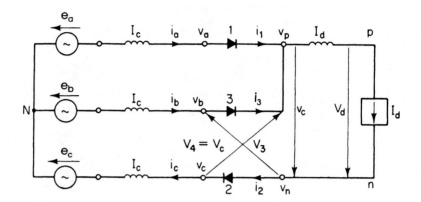

Figure 7-23. Bridge Converter with Valves 1, 2, and 3 Conducting.

We are thus led to conclude that the current $i_3(t)$ satisfies Eq. (7.80) obtained by combining Eqs. (7.76), (7.77), (7.78), and (7.79). Therefore,

$$\frac{di_3}{dt} = \frac{\sqrt{3}}{2L_c}(E_m)\sin \omega t \qquad (7.80)$$

Let us define

$$I_{s2} = \frac{\sqrt{3}}{2\omega}\left(\frac{E_m}{L_c}\right) \qquad (7.81)$$

As a result, Eq. (7.80) is rewritten as

$$\frac{di_3}{dt} = \omega I_{s2}\sin \omega t \qquad (7.82)$$

Integrating Eq. (7.82) we get

$$i_3(t) = I_c - I_{s2}\cos \omega t \qquad (7.83)$$

7.8 Analysis of the Three-Phase Bridge Converter

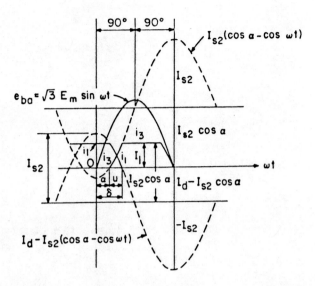

Figure 7-24. Currents i_1 and i_2 during Commutation of Valve 1 to Valve 3 as Arcs of Offset Sinusoidal Waves of Line-to-Line Short-Circuit of Amplitude I_{s2}, with i_3 Lagging 90° Behind the Commutation Voltage e_{ba}.

The constant of integration I_c is obtained from the boundary condition at $t = \alpha/\omega$. The result is

$$i_3(t) = I_{s2}(\cos\alpha - \cos\omega t) \qquad (7.84)$$

We can thus conclude that i_3, the current of the incoming valve during commutation, consists of a dc (constant) term and a sinusoidal term. The latter lags the commutating voltage by 90°, which is the characteristic of a purely inductive circuit and has a peak value I_{s2} equal to the current in a line-to-line short circuit on the ac source. The constant term, which makes $i_3 = 0$ at the beginning of commutation, will depend on α; and for $\alpha = 0$, it shifts the sine wave upward by its peak value. The current i_1 of the outgoing valve has a sine term of the same amplitude as that of i_3 but of opposite phase, and its constant term makes $i_1 = I_d$ at the beginning. The last observations are based on the KCL relation given in Eq. (7.77). Thus,

$$i_1(t) = I_d - i_3(t)$$

The currents i_1 and i_3 are shown in Figure 7-24.

The value of the dc current I_d can be obtained by noting that according to Eq. (7.75),

$$i_3\left(\frac{\delta}{\omega}\right) = I_d$$

Thus we have from Eq. (7.84),

$$I_d = I_{s2}(\cos \alpha - \cos \delta) \tag{7.85}$$

This gives the value of the dc current in terms of delay and overlap angles.

Voltage Drop Due to Overlap

The effect of the overlap on the voltage is to subtract an area A from the area A_0 of Figure 7-18 every sixth of a cycle ($\pi/3$ rad) as shown in Figure 7-25. As a result we have

$$A = \int_\alpha^\delta \left(e_b - \frac{e_a + e_b}{2}\right) d\theta = \int_\alpha^\delta \frac{(e_b - e_a)}{2}(d\theta) = \frac{\sqrt{3} E_m}{2} \int_\alpha^\delta \sin \theta \, d\theta$$

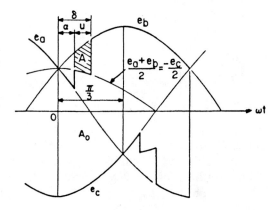

Figure 7-25. Details for the Derivation of Voltage Drop Caused by Overlap.

7.8 Analysis of the Three-Phase Bridge Converter

Thus

$$A = \frac{\sqrt{3}\,E_m}{2}(\cos\alpha - \cos\delta) \qquad (7.86)$$

Now the average voltage drop is obtained by dividing A by the duration $(\pi/3)$ to give

$$\Delta V_d = \frac{3}{\pi} A$$

or

$$\Delta V_d = \frac{V_{do}}{2}(\cos\alpha - \cos\delta) \qquad (7.87)$$

Thus the direct voltage accounting for the overlap is given by

$$V_d = V_{do}(\cos\alpha) - \Delta V_d$$

Hence

$$V_d = \frac{V_{do}}{2}(\cos\alpha + \cos\delta) \qquad (7.88)$$

With no overlap, $\delta = \alpha$; thus $V_d = V_{do}\cos\alpha$, which is the same as before. Recall from Eq. (7.73) that

$$\delta = \alpha + \mu$$

We thus have the expression for the dc voltage with overlap taken into account given by

$$V_d = \frac{V_{do}}{2}[\cos\alpha + \cos(\alpha + \mu)] \qquad (7.89)$$

Example 7-6

The transformer secondary line voltage to a three-phase, bridge-connected rectifier is 34.5 kV. Calculate the direct-voltage output when the overlap (commutation) angle is 15° and the delay angle is:

A. 0°
B. 15°
C. 30°
D. 45°

Solution

Given that $\mu = 15°$ and $V_L = 34.5$ kV, then we have by Eq. (7.64),

$$V_{do} = \frac{3\sqrt{2}}{\pi}(V_L) = 46.59 \text{ kV}$$

Using Eq. (7.89),

$$V_d = \frac{V_{do}}{2}[\cos\alpha + \cos(\alpha + \mu)]$$
$$= 23.296[\cos\alpha + \cos(\alpha + 15°)]$$

A. For $\alpha = 0°$,

$$V_d = 23.296(1 + \cos 15°) = 45.80 \text{ kV}$$

B. For $\alpha = 15°$,

$$V_d = 23.296(\cos 15° + \cos 30°) = 42.68 \text{ kV}$$

C. For $\alpha = 30°$,

$$V_d = 23.296(\cos 30° + \cos 45°) = 36.65 \text{ kV}$$

D. For $\alpha = 45°$

$$V_d = 23.296(\cos 45° + \cos 60°) = 28.12 \text{ kV}$$

Alternating-Current Relations Including Overlap

Assuming that the power on the ac side is equal to that on the dc side, we can write

$$V_d I_d = \sqrt{3}\, V_L I_{L_1} \cos\phi$$

Eliminating the voltages from the above equation using Eqs. (7.64) and (7.89) to show the relation between V_d and V_L, we conclude that

$$I_{L_1}\cos\phi = \frac{\sqrt{6}}{\pi}(I_d)\frac{\cos\alpha + \cos\delta}{2} \qquad (7.90)$$

Let us assume that the following approximation is true:

$$I_{L_1} \cong \frac{\sqrt{6}}{\pi}(I_d) \qquad (7.91)$$

This assumes that only the first harmonic appears on the ac side, and thus the rms value of the ac line current is equal to the rms of the first harmonic as given in Eq. (7.67).

With this assumption we find that the power factor is

$$\cos\phi \cong \frac{\cos\alpha + \cos\delta}{2} \qquad (7.92)$$

Recalling that $\delta = \alpha + \mu$, we can thus conclude that the dc current and voltage given by Eqs. (7.85), (7.81), (7.89), and (7.64) can be written as

$$I_d = \frac{V_L}{\sqrt{2}\, X_c}[\cos\alpha - \cos(\alpha+\mu)] \tag{7.93}$$

$$V_d = \frac{3\sqrt{2}\, V_L}{2\pi}[\cos\alpha + \cos(\alpha+\mu)] \tag{7.94}$$

where

$$X_c = \omega L_c \tag{7.95}$$

The magnitude of the ac line current is related to the dc current by Eq. (7.91), which is substituted in Eq. (7.93) to conclude that

$$I_{L_1} = \frac{\sqrt{3}\, V_L}{\pi X_c}[\cos\alpha - \cos(\alpha+\mu)] \tag{7.96}$$

Expanding the cosine terms we get

$$I_{L_1} = \frac{2\sqrt{3}\, V_L}{\pi X_c}\sin\left(\frac{2\alpha+\mu}{2}\right)\sin\left(\frac{\mu}{2}\right) \tag{7.97}$$

The active power component of the ac line current is denoted by I_p and is given on the basis of equal power on the ac and dc sides by

$$I_p = \frac{V_d I_d}{\sqrt{3}\, V_L} \tag{7.98}$$

Note that

$$I_p = I_{L1}\cos\phi \tag{7.99}$$

Using Eqs. (7.92) and (7.96), we obtain

$$I_p = \frac{\sqrt{3}\, V_L}{4\pi X_c}\{\cos 2\alpha - \cos[2(\alpha+\mu)]\} \tag{7.100}$$

or in terms of the expanded cosines,

$$I_p = \frac{\sqrt{3}\, V_L}{2\pi X_c}\sin\mu \sin(2\alpha+\mu) \tag{7.101}$$

To derive the reactive component of the ac line current, put

$$\tilde{A} = \frac{\sqrt{3}\, V_L}{2\pi X_c} \tag{7.102}$$

and
$$\theta = \frac{2\alpha + \mu}{2}, \qquad \Psi = \frac{\mu}{2} \qquad (7.103)$$

As a result, Eqs. (7.97) and (7.101) can be written as
$$I_{L_1} = 4\tilde{A} \sin\theta \sin\Psi \qquad (7.104)$$
$$I_p = \tilde{A} \sin 2\theta \sin 2\Psi \qquad (7.105)$$

The power factor according to Eq. (7.99) is given by
$$\cos\phi = \frac{I_p}{I_{L_1}} \qquad (7.106)$$

As a result, substituting for I_p from Eq. (7.105) and I_{L1} from Eq. (7.104) in Eq. (7.106), we obtain
$$\cos\phi = \cos\theta \cos\Psi \qquad (7.107)$$

Using the identity $\sin^2\phi = 1 - \cos^2\phi$, we get from Eq. (7.107),
$$\sin\phi = \sqrt{1 - \cos^2\theta \cos^2\Psi}$$

or
$$\sin\phi = \cos\Psi \sqrt{\frac{1}{\cos^2\Psi} - \cos^2\theta}$$

We make the approximation
$$\frac{1}{\cos^2\Psi} \simeq 1$$

Consequently,
$$\sin\phi = \sin\theta \cos\Psi \qquad (7.108)$$

We can now write an alternate form of the reactive component of the ac line current I_r given by
$$I_r = I_{L1} \sin\phi \qquad (7.109)$$

using the expressions of Eqs. (7.104) and (7.108). As a result,
$$I_r = \tilde{A}(\sin 2\psi)(1 - \cos 2\theta) \qquad (7.110)$$

In terms of α and μ, we use Eq. (7.103) to get
$$I_r = \tilde{A}[\sin\mu - \sin\mu \cos(2\alpha + \mu)] \qquad (7.111)$$

A reasonable approximation is made on the first sine term:
$$\sin\mu \cong \mu \qquad (7.112)$$

7.8 Analysis of the Three-Phase Bridge Converter

Thus Eq. (7.111) can be written as

$$I_r = \tilde{A}[\mu - \sin\mu\cos(2\alpha + \mu)] \tag{7.113}$$

Substituting for $\tilde{A}$, from Eq. (7.102) we get

$$I_r = \frac{\sqrt{3}\,V_L}{2\pi X_c}[\mu - \sin\mu\cos(2\alpha + \mu)] \tag{7.114}$$

The reactive power on the ac side given by

$$Q = \sqrt{3}\,V_L I_r \tag{7.115}$$

can be calculated using Eq. (7.114). Clearly the active power is calculated from Eq. (7.101) using

$$P = \sqrt{3}\,V_L I_p \tag{7.116}$$

Equivalent Circuit of the Bridge Rectifier

It is possible to eliminate reference to the overlap angle using the following argument. The dc output voltage V_d has been shown to be given by Eq. (7.88):

$$V_d = V_{do}(\cos\alpha) - \Delta V_d \tag{7.117}$$

where the voltage drop due to overlap is given by Eq. (7.87):

$$\Delta V_d = \frac{V_{do}}{2}(\cos\alpha - \cos\delta) \tag{7.118}$$

We would like to relate ΔV_d to the dc current I_d. Recall Eq. (7.85):

$$I_d = I_{s2}(\cos\alpha - \cos\delta) \tag{7.119}$$

Thus substituting Eq. (7.119) into Eq. (7.118), we obtain

$$\Delta V_d = \frac{V_{do}}{2} \cdot \frac{I_d}{I_{s2}} \tag{7.120}$$

Now by the definition of Eq. (7.81), we have

$$I_{s2} = \frac{\sqrt{3}\,E_m}{2X_c} \tag{7.121}$$

where $X_c = \omega L_c$. Since

$$V_{do} = \frac{3\sqrt{3}\,E_m}{\pi}$$

we conclude that Eq. (7.121) can be rewritten as

$$I_{s2} = \frac{V_{do}}{2} \cdot \frac{1}{\left(\dfrac{3X_c}{\pi}\right)} \tag{7.122}$$

Let us define the equivalent commutation resistance R_c by

$$R_c = \left(\frac{3}{\pi}\right) X_c \qquad (7.123)$$

Thus Eq. (7.122) reduces to

$$I_{s2} = \frac{V_{do}}{2} \cdot \frac{1}{R_c} \qquad (7.124)$$

As a result, Eq. (7.120) is rewritten with the help of Eq. (7.124) as

$$\Delta V_d = R_c I_d \qquad (7.125)$$

The voltage drop due to commutation is thus proportional to I_d.

An *equivalent circuit* of the bridge rectifier, operating at constant alternating voltage and constant ignition angle based on the above analysis, is given in Figure 7-26. The direct voltages and current in this circuit are the average without ripple. Note that the overlap angle has been eliminated and in its place we have R_c. The output voltage is thus obtained from Eqs. (7.117) and (7.125) as

$$V_d = V_{do}(\cos \alpha) - R_c I_d \qquad (7.126)$$

Example 7-7

Calculate the necessary secondary line voltage of the transformer for a three-phase bridge rectifier to provide a direct voltage of 120 kV. Assume $\alpha = 30°$ and $\mu = 15°$. Calculate the effective reactance X_c if the rectifier is delivering 800 amperes dc.

Solution

We use

$$V_d = \frac{V_{do}}{2}[\cos \alpha + \cos(\alpha + \mu)]$$

Thus

$$120 = \frac{V_{do}}{2}[\cos(30°) + \cos(45°)]$$

Hence,

$$V_{do} = 152.56 \text{ kV}$$

But

$$V_{do} = \frac{3\sqrt{2}}{\pi}(V_L)$$

Hence the required transformer voltage is

$$V_L = 112.97 \text{ kV}$$

We have,

$$V_d = V_{do}(\cos \alpha) - R_c I_d$$

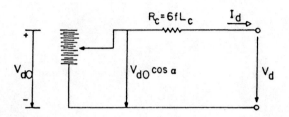

Figure 7-26. Equivalent Circuit of Bridge Rectifier.

Thus,
$$120 \times 10^3 = (152.56)(10^3)\cos 30° - (800)R_c$$

As a result,
$$R_c = 15.151 \text{ ohms}$$

But
$$R_c = \frac{3}{\pi}(X_c)$$

Thus the required transformer reactance is given by
$$X_c = 15.8661 \text{ ohms}$$

7.9 INVERSION IN THREE-PHASE BRIDGE CONVERTER

The direct voltage of a converter has been seen to be given by Eq. (7.89) as

$$V_d = \frac{V_{do}}{2}[\cos \alpha + \cos(\alpha + \mu)] \qquad (7.127)$$

This voltage is positive when the converter is operating as a rectifier. With no overlap, this occurs for $0 < \alpha < 90°$. Increasing α results in decreasing the output voltage. The dc voltage becomes negative for $90° < \alpha < 180°$, with no overlap. The valve current cannot be reversed since conduction occurs in only one direction. Thus reversal of the direct voltage V_d implies a reversal of power. In this case, the converter is said to be in the *inversion mode*.

With overlap, inversion starts at a delay angle α_t at which the direct voltage is zero. This takes place for

$$\cos \alpha_t + \cos(\alpha_t + \mu) = 0$$

or

$$2\alpha_t + \mu = \pi$$

Thus the delay angle corresponding to start of inversion is

$$\alpha_t = \frac{\pi - \mu}{2} \tag{7.128}$$

This is less than 90°.

For operation as a rectifier, the angles of ignition α and extinction δ are measured by the delay from the instant at which the commutating voltage is zero and increasing ($\omega t = 0$). Angles defined in the same way and having values between 90° and 180° could be used for inverter operation. Common practice, however, is to define the ignition angle β and the extinction angle γ by their advance with respect to the instant when the commutation voltage is zero and decreasing ($\omega t = 180°$ for ignition of valve 3 and extinction of valve 1).

Figure 7-27 shows the relations between the various angles defined. From the figure we have

$$\beta = \pi - \alpha_i \tag{7.129}$$

$$\gamma = \pi - \delta_i \tag{7.130}$$

$$\mu = \delta - \alpha_t \tag{7.131}$$

$$\mu_i = \beta - \gamma \tag{7.132}$$

$$\delta_i = \alpha_i + \mu_i \tag{7.133}$$

The general converter equations are given by Eq. (7.119) and Eq. (7.89) as

$$I_d = I_{s2}(\cos \alpha - \cos \delta) \tag{7.134}$$

$$V_d = \frac{V_{do}}{2}(\cos \alpha + \cos \delta) \tag{7.135}$$

To obtain the equations for inverter operation, we change the sign of V_d and put $\cos \alpha = -\cos \beta$ and $\cos \delta = -\cos \gamma$. Thus Eqs. (7.134) and (7.135) become

$$I_d = I_{s2}(\cos \gamma - \cos \beta) \tag{7.136}$$

$$V_d = \frac{V_{do}(\cos \gamma + \cos \beta)}{2} \tag{7.137}$$

Note that inverter voltage, considered negative in the general converter equations, is usually taken as positive when written specifically for an inverter. In the case of operation with constant ignition advance angle β,

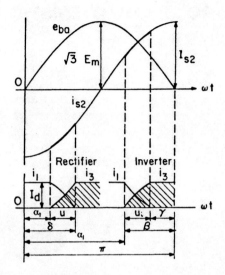

Figure 7-27. Angles Defined for Converter Operation.

Eq. (7.126) becomes

$$V_d = V_{do}(\cos \beta) + R_c I_d \qquad (7.138)$$

For the more commonly used mode of constant extinction angle γ, elimination of $\cos \beta$ from Eqs. (7.136) and (7.137) yields

$$V_d = V_{do}(\cos \gamma) - R_c I_d \qquad (7.139)$$

The above two equations result in two possible equivalent circuits for inverter operation as shown in Figure 7-28.

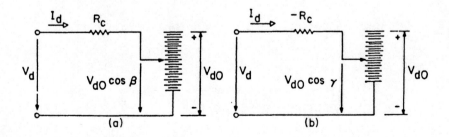

Figure 7-28. Equivalent Circuits of Inverter.

7.10 HVDC LINK AND CONVERTER CONTROL CHARACTERISTICS

A high-voltage dc link is an ac/dc/ac interconnection where ac power is converted to dc using a three-phase bridge rectifier. The dc power is fed through a cable or a long transmission line. At the end of the link there is a three-phase bridge inverter where dc power is converted to ac, which is then fed to the ac system. The equivalent circuit of Figure 7-29 represents the dc circuit. Subscripts r and i refer to rectifier and inverter respectively.

The dc current flowing from rectifier to inverter is given by Ohm's law as either

$$I_d = \frac{V_{do_r} \cos\alpha - V_{do_i} \cos\gamma}{R_{cr} + R_l - R_{ci}} \qquad (7.140)$$

for constant extinction angle γ, or

$$I_d = \frac{V_{do_r} \cos\alpha - V_{do_i} \cos\beta}{R_{cr} + R_l + R_{ci}} \qquad (7.141)$$

for constant ignition angle β.

The direct voltage and current can be controlled by either of two different methods: (1) grid control (α and β) or (2) control of alternating voltage (and hence V_{do_r} and V_{do_i}). The latter is done through transformer tap changing, which is slower (in the neighborhood of 5 seconds per step) in

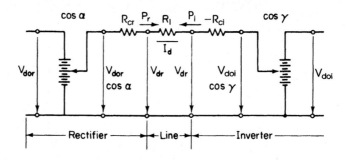

Figure 7-29. Equivalent Circuit of HVDC Link.

7.10 HVDC Link and Converter Control Characteristics

comparison with grid control (1 to 10 ms). Both means are applied with grid control initially for fast action and are followed by tap changing.

The operation of the converters follows selected characteristics in order to meet the requirements of control and protection. This is termed "compounding the converter." The rectifier's most desirable mode of control is constant current control (abbreviated to C.C. control), which requires a control action that adjusts the angle of ignition so that the current is maintained at a set value. If the control functions ideally, the rectifier characteristic is a vertical line. In practice it has a high negative slope and is limited by a direct voltage ceiling corresponding to point B in Figure 7-30.

Suppose that the ac voltage decreased at the rectifier. If α stayed constant, then the direct voltage would decrease. The C.C. control raises the direct voltage either to its initial value or until the minimum α limit is reached. When the latter happens, the rectifier operates on the constant ignition angle (C.I.A.) characteristic corresponding to the minimum ignition angle. The voltage in this case is given by

$$V_d = V_{dor} \cos \alpha_{min} - I_d R_c$$

This is shown as line AB in Figure 7-30. Thus the rectifier characteristics consist of two line segments.

When the converter operates as an inverter, commutation must be completed prior to reversal of the commutating voltage (at $\gamma = 0$) to avoid

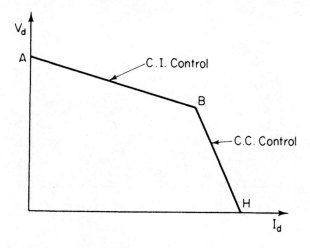

Figure 7-30. Rectifier Characteristics.

commutation failure. The minimum angle γ_0 corresponds to the time required for deionization of the arc (1 to 8°). This means that the angle of advance ($\beta = \gamma + \mu$) should be made large. But on the other hand, if β is too large, then the power factor becomes poor. Thus we should keep β and hence γ to a minimum. This leads to the requirement of a fast-acting control that maintains the extinction angle γ to a definite minimum. Such a control is referred to as the "constant extinction angle control" (abbreviated to C.E.A. control). Thus the $V_d I_d$ characteristic will be represented by the line *FD* in Figure 7-31. In addition, a C.C. control that advances the angle β more than that required for the C.E.A. control is required to ensure that current does not decrease below a set value. This gives rise to the segment *GF* in Figure 7-31.

The actual characteristics of the converter in both modes of operation are now seen to be as shown in Figure 7-32. Note that the difference between the current setting for operation as an inverter from that for operation as a rectifier is denoted by ΔI_d (in the neighborhood of 15 percent of rated current). The control scheme for a HVDC link is shown in Figure 7-33.

Consider a system of two converters as shown in Figure 7-33. If the C.C. control setting of converter 1 is higher than the setting of converter 2 by a margin ΔI_d, the operating characteristic is given by Figure 7-34. The current in the dc line is given by the point of intersection P_1 of the characteristics of the two converters. Converter 1 operates as a rectifier, and converter 2 as an inverter. If the C.I.A. characteristic of converter 2 is as shown, then converter 1 operates on C.C. control and converter 2 on C.E.A.

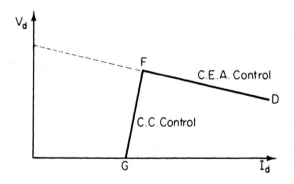

Figure 7-31.   Inverter Characteristics.

7.10 HVDC Link and Converter Control Characteristics

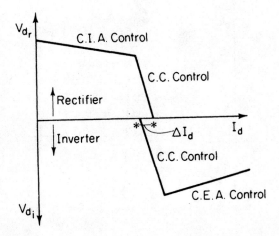

Figure 7-32. Complete Control of a Converter, from Inversion to Rectification.

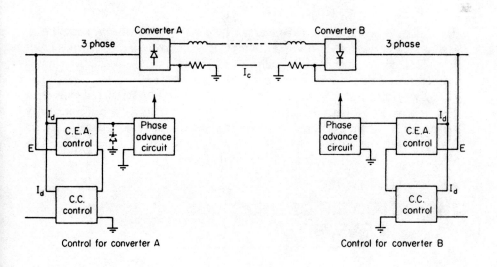

Figure 7-33. Schematic Diagram of Control of a HVDC System: C.E.A. = constant extinction angle; and C.C. = constant current.

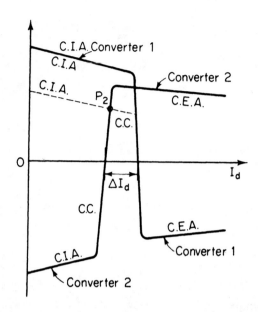

Figure 7-34. Control Characteristic with Converter 1 Operating as a Rectifier and Converter 2 as an Inverter.

control at point P_1 in Figure 7-34. If the C.I.A. characteristic of 1 falls below the C.E.A. characteristic of converter 2, as shown by the broken line in Figure 7-34, converter 1 will operate on C.E.A. control, and converter 2 will be on C.C. control as shown by point P_2 in Figure 7-34. Converter 1 remains to operate as a rectifier and converter 2 as an inverter.

By reversing the "margin setting," i.e., making the setting of converter 2 to exceed that of converter 1, the flow of power can be automatically reversed. Converter 2 will now operate as a rectifier, and converter 2 will be the inverter. The operation point is now P_3 as shown in Figure 7-35. The reversal of power is a result of the reversal of polarity.

7.11 ANALYSIS OF HVDC LINK PERFORMANCE

Consider the basic model of a dc link shown schematically in Figure 7-36. Its equivalent circuit is shown in Figure 7-37. On the ac side, the link is represented by two nodes i and j, whose voltages are V_i/ψ_i and V_j/ψ_j respectively. The ac currents at the rectifier and inverter terminals are denoted by I_i/η_i and I_j/η_j respectively. The tap-changing transformers have tap ratios a_i and a_j. We have already at our disposal all the relations necessary to model the link for performance analysis purposes.

The line voltage on the secondary side of the rectifier transformer is $a_i V_i$, and the line current is I_i/a_i. The direct voltage at the rectifier terminal V_{d_i} is given according to Eq. (7.126) as modified by Eqs. (7.123) and (7.64):

$$V_{d_i} = \frac{3\sqrt{2}}{\pi}(a_i V_i)\cos\alpha_i - \frac{3}{\pi}(X_{c_i} I_d) \qquad (7.142)$$

The direct current I_d is related to the secondary ac line current by Eq. (7.91). Thus

$$I_{s_i} = \frac{\sqrt{6}}{\pi}(I_d)$$

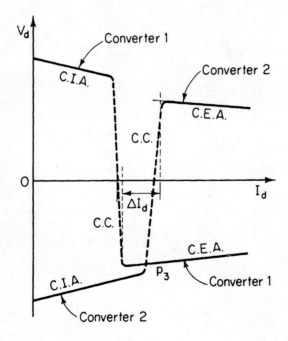

Figure 7-35. Control Characteristic with Converter 1 Operating as an Inverter and Converter 2 as a Rectifier.

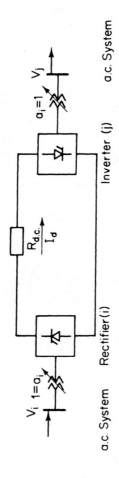

Figure 7-36. Schematic of a dc Link.

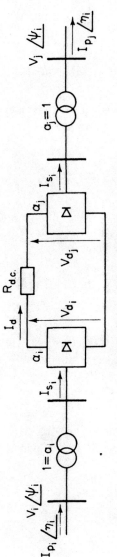

Figure 7-37. Single-Line Diagram of dc Link.

where

$$I_{s_i} = \frac{I_i}{a_i} \qquad (7.143)$$

Here α_i is the rectifier's delay angle, and X_{c_i} is its commutating reactance. The application of KVL in the dc link gives

$$V_{d_i} - V_{d_j} = I_d R_{dc} \qquad (7.144)$$

where V_{d_j} is the direct voltage at the inverter terminals, and R_{dc} is the link's resistance. The inverter voltage equation is given by Eq. (7.139) as

$$V_{d_j} = \frac{3\sqrt{2}}{\pi}(a_j V_j)\cos\gamma_j - \frac{3}{\pi}(X_{c_j} I_d) \qquad (7.145)$$

Here γ_j is the inverter's extinction angle, and X_{c_j} is its commutating reactance.

The power relations are given by

$$V_{d_i} I_d = \sqrt{3}\, V_i I_i \cos\phi_i \qquad (7.146)$$

$$V_{d_j} I_d = \sqrt{3}\, V_j I_j \cos\phi_j \qquad (7.147)$$

The ac line current on the inverter side is

$$I_j = a_j \frac{\sqrt{6}}{\pi}(I_d) \qquad (7.148)$$

Problems 7-A-15, 7-A-16, and 7-A-17 illustrate the use of the analysis equations.

A Per Unit System

There are a number of per unit (p.u.) systems in use in connection with HVDC systems. We illustrate one such a system. The system is chosen such that the same base MVA and voltage are used on the ac and dc sides.

Thus,

$$P_{b_{dc}} = P_{b_{ac}} = P_b \qquad (7.149)$$

$$V_{b_{dc}} = V_{b_{ac}} = V_b \qquad (7.150)$$

The base ac current can thus be calculated as

$$I_{b_{ac}} = \frac{P_b}{\sqrt{3}\, V_b} \qquad (7.151)$$

The base dc current is

$$I_{b_{dc}} = \frac{P_b}{V_b} \qquad (7.152)$$

As a result, we conclude that

$$I_{b_{dc}} = \sqrt{3}\, I_{b_{ac}} \qquad (7.153)$$

The base dc current is $\sqrt{3}$ times the base ac current.

This system enables us to write the condition that power on the ac side is equal to the power on the dc side in a computationally simple way. In terms of volts and amperes, we have

$$V_d I_d = \sqrt{3}\, V_{ac} I_{ac} \cos\phi \qquad (7.154)$$

In terms of per unit quantities, the above equation is

$$(V_{d_{p.u.}} V_b)(I_{d_{p.u.}} I_{b_{dc}}) = \sqrt{3}\,(V_{ac_{p.u.}} V_b)(I_{ac_{p.u.}} I_{b_{ac}}) \cos\phi$$

or

$$V_{d_{p.u.}} I_{d_{p.u.}} = \frac{\sqrt{3}\, I_{b_{ac}}}{I_{b_{dc}}} (V_{ac_{p.u.}} I_{ac_{p.u.}}) \cos\phi$$

As a result,

$$V_{d_{p.u.}} I_{d_{p.u.}} = V_{ac_{p.u.}} I_{ac_{p.u.}} \cos\phi \qquad (7.155)$$

The relation between ac and dc currents in amperes is given by

$$I_L = \frac{\sqrt{6}}{\pi}(I_d) \qquad (7.156)$$

In terms of per unit quantities we have

$$I_{L_{p.u.}} I_{b_{ac}} = \frac{\sqrt{6}}{\pi}(I_{d_{p.u.}} I_{b_{dc}})$$

Thus,

$$I_{L_{p.u.}} = \frac{I_{b_{dc}}}{I_{b_{ac}}} \frac{\sqrt{6}}{\pi}(I_{d_{p.u.}})$$

This gives the per unit relationship:

$$I_{L_{p.u.}} = \frac{3\sqrt{2}}{\pi}(I_{d_{p.u.}}) \qquad (7.157)$$

Define the constant K_1 by

$$K_1 = \frac{3\sqrt{2}}{\pi} \qquad (7.158)$$

We thus have from Eqs. (7.157) and (7.158),

$$I_{L_{p.u.}} = K_1 I_{d_{p.u.}} \qquad (7.159)$$

Analysis in Terms of Per Unit Quantities

Proceeding from the primary side of the rectifier's transformer we can write the approximate relation:

$$I_{p_i} = a_i I_{s_i} \qquad (7.160)$$

This simply states that the fundamental current magnitudes on both sides of the lossless transformer are related by the off-nominal tap ratio.

The fundamental current magnitude on the converter side is related to the direct current by Eq. (7.159). Thus,

$$I_{s_i} = K_1 I_d \qquad (7.161)$$

The direct voltage (in terms of the commutating reactance and the voltage on the system side of the rectifier's transformer) is according to Eq. (7.142),

$$V_{d_i} = \frac{3\sqrt{2}}{\pi}(a_i V_i)\cos \alpha_i - \frac{3}{\pi}(I_d X_{c_i}) \qquad (7.162)$$

Let

$$K_2 = \frac{3}{\pi} \qquad (7.163)$$

Then

$$V_{d_i} - K_1 a_i V_i \cos \alpha_i + K_2 I_d X_{c_i} = 0 \qquad (7.164)$$

where X_{c_i} is the commutating reactance of the rectifier's ac source. The active power balance between the ac and dc sides gives

$$V_{d_i} I_{d_i} = V_i I_{p_i} \cos \phi_i \tag{7.165}$$

Using Eqs. (7.160) and (7.161) in Eq. (7.165), we get

$$V_{d_i} - K_1 a_i V_i \cos \phi_i = 0 \tag{7.166}$$

The rectifier voltage and inverter voltage on the dc part of the system are related by

$$V_{d_i} - V_{d_j} = R_{dc} I_d \tag{7.167}$$

The model of the inverter is similar to that of the rectifier. Thus the following equations apply:

$$V_{d_j} - K_1 a_j V_j \cos \gamma_j + K_2 I_d X_{c_j} = 0 \tag{7.168}$$

$$V_{d_j} - K_1 a_j V_j \cos \phi_j = 0 \tag{7.169}$$

The model developed so far consists of five independent equations—Eqs. (7.162), (7.166), (7.167), (7.168), and (7.169)—in terms of the following nine variables:

$$V_{d_i}, V_{d_j}, a_i, a_j, \cos \alpha_i, \cos \gamma_j, \phi_i, \phi_j, \text{ and } I_d$$

To solve for nine variables, four equations giving the control specifications are needed. Thus,

$$V_{d_i}^{sp} - V_{d_i} = 0$$
$$\cos \alpha_i^{sp} - \cos \alpha_i = 0$$
$$\cos \gamma_j^{sp} - \cos \gamma_j = 0$$
$$V_{d_j} I_d - P_{d_j}^{sp} = 0$$

Problem 7-A-18 illustrates the per unit system and provides an analysis of a system in its terms.

SOME SOLVED PROBLEMS

Problem 7-A-1

Consider an existing three-phase, double-circuit, ac line in relation to its conversion to dc with three circuits. Assume the same insulation level. Show that the ratio of power transmitted by dc to that by ac is given by

$$\frac{\text{Power by dc}}{\text{Power by ac}} = \sqrt{2}$$

Show also that the percentage loss ratio is

$$\frac{\text{Percentage losses by dc}}{\text{Percentage losses by ac}} = \frac{1}{\sqrt{2}} = 0.707$$

Solution

Power transmitted by ac $= (2)(3)V_{ph}I_L$. The ac line is converted to three dc circuits, each having two conductors at plus and minus $(V_d/2)$ to ground.

Power transmitted by dc $= 6V_d I_d$

For
$$I_L = I_d,$$

with the same insulation level of

$$V_d = \sqrt{2}\, V_{ph}.$$

The first required ratio is

$$\frac{\text{Power by dc}}{\text{Power by ac}} = \frac{6V_d I_d}{6V_{ph} I_L}$$

$$= \frac{V_d}{V_{ph}}$$

$$= \sqrt{2}$$

We also have

$$\frac{\text{Percentage losses by dc}}{\text{Percentage losses by ac}} = \frac{\text{Losses by dc}}{\text{Power by dc}} \cdot \frac{\text{Power by ac}}{\text{Losses by ac}}$$

$$= \frac{\text{Power by ac}}{\text{Power by dc}} \cdot \frac{\text{Losses by dc}}{\text{Losses by ac}}$$

$$= \frac{1}{\sqrt{2}} = 0.707$$

Problem 7-A-2

Assume a particular design criterion specifies that the ratio of insulation level for a dc bipolar line to the insulation level for the equivalent ac three-phase line (equal power transfer) is x. Show that the corresponding losses are related by

$$\frac{P_{L_{ac}}}{P_{L_{dc}}} = \frac{4}{3} x^2$$

Assume that the insulation level varies directly with the peak value of the voltage.

Solution

The insulation level ratio assumed is

$$x = \frac{V_d}{\sqrt{2}\, V_{ph}}$$

For equal power transfer,

$$P_{ac} = P_{dc}$$

Therefore,

$$3 V_{ph} I_L = 2 V_d I_d$$

Eliminating the voltages from both sides using the insulation ratio, we get

$$\frac{I_L}{I_d} = \left(\frac{2\sqrt{2}}{3}\right) x$$

The loss ratio is

$$\frac{P_{L_{ac}}}{P_{L_{dc}}} = \frac{3 I_L^2 R}{2 I_d^2 R}$$

$$= \frac{3}{2}\left(\frac{2\sqrt{2}}{3} x\right)^2 = \frac{4}{3} x^2$$

Problem 7-A-3

The losses for a proposed ac line are 60 MW. Find the corresponding losses if a dc line is designed such that the ratio of insulation level for the dc line to that for the equivalent ac three-phase line is 0.87.

Solution

We have from Problem 7-A-2,

$$x = 0.87$$

It has been shown earlier that

$$\frac{P_{L_{ac}}}{P_{L_{dc}}} = \frac{4}{3} x^2$$

Thus,

$$P_{L_{dc}} = \frac{3}{4 x^2} P_{L_{ac}}$$

$$= \frac{3}{4(0.87)^2}(60)$$

$$= 59.45 \text{ MW}$$

Problem 7-A-4

The annual savings to the utility due to insulation and loss reduction by using the dc option over the ac option depend on the choice of the ratio x defined in Problem 7-A-2.

Sketch the variation of the total annual savings with x. Obtain the optimum value of x for maximum total annual savings.

Solution

The total savings are given by the sum of the reduction in annual cost of losses and the reduction in the capital cost of insulation calculated on a yearly basis.

- Cost of the ac insulation-related equipment $= A$.
- Cost of the dc insulation-related equipment $= x\tilde{A}$ where $\tilde{A}$ is as defined in text.

Thus,

$$\text{Savings in insulation cost} = A - x\tilde{A}$$

$$\text{Annual savings in loss cost} = B\left(1 - \frac{P_{L_{dc}}}{P_{L_{ac}}}\right)$$

$$= B\left(-\frac{3}{4x^2}\right)$$

Thus we have the total annual savings given by

$$S = A - \tilde{A}x + B\left(1 - \frac{3}{4x^2}\right)$$

A sketch of the annual savings is shown in Figure 7-38. To maximize the savings,

$$\frac{\partial S}{\partial x} = 0$$

Thus,

$$-\tilde{A} - \frac{3}{4}B\left(\frac{-2}{x^3}\right) = 0$$

The optimum value of x is thus

$$x^* = \left(\frac{3B}{2\tilde{A}}\right)^{1/3}$$

The maximum savings are thus

$$S_{\max} = A + B - \left(\frac{3B + 4\tilde{A}x^3}{4x^2}\right)$$

$$= A + B - 1.72 B^{1/3}\tilde{A}^{2/3}$$

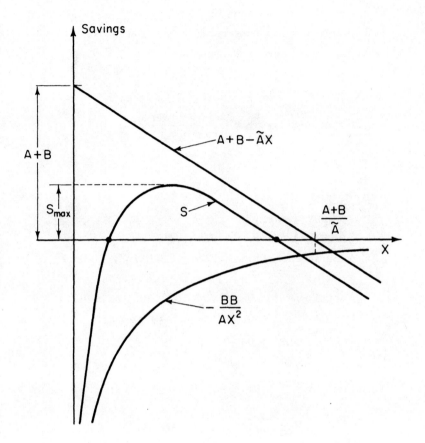

Figure 7-38. Annual Savings Variation with x.

Problem 7-A-5

High-voltage dc is used for the transmission of 800-MW power from a remote generating site to the load center. Assume that transmission losses are 4 percent of total using the ac alternative and that this loss costs 35 mils/kWh. A loss load factor of 0.56 is commonly assumed. The line design is such that the losses using dc are 80 percent of their ac counterpart.

Assuming that this ratio is the optimal choice such that total annual cost of losses and insulation are a minimum, calculate:

A. The annual cost of the losses using the ac alternative.
B. The annual cost of the losses using dc.
C. The annual cost of insulation amortized annually.

Hint:

$$\text{Loss load factor} = \frac{\text{average power loss}}{\text{peak power loss}}$$

Solution

A. For the ac alternative, we calculate as follows:

$$\text{Peak power loss} = (0.04)(800) = 32 \text{ MW}$$
$$\text{Average power loss} = (0.56)(32) = 17.92 \text{ MW}$$
$$\text{Annual energy loss} = (17.92)(8760)(10^3)$$
$$= 1.569792 \times 10^8 \text{ kWh}$$
$$\text{Annual cost of losses} = (1.569792 \times 10^8)(35 \times 10^{-3})$$
$$= 5.49427 \times 10^6 \text{ \$/year}$$

B. For the cost of losses we have

$$\frac{P_{L_{dc}}}{P_{L_{ac}}} = 0.8$$

Thus,

$$\text{Annual cost of dc loss} = (0.8)(5.49427 \times 10^6)$$
$$= 4.39542 \times 10^6 \text{ \$/year}$$

C. For optimal design we have from example 7-4 that

$$\text{Cost of dc insulation} = 2 \text{ (annual cost of dc loss)}$$
amortized annually
$$= 2(4.39542 \times 10^6)$$
$$= 8.79084 \times 10^6 \text{ \$/year}$$

Problem 7-A-6

Consider the cascade of two three-phase, one-way rectifiers shown in Figure 7-39. Show that

A. Average valve current is $0.333 I_d$.
B. Peak inverse voltage on a valve is $1.047 V_d$.
C. Direct-current voltage ripple, peak to peak, is $0.14 V_d$.

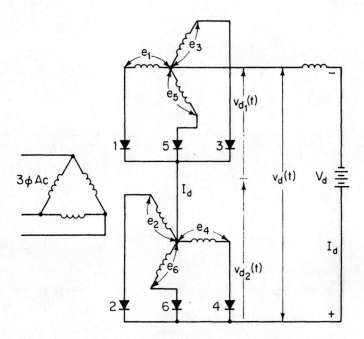

Figure 7-39. Cascade of Two Three-Phase, One-Way Rectifiers.

Solution

We assume the secondary voltages to vary according to the following:

$$e_1(t) = E_m \cos \omega t$$
$$e_2(t) = E_m \cos\left(\omega t - \frac{\pi}{3}\right)$$
$$e_3(t) = E_m \cos\left(\omega t - \frac{2\pi}{3}\right)$$
$$e_4(t) = E_m \cos(\omega t - \pi)$$
$$e_5(t) = E_m \cos\left(\omega t - \frac{4\pi}{3}\right)$$
$$e_6(t) = E_m \cos\left(\omega t - \frac{5\pi}{3}\right)$$

The first group of valves $(1, 3, 5)$ produces an unfiltered voltage $v_{d1}(t)$, and the second group $(2, 4, 6)$ produces $v_{d2}(t)$. Each of the secondaries in the second group is phase-displaced from the corresponding first by $\pi/3$ electrical radians.

Each valve in the group conducts for one-third of the cycle. The current waveforms are shown in Figure 7-40. The pulse height is I_d. The

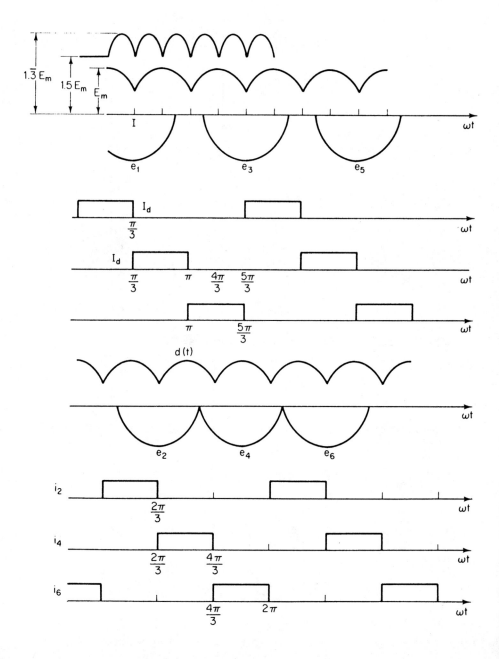

Figure 7-40. Waveforms for the Cascade of Two Three-Phase, One-Way Rectifiers.

average current is thus

$$I_{av} = \frac{I_d}{3}$$
$$= 0.333 I_d$$

The unfiltered voltage $v_d(t)$ is the sum of the voltages due to the two groups; that is,

$$v_d(t) = v_{d1}(t) + v_{d2}(t)$$

Since the two groups are similar, the average (dc voltage) is given by

$$V_d = V_{d1} + V_{d2}$$
$$= 2V_{d1}$$

We have for a three-phase, one-way circuit,

$$V_{d1} = 0.827 E_m$$

Thus,

$$V_d = 1.654 E_m$$

The inverse relation is thus

$$E_m = 0.6046 V_d$$

The rms secondary voltage is

$$E_{rms} = \frac{E_m}{\sqrt{2}}$$

Thus in terms of the dc voltage, we have

$$E_{rms} = 0.42752 V_d$$

The peak inverse voltage for a three-phase, one-way circuit is

$$\text{PIV} = \sqrt{3}\, E_m$$

In terms of the dc voltage, we get

$$\text{PIV} = (\sqrt{3})(0.6046 V_d)$$
$$= 1.0472 V_d$$

The total voltage waveform $v_d(t)$ is obtained using Figure 7-40 in analytic form as:

$$v_d(t) = E_m \left[\cos(\omega t) + \cos\left(\omega t - \frac{\pi}{3}\right) \right]$$
$$= \sqrt{3}\, E_m \cos\left(\omega t - \frac{\pi}{6}\right) \qquad \left(0 \leqslant \omega t \leqslant \frac{\pi}{3}\right)$$

$$v_d(t) = E_m \left[\cos\left(\omega t - \frac{2\pi}{3}\right) + \cos\left(\omega t - \frac{\pi}{3}\right) \right]$$
$$= \sqrt{3}\, E_m \cos\left(\omega t - \frac{\pi}{2}\right) \qquad \left(\frac{\pi}{3} \leqslant \omega t \leqslant \frac{2\pi}{3}\right)$$

$$v_d(t) = E_m\left[\cos\left(\omega t - \frac{2\pi}{3}\right) + \cos(\omega t - \pi)\right]$$

$$= \sqrt{3}\,E_m\cos\left(\omega t - \frac{5\pi}{6}\right) \qquad \left(\frac{2\pi}{3} \leqslant \omega t \leqslant \pi\right)$$

$$v_d(t) = E_m\left[\cos\left(\omega t - \frac{4\pi}{3}\right) + \cos(\omega t - \pi)\right]$$

$$= \sqrt{3}\,E_m\cos\left(\omega t - \frac{7\pi}{6}\right) \qquad \left(\pi \leqslant \omega t \leqslant \frac{4\pi}{3}\right)$$

$$v_d(t) = E_m\left[\cos\left(\omega t - \frac{4\pi}{3}\right) + \cos\left(\omega t - \frac{5\pi}{3}\right)\right]$$

$$= \sqrt{3}\,E_m\cos\left(\omega t - \frac{3\pi}{2}\right) \qquad \left(\frac{4\pi}{3} \leqslant \omega t \leqslant \frac{5\pi}{3}\right)$$

$$v_d(t) = E_m\left[\cos(\omega t) + \cos\left(\omega t - \frac{5\pi}{3}\right)\right]$$

$$= \sqrt{3}\,E_m\cos\left(\omega t - \frac{11\pi}{6}\right) \qquad \left(\frac{5\pi}{3} \leqslant \omega t \leqslant 2\pi\right)$$

The peak-to-peak ripple is obtained as

$$\text{PPR} = v_d\left(\frac{\pi}{6}\right) - v_d(0)$$

$$= \sqrt{3}\,E_m - \sqrt{3}\,E_m\cos\frac{\pi}{6}$$

$$= \sqrt{3}\,E_m\left(1 - \cos\frac{\pi}{6}\right)$$

$$= 0.2321\,E_m$$

In terms of the direct voltage, we have

$$\text{PPR} = (0.2321)(0.6046V_d)$$

$$= 0.1403V_d$$

Problem 7-A-7

Consider the Y-Y interphase rectifier circuit shown in Figure 7-41. Show that

A. Average valve current is $0.167I_d$.
B. Peak inverse voltage on a valve is $2.094V_d$.
C. Direct-current voltage ripple, peak to peak, is $0.14V_d$.

Solution

The rectifier circuit is redrawn in Figure 7-42 with each group composed of a three-phase secondary with the three valves replaced by a single block. This clearly shows the voltages in the circuit. Each valve conducts for

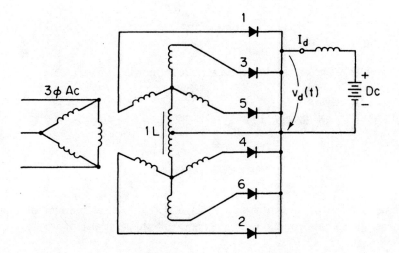

Figure 7-41. Y-Y Interphase Rectifier Circuit.

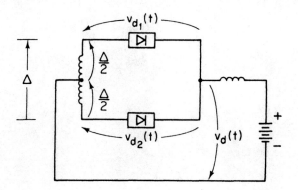

Figure 7-42. Equivalent of Y-Y Interphase Rectifier Circuit.

one-sixth of a cycle. Hence

$$I_{av} = \frac{I_d}{6}$$
$$= 0.167 I_d$$

The voltage across the interphase transformer is Δ. Thus,

$$\Delta = v_{d1}(t) - v_{d2}(t)$$

The output voltage $v_d(t)$ is obtained as

$$v_d(t) = v_{d2}(t) + \left(\frac{\Delta}{2}\right)$$
$$= \frac{v_{d1}(t) + v_{d2}(t)}{2}$$

Thus the output (unfiltered) voltage $v_d(t)$ is the average of the two three-phase, one-way circuits' voltages. For the first circuit we have

$$v_{d1}(t) = E_m \cos(\omega t) \qquad \left(-\frac{\pi}{3} \leqslant \omega t \leqslant \frac{\pi}{3}\right)$$

The second circuit lags the first by 60°; thus,

$$v_{d2}(t) = E_m \cos\left(\omega t - \frac{\pi}{3}\right) \qquad \left(0 \leqslant \omega t \leqslant \frac{2\pi}{3}\right)$$

We then have

$$v_d(t) = 0.5 E_m \left[\cos(\omega t) + \cos\left(\omega t - \frac{\pi}{3}\right)\right]$$
$$= 0.867 E_m \cos\left(\omega t - \frac{\pi}{6}\right) \qquad \left(0 \leqslant \omega t \leqslant \frac{\pi}{3}\right)$$

The average output voltage is

$$v_d = \frac{0.867 E_m}{\frac{\pi}{3}} \int_0^{\pi/3} \cos\left(\theta - \frac{\pi}{6}\right) d\theta$$
$$= 0.827 E_m (2 \sin 30°)$$
$$= 0.827 E_m$$

With the inverse relation,

$$E_m = 1.2092 V_d$$

The inverse voltage is obtained as

$$v_i(t) = e_a(t) - e_b(t)$$
$$= \sqrt{3} E_m \cos\left(\omega t + \frac{\pi}{6}\right)$$

Its peak is

$$PIV = \sqrt{3} E_m$$
$$= 2.0944 V_d$$

The peak-to-peak ripple is

$$\text{PPR} = v_d\left(\frac{\pi}{6}\right) - v_d\left(\frac{\pi}{3}\right)$$

$$= 0.867 E_m \left[1 - \cos\left(\frac{\pi}{6}\right)\right]$$

$$= 0.116 E_m$$

Thus in terms of the dc voltage, we have

$$\text{PPR} = (0.116)(1.2092) V_d$$

$$= 0.1403 V_d$$

Problem 7-A-8

Consider the six-phase diametrical converter circuit shown in Figure 7-43. Show that

A. Average current in a valve is $0.167 I_d$.
B. Peak inverse voltage on a valve is $2.094 V_d$.
C. Direct-current ripple, peak to peak, is $0.14 V_d$.

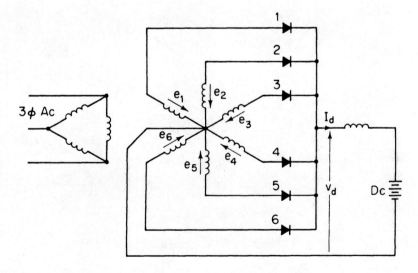

Figure 7-43. Six-Phase Diametrical Converter.

Solution

The secondary transformer voltages are assumed to vary as shown below:

$$e_1(t) = E_m \cos \omega t$$
$$e_2(t) = E_m \cos\left(\omega t - \frac{\pi}{3}\right)$$
$$e_3(t) = E_m \cos\left(\omega t - \frac{2\pi}{3}\right)$$
$$e_4(t) = E_m \cos(\omega t - \pi) = -e_1(t)$$
$$e_5(t) = E_m \cos\left(\omega t - \frac{4\pi}{3}\right) = -e_2(t)$$
$$e_6(t) = E_m \cos\left(\omega t - \frac{5\pi}{3}\right) = -e_3(t)$$

These voltages are shown in Figure 7-44. Since the cathodes of all valves are connected to a common node, the common voltage $v_d(t)$ is the most positive voltage. This is shown in the heavy line. For example, $e_1(t)$ is the most positive in the range

$$\left(-\frac{\pi}{6} \leqslant \omega t \leqslant \frac{\pi}{6}\right)$$

and in this case, valve 1 conducts. For the next range

$$\left(\frac{\pi}{6} \leqslant \omega \leqslant \frac{\pi}{2}\right)$$

valve 2 conducts, and so on. Thus,

$$i_1(t) = I_d \qquad \left(-\frac{\pi}{6} \leqslant \omega t \leqslant \frac{\pi}{6}\right)$$
$$= 0 \qquad \left(\frac{\pi}{6} \leqslant \omega t \leqslant \frac{11\pi}{6}\right)$$

The average valve current is

$$I_{av} = \frac{I_d}{6} = 0.167 I_d$$

The direct voltage $v_d(t)$ has the value

$$v_d(t) = e_1(t) = E_m \cos \omega t \qquad \left(-\frac{\pi}{6} \leqslant \omega t \leqslant \frac{\pi}{6}\right)$$

The average or dc value of the voltage is thus

$$V_d = \frac{1}{\frac{\pi}{3}} \int_{-\pi/6}^{\pi/6} (E_m) \cos\theta\, d\theta$$

$$= \frac{E_m}{\frac{\pi}{3}} (2 \sin 30°) = \frac{3 E_m}{\pi}$$

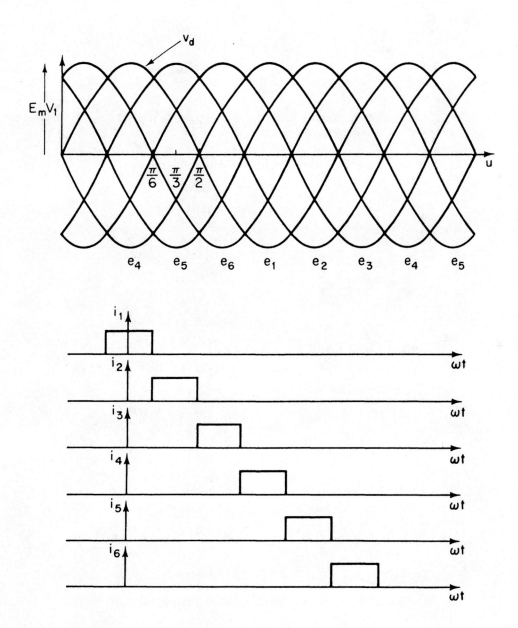

Figure 7-44. Waveforms for the Six-Phase Diametrical Converter.

Thus,

$$E_m = \frac{\pi}{3}(V_d) = 1.0472 V_d$$

$$E_{rms} = \frac{E_m}{\sqrt{2}} = 0.7405 V_d$$

The peak inverse voltage is obtained by considering $v_1(t)$, the voltage on valve 1. When valve 2 conducts,

$$v_1(t) = v_2(t) = e_1(t) - e_2(t)$$
$$= E_m \cos \omega t - E_m \cos(\omega t - 60°)$$
$$= E_m \cos(\omega t + 60°)$$

This is not the severest case, since when valve 4 conducts,

$$v_1(t) = v_4(t) = e_1(t) - e_4(t)$$
$$= 2 e_1(t)$$
$$= 2 E_m \cos(\omega t)$$

Thus the PIV is $2 E_m$.

$$\text{PIV} = 2 E_m$$
$$= 2.0944 V_d$$

The dc voltage ripple, peak-to-peak, is obtained as

$$\text{PPR} = E_m - e_a\left(\frac{\pi}{6}\right)$$
$$= E_m(1 - \cos 30°)$$
$$= 0.134 E_m$$
$$= 0.1403 V_d$$

Problem 7-A-9

Consider the cascade of three single-phase, full-wave rectifiers shown in Figure 7-45. Show that

A. Average valve current is $0.5 I_d$.
B. Peak inverse voltage on a valve is $1.047 V_d$.
C. Direct-current voltage ripple, peak to peak, is $0.14 V_d$.

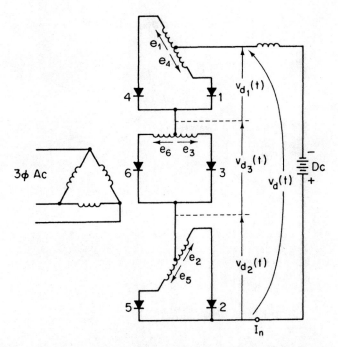

Figure 7-45. Cascade of Three Single-Phase, Full-Wave Rectifiers.

Solution

The transformer secondary voltages are assumed to be given by

$$e_1(t) = E_m \cos \omega t$$
$$e_2(t) = E_m \cos(\omega t - 120°)$$
$$e_3(t) = E_m \cos(\omega t - 240°)$$
$$e_4(t) = E_m \cos(\omega t - 180°)$$
$$= -e_1(t)$$
$$e_5(t) = E_m \cos(\omega t - 300°)$$
$$= -e_2(t)$$
$$e_6(t) = E_m \cos(\omega t - 60°)$$
$$= -e_3(t)$$

Valves 1 and 4 operate as a single-phase, full-wave rectifier with each valve conducting for one-half a cycle of the ac voltage. The output $v_{d_1}(t)$ is the positive values of $e_1(t)$ and $e_4(t)$. Valves 2 and 5 operate similarly, but the output is displaced from that of the first set by $2\pi/3$ electrical radians. The output is denoted $v_{d_2}(t)$. Valves 3 and 6 operate similarly, with a further $2\pi/3$ electrical radians displacement. Thus the voltage components are

$$v_{d_1}(t) = E_m \cos \omega t \qquad \left(-\frac{\pi}{2} \leq \omega t \leq \frac{\pi}{2}\right)$$

$$= -E_m \cos \omega t \qquad \left(\frac{\pi}{2} \leq \omega t \leq \frac{3\pi}{2}\right)$$

$$v_{d_2}(t) = E_m \cos\left(\omega t - \frac{2\pi}{3}\right) \qquad \left(\frac{\pi}{6} \leq \omega t \leq \frac{7\pi}{6}\right)$$

$$= -E_m \cos\left(\omega t - \frac{2\pi}{3}\right) \qquad \left(\frac{7\pi}{6} \leq \omega t \leq \frac{13\pi}{6}\right)$$

$$v_{d_3}(t) = E_m \cos\left(\omega t - \frac{4\pi}{3}\right) \qquad \left(\frac{5\pi}{6} \leq \omega t \leq \frac{11\pi}{6}\right)$$

$$= -E_m \cos\left(\omega t - \frac{4\pi}{3}\right) \qquad \left(\frac{11\pi}{6} \leq \omega t \leq \frac{17\pi}{6}\right)$$

The output voltage is clearly

$$v_d(t) = v_{d_1}(t) + v_{d_2}(t) + v_{d_3}(t)$$

The filtered output V_d is

$$V_d = \frac{1}{2\pi} \int_0^{2\pi} \left[v_{d_1}(t) + v_{d_2}(t) + v_{d_3}(t)\right] d\omega t$$

Thus,

$$V_d = V_{d_1} + V_{d_2} + V_{d_3} = 3V_{d_1}$$

where

$$V_{d_1} = \frac{1}{2\pi} \int_0^{2\pi} v_{d_1}(t)\, dt = \frac{2E_m}{\pi}$$

Therefore,

$$V_d = \frac{6E_m}{\pi} = 1.9099 E_m$$

with the inverse relation

$$E_m = 0.5236 V_d$$

The valve current is a pulse of height I_d and duration of one-half cycle. Thus,

$$I_{av} = \frac{I_d}{2}$$

The peak inverse voltage (PIV) from the full-wave rectifier discussion is

$$\text{PIV} = 2E_m$$
$$= 1.0472 V_d$$

The unfiltered output voltage expression $v_d(t)$ is obtained as follows:

$$v_d(t) = E_m\left[\cos\omega t - \cos\left(\omega t - \frac{2\pi}{3}\right) - \cos\left(\omega t - \frac{4\pi}{3}\right)\right]$$
$$= 2E_m\cos\omega t \qquad \left(-\frac{\pi}{6} \leqslant \omega t \leqslant \frac{\pi}{6}\right)$$

$$v_d(t) = E_m\left[\cos\omega t + \cos\left(\omega t - \frac{2\pi}{3}\right) - \cos\left(\omega t - \frac{4\pi}{3}\right)\right]$$
$$= 2E_m\cos\left(\omega t - \frac{\pi}{3}\right) \qquad \left(\frac{\pi}{6} \leqslant \omega t \leqslant \frac{\pi}{2}\right)$$

$$v_d(t) = E_m\left[-\cos\omega t + \cos\left(\omega t - \frac{2\pi}{3}\right) - \cos\left(\omega t - \frac{4\pi}{3}\right)\right]$$
$$= 2E_m\cos\left(\omega t - \frac{2\pi}{3}\right) \qquad \left(\frac{\pi}{2} \leqslant \omega t \leqslant \frac{5\pi}{6}\right)$$

$$v_d(t) = E_m\left[-\cos\omega t + \cos\left(\omega t - \frac{2\pi}{3}\right) + \cos\left(\omega t - \frac{4\pi}{3}\right)\right]$$
$$= 2E_m\cos(\omega t - \pi) \qquad \left(\frac{5\pi}{6} \leqslant \omega t \leqslant \frac{7\pi}{6}\right)$$

$$v_d(t) = E_m\left[-\cos\omega t - \cos\left(\omega t - \frac{2\pi}{3}\right) + \cos\left(\omega t - \frac{4\pi}{3}\right)\right]$$
$$= 2E_m\cos\left(\omega t - \frac{4\pi}{3}\right) \qquad \left(\frac{7\pi}{6} \leqslant \omega t \leqslant \frac{3\pi}{2}\right)$$

$$v_d(t) = E_m\left[\cos\omega t - \cos\left(\omega t - \frac{2\pi}{3}\right) + \cos\left(\omega t - \frac{4\pi}{3}\right)\right]$$
$$= 2E_m\cos\left(\omega t - \frac{5\pi}{3}\right) \qquad \left(\frac{3\pi}{2} \leqslant \omega t \leqslant \frac{11\pi}{6}\right)$$

The output voltage is shown in Figure 7-46. The peak-to-peak ripple is therefore obtained as

$$\text{PPR} = v_d(0) + v_d\left(\frac{\pi}{6}\right)$$
$$= 2E_m\left[1 - \cos\left(\frac{\pi}{6}\right)\right] = 0.2679 E_m$$

In terms of the dc voltage, we get
$$\text{PPR} = (0.2679)(0.5236)V_d = 0.1403V_d$$

Problem 7-A-10

Consider the cascade of two three-phase bridge rectifers shown in Figure 7-47. Show that

A. Average valve current is $0.33I_d$.

B. Peak inverse voltage on a valve is $0.524V_d$.

C. Direct-current voltage ripple, peak to peak, is $0.035V_d$.

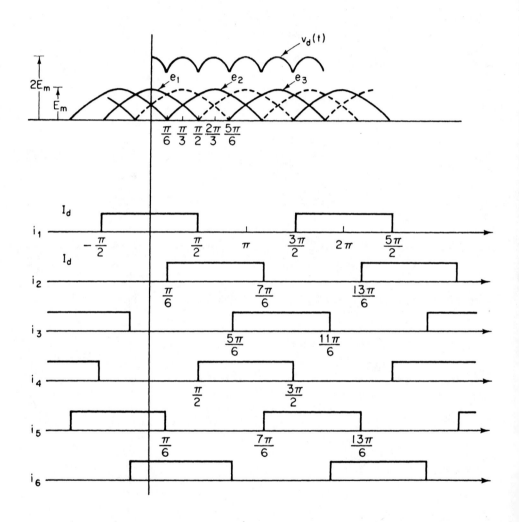

Figure 7-46. Waveforms for Cascade of Three Single-Phase, Full-Wave Rectifiers.

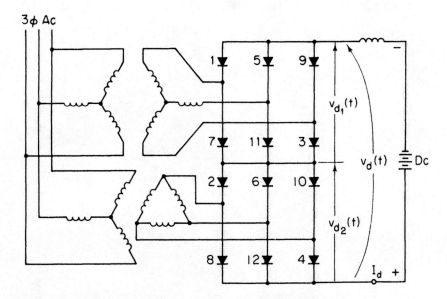

Figure 7-47. Cascade of Two Three-Phase Bridge Rectifiers.

Solution

The ac voltage of the lower bridge lags that of the upper by 30° electrical. For the top bridge, we have

$$v_{d_1}(t) = \sqrt{3}\, E_m \cos\left(\omega t - \frac{\pi}{6}\right) \qquad \left(0 \leqslant \omega t \leqslant \frac{\pi}{3}\right)$$

With the 30° shift, the bottom bridge's output unfiltered voltage is

$$v_{d_2}(t) = \sqrt{3}\, E_m \cos\left(\omega t - \frac{\pi}{3}\right) \qquad \left(\frac{\pi}{6} \leqslant \omega t \leqslant \frac{\pi}{2}\right)$$

The output voltage of the combination is thus

$$v_d(t) = \sqrt{3}\, E_m \left[\cos\left(\omega t - \frac{\pi}{6}\right) + \cos\left(\omega t - \frac{\pi}{3}\right)\right]$$

$$= 1.9319\sqrt{3}\, E_m \cos\left(\omega t - \frac{\pi}{4}\right) \qquad \left(\frac{\pi}{6} \leqslant \omega t \leqslant \frac{\pi}{3}\right)$$

The dc voltage V_d is thus

$$V_d = 1.9319\sqrt{3}\, E_m \left\{\left[\int_{\pi/6}^{\pi/3} \cos\left(\theta - \frac{\pi}{4}\right) d\theta\right] \Big/ \frac{\pi}{6}\right\}$$

$$= \frac{(6)(1.9319\sqrt{3})\, E_m}{\pi} 2\sin\left(\frac{\pi}{12}\right) = 3.3042\, E_m$$

This can also be obtained from the dc value of one bridge V_{d_1} as

$$V_d = 2V_{d_1} = 2\left[\frac{3\sqrt{3}}{\pi}(E_m)\right]$$

The transformer secondary voltage is given by the inverse relation

$$E_m = \frac{1}{3.3042} V_d$$

or

$$E_m = 0.3026 V_d$$

The peak inverse voltage and average current in a valve are given by the same expressions as for a single bridge:

$$I_{av} = \frac{I_d}{3}$$
$$= 0.333 I_d$$
$$\text{PIV} = \frac{\pi}{3}(V_{d_1})$$
$$= \frac{\pi}{3}\left(\frac{V_d}{2}\right)$$

Thus,

$$\text{PIV} = 0.5236 V_d$$

The peak-to-peak ripple is

$$\text{PPR} = v_d\left(\frac{\pi}{4}\right) - v_d\left(\frac{\pi}{3}\right)$$
$$= 1.9319\sqrt{3}\, E_m\left(1 - \cos\frac{\pi}{12}\right)$$
$$= 0.114 E_m$$

In terms of the dc voltage, we thus have

$$\text{PPR} = (0.114)(0.3026 V_d)$$

or

$$\text{PPR} = 0.0345 V_d$$

Problem 7-A-11

It is required to obtain a direct voltage of 40 kV from a bridge-connected rectifier operating with $\alpha = 30°$ and $\mu = 15°$. Calculate the necessary secondary voltage of the rectifier transformer, which is nominally rated at 230 kV/34.5 kV; and calculate the tap ratio required.

Solution

$$V_d = \frac{V_{d_0}}{2}[\cos(\alpha) + \cos(\alpha + \mu)]$$

$$40 = \frac{V_{d_0}}{2}(\cos 30° + \cos 45°)$$

Thus,

$$V_{d_0} = 50.85 \text{ kV}$$

but

$$V_{d_0} = \frac{3\sqrt{2}}{\pi}(V_L)$$

Thus,

$$V_L = 37.66 \text{ kV}$$

$$\text{Tap ratio} = \frac{37.66}{34.5} = 1.091$$

Problem 7-A-12

Assume for a three-phase bridge rectifier that the transformer secondary leakage reactance per phase is 0.4 ohm, and the secondary line voltage is 400 V. If the output current is 200 A, find the angle of overlap and the dc output voltage. Assume a delay angle of 15°.

Solution

We have

$$V_{d0} = \frac{3\sqrt{2}}{\pi}(V_L) = \frac{3\sqrt{2}}{\pi}(400) = 540.19 \text{ V}$$

$$R_c = \frac{3X_c}{\pi} = \frac{(3)(0.4)}{\pi} = 0.382 \text{ ohm}$$

$$I_{s_2} = \frac{V_{d0}}{2}\left(\frac{1}{R_c}\right) = 707.107 \text{ A}$$

$$I_d = I_{s_2}(\cos \alpha - \cos \delta)$$

$$200 = 707.107(\cos 15° - \cos \delta)$$

$$\delta = 46.915°$$

But

$$\delta = \alpha + \mu$$

Thus the overlap angle is obtained as
$$\mu = 31.915°$$
The dc output voltage is now calculated as
$$V_d = V_{d0}(\cos\alpha + \cos\delta)$$
$$= \frac{540.19}{2}(\cos 15° + \cos 46.915°)$$
$$= 445.389 \text{ V}$$

Problem 7-A-13

A three-phase bridge inverter has a commutating reactance of 150 ohms. The current and voltage at the dc side are 1053 A and 285 kV respectively. The ac line voltage is 345 kV. Calculate the extinction angle γ and the overlap angle.

Solution

Utilizing
$$V_d = V_{d0}\cos\gamma - \frac{3}{\pi}(I_d X_c)$$

we have
$$V_{d0} = \frac{3\sqrt{2}}{\pi}(V_L)$$
$$= 465.91 \text{ kV}$$

Thus
$$285 = 465.91 \cos\gamma - \frac{3}{\pi}(1053)(150)(10^{-3})$$

from which
$$\gamma = 20.70°$$

Now to find the overlap angle,
$$V_d = \frac{V_{d0}}{2}[\cos\gamma + \cos(\gamma + \mu)]$$
$$285 = \frac{465.91}{2}[\cos(20.7) + \cos(20.7 + \mu)]$$

As a result, we obtain the overlap angle
$$\mu = 52.56°$$

Problem 7-A-14

The ac line voltage of a three-phase bridge inverter is 160 kV when the extinction angle is 20° with an overlap of 20.342°. Calculate the dc voltage at the inverter. Calculate the necessary extinction angle to maintain the ac line voltage at 160 kV when the dc voltage drops to 180 kV. Assume the overlap angle to remain unchanged.

Solution

$$V_{do} = \frac{3\sqrt{2}}{\pi}(V_L) = 216.08 \text{ kV}$$

Now,

$$V_d = \frac{V_{do}}{2}[\cos\gamma + \cos(\gamma + \mu)]$$
$$= \frac{216.08}{2}(\cos 20° + \cos 40.342°)$$
$$= 183.87 \text{ kV}$$

which is the dc voltage required.

To maintain V_L constant, V_{do} will also be constant; thus

$$180 = \frac{216.08}{2}[\cos\gamma + \cos(\gamma + 20.342°)]$$

Thus,

$$\cos\gamma + \cos(\gamma + 20.342) = 1.67$$

Expanding the cosine term,

$$1.94 \cos\gamma - 0.35 \sin\gamma = 1.67$$

Let

$$\cos\theta = \frac{1.94}{\sqrt{(1.94)^2 + (0.35)^2}}$$

Thus

$$\theta = 10.17°$$

Then

$$\cos\theta \cos\gamma - \sin\theta \sin\gamma = 0.85$$

As a result,

$$\cos(\theta + \gamma) = 0.85$$

This gives

$$\theta + \gamma = 32.19$$

from which we calculate

$$\gamma = 22.01°$$

This is the new extinction angle required.

Problem 7-A-15

A HVDC link has the following parameters:

$$X_{cr} = 58 \text{ ohms}$$
$$R_{dc} = 3 \text{ ohms}$$
$$X_{ci} = 57 \text{ ohms}$$

The ac line voltage to the rectifier terminals is 320 kV when delivering 500 MW at 335 kV dc. The inverter operates with an extinction angle of 21.5°. Calculate:

A. The delay angle of the rectifier.
B. The ac line current and power factor at the rectifier terminals.
C. The ac line current, voltage, and power factor at the inverter terminals.

Solution

A. The dc current I_d is obtained as

$$I_d = \frac{500 \times 10^3}{335} = 1492.54 \text{ A}$$

The rectifier equation is applied. Thus,

$$335 = \left(\frac{3\sqrt{2}}{\pi}\right)(320)\cos\alpha - \frac{3}{\pi}(1492.54)(58)(10^{-3})$$

As a result, the delay angle is

$$\alpha = 14.88°$$

B. The ac line current is related to the dc current by

$$I_L = \frac{\sqrt{6}}{\pi}(I_d)$$

Thus,

$$I_{L_r} = 1163.73 \text{ A}$$

The ac power at the rectifier satisfies

$$P_{ac_r} = \sqrt{3}\,(V_{L_r} I_{L_r})\cos\phi_r$$
$$500 \times 10^6 = \sqrt{3}\,(320)(10^3)(1163.73)\cos\phi_r$$
$$\cos\phi_r = 0.775$$

C. The voltage equation of the dc line is

$$V_{dr} - V_{di} = I_d R_{dc}$$

Thus,
$$V_{di} = 335 - (1492.54)(3)(10^{-3})$$
$$= 330.522 \text{ kV}$$

The inverter equation is applied. Thus,
$$V_{d_i} = \frac{3\sqrt{2}}{\pi}(V_{L_i})\cos\gamma - \frac{3}{\pi}(I_d X_{c_i})$$

$$330.522 = \frac{3\sqrt{2}}{\pi}(V_{L_i})\cos(21.5) - \frac{3}{\pi}(1492.54)(57)(10^{-3})$$

The ac line voltage at the inverter is given by
$$V_{L_i} = 327.705 \text{ kV}$$

The ac line current is the same as that for the rectifier. The power at the inverter is given by
$$P_{d_i} = V_{d_i} I_d$$
$$= (330.522)(1492.54)(10^{-3})$$
$$= 493.139 \text{ MW}$$

For a loss-free inverter, the ac line power is
$$P_{ac_i} = 493.139 \text{ MW}$$

The power factor at the inverter ac terminals is obtained as
$$\cos\phi_i = \frac{493.139 \times 10^6}{(\sqrt{3})(327.705)(10^3)(1163.73)}$$
$$= 0.747$$

Problem 7-A-16

The output power of the dc link of Problem 7-A-15 is maintained at 500 MW at an ac line voltage of 330 kV and a power factor of 0.78. Calculate the necessary ac line voltage, current, and power factor at the rectifier terminals when operating with a 15° delay angle.

Solution

From the data at the inverter we have
$$I_{L_i} = \frac{500 \times 10^6}{(\sqrt{3})(330)(10^3)(0.78)} = 1121.50 \text{ A}$$

Thus
$$I_d = \frac{\pi}{\sqrt{6}}(I_L) = 1438.38 \text{ A}$$

The dc voltage at the inverter terminals is calculated using

$$P_{d_i} = V_{d_i} I_d$$

$$500 \times 10^6 = V_{d_i}(1438.38)$$

As a result,

$$V_{d_i} = 347.61 \text{ kV}$$

The rectifier voltage is now obtained using

$$V_{dr} = V_{d_i} + I_d R_{dc}$$

$$= 347.61 + (1438.38)(3)(10^{-3})$$

$$= 351.93 \text{ kV}$$

The rectifier voltage equation is

$$V_{dr} = \frac{3\sqrt{2}}{\pi}(V_{L_r})\cos\alpha - \frac{3}{\pi}(X_{c_r} I_d)$$

$$351.93 = \frac{3\sqrt{2}}{\pi}(V_{L_r})(\cos 15°) - \frac{3}{\pi}(58)(1438.38)(10^{-3})$$

As a result,

$$V_{L_r} = 330.86 \text{ kV}$$

This is the required line voltage at the rectifier's ac terminals. The power input to the rectifier is obtained from

$$P_{dc_r} = V_{d_r} I_d$$

$$= (351.93)(1438.38)(10^{-3})$$

$$= 506.21 \text{ MW}$$

The line current at the rectifier is the same as that at the inverter:

$$I_{L_r} = 1121.50 \text{ A}$$

For a loss-free rectifier,

$$P_{ac_r} = P_{dc_r}$$

$$506.21 \times 10^6 = \sqrt{3}\,(330.86 \times 10^3)(1121.50)\cos\phi_r$$

$$\cos\phi_r = 0.7876$$

Problem 7-A-17

Figure 7-48 shows a single-line diagram of a three-terminal dc system. Calculate the ac line voltage, current, and power factor at the inverter terminals labeled 3.

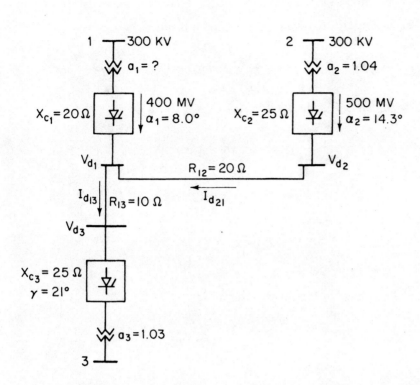

Figure 7-48. A Three-Terminal dc System for Problem 7-A-17.

Solution

We start at the rectifier labeled 2. The dc current is related to the voltage by

$$V_{d_2} I_{d_2} = 500 \times 10^6$$

As a result, the rectifier's voltage equation is written as

$$\frac{500 \times 10^3}{I_{d_2}} = \frac{3\sqrt{2}}{\pi}(300)(1.04)\cos 14.3° - \frac{3}{\pi}(I_{d_2})(25)(10^{-3})$$

Solving the quadratic equation, we get

$$I_{d_2} = 1327.68 \text{ A}$$

from which

$$V_{d_2} = \frac{500 \times 10^3}{1327.68} = 376.6 \text{ kV}$$

The current $I_{d_{21}}$ is the same as I_{d_2}. As a result,
$$V_{d_1} = V_{d_2} - I_{d_{21}} R_{12}$$
$$= 376.6 - (1327.68)(20)(10^{-3})$$
$$= 350.04 \text{ kV}$$

The current I_{d_1} is obtained from
$$P_{d_1} = V_{d_1} I_{d_1}$$
$$400 \times 10^3 = (350.04) I_{d_1}$$

Thus,
$$I_{d_1} = 1142.72 \text{ A}$$

The current $I_{d_{13}}$ is
$$I_{d_{13}} = I_{d_1} + I_{d_{21}}$$
$$= 1142.72 + 1327.68$$
$$= 2470.40 \text{ A}$$

The dc voltage at the inverter terminal is
$$V_{d_3} = V_{d_1} - I_{d_{13}} R_{13}$$
$$= 350.04 - (2470.40)(10)(10^{-3})$$
$$= 325.34 \text{ kV}$$

The current through the inverter is $I_{d_{13}}$. Thus the power to the inverter is
$$P_{d_3} = V_{d_3} I_{d_{13}}$$
$$= 803.71 \text{ MW}$$

The inverter's ac line voltage is obtained from
$$V_{d_3} = \frac{3\sqrt{2}}{\pi}(a_3 V_{L_3}) \cos \alpha - \frac{3}{\pi}(I_{d_3} X_{c_3})$$
$$325.34 = \frac{3\sqrt{2}}{\pi}(1.03)(V_{L_3}) \cos 21° - \frac{3}{\pi}(2470.4)(25)(10^{-3})$$

As a result, we obtain
$$V_{L_3} = 295.95 \text{ kV}$$

The line ac current $I_{L_{3'}}$ is related to I_{d_3} by
$$I_{L_{3'}} = \frac{\sqrt{6}}{\pi}(I_{d_3})$$

Thus the line current on the bus 3 side is given by
$$I_{L_3} = a_3 I_{L_{3'}}$$
$$I_{L_3} = 1983.94 \text{ A}$$

The power factor is now calculated from
$$P_{d_3} = \sqrt{3}\,(V_{L_3}I_{L_3})\cos\phi_3$$
$$803.71 \times 10^3 = \sqrt{3}\,(295.95)(1983.94)\cos\phi_3$$
from which
$$\cos\phi_3 = 0.7903$$

Problem 7-A-18

Assume that a per unit system is adopted for the three-terminal dc system of Problem 7-A-17. The following base values are given:
$$P_b = 500 \text{ MW}$$
$$V_b = 300 \text{ kV}$$
Convert the data of the problem into the new per unit system and repeat the solution procedure.

Solution

The base dc current is
$$I_{b_{dc}} = \frac{P_b}{V_b} = \frac{500 \times 10^3}{300} = 1666.67 \text{ A}$$
The base resistance on the dc side is
$$Z_{b_{dc}} = \frac{V_b}{I_{b_{dc}}} = 180 \text{ ohms}$$
The base ac current is
$$I_{b_{ac}} = \frac{I_{b_{dc}}}{\sqrt{3}} = 962.25 \text{ A}$$

In the per unit (p.u.) system, the given data are as follows:
$$P_{1_{p.u.}} = \frac{P_1}{P_b} = \frac{400}{500} = 0.8 \text{ p.u.}$$
$$P_{2_{p.u.}} = \frac{500}{500} = 1 \text{ p.u.}$$
$$V_{L_{1_{p.u.}}} = V_{L_{2_{p.u.}}} = \frac{300}{300} = 1 \text{ p.u.}$$
$$X_{c_{1_{p.u.}}} = \frac{20}{180} = 0.111$$
$$X_{c_{2_{p.u.}}} = X_{c_{3_{p.u.}}} = \frac{25}{180} = 0.1389$$
$$R_{12_{p.u.}} = \frac{20}{180} = 0.1111 \text{ p.u.}$$
$$R_{13_{p.u.}} = \frac{10}{180} = 0.0556 \text{ p.u.}$$

The solution procedure in per unit values is as follows. (We drop the subscript *pu* for simplicity.)

$$\frac{1}{I_{d_2}} = \frac{3\sqrt{2}}{\pi}(1)(1.04)\cos 14.3° - \frac{3}{\pi}(I_{d_2})(0.1389)$$

The solution for I_{d_2} is

$$I_{d_2} = 0.7966 \text{ p.u.}$$

From this we get

$$V_{d_2} = \frac{1}{0.7966} = 1.2553 \text{ p.u.}$$

Now the rectifier terminal voltage is

$$V_{d_1} = 1.2553 - (0.7966)(0.111)$$
$$= 1.1668 \text{ p.u.}$$

The current in rectifier 1 is obtained from the power and voltages as

$$I_{d_1} = \frac{0.8}{1.1668} = 0.6856 \text{ p.u.}$$

The current $I_{d_{13}}$ is the sum of I_{d_2} and I_{d_1}. Thus,

$$I_{d_{13}} = 0.6856 + 0.7966$$
$$= 1.4822 \text{ p.u.}$$

The inverter's dc voltage is obtained from the voltage drop equation:

$$V_{d_3} = 1.1668 - (1.4822)(0.0556)$$
$$= 1.0844 \text{ p.u.}$$

The dc power to the inverter is now given by

$$P_{d_3} = (1.0844)(1.4822)$$
$$= 1.6073 \text{ p.u.}$$

The inverter's ac line voltage is obtained using

$$V_{d_3} = \frac{3\sqrt{2}}{\pi}(a_3 V_{L_3})\cos\gamma - \frac{3}{\pi}(I_{d_3} X_{c_3})$$

$$1.0844 = \frac{3\sqrt{2}}{\pi}(1.03)(V_{L_3})\cos 21° - \frac{3}{\pi}(1.4822)(0.1389)$$

$$V_{L_3} = 0.986 \text{ p.u.}$$

The ac current in this per unit system is related to the dc current. Thus,

$$I_{L_{3'}} = \frac{3\sqrt{2}}{\pi}(I_{d_3}) = 2.00 \text{ p.u.}$$

$$I_{L_3} = a_3 I_{L_{3'}} = 2.06$$

$$P_{d_3} = V_3 I_{L_3} \cos\phi_3$$

$$1.6073 = (0.986)(2.06)\cos\phi_3$$

$$\cos\phi_3 = 0.7903$$

PROBLEMS

Problem 7-B-1

A 400-kV, direct-current transmission option is compared with a 760-kV, three-phase alternating current option for equal power transmission. Calculate the ratio of insulation levels and hence the ratio of transmission losses.

Problem 7-B-2

A direct-current transmission option is desired such that the dc power loss is 150 percent of the power losses when ac of 345 kV is used to transmit the same power. Calculate the required dc voltage and the ratio of insulation level.

Problem 7-B-3

The losses for a proposed ac line are 120 MW. Find the corresponding losses if a dc line is designed such that the ratio of insulation level for the dc line to that for the equivalent ac three-phase line is 0.87. Assuming that the dc line resistance is 40 ohms, find the ac and dc line currents.

Problem 7-B-4

Repeat Problem 7-B-2 if the ac voltage is 760 kV.

Problem 7-B-5

Repeat Problem 7-B-3 if the ac line losses are 100 MW.

Problem 7-B-6

A. Assume that a design choice calls for a ratio y of the losses in the bipolar dc case to losses in the three-phase ac alternative. Thus,

$$\frac{P_{Ldc}}{P_{Lac}} = y$$

Assume that the power transfer for each alternative is not equal, with the ratio of dc power transmitted to that of the ac by

$$\frac{P_{dc}}{P_{ac}} = \beta$$

Further, we account for skin effects by assuming that the line resistance in the ac operation is related to the dc line resistance by

$$\frac{R_{ac}}{R_{dc}} = \gamma$$

Show that under these conditions the ratio of insulation levels is given by

$$\frac{\text{Insulation level for dc}}{\text{Insulation level for ac}} = \frac{0.87\beta \cos\phi}{\sqrt{\gamma y}}$$

where $\cos\phi$ is the power factor in the ac option.

B. Assume that for a certain conductor type,

$$R_{ac} = 1.1 R_{dc}$$

Find the ratio of ac line losses to the corresponding dc losses if the dc option is designed to carry 120 percent of the corresponding ac power while maintaining the same insulation levels. Assume $\cos\phi = 0.85$ for ac operation.

Problem 7-B-7

A. Assume that for a certain conductor type,

$$R_{ac} = 1.1 R_{dc}$$

Find the ratio of ac line losses to the corresponding dc losses if the dc option is designed to carry 150 percent of the corresponding ac power while maintaining the same insulation levels. Assume ac power factor of unity. (Use results given in Problem 7-B-6.)

B. For the conductors of part (a) and the same power-carrying capacity indicated, find the ratio of insulation levels for equal transmission losses in both ac and dc options. (Use results given in Problem 7-B-6.)

Problem 7-B-8

Show that for the minimum cost of operating the HVDC option under the conditions outlined in Problem 7-B-6, the minimization procedure gives

$$By = \frac{0.87\beta \tilde{A} \cos \phi}{2\sqrt{\gamma}\sqrt{y}}$$

where B and $\tilde{A}$ are as defined in text.

Problem 7-B-9

The ratio $(\tilde{A}/B)$ is 1.6 for a contemplated HVDC option. Assume that the ac alternative to transmit the same power operates at a power factor of 0.8 and that the ac line resistance is 110 percent of the dc resistance. Find the ratio of dc power losses to that of the alternative ac option.

Problem 7-B-10

A three-phase, two-way bridge converter (Graetz circuit) delivers 2000 A when the dc voltage is 40 kV. Calculate:

A. The peak inverse voltage.

B. Peak-to-peak ripple.

C. Valve rating.

D. Transformer secondary voltampere rating.

Assume ideal rectifier characteristics.

Problem 7-B-11

Repeat Problem 7-B-10 for

$$V_d = 250 \text{ kV}$$
$$I_{dc} = 1000 \text{ A}$$

Problem 7-B-12

Repeat Problem 7-B-10 for

$$V_d = 390 \text{ kV}$$
$$I_{dc} = 1000 \text{ A}$$

Problem 7-B-13

Repeat Problem 7-B-10 for

$$V_d = 250 \text{ kV}$$
$$P_{dc} = 140 \text{ MW}$$

Problem 7-B-14

Repeat Problem 7-B-10 if a three-phase, one-way converter circuit is used. Compare the results.

Problem 7-B-15

Assume that the line-to-line-secondary transformer voltage in a three-phase, two-way bridge converter is 34.5 kV while delivering dc power of 80 MW. Assume ideal rectifier characteristics, and calculate:

A. The dc voltage.
B. The peak inverse voltage.
C. Peak-to-peak ripple.
D. Valve rating.
E. Transformer secondary voltampere rating.

Problem 7-B-16

A three-phase bridge converter operates as a rectifier with a delay angle of 8° when the line voltage on the transformer secondary is 170 kV. Assuming that the dc power delivered is 200 MW, calculate the direct voltage and current at the rectifier terminals. (Neglect overlap.)

Problem 7-B-17

A three-phase bridge rectifier operates with zero delay angle to deliver dc power and current P_{d_1} and I_{d_1} respectively. If the delay angle is α_2, the dc power and current are P_{d_2} and I_{d_2} respectively. Find α_2, given that

$$\frac{P_{d_1}}{P_{d_2}} = 1.2 \quad \text{and} \quad \frac{I_{d_1}}{I_{d_2}} = 1.1$$

(Neglect overlap.)

Problem 7-B-18

The transformer secondary line voltage to a three-phase bridge rectifier is 180 kV. Calculate the dc voltage output for a delay angle of 15° and an overlap angle of 20°.

Problem 7-B-19

Calculate the delay angle α for a three-phase bridge rectifier if the transformer secondary line voltage is 175 kV when the dc output voltage is 200 kV for an overlap angle of 21°.

Problem 7-B-20

The transformer secondary line voltage to a three-phase rectifier is 169.46 kV when the dc terminal voltage is 195 kV at a delay angle of 19°. Calculate the angle of overlap.

Problem 7-B-21

A three-phase bridge rectifier operates with a delay angle of 18.97° when delivering a dc current of 1025.5 A at a dc voltage of 195 kV. Assuming the commutating reactance to be 21.8 ohms, find the necessary ac line voltage.

Problem 7-B-22

Calculate the delay angle for a three-phase bridge rectifier supplying a load of 160 MW when the dc output voltage is 200 kV for an ac line voltage of 180 kV. Assume commutating reactance of 20 ohms.

Problem 7-B-23

The ac line voltage to a three-phase bridge rectifier is 170 kV when delivering dc power of 200 MW. Assuming that the delay angle is 20° and that the commutating reactance is 20 ohms, calculate the dc current.

Problem 7-B-24

The dc voltage and current at the terminals of a three-phase bridge converter operating as an inverter are 170 kV and 900 A respectively. Assuming that the commutating reactance of the inverter is 20 ohms and that the extinction angle is 20°, find the line voltage at the ac terminals of the inverter.

Problem 7-B-25

The line voltage at the ac terminals of a three-phase bridge inverter is 158.5 kV when receiving dc current of 1025 A at 183.36 kV dc. Assume that the commutating reactance is 20.643 ohms. Calculate the extinction angle γ.

Problem 7-B-26

The dc power into a three-phase bridge inverter is 200 MW when the output ac line voltage is 160 kV at an extinction angle of 18°. Assuming that the commutating reactance is 20 ohms, find the dc current into the inverter.

Problem 7-B-27

A HVDC link is operating with an ac line voltage to the rectifier of 175 kV. When the dc current is 1100 A, the rectifier delay angle is 10° and the inverter's extinction angle is 20°. The resistance of the line is 23 ohms, commutating reactance of the rectifier is 22 ohms and the commutating reactance of the inverter is 21 ohms. Calculate the dc voltages at the rectifier and inverter terminals as well as the ac line voltage at the inverter.

Problem 7-B-28

Assume that the link of Problem 7-B-27 is delivering 1000 A dc with the ac line voltage to the rectifier being 180 kV, and that of the inverter is 165 kV. Assume that the delay angle is now 15°. Calculate the dc voltages at the rectifier and inverter terminals as well as the inverter's extinction angle.

Problem 7-B-29

The link of Problem 7-B-27 is operating with a delay angle of 15° and extinction angle of 18°. If the output ac line voltage is 170 kV and the dc current is 1000 A, calculate the dc voltages at both ends and the ac line voltage at the rectifier terminals.

Problem 7-B-30

The link of Problem 7-B-27 is delivering a dc current of 1050 A when the ac line voltage at the rectifier is 190 kV and that at the inverter is 175 kV. Assume the inverter's extinction angle is 18°. Calculate the dc voltages and the rectifier's delay angle.

Problem 7-B-31

The link of Problem 7-B-27 is operating with a delay angle of 15° and an extinction angle of 20° when the ac line voltages at the rectifier and inverter are 185 kV and 170 kV respectively. Find the dc current delivered.

Problem 7-B-32

The link of Problem 7-B-27 is operating with a delay angle of 20° and an extinction angle of 18° when the dc voltage at the rectifier is 200 kV and that of the inverter is 180 kV. Calculate the dc current and the ac line voltages.

Problem 7-B-33

Repeat Problem 7-B-27 if the dc power is specified as 200 MW instead of the dc current specified in that problem.

Problem 7-B-34

Repeat Problem 7-B-28 for the dc power specified as 200 MW instead of the dc current specification.

Problem 7-B-35

Repeat Problem 7-B-29 for 200 MW dc power replacing the current specification.

Problem 7-B-36

Repeat Problem 7-B-30 for 200 MW dc power replacing the current specification.

Problem 7-B-37

Consider the conversion of Problem 7-B-27 to a per unit representation. Assume the power base to be

$$P_b = 200 \text{ MW}$$

and voltage base to be

$$V_b = 200 \text{ kV}$$

for both ac and dc sides.

A. Find the base current and resistance on the dc side as well as on the ac side.

B. Repeat the solution to the problem in the given per unit system.

Problem 7-B-38

Repeat Problem 7-B-37 assuming the power base is

$$P_b = 100 \text{ MW}$$

and voltage base is

$$V_b = 172.5 \text{ kV}$$

Problem 7-B-39

Consider the dc link shown in Figure 7-49. Assume that the ac voltage at transformer primary is given by

$$|V_1| = 1.02 \text{ p.u.}$$

The rectifier's commutating reactance is 0.10 p.u. and that of the inverter is 0.08 p.u. The tap ratio of the rectifier transformer is 1.05 and that for the inverter transformer is 1.03. The dc link's line resistance is 0.004 p.u. Assume that the rectifier's dc voltage output is 1.280 p.u. when delivering dc

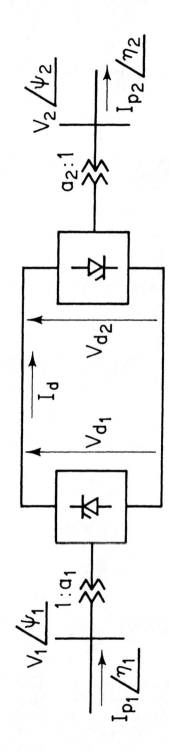

Figure 7-49. Schematic for Problem 7-B-39.

current of 0.500 p.u. Calculate:

A. The delay angle of the rectifier.
B. The power factor on the primary side of the rectifier's transformer.
C. The dc voltage at the inverter terminals.
D. The inverter's extinction angle if ac voltage is maintained at 1.00 p.u.
E. The power factor at the receiving end.

Problem 7-B-40

Consider the dc link shown in Figure 7-49. Assume that the dc voltage at the transformer primary is given by

$$|V_1| = 1.03 \text{ p.u.}$$

The rectifier's commutating reactance is 0.126 p.u. and that of the inverter is 0.07275 p.u. The rectifier is operating with a 9° delay angle, and the inverter is operating with a 10° extinction angle. The dc link's line resistance is 0.00334 p.u. Assume that the rectifier's dc voltage output is 1.286 p.u. when delivering dc current of 0.456 p.u. Calculate:

A. The tap ratio of the rectifier's transformer.
B. The power factor on the primary side of the rectifier's transformer.
C. The dc voltage at the inverter terminals.
D. The input active power to the link, dc line loss, and output active power.
E. The inverter's transformer tap ratio if the ac voltage is maintained at 1.06 p.u.
F. The power factor at the receiving end.

CHAPTER VIII

Faults on Electric Energy Systems

8.1 INTRODUCTION

A fault occurs when two or more conductors that normally operate with a potential difference come in contact with each other. The contact may be a physical metallic one, or it may occur through an arc. In the metal-to-metal contact case, the voltage between the two parts is reduced to zero. On the other hand, the voltage through an arc will be of a very small value. Faults in three-phase systems are classified as:

1. Balanced or symmetrical three-phase faults.
2. Single line-to-ground faults.
3. Line-to-line faults.
4. Double line-to-ground faults.

Generators may fail due to insulation breakdown between turns in the same slot or between the winding and the steel structure of the machine. The same is true for transformers. The breakdown is the result of insulation deterioration combined with switching and/or lightning overvoltages. Overhead lines are constructed of bare conductors. Wind, sleet, trees, cranes, kites, airplanes, birds, or damage to supporting structure are causes for accidental faults on overhead lines. Contamination of insulators and lightning overvoltages will in general result in faults. Deterioration of insulation in underground cables results in short circuits. This is mainly attributed to aging combined with overloading. About 75 percent of the energy system's faults are due to the second category and result from insulator flashover during electrical storms. Only one in twenty faults is due to the balanced category.

As a result of a fault, currents of high value will flow through the network to the faulted point. The amount of current will be much greater than the designed thermal ability of the conductors in the power lines or machines feeding the fault. As a result, temperature rise may cause damage by annealing of conductors and insulation charring. In addition to this, the low voltage in the neighborhood of the fault will render the equipment inoperative.

Fault or short-circuit studies are obviously an essential tool for the electric energy systems engineer. The task here is to be able to calculate the fault conditions and to provide protective equipment designed to isolate the faulted zone from the remainder of the system in the appropriate time. The least complex category computationally is the balanced fault. This tempts the engineer to base his decisions on its results. The balanced fault could (in some locations) result in currents smaller than that due to any other type of fault. However, the interrupting capacity of breakers should be chosen to accommodate the largest of fault currents.

8.2 TRANSIENTS DURING A BALANCED FAULT

The dependence of the value of the short-circuit current in the electric power system on the instant in the cycle at which the short circuit occurs can be verified using a very simple model. The model is a generator with series resistance R and inductance L as shown in Figure 8-1. The voltage of the generator is assumed to vary as

$$e(t) = E_m \sin(\omega t + \alpha) \tag{8.1}$$

With a balanced fault placed on the generator terminals at $t = 0$, we can show that a dc term will in general exist. Its initial magnitude may be equal to the magnitude of the steady-state current term.

The transient current $i(t)$ is given by

$$i(t) = \left(\frac{E_m}{Z}\right)\left[\sin(\omega t + \alpha - \Theta) - \sin(\alpha - \Theta)e^{-RT/L}\right] \tag{8.2}$$

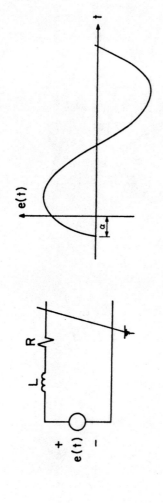

Figure 8-1. (a) Generator Model. (b) Voltage Waveform.

where

$$Z = (R^2 + \omega^2 L^2)^{1/2}$$
$$\Theta = \tan^{-1}\left(\frac{\omega L}{R}\right)$$

The worst possible case occurs for the value of α given by

$$\tan \alpha = -\frac{R}{\omega L}$$

In this case, the current magnitude will approach twice the steady-state maximum value immediately after the short circuit. The transient current is given by

$$i(t) = \frac{E_m}{Z}(-\cos \omega t + e^{-Rt/L}) \tag{8.3}$$

For small t,

$$e^{-Rt/L} \cong 1$$

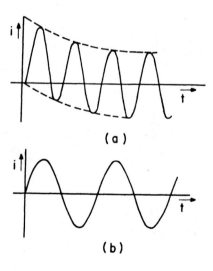

(a)

(b)

Figure 8-2. (a) Short-Circuit Current Wave Shape for tan $\alpha = -(R/\omega L)$. (b) Short-Circuit Current Wave Shape for tan $\alpha = (\omega L/R)$.

8.2 Transients During a Balanced Fault

Thus,

$$i(t) = \frac{E_m}{Z}(1 - \cos \omega t) \tag{8.4}$$

It is clear that

$$i_{max} = \frac{2E_m}{Z}$$

This waveform is shown in Figure 8-2(a).

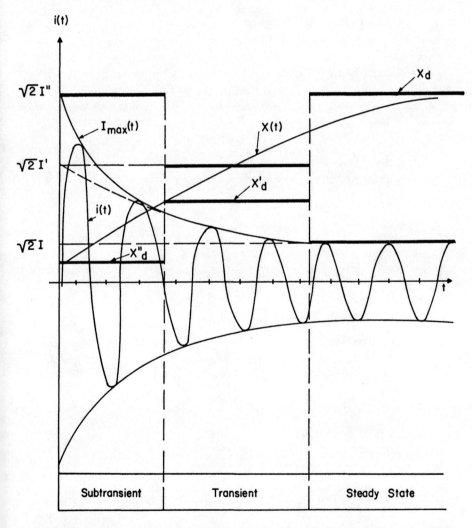

Figure 8-3. Symmetrical Short-Circuit Current and Reactances for a Synchronous Machine.

For the case

$$\tan \alpha = \frac{\omega L}{R}$$

we have

$$i(t) = \frac{E_m}{Z} \sin \omega t \qquad (8.5)$$

This waveform is shown in Figure 8-2(b).

It is clear from inspection of either the expression for the short-circuit current given by Eq. (8.2) or the response waveform given in Figure 8-3 that the reactance of the machine appears to be time-varying. This is so if we assume a fixed voltage source E. For our power system purposes, we let the reactance vary in a stepwise fashion X_d'', X_d', and X_d as shown in Figure 8-3.

The current history $i(t)$ can be approximated in three time zones by three different expressions. In the first, denoted the subtransient interval and lasting up to two cycles, the current is I''. This defines the direct-axis subtransient reactance:

$$X_d'' = \frac{E}{I''} \qquad (8.6)$$

The second, denoted the transient interval, gives rise to

$$X_d' = \frac{E}{I'} \qquad (8.7)$$

where I' is the transient current and X_d' is direct-axis transient reactance. The transient interval lasts for about 30 cycles.

The steady-state condition gives the direct-axis synchronous reactance:

$$X_d = \frac{E}{I} \qquad (8.8)$$

Table 8-1 lists typical values of the reactances defined in Eqs. (8.6), (8.7), and (8.8). Note that the subtransient reactance can be as low as 7 percent of the synchronous reactance.

8.3 THE METHOD OF SYMMETRICAL COMPONENTS

The method of symmetrical components is used to transform an unbalanced three-phase system into three sets of balanced three-phase phasors. The basic idea of the transformation is simple. Given three voltage phasors V_A, V_B, and V_C, it is possible to express each as the sum of three

TABLE 8-1
Typical Sequence Reactance Values for Synchronous Machines

	Two-Pole Turbine Generator			Four-Pole Turbine Generator			Salient-Pole Machine with Dampers			Salient-Pole Generator without Dampers			Synchronous Condensers		
	Low	Avg.	High	Low	Avg.	High	Low	Avg.	High	Low	Avg.	High	Low	Avg.	High
X_d	0.95	1.2	1.45	1.00	1.2	1.45	0.6	1.25	1.50	0.6	1.25	1.5	1.25	2.2	2.65
X_d'	0.12	0.15	0.21	0.20	0.23	0.28	0.20	0.30	0.50	0.20	0.30	0.50	0.30	0.48	0.60
X_d''	0.07	0.09	0.14	0.12	0.14	0.17	0.13	0.2	0.32	0.20	0.30	0.50	0.19	0.32	0.36
X_-	0.07	0.09	0.14	0.12	0.14	0.17	0.13	0.2	0.32	0.35	0.48	0.65	0.18	0.31	0.48
X_0	0.01	0.03	0.08	0.015	0.08	0.14	0.03	0.18	0.23	0.03	0.19	0.24	0.025	0.14	0.18

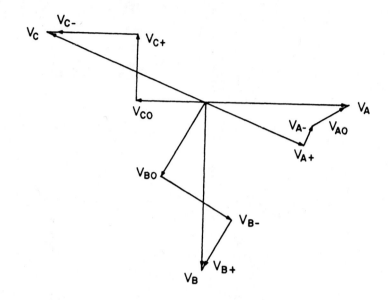

Figure 8-4. An Unbalanced Set of Voltage Phasors and a Possible Decomposition.

phasors as follows:

$$V_A = V_{A+} + V_{A-} + V_{A0} \tag{8.9}$$
$$V_B = V_{B+} + V_{B-} + V_{B0} \tag{8.10}$$
$$V_C = V_{C+} + V_{C-} + V_{C0} \tag{8.11}$$

Figure 8-4 shows the phasors V_A, V_B, and V_C as well as a particular possible choice of the decompositions.

Obviously the possible decompositions are many. We require that the sequence voltages V_{A+}, V_{B+}, and V_{C+} form a balanced positively rotating system. Thus the phasor magnitudes are equal, and the phasors are 120° apart in a sequence $A-B-C$, as shown in Figure 8-5(a). As a result, we have

$$V_{B+} = e^{-j120°} V_{A+}$$
$$V_{C+} = e^{j120°} V_{A+}$$

Similarly, we require that the sequence voltages V_{A-}, V_{B-}, and V_{C-} form a balanced negatively rotating system. This differs from the first

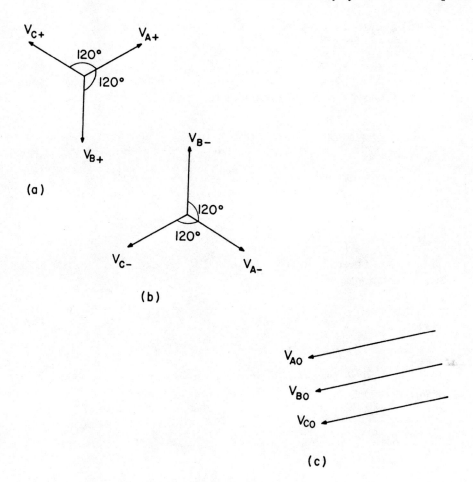

Figure 8-5. (a) Positive Sequence Voltage Phasors.
(b) Negative Sequence Voltage Phasors.
(c) Zero Sequence Voltage Phasors.

requirement in that the sequence is C–B–A, as shown in Figure 8-5(b). Thus,

$$V_{B-} = e^{j120} V_{A-}$$
$$V_{C-} = e^{-j120} V_{A-}$$

The sequence voltages $V_{A_0}, V_{B_0}, V_{C_0}$ are required to be equal in magnitude and phase. Thus,

$$V_{B0} = V_{A0}$$
$$V_{C0} = V_{A0}$$

478   Faults on Electric Energy Systems

For notational simplicity, we introduce the complex operator α defined by

$$\alpha = e^{j120°} \tag{8.12}$$

We can thus rewrite our requirements as

$$V_{B+} = \alpha^2 V_{A+} \tag{8.13}$$
$$V_{C+} = \alpha V_{A+} \tag{8.14}$$
$$V_{B-} = \alpha V_{A-} \tag{8.15}$$
$$V_{C-} = \alpha^2 V_{A-} \tag{8.16}$$
$$V_{B0} = V_{A0} \tag{8.17}$$
$$V_{C0} = V_{A0} \tag{8.18}$$

The original phasor voltages V_A, V_B, and V_C are expressed in terms of the sequence voltages as

$$V_A = V_{A+} + V_{A-} + V_{A0} \tag{8.19}$$
$$V_B = \alpha^2 V_{A+} + \alpha V_{A-} + V_{A0} \tag{8.20}$$
$$V_C = \alpha V_{A+} + \alpha^2 V_{A-} + V_{A0} \tag{8.21}$$

The inverse relation giving the positive sequence voltage V_{A+}, the negative sequence voltage V_{A-}, and the zero sequence voltage V_{A0} is obtained by solving the above three simultaneous equations to give

$$V_{A+} = \frac{1}{3}(V_A + \alpha V_B + \alpha^2 V_C) \tag{8.22}$$

$$V_{A-} = \frac{1}{3}(V_A + \alpha^2 V_B + \alpha V_C) \tag{8.23}$$

$$V_{A0} = \frac{1}{3}(V_A + V_B + V_C) \tag{8.24}$$

Some of the properties of the operator α are as follows:

$$\alpha^2 = \alpha^{-1}$$
$$\alpha^3 = 1$$
$$1 + \alpha + \alpha^2 = 0$$
$$\alpha^* = \alpha^2$$
$$(\alpha^2)^* = \alpha$$
$$1 + \alpha^* + (\alpha^*)^2 = 0$$

where as usual the asterisk ($*$) denotes complex conjugation.

8.3 The Method of Symmetrical Components

The first property is obtained as follows:

$$\alpha^2 = (e^{j120})^2$$
$$= e^{j240}$$
$$= e^{-j120°}$$
$$= (e^{j120°})^{-1}$$
$$= \alpha^{-1}$$

The second property follows by multiplying both sides of the first by α.

For clarity, we will drop the suffix A from the sequence voltage symbols. Thus,

$$V_+ = V_{A+} \qquad (8.25)$$
$$V_- = V_{A-} \qquad (8.26)$$
$$V_0 = V_{A0} \qquad (8.27)$$

Our relations are thus

$$V_A = V_+ + V_- + V_0 \qquad (8.28)$$
$$V_B = \alpha^2 V_+ + \alpha V_- + V_0 \qquad (8.29)$$
$$V_C = \alpha V_+ + \alpha^2 V_- + V_0 \qquad (8.30)$$

and

$$V_+ = \frac{1}{3}(V_A + \alpha V_B + \alpha^2 V_C) \qquad (8.31)$$
$$V_- = \frac{1}{3}(V_A + \alpha^2 V_B + \alpha V_C) \qquad (8.32)$$
$$V_0 = \frac{1}{3}(V_A + V_B + V_C) \qquad (8.33)$$

We have the following two examples:

Example 8-1

Given the system of unbalanced voltages:

$$V_A = 1\underline{/0°} = 1$$
$$V_B = 1\underline{/-120°} = \alpha^2$$
$$V_C = 0$$

find the positive, negative, and zero sequence voltages.

Solution

We can obtain the sequence voltages using Eq. (8.31), (8.32), and (8.33) as follows:

$$V_+ = \frac{1}{3}(1 + \alpha^3) = \frac{2}{3}$$

$$V_- = \frac{1}{3}(1 + \alpha^4) = \frac{1 + \alpha}{3}$$

$$= -\frac{\alpha^2}{3}$$

$$= \frac{(e^{j180})(e^{j240})}{3}$$

$$= \frac{e^{j60}}{3}$$

$$V_0 = \frac{1}{3}(1 + \alpha^2) = -\frac{\alpha}{3}$$

$$= \frac{e^{j300}}{3}$$

$$= \frac{e^{-j60}}{3}$$

The decomposition of V_A is as shown in Figure 8-6.

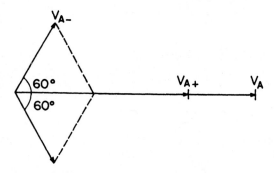

Figure 8-6. Decomposition of V_A for Example 8-1.

Example 8-2

Given that

$$V_0 = 100$$
$$V_+ = 200 - j100$$
$$V_- = -100$$

find the phase voltages V_A, V_B, and V_C.

Solution

$$\begin{aligned}
V_A &= V_+ + V_- + V_0 \\
&= 100 + 200 - j100 - 100 = 200 - j100 \\
V_B &= \alpha^2 V_+ + \alpha V_- + V_0 \\
&= \left(1\underline{/240°}\right)\left(223.61\underline{/-26.57°}\right) + \left(1\underline{/120°}\right)(-100) + 100 \\
&= 212.98\underline{/-99.9°} \\
V_C &= \alpha V_+ + \alpha^2 V_- + V_0 \\
&= \left(1\underline{/120°}\right)\left(223.61\underline{/-26.57°}\right) + \left(1\underline{/240°}\right)\left(100\underline{/180°}\right) + 100 \\
&= 338.59\underline{/66.21°}
\end{aligned}$$

Just to emphasize that the method of symmetrical components applies to currents, as well as voltages we give the following example:

Example 8-3

The following currents were recorded under fault conditions in a three-phase system:

$$I_A = 150\underline{/45°}\,A$$
$$I_B = 250\underline{/150°}\,A$$
$$I_C = 100\underline{/300°}\,A$$

Calculate the values of the positive, negative, and zero phase sequence components for each line.

Solution

$$I_0 = \frac{1}{3}(I_A + I_B + I_c)$$

$$= \frac{1}{3}(106.07 + j106.07 + j106.07 - 216.51 + j125.00 + 50 - j86.6)$$

$$= \frac{(-60.44 + j144.46)}{3}$$

$$= 52.2\underline{/112.7°}$$

$$I_+ = \frac{1}{3}(I_A + \alpha I_B + \alpha^2 I_C) = \frac{1}{3}\left(150\underline{/45°} + 250\underline{/270°} + 100\underline{/180°}\right)$$

$$= 48.02\underline{/-87.6°}$$

$$I_- = \frac{1}{3}(I_A + \alpha^2 I_B + \alpha I_C)$$

$$= 163.21\underline{/40.45°}$$

Power in Symmetrical Components

The total power in a three-phase network is given by

$$S = V_A I_A^* + V_B I_B^* + V_C I_C^*$$

where the asterisk denotes complex conjugation.

Each of the terms using the symmetrical component notation is given by

$$V_A I_A^* = (V_+ + V_- + V_0)(I_+ + I_- + I_0)^*$$
$$V_B I_B^* = (\alpha^2 V_+ + \alpha V_- + V_0)(\alpha^2 I_+ + \alpha I_- + I_0)^*$$
$$V_C I_C^* = (\alpha V_+ + \alpha^2 V_- + V_0)(\alpha I_+ + \alpha^2 I_- + I_0)^*$$

Performing the conjugation operation, we obtain the following:

$$V_A I_A^* = (V_+ + V_- + V_0)(I_+^* + I_-^* + I_0^*)$$
$$V_B I_B^* = (\alpha^2 V_+ + \alpha V_- + V_0)(\alpha I_+^* + \alpha^2 I_-^* + I_0^*)$$
$$V_C I_C^* = (\alpha V_+ + \alpha^2 V_- + V_0)(\alpha^2 I_+^* + \alpha I_-^* + I_0^*)$$

Use has been made of the properties

$$(\alpha^2)^* = \alpha$$
$$(\alpha)^* = \alpha^2$$
$$\alpha^3 = 1$$

Expanding each of the powers gives

$$V_A I_A^* = V_+ I_+^* + V_+(I_-^* + I_0^*) + V_- I_-^*$$
$$+ V_-(I_+^* + I_0^*) + V_0 I_0^* + V_0(I_+^* + I_-^*)$$
$$V_B I_B^* = V_+ I_+^* + V_+(\alpha I_-^* + \alpha^2 I_0^*) + V_- I_-^*$$
$$+ V_-(\alpha^2 I_+^* + \alpha I_0^*) + V_0 I_0^* + V_0(\alpha I_+^* + \alpha^2 I_-^*)$$
$$V_C I_C^* = V_+ I_+^* + V_+(\alpha^2 I_-^* + \alpha I_0^*) + V_- I_-^* + V_-(\alpha I_+^* + \alpha^2 I_0^*)$$
$$+ V_0 I_0^* + V_0(\alpha^2 I_+^* + \alpha I_-^*)$$

To obtain the total power, we add the above expressions. Note that all even terms in the expressions add to zero since

$$\alpha^2 + \alpha + 1 = 0$$

As a result, we have

$$S = 3(V_+ I_+^* + V_- I_-^* + V_0 I_0^*) \qquad (8.34)$$

We conclude that the total power is three times the sum of powers in individual sequence networks.

8.4 SEQUENCE NETWORKS

Positive Sequence Networks

The positive sequence network for a given power system shows all the paths for the flow of positive sequence currents in the system. The one-line diagram of the system is converted to an impedance diagram that shows the equivalent circuit of each component under balanced operating conditions.

Each generator in the system is represented by a generated voltage in series with the appropriate reactance and resistance. Current-limiting impedances between the generator's neutral and ground pass no positive sequence current and hence are not included in the positive sequence network.

To simplify the calculations, all resistance and the magnetizing current for each transformer are neglected. For transmission lines, the line's shunt capacitance is neglected, and in many instances, so is the resistance.

Motor loads, whether synchronous or induction, are included in the network as generated EMF's in series with the appropriate reactance. Static loads are mostly neglected in fault studies.

Negative Sequence Networks

Having obtained the positive sequence network, the process of finding the negative sequence network follows easily. Three-phase generators and motors have only positive sequence-generated voltages. Thus the negative sequence network model will not contain EMF sources associated with rotating machinery. Note that the negative sequence impedance for this type of device will in general be different from the positive sequence values. For static devices such as transmission lines and transformers, the negative sequence impedances have the same values as the corresponding positive sequence impedances.

The current-limiting impedances between the generator's neutral and ground will not appear in the negative sequence network. This arises simply because negative sequence currents are balanced.

Zero Sequence Networks

The zero sequence network of a system depends on the nature of the connections of the three-phase windings for each of the system's components.

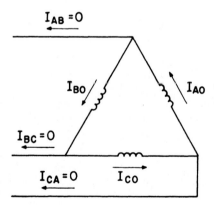

Figure 8-7. Delta-Connected Winding and Zero Sequence Currents.

Delta-Connected Winding

Zero sequence currents can exist in the phase windings of the delta connection. However, since we have the requirement

$$I_{A0} = I_{B0} = I_{C0} = I_0$$

we conclude that the line currents are zero. For example,

$$I_{AB} = I_{A0} - I_{B0} = 0$$

This situation is shown in Figure 8-7.

The single-phase equivalent zero sequence network for a delta-connected load with zero sequence impedance Z_0 is shown in Figure 8-8.

Wye-Connected Winding

In the presence of a neutral return wire, zero sequence currents will exist both in the phase windings as well as on the lines. The neutral current I_N will be

$$I_N = I_{A0} + I_{B0} + I_{C0}$$
$$= 3I_0$$

This is shown in Figure 8-9(a).

In the case of a system with no neutral return, $I_N = 0$ shows that no zero sequence currents can exist. This is shown in Figure 8-9(b). Zero sequence equivalents are shown in Figure 8-10.

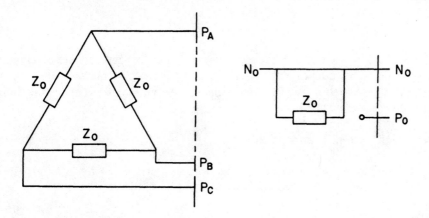

Figure 8-8. Zero Sequence Equivalent of a Delta-Connected Load.

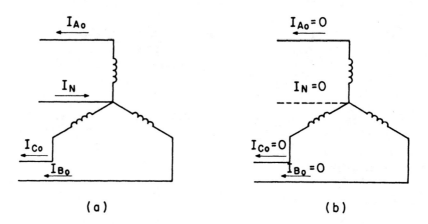

Figure 8-9. Wye-Connected Winding with and without Neutral Return.

Transformer's Zero Sequence Equivalents

There are various possible combinations of the primary and secondary connections for three-phase transformers. These alter the corresponding zero sequence network.

Delta-delta Bank

Since for a delta circuit no return path for zero sequence current exists, no zero sequence current can flow into a delta-delta bank, although it can circulate within the delta windings. The equivalent circuit connections are shown in Figure 8-11.

Wye-delta Bank, Ungrounded Wye

For an ungrounded wye connection, no path exists for zero sequence current to the neutral. The equivalent circuit is shown in Figure 8-12.

Wye-delta Bank, Grounded wye

Zero sequence currents will pass through the wye winding to ground. As a result, secondary zero sequence currents will circulate through the

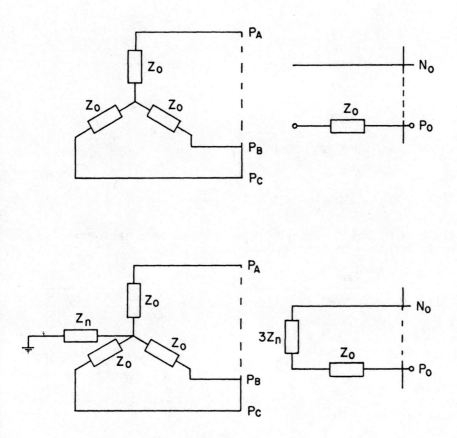

Figure 8-10. Zero Sequence Networks for Y-Connected Loads.

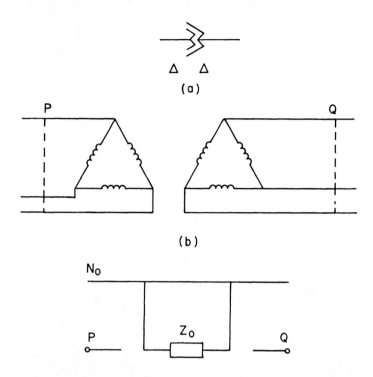

Figure 8-11. Zero Sequence Equivalent Circuits for a Three-Phase Transformer Bank Connected in delta-delta.

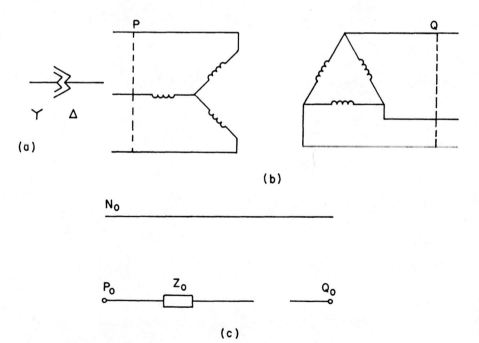

Figure 8-12. Zero Sequence Equivalent Circuits for a Three-Phase Transformer Bank Connected in Wye-delta.

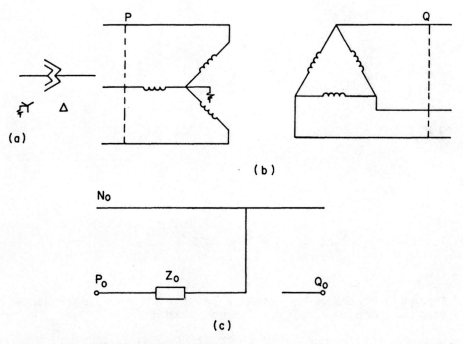

Figure 8-13. Zero Sequence Equivalent Circuit for a Three-Phase Transformer Bank Connected in Wye-Delta Bank with Grounded Y.

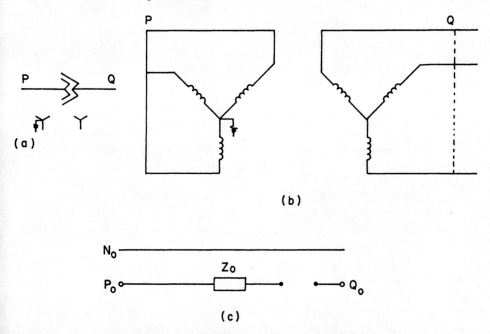

Figure 8-14. Zero Sequence Equivalent Circuit for a Three-Phase Transformer Bank Connected in Wye-Wye with One Grounded Neutral.

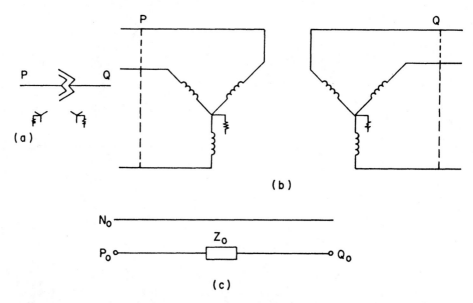

Figure 8-15. Zero Sequence Equivalent Circuit for a Three-Phase Transformer Bank Connected in Wye-Wye with Neutrals Grounded.

delta winding. No zero sequence current will exist on the lines of the secondary. The equivalent circuit is shown in Figure 8-13.

Wye-wye Bank, One Neutral Grounded

With an ungrounded wye, no zero sequence current can flow. The nonpresence of the current in one winding means that no current exists in the other. Figure 8-14 illustrates the situation.

Wye-wye Bank, Both Neutrals Grounded

With both wyes grounded, zero sequence current can flow. The presence of the current in one winding means that secondary current exists in the other. Figure 8-15 illustrates the situation.

A hypothetical system is shown in Figure 8-16(a) to illustrate the construction of the zero sequence network with different transformer connections. Figure 8-16(b) shows the zero sequence network connections using the principles illustrated in this section.

Sequence Impedances for Synchronous Machines

For a synchronous machine, sequence impedances are essentially reactive. The positive, negative, and zero sequence impedances have in general different values.

Positive Sequence Impedance

Depending on the time interval of interest, one of three reactances may be used:

1. For the subtransient interval, we use the subtransient reactance:
$$Z_+ = jX_d''$$

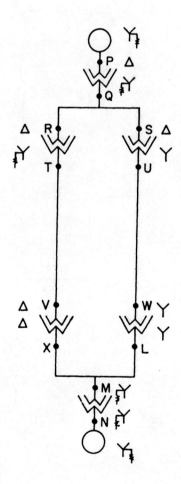

Figure 8-16a. Example System for Zero Sequence Network Illustration.

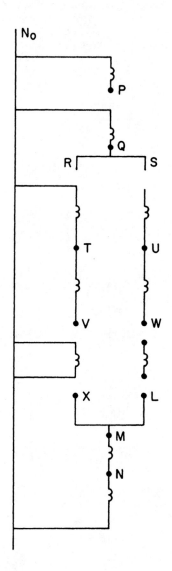

Figure 8-16b.

2. For the transient interval, use is made of the corresponding reactance:
$$Z_+ = jX'_d$$

3. In the steady state, we have
$$Z_+ = jX_d$$

Negative Sequence Impedance

The MMF produced by negative sequence armature current rotates in a direction opposite to the rotor and hence opposite to the dc field winding. As a result, the reactance of the machine will be different from that for the positively rotating sequence. The negative sequence reactance for a synchronous machine is often defined as

$$Z_- = j\left(\frac{X''_d + X''_q}{2}\right)$$

Zero Sequence Impedance

The zero sequence impedance of a synchronous machine is quite variable and depends on the nature of the stator windings. In general, these will be much smaller than the corresponding positive and negative sequence reactances. Table 8-1 shows typical values for synchronous machine sequence reactances.

Sequence Impedances for a Transmission Link

Consider a three-phase transmission link of impedance Z_L per phase as shown in Figure 8-17. The return (or neutral) impedance is Z_N. If the system voltages are unbalanced, we have a neutral current I_N. Thus,

$$I_N = I_A + I_B + I_C$$

The voltage drops ΔV_A, ΔV_B, and ΔV_C across the link are as shown below:

$$\Delta V_A = I_A Z_L + I_N Z_N$$
$$\Delta V_B = I_B Z_L + I_N Z_N$$
$$\Delta V_C = I_C Z_L + I_N Z_N$$

In terms of sequence voltages and currents, we thus have

$$\Delta V_+ + \Delta V_- + \Delta V_0 = \left[(I_+ + I_- + I_0)Z_L\right] + 3I_0 Z_N$$
$$\alpha^2 \Delta V_+ + \alpha \Delta V_- + \Delta V_0 = \left[(\alpha^2 I_+ + \alpha I_- + I_0)Z_L\right] + 3I_0 Z_N$$
$$\alpha \Delta V_+ + \alpha^2 \Delta V_- + \Delta V_0 = \left[(\alpha I_+ + \alpha^2 I_- + I_0)Z_L\right] + 3I_0 Z_N$$

494 Faults on Electric Energy Systems

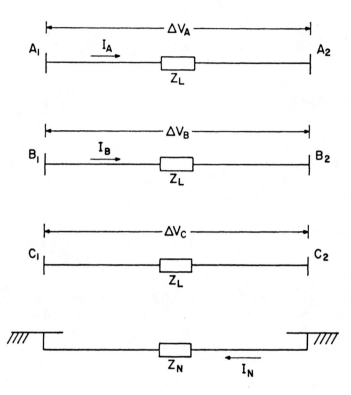

Figure 8-17. Three-Phase Transmission Link.

The above equations give

$$\Delta V_+ = I_+ Z_L$$
$$\Delta V_- = I_- Z_L$$
$$\Delta V_0 = I_0 (Z_L + 3Z_N)$$

We identify the following sequence impedances:

$$Z_0 = Z_L + 3Z_N$$

$$Z_- = Z_L$$
$$Z_+ = Z_L$$

It is thus clear that the impedance of the neutral path will enter into the zero sequence impedance in addition to the link's impedance Z_L. However, for the positive and negative sequence impedances, only the link's impedance appears.

Example 8-4

Obtain the sequence networks for the system shown in Figure 8-18 in the case of a fault at F. Assume the following data in pu on the same base are given:

Generator G_1:	$X_+ = 0.2$ p.u.
	$X_- = 0.12$ p.u.
	$X_0 = 0.06$ p.u.
Generator G_2:	$X_+ = 0.33$ p.u.
	$X_- = 0.22$ p.u.
	$X_0 = 0.066$ p.u.
Transformer T_1:	$X_+ = X_- = X_0 = 0.2$ p.u.
Transformer T_2:	$X_+ = X_- = X_0 = 0.225$ p.u.
Transformer T_3:	$X_+ = X_- = X_0 = 0.27$ p.u.
Transformer T_4:	$X_+ = X_- = X_0 = 0.16$ p.u.
Line L_1:	$X_+ = X_- = 0.14$ p.u.
	$X_0 = 0.3$ p.u.
Line L_2:	$X_+ = X_- = 0.35$ p.u.
	$X_0 = 0.6$ p.u.

Solution

The positive sequence network and steps in its reduction to a single source impedance by Thévenin's equivalent are shown in Figure 8-19. The negative sequence network and its reduction are given in Figure 8-20. The zero sequence network is treated similarly in Figure 8-21. The reader is invited to examine Problem 8-A-10, which involves essentially the same system as that of this example except for the existence of a load. We will find in that problem that the use of Thévenin's theorem is a prerequisite.

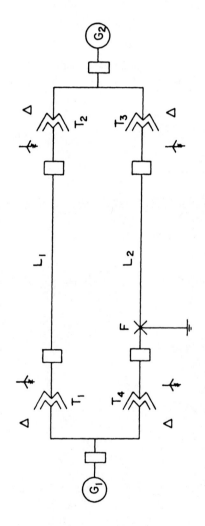

Figure 8-18. System for Example 8-4.

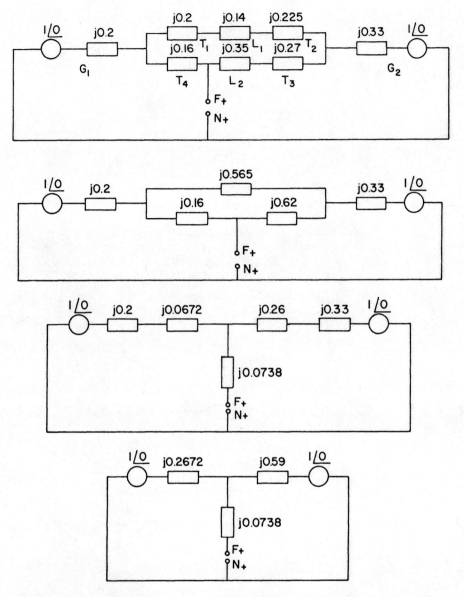

Figure 8-19. Positive Sequence Network and Steps in Its Reduction for Example 8-4.

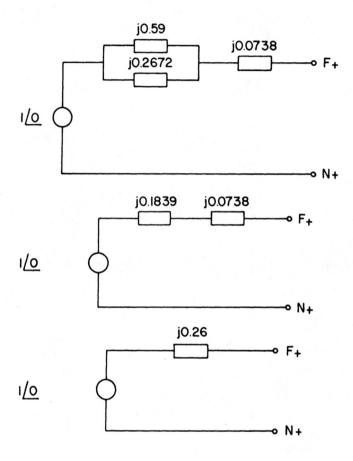

Figure 8-19 (Cont.)

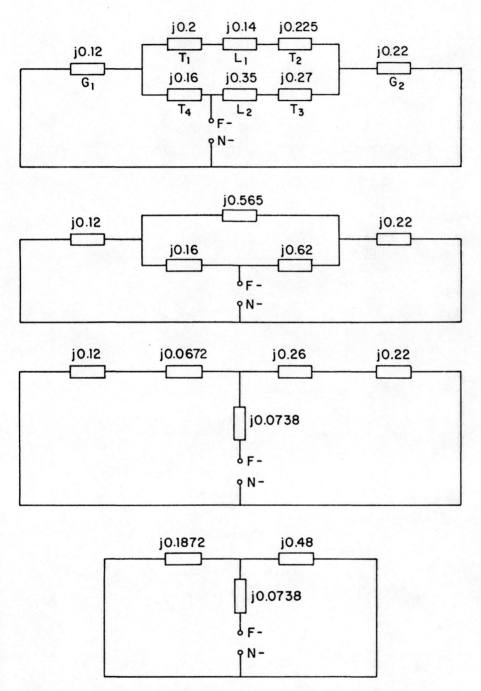

Figure 8-20. Negative Sequence Network and Its Reduction for Example 8-4.

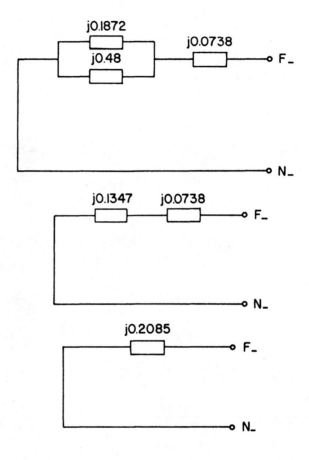

Figure 8-20 (*Cont.*)

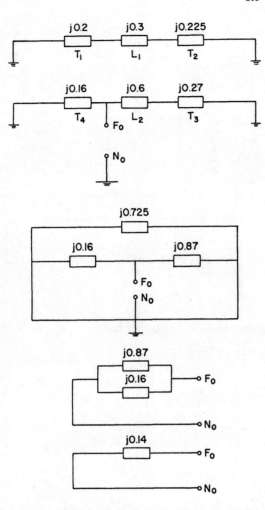

Figure 8-21. Reduction of the Zero Sequence Network for Example 8-4.

8.5 LINE-TO-GROUND FAULT

Assume that phase A is shorted to ground at the fault point F as shown in Figure 8-22. The phase B and C currents are assumed negligible, and we can thus write $I_B = 0$, $I_C = 0$. The sequence currents are obtained as follows:

For the positive sequence value we have

$$I_+ = \frac{1}{3}(I_A + \alpha I_B + \alpha^2 I_C)$$

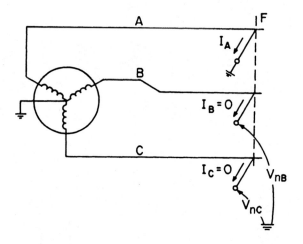

Figure 8-22. Line-to-Ground Fault Schematic.

This gives

$$I_+ = \frac{I_A}{3}$$

For the negative sequence current, we have

$$I_- = \frac{1}{3}(I_A + \alpha^2 I_B + \alpha I_C)$$

This gives

$$I_- = \frac{I_A}{3}$$

Likewise for the zero sequence current, we get

$$I_0 = \frac{I_A}{3}$$

We conclude then that in the case of a single line-to-ground fault, the sequence currents are equal, and we write

$$I_+ = I_- = I_0 = \frac{I_A}{3} \qquad (8.35)$$

8.5 Line-to-Ground Fault

With the generators normally producing balanced three-phase voltages, which are positive sequence only, we can write

$$E_+ = E_A \tag{8.36}$$
$$E_- = 0 \tag{8.37}$$
$$E_0 = 0 \tag{8.38}$$

Let us assume that the sequence impedances to the fault are given by Z_+, Z_-, Z_0. We can write the following expressions for sequence voltages at the fault:

$$V_+ = E_+ - I_+ Z_+ \tag{8.39}$$
$$V_- = 0 - I_- Z_- \tag{8.40}$$
$$V_0 = 0 - I_0 Z_0 \tag{8.41}$$

The fact that phase A is shorted to the ground is used. Thus

$$V_A = 0$$

we also recall that

$$V_A = V_+ + V_- + V_0$$

Our conclusion is

$$0 = E_+ - I_0(Z_+ + Z_- + Z_0)$$

or

$$I_0 = \frac{E_+}{Z_+ + Z_- + Z_0} \tag{8.42}$$

The resulting equivalent circuit is shown in Figure 8-23.

We can now state the solution in terms of phase currents:

$$I_A = \frac{3E_+}{Z_+ + Z_- + Z_0}$$
$$I_B = 0$$
$$I_C = 0 \tag{8.43}$$

For phase voltages we have

$$V_A = 0$$
$$V_B = \frac{E_B(1-\alpha)[Z_0 + (1+\alpha)Z_-]}{Z_0 + Z_- + Z_+}$$
$$V_C = \frac{E_C(1-\alpha)[(1+\alpha)Z_0 + Z_-]}{Z_0 + Z_- + Z_+} \tag{8.44}$$

The last two expressions can be derived easily from the basic relations. For phase B, we have

$$V_B = \alpha^2 V_+ + \alpha V_- + V_0$$

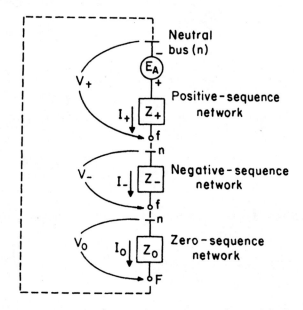

Figure 8-23. Equivalent Circuit for Single Line-to-Ground Fault.

Using Eqs. (8.39), (8.40), and (8.41), we have

$$V_B = \alpha^2(E_+ - I_+ Z_+) + \alpha(0 - I_- Z_-) + 1(0 - I_0 Z_0)$$

Inserting Eq. (8.35), we thus have

$$V_B = \alpha^2 E_+ - \frac{E_+(\alpha^2 Z_+ + \alpha Z_- + Z_0)}{[Z_+ + Z_- + Z_0]}$$

which reduces to

$$V_B = \frac{E_+[(\alpha^2 - \alpha)Z_- + (\alpha^2 - 1)Z_0]}{Z_+ + Z_- + Z_0}$$

Remembering that

$$E_B = \alpha^2 E_+$$

we obtain

$$V_B = \frac{E_B(1-\alpha)[Z_0 + (1+\alpha)Z_-]}{[Z_0 + Z_- + Z_+]}$$

Similarly, we get the result for phase C.

Example 8-5

For the system of Example 8-4 find the voltages and currents at the fault point for a single line-to-ground fault.

Solution

The sequence networks are connected in series for a single line-to-ground fault. This is shown in Figure 8-24.

The sequence currents are given by

$$I_+ = I_- = I_0 = \frac{1}{j(0.2577 + 0.2085 + 0.14)}$$
$$= 1.65\underline{/-90°} \text{ p.u.}$$

Therefore,

$$I_A = 3I_+ = 4.95\underline{/-90°} \text{ p.u.}$$
$$I_B = I_C = 0$$

The sequence voltages are as follows:

$$V_+ = E_+ - I_+ Z_+$$
$$= 1\underline{/0} - (1.65\underline{/-90°})(0.269\underline{/90°})$$
$$= 0.57 \text{ p.u.}$$
$$V_- = -I_- Z_-$$
$$= -(1.65\underline{/-90°})(0.2085\underline{/90°})$$
$$= -0.34 \text{ p.u.}$$
$$V_0 = -I_0 Z_0$$
$$= -(1.65\underline{/-90°})(0.14\underline{/90°})$$
$$= -0.23 \text{ p.u.}$$

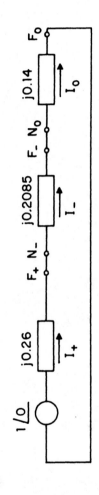

Figure 8-24. Sequence Network Connection for Example 8-5.

The phase voltages are thus

$$V_A = V_+ + V_- + V_0 = 0$$
$$V_B = \alpha^2 V_+ + \alpha V_- + V_0$$
$$= (1/240°)(0.57) + (1/120°)(-0.34) + (-0.23)$$
$$= 0.86/-113.64° \text{ p.u.}$$
$$V_C = \alpha V_+ + \alpha^2 V_- + V_0$$
$$= (1/120°)(0.57) + (1/240°)(-0.34) + (-0.23)$$
$$= 0.86/113.64° \text{ p.u.}$$

8.6 DOUBLE LINE-TO-GROUND FAULT

We will consider a general fault condition. In this case we assume that phase B has fault impedance of Z_f; phase C has a fault impedance of Z_f; and the common line-to-ground fault impedance is Z_g. This is shown in Figure 8-25.

The boundary conditions are as follows:

$$I_A = 0$$
$$V_{Bn} = I_B(Z_f + Z_g) + I_C Z_g$$
$$V_{Cn} = I_B Z_g + (Z_f + Z_g) I_C$$

The potential difference between phases B and C is thus

$$V_{Bn} - V_{Cn} = I_B Z_f - I_C Z_f$$

Substituting in terms of sequence currents and voltages, we have

$$(\alpha^2 - \alpha)(V_+ - V_-) = (\alpha^2 - \alpha)(I_+ - I_-)Z_f$$

As a result, we get

$$V_+ - I_+ Z_f = V_- - I_- Z_f \quad (8.45)$$

The sum of phase voltages is

$$V_{Bn} + V_{Cn} = (I_B + I_C)(Z_f + 2Z_g)$$

In terms of sequence quantities this gives

$$2V_0 - V_+ - V_- = (2I_0 - I_+ - I_-)(Z_f + 2Z_g)$$

Recall that, since $I_A = 0$, we have

$$I_+ + I_- + I_0 = 0$$

We can thus assert that

$$2V_0 - V_+ - V_- = 3I_0(Z_f + 2Z_g) \quad (8.46)$$

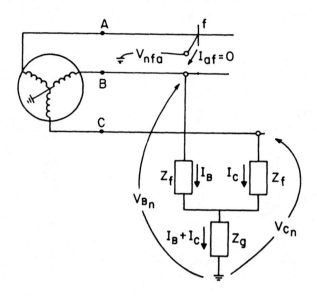

Figure 8-25. Circuit with Double Line-to-Ground fault.

Substituting for V_- from Eq. (8.45), we get

$$2V_0 - 2V_+ + I_+ Z_f + (I_+ + I_0)Z_f = 3I_0(Z_f + 2Z_g)$$

The above reduces to

$$V_0 - I_0(Z_f + 3Z_g) = V_+ - I_+ Z_f$$

Now we have

$$V_+ = E_+ - I_+ Z_+$$
$$V_- = -I_- Z_-$$
$$V_0 = -I_0 Z_0$$

Consequently,

$$E_+ - I_+(Z_+ + Z_f) = -I_-(Z_- + Z_f)$$
$$= -I_0(Z_0 + Z_f + 3Z_g) \qquad (8.47)$$

8.6 Double Line-to-Ground Fault

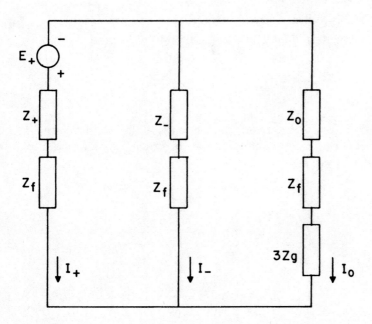

Figure 8-26. Sequence Network for Double Line-to-Ground Fault.

The equivalent circuit is shown in Figure 8-26. It is clear from Eq. (8.47) that the sequence networks are connected in parallel.

From the equivalent circuit we obtain the positive sequence current

$$I_+ = \frac{E_+}{Z_+ + Z_f + \left[\dfrac{(Z_- + Z_f)(Z_0 + Z_f + 3Z_g)}{Z_- + Z_0 + 2Z_f + 3Z_g}\right]} \qquad (8.48)$$

The negative sequence current is

$$I_- = -I_+ \left(\frac{Z_0 + Z_f + 3Z_g}{Z_- + Z_0 + 2Z_f + 3Z_g}\right) \qquad (8.49)$$

Finally,

$$I_0 = -(I_+ + I_-) \qquad (8.50)$$

Example 8-6

For the system of Example 8-4 find the voltages and currents at the fault point for a double line-to-ground fault. Assume
$$Z_f = j0.05 \text{ p.u.}$$
$$Z_g = j0.033 \text{ p.u.}$$

Solution

The sequence network connection is as shown in Figure 8-27. Steps of the network reduction are also shown. From the figure, sequence currents are as follows:

$$I_+ = \frac{1/0}{0.45/90°} = 2.24/-90°$$

$$I_- = -I_+ \left(\frac{0.29}{0.29 + 0.2585} \right)$$
$$= -1.18/-90°$$

$$I_0 = -1.06/-90°$$

The sequence voltages are calculated as follows.

$$V_+ = E_+ - I_+ Z_+$$
$$= 1/0 - (2.24/-90°)(0.26/90°)$$
$$= 0.42$$

$$V_- = -I_- Z_-$$
$$= +(1.18)(0.2085) = 0.25$$

$$V_0 = -I_0 Z_0$$
$$= (1.06)(0.14) = 0.15$$

The phase currents are obtained as

$$I_A = 0$$
$$I_B = \alpha^2 I_+ + \alpha I_- + I_0$$
$$= (1/240)(2.24/-90°) + (1/120)(-1.18/-90°)$$
$$+ (-1.06/-90°)$$
$$= 3.36/151.77°$$

$$I_C = \alpha I_+ + \alpha^2 I_- + I_0$$
$$= (1/120°)(2.24/-90°) + (1/240°)(-1.18/-90°)$$
$$+ (-1.06/-90°)$$
$$= 3.36/28.23°$$

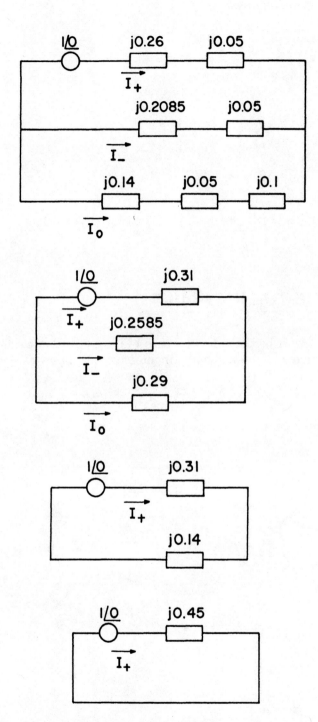

Figure 8-27. Sequence Network for Example 8-6.

Phase voltages are

$$V_A = V_+ + V_- + V_0$$
$$= 0.42 + 0.25 + 0.15$$
$$= 0.82$$
$$V_B = \alpha^2 V_+ + \alpha V_- + V_0$$
$$= (1\underline{/240°})(0.42) + (1\underline{/120°})(0.25) + (0.15)$$
$$= 0.24\underline{/-141.49°}$$
$$V_C = \alpha V_+ + \alpha^2 V_- + V_0$$
$$= (1\underline{/120°})(0.42) + (1\underline{/240°})(0.25) + 0.15$$
$$= 0.24\underline{/141.49°}$$

8.7 LINE-TO-LINE FAULT

Let phase A be the unfaulted phase. Figure 8-28 shows a three-phase system with a line-to-line short circuit between phases B and C. The

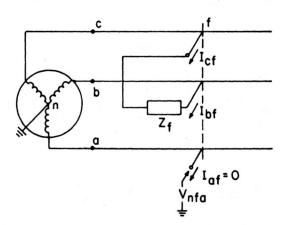

Figure 8-28. Example of a Line-to-Line Fault.

8.7 Line-to-Line Fault

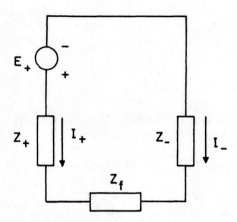

Figure 8-29. Line-to-Line Equivalent Circuit.

boundary conditions in this case are

$$I_A = 0$$
$$I_B = -I_C$$
$$V_B - V_C = I_B Z_f$$

The first two conditions yield

$$I_0 = 0$$
$$I_+ = -I_- = \frac{1}{3}(\alpha - \alpha^2)I_B$$

The voltage conditions give

$$(\alpha^2 - \alpha)(V_+ - V_-) = Z_f(\alpha^2 - \alpha)I_+$$

which reduces to

$$V_+ - V_- = Z_f I_+ \tag{8.51}$$

The equivalent circuit will take on the form shown in Figure 8-29. Note that the zero sequence network is not included since $I_0 = 0$.

Example 8-7

For the system of Example 8-4, find the voltages and currents at the fault point for a line-to-line fault through an impedance $Z_f = j0.05$ p.u.

Solution

The sequence network connection is as shown in Figure 8-30. From the diagram,

$$I_+ = -I_- = \frac{1\underline{/0}}{0.5185\underline{/90°}}$$

$$= 1.93\underline{/-90°} \text{ p.u.}$$

$$I_0 = 0$$

The phase currents are thus

$$I_A = 0$$
$$I_B = -I_C$$
$$= (\alpha^2 - \alpha)I_+$$
$$= \left(1\underline{/240°} - 1\underline{/120°}\right)\left(1.93\underline{/-90°}\right)$$
$$= 3.34\underline{/-180°} \text{ p.u.}$$

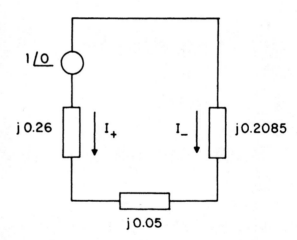

Figure 8-30. Sequence Network Connection for Example 8-7.

The sequence voltages are

$$V_+ = E_+ - I_+ Z_+$$
$$= 1\underline{/0°} - (1.93\underline{/-90°})(0.26\underline{/90°})$$
$$= 0.5 \text{ p.u.}$$
$$V_- = -I_- Z_-$$
$$= -(-1.93\underline{/-90°})(0.2085\underline{/90°})$$
$$= 0.4 \text{ p.u.}$$
$$V_0 = -I_0 Z_0$$
$$= 0$$

The phase voltages are obtained as shown below:

$$V_A = V_+ + V_- + V_0$$
$$= 0.9 \text{ p.u.}$$
$$V_B = \alpha^2 V_+ + \alpha V_- + V_0$$
$$= (1\underline{/240°})(0.5) + (1\underline{/120°})(0.4)$$
$$= 0.46\underline{/-169.11°}$$
$$V_C = \alpha V_+ + \alpha^2 V_- + V_0$$
$$= (1\underline{/120°})(0.5) + (1\underline{/240°})(0.4)$$
$$= 0.46\underline{/169.11°}$$

As a check we calculate

$$V_B - V_C = 0.17\underline{/-90°}$$
$$I_B Z_f = (3.34\underline{/-180°})(0.05\underline{/90°})$$
$$= 0.17\underline{/-90°}$$

Hence,

$$V_B - V_C = I_B Z_f$$

8.8 THE BALANCED THREE-PHASE FAULT

Let us now consider the situation with a balanced three-phase fault on phases A, B, and C, all through the same fault impedance Z_f. This fault condition is shown in Figure 8-31. It is clear from inspection of Figure 8-31

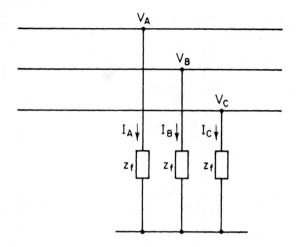

Figure 8-31. A Balanced Three-Phase Fault.

that the phase voltages at the fault are given by

$$V_A = I_A Z_f \tag{8.52}$$

$$V_B = I_B Z_f \tag{8.53}$$

$$V_C = I_C Z_f \tag{8.54}$$

The positive sequence voltages are obtained using the following

$$V_+ = \frac{1}{3}(V_A + \alpha V_B + \alpha^2 V_C)$$

Using Eqs. (8.52), (8.53), and (8.54), we thus conclude

$$V_+ = \frac{1}{3}(I_A + \alpha I_B + \alpha^2 I_C) Z_f$$

However, in view of Eq. (8.31) for currents, we get

$$V_+ = I_+ Z_f \tag{8.55}$$

The negative sequence voltage is similarly given by

$$V_- = I_- Z_f \tag{8.56}$$

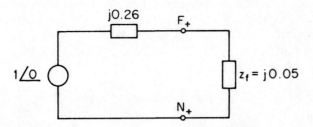

Figure 8-32. Positive Sequence Network for Example 8-8.

The zero sequence voltage is also

$$V_0 = I_0 Z_f \tag{8.57}$$

For a balanced source we have

$$V_+ = E - I_+ Z_+ \tag{8.58}$$

$$V_- = 0 - I_- Z_- \tag{8.59}$$

$$V_0 = 0 - I_0 Z_0 \tag{8.60}$$

Combining Eqs. (8.55) and (8.58), we conclude that

$$V_+ = E - I_+ Z_+ = I_+ Z_f \tag{8.61}$$

As a result,

$$I_+ = \frac{E}{Z_+ + Z_f} \tag{8.62}$$

Combining Eqs. (8.56) and (8.59) gives

$$I_- = 0 \tag{8.63}$$

Finally Eqs. (8.57) and (8.60) give

$$I_0 = 0 \tag{8.64}$$

Faults on Electric Energy Systems

The implications of Eqs. (8.63) and (8.64) are obvious. No zero sequence nor negative sequence components of the current exist. Instead, only positive sequence quantities are obtained in the case of a balanced three-phase fault.

Example 8-8

For the system of Example 8-4, find the short-circuit currents at the fault point for a balanced three-phase fault through three impedances of value $Z_f = j0.05$ p.u., each.

Solution

We need only to recall the reduced positive sequence network of Figure 8-19 given again in Figure 8-32. From Figure 8-32 we assert that

$$I_{A_{sc}} = I_+ = \frac{1\angle 0}{j(0.26 + 0.05)} = 3.23\angle -90°$$

SOME SOLVED PROBLEMS

Problem 8-A-1

The zero and positive sequence components of an unbalanced set of voltages are

$$V_+ = 2$$
$$V_0 = 0.5 - j0.866$$

The phase A voltage is

$$V_A = 3$$

Obtain the negative sequence component and the B and C phase voltages.

Solution

We have
$$V_A = V_+ + V_- + V_0$$
$$3 = 2 + V_- + (0.5 - j0.866)$$

Thus,
$$V_- = 0.5 + j0.866 = 1\underline{/60°}$$

In polar form, we have
$$V_0 = 0.5 - j0.866 = 1\underline{/-60°}$$

Now for phase B, we have
$$V_B = \alpha^2 V_+ + \alpha V_- + V_0$$
$$= 2\underline{/240°} + 1\underline{/180°} + 1\underline{/-60°}$$
$$= 3\underline{/-120°}$$

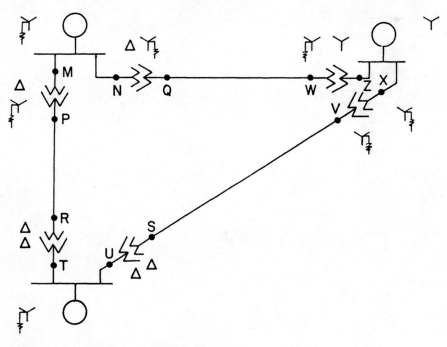

Figure 8-33. System for Problem 8-A-2.

For phase C, we have

$$V_C = \alpha V_+ + \alpha^2 V_- + V_0$$
$$= 2\underline{/120°} + 1\underline{/300°} + 1\underline{/-60°}$$
$$= 0$$

Problem 8-A-2

Draw the zero sequence network for the system shown in Figure 8-33.

Solution

The zero sequence network is shown in Figure 8-34.

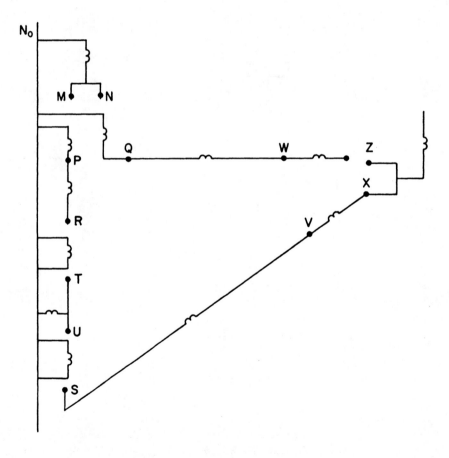

Figure 8-34. Zero Sequence Network for Problem 8-A-2.

Problem 8-A-3

Suppose that an unsymmetrical fault condition gives the following data at the fault point.

$$E = 1.0$$
$$I_+ Z_+ = 0.2$$
$$I_- Z_- = 0.2$$
$$I_0 Z_0 = 0.6$$

A. Find the phase and line-to-line voltages at the fault.
B. Identify the type of fault.

Solution

A. From the given information, we have

$$V_+ = E_+ - I_+ Z_+$$
$$= 1.0 - 0.2 = 0.8$$
$$V_- = -I_- Z_-$$
$$= -0.2$$
$$V_0 = -I_0 Z_0$$
$$= -0.6$$

The phase voltages are thus obtained as

$$V_A = V_+ + V_- + V_0$$
$$= 0.8 - 0.2 - 0.6 = 0$$
$$V_B = \alpha^2 V_+ + \alpha V_- + V_0$$
$$= 0.8 \underline{/240°} - 0.2 \underline{/120°} - 0.6$$
$$= 1.25 \underline{/-136.1°}$$
$$V_C = \alpha V_+ + \alpha^2 V_- + V_0$$
$$= 0.8 \underline{/120°} - 0.2 \underline{/240°} - 0.6$$
$$= 1.25 \underline{/136.1°}$$

The line-to-line voltages are

$$V_{AB} = V_A - V_B = 0 - 125 \underline{/-136.1°} = 1.25 \underline{/43.9°}$$
$$V_{BC} = V_B - V_C = 1.25 \underline{/-136.1°} - 1.25 \underline{/136.1°}$$
$$= 1.73 \underline{/-90°}$$
$$V_{CA} = V_C - V_A = 1.25 \underline{/136.1°} - 0 = 1.25 \underline{/136.1°}$$

B. From part (a), $V_A = 0$, indicating that this condition is a single line-to-ground fault on phase A.

Problem 8-A-4

The following sequence impedances exist between the source and the point of fault on a radial transmission system:

$$Z_+ = 0.3 + j0.6 \text{ p.u.}$$
$$Z_- = 0.3 + j.055 \text{ p.u.}$$
$$Z_0 = 1 + j0.78 \text{ p.u.}$$

The fault path to earth on a single line-to-ground fault has a resistance of 0.66 p.u. Determine the fault current and the voltage at the point of fault.

Solution

We have with reference to Figure 8-35,

$$Z_t = 3.6 + j1.93 = 4.08 \underline{/28.2°}$$
$$I_+ = I_- = I_0 = \frac{E}{Z_t} = 0.24 \underline{/-28.2°}$$

Thus,

$$I_A = 3I_+ = 0.73 \underline{/-28.2°}$$
$$V_f = 3Z_f I_0 = 0.48 \underline{/-28.2°}$$
$$= 0.43 - j0.23$$

Problem 8-A-5

For the radial transmission system of Problem 8-A-4, calculate the three-phase fault current. Compare with the single line-to-ground fault current assuming the fault path to ground has a negligible impedance.

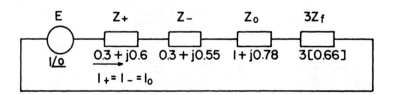

Figure 8-35. Network for Problem 8-A-4.

Solution

The three-phase fault current is

$$I_3 = \frac{E}{Z_+} = \frac{1}{(0.3 + j0.6)}$$
$$= 1.49\underline{/-63.435°}$$

The single line-to-ground fault current is

$$I_S = \frac{3E}{Z_+ + Z_- + Z_0}$$
$$= \frac{3}{1.6 + j1.435}$$
$$= 1.40\underline{/-41.89°}$$

Problem 8-A-6

A turbine generator has the following sequence reactances:

$$X_+ = 0.1$$
$$X_- = 0.13$$
$$X_0 = 0.04$$

Compare the fault currents for a three-phase fault and a single line-to-ground fault. Find the value of an inductive reactance to be inserted in the neutral connection to limit the current for a single line-to-ground fault to that for a three-phase fault.

Solution

$$I_3 = \frac{1}{0.1} = 10$$
$$I_s = \frac{3}{0.1 + 0.13 + 0.04}$$
$$= 11.11$$

The single line-to-ground fault current is higher than that for a three-phase fault.

With a neutral reactance X_n, we have

$$I_s = \frac{3}{0.1 + 0.13 + 0.04 + 3X_n}$$

For $I_s = 10$, we get

$$X_n = 0.01$$

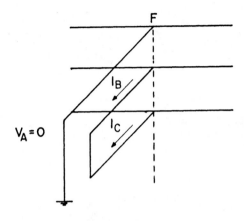

Figure 8-36. Fault for Problem 8-A-7.

Problem 8-A-7

Obtain the sequence network connection for a simultaneous single line-to-ground fault on phase A and line-to-line fault between phases B and C as shown in Figure 8-36.

Solution

The boundary conditions are

$$V_A = 0$$
$$V_B = V_C$$
$$I_B + I_C = 0$$

The sequence currents are thus

$$I_+ = \frac{1}{3}\left[I_A + (\alpha - \alpha^2)I_B\right]$$
$$I_- = \frac{1}{3}\left[I_A + (\alpha^2 - \alpha)I_B\right]$$
$$I_0 = \frac{1}{3}I_A$$

We conclude that

$$I_+ + I_- = 2I_0$$

This necessitates the connection shown in Figure 8-37.

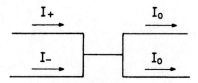

Figure 8-37. Sequence Network Connection to Satisfy Current Relation for Problem 8-A-7.

The sequence voltages are obtained as follows:

$$V_+ = \frac{1}{3}[(\alpha + \alpha^2)V_B] = -\frac{1}{3}V_B$$

$$V_- = \frac{1}{3}[(\alpha + \alpha^2)V_B] = -\frac{1}{3}V_B$$

$$V_0 = \frac{1}{3}(2V_B)$$

Thus,

$$V_+ = V_-$$
$$V_+ + V_- + V_0 = 0$$

Consequently we have the sequence network connection shown in Figure 8-38.

Problem 8-A-8

A simultaneous fault occurs at the load end of a radial line. The fault consists of a line-to-ground fault on phase A and a line-to-line fault on phases B and C. The current in phase A is $-j5$ p.u., whereas that in phase B is $I_B = -3.46$ p.u. Given that $E = 1/\underline{0}$ and $Z_+ = j0.25$, find Z_- and Z_0.

Solution

Noting that $I_C = -I_B$, we have the following sequence currents:

$$I_+ = \frac{1}{3}[I_A + (\alpha - \alpha^2)I_B]$$

$$= \frac{1}{3}\left[-j5 + (1.732/\underline{90°})(-3.46)\right]$$

$$= -\frac{j11}{3} = -j3.67$$

$$I_- = \frac{1}{3}[I_A + (\alpha^2 - \alpha)I_B]$$

$$= \frac{1}{3}\left[-j5 + (1.732/\underline{-90°})(-3.46)\right]$$

$$= j0.333$$

$$I_0 = \frac{I_A}{3} = -j1.667$$

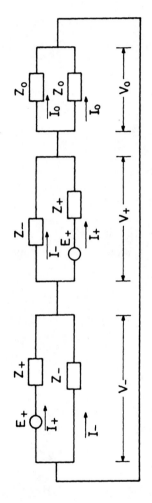

Figure 8-38. Sequence Network for System of Problem 8-A-7.

The positive sequence voltage is thus
$$V_+ = E - I_+ Z_+$$
$$= 1\underline{/0} + (j3.67)(j0.25) = 0.0825$$

The negative sequence voltage is obtained as
$$V_- = V_+ = 0.0825.$$

However, since
$$V_- = -I_- Z_-,$$

we conclude that
$$0.0825 = -j0.333 Z_-.$$

Thus,
$$Z_- = j0.2475.$$

Now, since we have
$$V_+ + V_- + V_0 = 0,$$

we obtain
$$V_0 = -2V_+ = -0.165$$
$$= -I_0 Z_0.$$

As a result we have
$$Z_0 = \frac{0.165}{-j1.667} = j0.099 \text{ pu}$$

Problem 8-A-9

Obtain the sequence networks for the system shown in Figure 8-39. Assume the following data in p.u. on the same base.

Generator G_1:	$X_+ = 0.2$ p.u.
	$X_- = 0.12$ p.u.
	$X_0 = 0.06$ p.u.
Generator G_2:	$X_+ = 0.33$ p.u.
	$X_- = 0.22$ p.u.
	$X_0 = 0.066$ p.u.
Transformer T_1:	$X_+ = X_- = X_0 = 0.2$ p.u.
T_2:	$X_+ = X_- = X_0 = 0.225$ p.u.
T_3:	$X_+ = X_- = X_0 = 0.27$ p.u.
T_4:	$X_+ = X_- = X_0 = 0.16$ p.u.
Line L_1:	$X_+ = X_- = 0.14$ p.u.
	$X_0 = 0.3$ p.u.

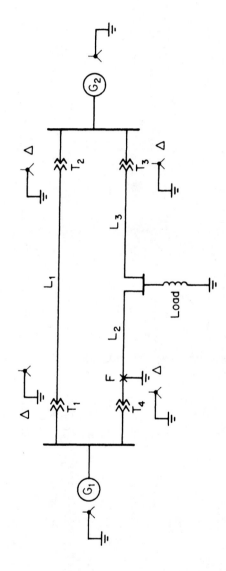

Figure 8-39. Network for Problem 8-A-9.

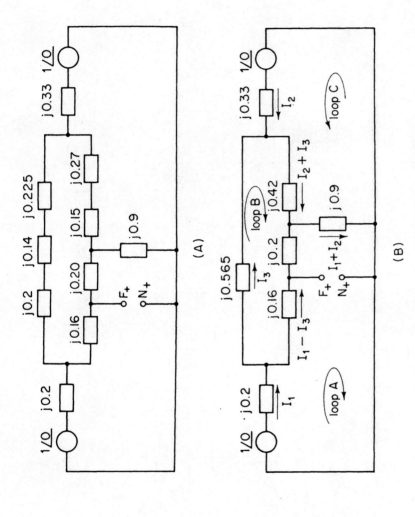

Figure 8-40. Positive Sequence Network for Problem 8-A-9.

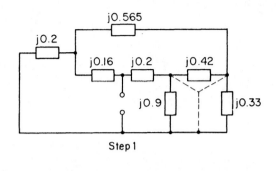

Step 1

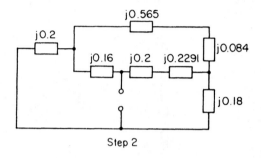

Step 2

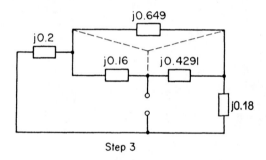

Step 3

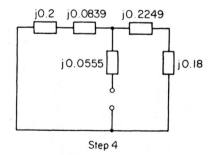

Step 4

Figure 8-41. Steps in Positive Sequence Impedance Reduction.

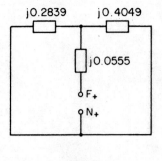

Step 5

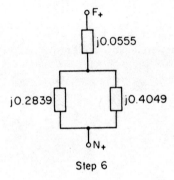

Step 6

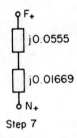

Step 7

Step 8

Figure 8-41 *(Cont.)*

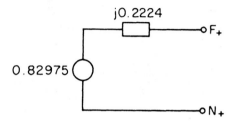

Figure 8-42. Positive Sequence Network Equivalent for Problem 8-A-9.

$$\text{Line } L_2: \qquad X_+ = X_- = 0.20 \text{ p.u.}$$

$$X_0 = 0.4 \text{ p.u.}$$

$$\text{Line } L_3: \qquad X_+ = X_- = 0.15 \text{ p.u.}$$

$$X_0 = 0.2 \text{ p.u.}$$

$$\text{Load:} \qquad X_+ = X_- = 0.9 \text{ p.u.}$$

$$X_0 = 1.2 \text{ p.u.}$$

Assume an unbalanced fault occurs at F. Find the equivalent sequence networks for this condition.

Solution

The positive sequence network is as shown in Figure 8-40(a). One step in the reduction can be made, the result of which is shown in Figure 8-40(b). To avoid tedious work we utilize Thévenin's theorem to obtain the positive sequence network in reduced form. We assign currents I_1, I_2, and I_3 as shown in Figure 8-40(b) and proceed to solve for the open-circuit voltage between F_+ and N_+.

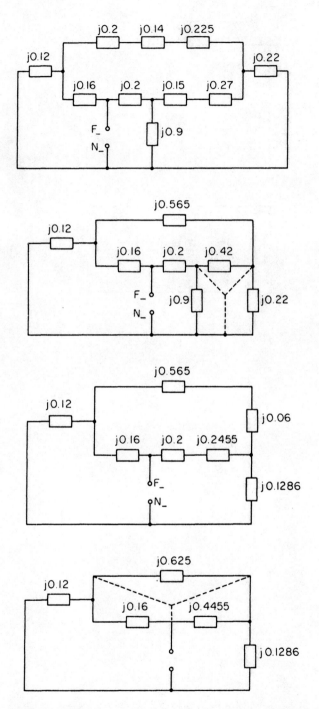

Figure 8-43. Steps in Reduction of the Negative Sequence Network for Problem 8-A-9.

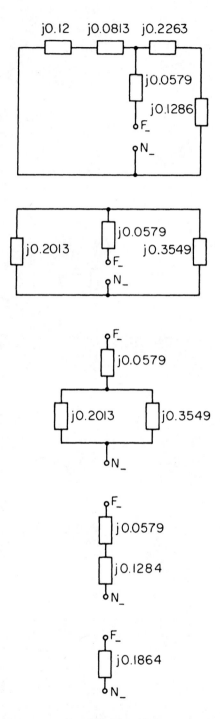

Figure 8-43 (*Cont.*)

Consider loop A. We can write

$$1\underline{/0} = j[0.2I_1 + 0.36(I_1 - I_3) + 0.9(I_1 + I_2)]$$

For loop B, we have

$$0 = j[0.565I_3 + 0.42(I_2 + I_3) - 0.36(I_1 - I_3)]$$

For loop C, we have

$$1\underline{/0} = j[0.33I_2 + 0.42(I_2 + I_3) + 0.9(I_1 + I_2)]$$

The above three equations are rearranged to give

$$1\underline{/0} = j(1.46I_1 + 0.9I_2 - 0.36I_3)$$
$$0 = 0.36I_1 - 0.42I_2 - 1.345I_3$$
$$1\underline{/0} = j(0.9I_1 + 1.65I_2 + 0.42I_3)$$

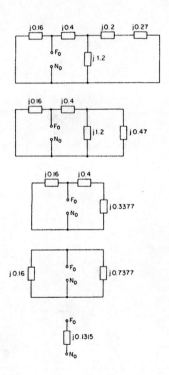

Figure 8-44. Steps in Reducing the Zero Sequence Network for Problem 8-A-9.

Solving, we obtain

$$I_1 = -j0.4839$$
$$I_2 = -j0.3357$$
$$I_3 = -j0.0247$$

As a result, we get

$$\begin{aligned}V_{F_+N_+} = V_{TH} &= 1 - j0.2I_1 - j0.16(I_1 - I_3)\\ &= 1 - (0.2)(0.4839) - (0.16)(0.4839 - 0.0247)\\ &= 0.82975\end{aligned}$$

We now turn our attention to the Thévenin's equivalent impedance, which is obtained by shorting out the sources and using network reduction. The steps are shown in Figure 8-41. As a result, we get

$$Z_+ = j0.224$$

The positive sequence equivalent is shown in Figure 8-42.

The negative sequence and zero sequence impedance networks and steps in their reduction are shown in Figure 8-43 and Figure 8-44. As a result, we get

$$Z_- = j0.1864$$
$$Z_0 = j0.1315$$

PROBLEMS

Problem 8-B-1

Consider the case of an open-line fault on phase A of a three-phase system, such that

$$I_A = 0$$
$$I_B = \alpha^2 I$$
$$I_C = \alpha I$$

Find the sequence currents I_+, I_-, and I_0.

Problem 8-B-2

Consider the case of a three-phase system supplied by a two-phase source such that

$$V_A = V$$
$$V_B = -jV$$
$$V_C = 0$$

Find the sequence voltages V_+, V_-, and V_0.

Problem 8-B-3

Calculate the phase currents and voltages for an unbalanced system with the following sequence values:

$$I_+ = I_- = I_0 = -j1.2737$$
$$V_+ = 0.597$$
$$V_- = -0.343$$
$$V_0 = -0.254$$

Problem 8-B-4

Calculate the apparent power consumed in the system of Problem 8-B-3 using sequence quantities and phase quantities.

Problem 8-B-5

Assume an unbalanced fault occurs on the line bus of transformer T_3 in the system of Example 8-4. Find the equivalent sequence networks for this condition.

Problem 8-B-6

Repeat Problem 8-B-5 for a fault on the generator bus of G_2.

Problem 8-B-7

Repeat Problem 8-B-5 for the fault in the middle of the line L_1.

Problem 8-B-8

Calculate the fault current for a single line-to-ground fault on phase A for a fault location as in Problem 8-B-5.

Problem 8-B-9

Repeat Problem 8-B-8 for a fault location as in Problem 8-B-5.

Problem 8-B-10

Repeat Problem 8-B-8 for a fault location as in Problem 8-B-7.

Problem 8-B-11

Calculate the fault current in phase B for a double line-to-ground fault for a fault location as in Problem 8-B-5.

Problem 8-B-12

Repeat Problem 8-B-11 for a fault location as in Problem 8-B-6.

Problem 8-B-13

Repeat Problem 8-B-11 for a fault location as in Problem 8-B-7.

Problem 8-B-14

Calculate the fault current in phase B for a line-to-line fault for a fault location as in Problem 8-B-5.

Problem 8-B-15

Repeat Problem 8-B-14 for a fault location as in Problem 8-B-6.

Problem 8-B-16

Repeat Problem 8-B-14 for a fault location as in Problem 8-B-7.

Problem 8-B-17

Calculate the fault current for a single line-to-ground fault on phase A for a fault location as in Problem 8-A-9.

Problem 8-B-18

Calculate the fault current in phase B for a double line-to-ground fault for a fault location at F, as in Problem 8-A-9.

Problem 8-B-19

Calculate the fault current in phase B for a line-to-line fault for a fault at F, as in Problem 8-A-9.

Problem 8-B-20

Repeat Problem 8-A-9 with the transformer T_4 connected in wye/delta, with the wye grounded.

Problem 8-B-21

Repeat Problem 8-B-17 for the conditions of Problem 8-B-20.

Problem 8-B-22

Repeat Problem 8-B-18 for the conditions of Problem 8-B-20.

Problem 8-B-23

Repeat Problem 8-B-19 for the conditions of Problem 8-B-20.

Problem 8-B-24

The following sequence voltages were recorded on an unbalanced fault:
$$V_+ = 0.5 \text{ p.u.}$$
$$V_- = -0.4 \text{ p.u.}$$
$$V_0 = -0.1 \text{ p.u.}$$

Given that the positive sequence fault current is $-j1$, calculate the sequence impedances. Assume $E = 1$.

Problem 8-B-25

The following sequence currents were recorded on an unbalanced fault condition:
$$I_+ = -j1.653 \text{ p.u.}$$
$$I_- = j0.5 \text{ p.u.}$$
$$I_0 = j1.153 \text{ p.u.}$$

Identify the type of fault. Assuming that $E = 1$ p.u. and $V_+ = 0.175$ p.u., find the sequence impedances for the system under these conditions.

Problem 8-B-26

The positive sequence current for a double line-to-ground fault in a system is $-j1$ p.u., and the corresponding negative sequence current is $j0.333$ p.u. Given that the positive sequence impedance is 0.8 p.u., find the negative and zero sequence impedances.

Problem 8-B-27

The zero sequence current in a system on a certain fault is zero. The positive sequence voltage is 0.45 p.u., and the positive sequence current is $-j1$ p.u. Calculate the positive and negative sequence impedances.

Problem 8-B-28

The positive sequence current on a single line-to-ground fault on phase A at the load end of a radial transmission system is $-j2$ p.u. For a double line-to-ground fault on phases B and C, the positive sequence current is $-j3.57$ p.u., and for a double-line fault between phases B and C, its value is $-j2.67$. Assuming the sending-end voltage $E = 1.2$, find the sequence impedances for this system.

Problem 8-B-29

A single line-to-ground fault on phase A at the load end of a radial line results in a short-circuit current of $-j5$ p.u. A line-to-line fault on phases B

and C results in a current $I_B = -3.46$ p.u. Given that $E = 1\underline{/0}$ and $Z_+ = j0.3$, find Z_- and Z_0.

Problem 8-B-30

The equivalent zero sequence impedance in a system is $Z_0 = j0.05$ p.u. The positive sequence current for a double line-to-ground fault is $-j2.26$ p.u., and for a single line-to-ground its value is $-j1.33$ p.u. Calculate the positive and negative sequence impedances of the system and hence predict the positive sequence current for a line-to-line fault.

Problem 8-B-31

The positive sequence voltage and current are 0.5 p.u. and $-j1.25$ p.u. respectively, for a double-line fault. The positive sequence current is $-j1$ p.u. for a single line-to-ground fault. Find the sequence impedances. Assume $E = 1$.

CHAPTER IX

System Protection

9.1 INTRODUCTION

The previous chapter treated the problems of system analysis under fault conditions that range from the balanced three-phase short circuit to the various unbalanced faults that may occur on the system. The result of the analysis provides a basis to determine the conditions that exist in the system under fault conditions. It is important to take the necessary action to prevent the faults, and if they do occur, to minimize possible damage or possible power disruption. A protection system continuously monitors the power system to ensure maximum continuity of electrical supply with minimum damage to life, equipment, and property.

The consequences of faults are diverse and include the following:

1. Abnormally large currents are caused to flow in parts of the system with the associated overheating of components.
2. System voltages will be off their normal acceptable levels, resulting in possible equipment damage.

3. Parts of the system will be caused to operate as unbalanced three-phase systems, which will mean improper operation of the equipment.

In view of the possible consequences, a number of important requirements for protective systems provide the basis for design criteria. These include:

1. *Reliability*: A reliable system should provide both dependability (guaranteed correct operation in response to faults) and security (avoiding unnecessary operation). Reliability requires that relay systems perform correctly under adverse system and environmental conditions.
2. *Speed*: Relays should respond to abnormal conditions in the least possible time. This usually means that the operating time should not exceed three cycles on a 60-Hz base.
3. *Selectivity*: A relay system should provide maximum possible service continuity with minimum system disconnection.
4. *Simplicity and economy*: The requirements of simplicity and economy are common in any engineering design, and relay systems are no exception.

A protective system is based on detecting fault conditions by continuously monitoring the power system variables such as current, voltage, power, frequency, and impedance. Measuring currents and voltages is performed by instrument transformers of the potential type (P.T.) or current type (C.T.). Instrument transformers feed the measured variables to the relay system, which in turn, upon detecting a fault, commands a circuit-interrupting device known as the *circuit breaker* (C.B.) to disconnect the faulted section of the system.

An electric power system is divided into protective zones for

Generators.

Transformers.

Bus bars.

Transmission and distribution circuits.

Motors.

The division is such that zones are given adequate protection while keeping service interruption to a minimum. A single-line diagram of a part of a power system with its zones of protection is given in Figure 9-1. It is to be noted that each zone is overlapped to avoid unprotected (blind) areas. The connections of current transformers achieve the overlapping. Figure 9-2 shows two possible arrangements. Note that a fault in the area of the two current transformers will trip all the breakers in both zones.

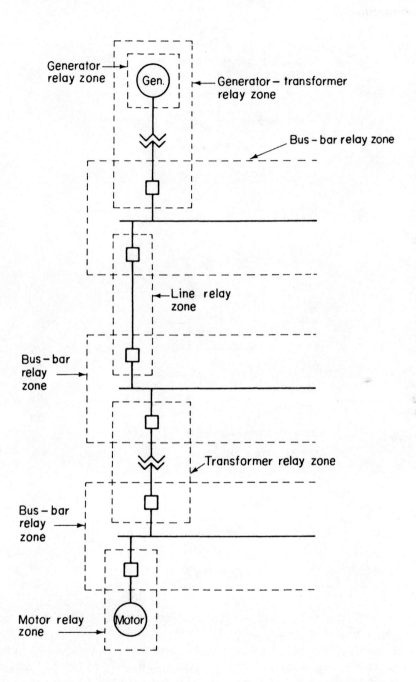

Figure 9-1. Typical Zones of Protection in Part of an Electric Power System.

544 System Protection

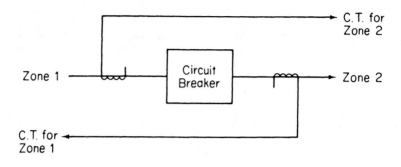

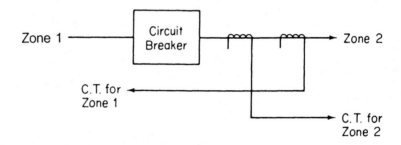

Figure 9-2. Overlapping around a Circuit Breaker.

9.2 PROTECTIVE RELAYS

A relay is a device that opens and closes electrical contacts to cause the operation of other devices under electric control. The action of a relay is essentially to detect intolerable or undesirable conditions within an assigned area. The relay acts to disconnect the area affected to prevent damage to personnel and property, by operating the appropriate circuit breakers.

One way for classifying relays is according to their function, that is, as *measuring* or *on-off* relays. The latter class is also known as *all-or-nothing* and includes relays such as time-lag relays, auxiliary relays, and tripping relays. The common feature in this class is that the relay does not have a

specified setting and is energized by a quantity that is either higher than
that at which it operates or lower than that at which it resets.

The class of measuring relays includes a number of types with the
common feature that they operate at a predetermined setting. Examples are
as follows:

- *Current relays*: Operate at a predetermined value of current. These
  include overcurrent and undercurrent relays.
- *Voltage relays*: Operate at a predetermined value of voltage. These
  include overvoltage and undervoltage relays.
- *Power relays*: Operate at a predetermined value of power. These
  include overpower and underpower relays.
- *Directional relays*:
  (i) Alternating current: Operate according to the phase relationship between alternating quantities.
  (ii) Direct current: Operate according to the direction of the current
  and are usually of the permanent-magnetic, moving-coil pattern.
- *Frequency relays*: Operate at a predetermined frequency. These
  include overfrequency and underfrequency relays.
- *Temperature relays*: Operate at a predetermined temperature in
  the protected component.
- *Differential relays*: Operate according to the scalar or vectorial
  difference between two quantities such as current, voltage, etc.
- *Distance relays*: Operate according to the "distance" between the
  relay's current transformer and the fault. The "distance" is measured in terms of resistance, reactance, or impedance.

Relays are made up of one or more fault-detecting units along with
the necessary auxiliary units. Basic units for relay systems can be classified
as being electromechanical units, sequence networks, or solid-state units.
The electromechanical types include those based on magnetic attraction,
magnetic induction, D'Arsonval, and thermal principles. Static networks
with three-phase inputs can provide a single-phase output proportional to
positive, negative, or zero sequence quantities. These are used as fault
sensors and are known as *sequence filters*. Solid-state relays use low power
components, which are designed into logic units used in many relays.

Electromechanical Relays

We consider first the magnetic attraction type, which can be classified
into three categories: the plunger unit, the clapper unit, and the polar unit.

The *plunger type* has cylindrical coils with an external magnetic
structure and a center plunger as shown in Figure 9-3. The plunger moves

546 System Protection

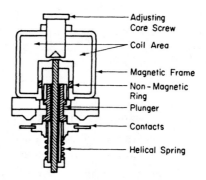

Figure 9-3. Plunger-Type Relay Unit.

upward to operate a set of contacts when the current or voltage applied to the coil exceeds a certain value. The moving force is proportional to the square of the current in the coil. These units are instantaneous since no delay is intentionally introduced. Typical operating times are 5 to 50 ms, with the longer times occurring near the threshold values of pickup. The unit shown in Figure 9-3 is used as a high dropout instantaneous overcurrent unit. The steel plunger floats in an air gap provided by a nonmagnetic ring in the center of the magnetic core. When the coil is energized, the plunger assembly moves upward, carrying a silver disc that bridges three stationary contacts (only two are shown). A helical spring absorbs the ac plunger's vibrations, producing good contact action.

Clapper units such as shown in Figure 9-4 have a U-shaped magnetic frame with a movable armature across the open end. The armature is hinged at one side and spring-restrained at the other. When the electrical coil is energized, the armature moves toward the magnetic core, opening or closing a set of contacts with a torque proportional to the square of the coil current. Clapper units are less accurate than plunger units and are primarily applied as auxiliary or "go/no go" units.

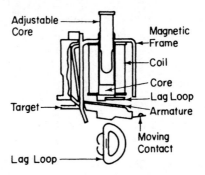

Figure 9-4. Clapper-Type Relay Unit.

Polar units use direct current applied to a coil wound around the hinged armature in the center of the magnetic structure. A permanent magnet across the structure polarizes the armature-gap poles. Two nonmagnetic spacers, located at the rear of the magnetic frame, are bridged by two adjustable magnetic shunts. This arrangement enables the magnetic flux paths to be adjusted for pickup and contact action. With balanced air gaps the armature will float in the center with the coil deenergized. With the gaps unbalanced, polarization holds the armature against one pole with the coil deenergized. The coil is arranged so that its magnetic axis is in line with the armature and at a right angle to the permanent magnet axis. Current in the coil magnetizes the armature either north or south, increasing or decreasing any prior polarization of the armature. If, as shown in Figure 9-5, the magnetic shunt adjustment normally makes the armature a north pole, it will move to the right. Direct current in the operating coil, which tends to make the contact end a south pole, will overcome this tendency, and the armature will move to the left to close the contacts.

Induction disc units are based on the watthour meter design and use the same operating principles. They operate by torque resulting from the interaction of fluxes produced by an electromagnet with those from induced currents in the plane of a rotatable aluminum disc. The unit shown in Figure 9-6 has three poles on one side of the disc and a common magnetic keeper on the opposite side. The main coil is on the center leg. Current (I) in the main coil produces flux (ϕ), which passes through the air gap and disc to the keeper. The flux ϕ is divided into ϕ_L through the left-hand leg and ϕ_R through the right-hand leg. A short-circuited lagging coil on the left leg causes ϕ_L to lag both ϕ_R and ϕ, producing a split-phase motor action. The flux ϕ_L induces a voltage V_s, and current I_s flows, in phase, in the shorted lag coil. The flux ϕ_T is the total flux produced by the main coil current (I). The three fluxes cross the disc air gap and produce eddy currents in the disc. As a result, the eddy currents set up counterfluxes, and

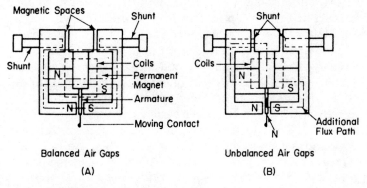

Figure 9-5. Polar-Type relay Unit. (a) Balanced Air Gaps. (b) Unbalanced Air Gaps.

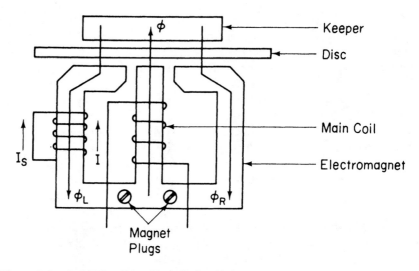

Figure 9-6. Induction Disc-Type Relay Unit.

the interaction of the two sets of fluxes produces the torque that rotates the disc.

A spiral spring on the disc shaft conducts current to the moving contact. This spring, together with the shape of the disc and the design of the electromagnet, provides a constant minimum operating current over the contact's travel range. A permanent magnet with adjustable keeper (shunt) damps the disc, and the magnetic plugs in the electromagnet control the degree of saturation. The spring tension, the damping magnet, and the magnetic plugs allow separate and relatively independent adjustment of the unit's inverse time overcurrent characteristics.

The operation of a *cylinder unit* is similar to that of an induction motor with salient poles for the stator windings. This unit has an inner steel core at the center of the square electromagnet with a thin-walled aluminum cylinder rotating in the air gap as shown in Figure 9-7. Cylinder travel is limited to a few degrees by the contact and the associated stops, and a spiral spring provides reset torque. Operating torque is a function of the product of the two operating quantities and the cosine of the angle between them. Different combinations of input quantities can be used for different applications, system voltages or currents, or network voltages.

A magnetic structure and an inner permanent magnet form a two-pole cylindrical core in the *D'Arsonval unit*, shown in Figure 9-8. A moving coil loop in the air gap is energized by direct current, which reacts with the air gap flux to create rotational torque. The D'Arsonval unit operates on very low energy input, available from dc shunts, bridge network, or rectified ac.

The balanced-beam relay operates on the balance principle by comparing two quantities, e.g., two currents, or one current and one voltage. The

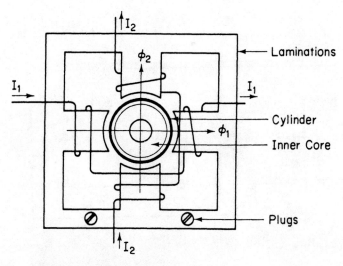

Figure 9-7. Cylinder-Type Relay Unit.

general form of the relay, comparing, for example, two currents, is illustrated in Figure 9-9. The beam is given a slight mechanical bias to the contact open position; this is achieved by spring or weight adjustment. Assuming that the two coils have an equal number of turns, operation is obtained when I_a^2 is equal to $I_b^2 + K$, where I_a is the current in the operating coil, I_b the current in the restraining coil, and K is a constant. The electromagnets may be provided with more than one coil, and some of the coils may be interconnected depending on the quantities to be measured.

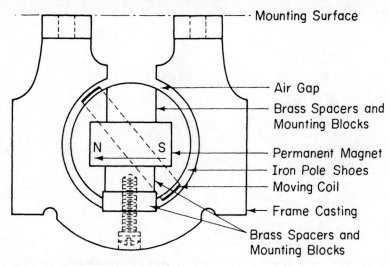

Figure 9-8. D'Arsonval-Type Unit.

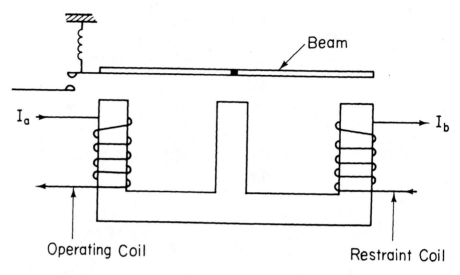

Figure 9-9. A Balanced-Beam Relay.

Sequence Filters

Static networks with three-phase current or voltage inputs can be arranged to provide a single-phase output proportional to positive, negative, or zero sequence quantities. The simplest sequence filter is the zero sequence current variety, which is composed of the secondaries of three current transformers connected in parallel to provide $3I_0$ from I_a, I_b, and I_c.

A basic circuit that provides many variations of sequence filters is shown in Figure 9-10. The open-circuit voltage V_F is seen to be

$$V_F = I_n R_0 + I_a R_1 + jX_m(I_c - I_b) \tag{9.1}$$

with

$$I_n = I_a + I_b + I_c \tag{9.2}$$

The open-circuit voltage in terms of sequence quantities is given by

$$V_F = \left(R_1 - \sqrt{3}\,X_m\right)I_+ + \left(R_1 + \sqrt{3}\,X_m\right)I_- + (R_1 + 3R_0)I_0 \tag{9.3}$$

It can be seen that different weights are given to each sequence current.

If we interchange I_b and I_c in the circuit, we get

$$V_F = \left(R_1 + \sqrt{3}\,X_m\right)I_+ + \left(R_1 - \sqrt{3}\,X_m\right)I_- + (R_1 + 3R_0)I_0 \tag{9.4}$$

If R_1 is chosen such that $X_m = R_1/\sqrt{3}$, we get

$$V_F = 2R_1 I_+ + (R_1 + 3R_0)I_0 \tag{9.5}$$

and the relay connected to the output terminals will respond to positive and zero sequence quantities.

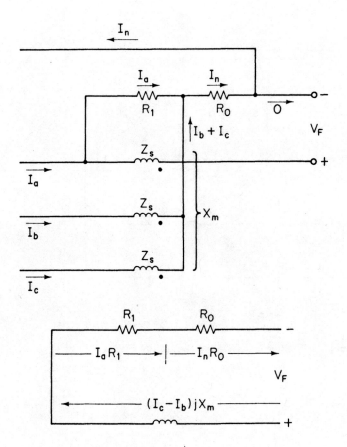

Figure 9-10. Sequence Filter.

Another type of circuit shown in Figure 9-11 provides the ability to respond to either positive or negative sequence quantities. Here we have

$$V_F = I_a\left(\frac{2R_1}{3}\right) + (I_c - I_b)(jX_m) - (I_b + I_c)\left(\frac{R_1}{3}\right) \quad (9.6)$$

Again this reduces to

$$V_F = \left(R_1 - \sqrt{3}\,X_m\right)I_+ + \left(R_1 + \sqrt{3}\,X_m\right)I_- \quad (9.7)$$

Now with the choice $X_m = R_1/\sqrt{3}$, we get

$$V_F = 2R_1 I_- \quad (9.8)$$

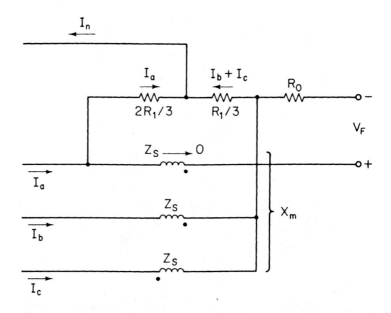

Figure 9-11. Positive or Negative Sequence Filter.

which gives the negative sequence filter characteristic. Interchanging I_b and I_c, we get

$$V_F = \left(R_1 + \sqrt{3}\,X_m\right)I_+ + \left(R_1 - \sqrt{3}\,X_m\right)I_- \qquad (9.9)$$

which gives the positive sequence characteristic for the choice $X_m = R_1/\sqrt{3}$:

$$V_F = 2R_1 I_+ \qquad (9.10)$$

Example 9-1

A. Consider the sequence filter for protective relaying of Figure 9-11. The open-circuit voltage V_F is given in terms of the three-phase currents by Eq. (9.6)

$$V_F = \left(\frac{2R_1}{3}\right)I_a + jX_m(I_c - I_b)$$
$$- \left(\frac{R_1}{3}\right)(I_b + I_c)$$

In this case V_F is considered as the Thévenin's equivalent voltage. By short-circuiting the terminals 1 and 2, we can calculate the Thévenin's equivalent impedance Z_{TH} using

$$Z_{TH} = \frac{V_{TH}}{I_{sc}}$$

Show that for this circuit

$$Z_{TH} = R_0 + R_1 + Z_S$$

B. Assume now that the sequence filter described above is connected at terminals 1 and 2 to a relay of resistance R_L. Assume that

$$R_1 = \sqrt{3}\, X_m$$

Show that the current in the relay is

$$I_L = \frac{2 R_1 I_-}{R_0 + R_1 + Z_S + R_L}$$

C. Given that $R_0 = 0.1\ \Omega$, $R_L = 0.1\ \Omega$, and $Z_S = 0.08\ \Omega$, it is required to adjust the sequence filter parameters so that for $I_- = 10$ A, the relay current $I_L = 5$ A. Find the values of R_1 and X_m.

Solution

A. Refer to Figure 9-12 showing the conditions with terminals 1 and 2 shorted. It is clear that KVL results in the following relation:

$$(R_0 + R_1 + Z_S) I_{sc} + \left[I_a\left(\frac{2 R_1}{3}\right) + (I_c - I_b) j X m \right.$$
$$\left. - (I_b + I_c)\left(\frac{R_1}{3}\right) \right] = 0$$

Substituting for the relation for V_F, we get

$$(R_0 + R_1 + Z_S) I_{sc} + V_F = 0$$

Thus,

$$Z_{TH} = R_0 + R_1 + Z_S$$

B. Put $X_m = R_1/\sqrt{3}$. Thus we get

$$V_F = 2 R_1 I_-$$

From the Thévenin's equivalent representation,

$$I_L = \frac{V_F}{R_{TH} + R_L}$$
$$= \frac{2 R_1}{R_{TH} + R_L} I_-$$
$$= \frac{2 R_1}{R_0 + R_1 + Z_S + R_L} I_-$$

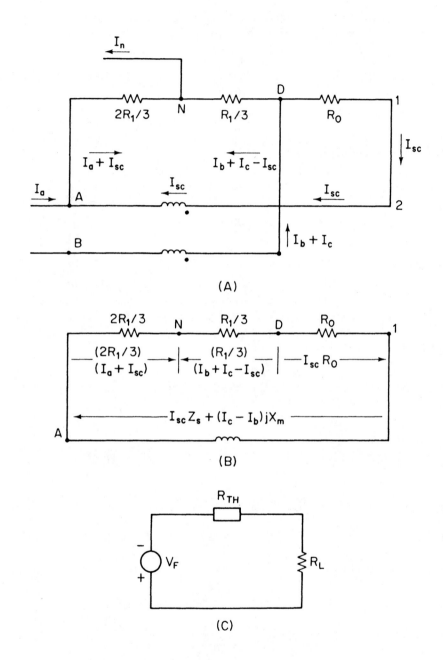

Figure 9-12. To Illustrate Steps in Solving Example 9-1.

C. With the following data:

$$I_- = 10 \text{ A},$$
$$I_L = 5 \text{ A}$$
$$R_L = 0.1 \text{ }\Omega$$
$$R_0 = 0.1 \text{ }\Omega$$
$$Z_S = 0.08 \text{ }\Omega$$

we obtain

$$5 = \frac{2R_1(10)}{0.1 + 0.08 + 0.1 + R_1}$$

The result is

$$R_1 = \frac{1.4}{15} = 0.0933 \text{ }\Omega, X_m = \frac{R_1}{\sqrt{3}} = 0.0539$$

Solid-State Units

Solid-state, linear, and digital-integrated circuit logic units are combined in a variety of ways to provide modules for relays and relay systems. Three major categories of circuits can be identified: (1) fault-sensing and data-processing logic units, (2) amplification logic units, and (3) auxiliary logic units.

Logic circuits in the fault-sensing and data-processing category employ comparison units to perform conventional fault-detection duties. Magnitude comparison logic units are used for overcurrent detection both of instantaneous and time overcurrent categories. For instantaneous overcurrent protection, a dc level detector, or a fixed reference magnitude comparator, is used. A variable reference magnitude comparator circuit is used for ground-distance protection. Phase-angle comparison logic circuits produce an output when the phase angle between two quantities is in the critical range. These circuits are useful for phase, distance, and directional relays.

9.3 THE X-R DIAGRAM

Consider a transmission line with series impedance Z_L and negligible shunt admittance. At the receiving end, a load of impedance Z_R is assumed. The schematic of the system is shown in Figure 9-13. The phasor diagram shown in Figure 9-14 is constructed with I taken as the reference. For our purpose the phasor diagram represents the relation

$$V_S = IZ_L + V_r \qquad (9.11)$$

giving rise to the heavy-lined diagram rather than the usual one shown by

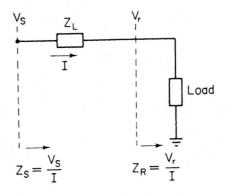

Figure 9-13. Schematic of Short Line.

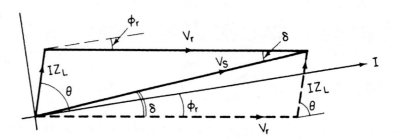

Figure 9-14. Voltage Phasor Diagram.

9.3 The X-R Diagram

the dashed line. On the diagram, δ is the torque angle, which is the angle between V_s and V_r.

If the phasor diagram, Eq. (9.11), is divided by the current I, we obtain the impedance equation

$$Z_s = Z_L + Z_r \qquad (9.12)$$

where

$$Z_s = \frac{V_s}{I}$$

$$Z_r = \frac{V_r}{I}$$

An impedance diagram is shown in Figure 9-15. This is called the X-R diagram since the real axis represents a resistive component (R), and the imaginary axis corresponds to a reactive component (X). The angle δ appears on the impedance diagram as that between Z_s and Z_r.

The evaluation of Z_r from complex power S_R and voltage V_r is straightforward. Here we utilize the following relationship valid for a series representation $Z_r = R_r + jX_r$:

$$R_r = \frac{|V_r|^2 P_r}{|S_r|^2} \qquad (9.13)$$

$$X_r = \frac{|V_r|^2 Q_r}{|S_r|^2} \qquad (9.14)$$

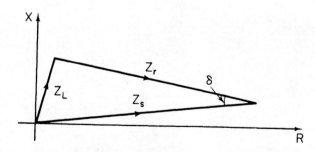

Figure 9-15. Impedance Diagram.

Example 9-2

Find Z_r, given that

$$|V_r| = 1 \text{ p.u.}$$
$$S_r = 2 + j0.8 \text{ p.u.}$$

Construct the impedance diagram for

$$Z_L = 0.1 + j0.3 \text{ p.u.}$$

Find Z_s for this condition, as well as the angle δ.

Solution

$$R_r = \frac{(1)^2(2)}{(2.154)^2} = 0.431 \text{ p.u.}$$

$$X_r = \frac{(1)^2(0.8)}{(2.154)^2} = 0.1724 \text{ p.u.}$$

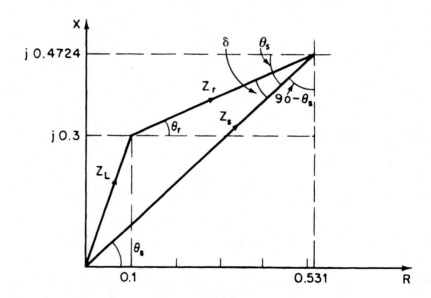

Figure 9-16. Construction of *X-R* Details for Example 9-2.

From the diagram, Z_s can be obtained graphically, or analytically. Thus

$$Z_s = Z_L + Z_r \qquad\qquad \delta = \theta_s - \theta_r$$
$$= 0.7107\,\underline{/41.6576°} \qquad\qquad = 26.54°$$
$$\theta_s = 41.6576°$$

Figure 9-16 illustrates the construction.

9.4 RELAY COMPARATORS

Relay comparators can have any number of input signals. However, we focus our attention here on the two-input comparator shown schematically in Figure 9-17. The input to the two transformer circuits 1 and 2 includes the line voltage V_L and current I_L. The output of transformer 1 is V_1, and that of transformer 2 is V_2. Both V_1 and V_2 are input to the comparator, which produces a trip (operate) signal whenever $|V_2|>|V_1|$ in an amplitude comparison mode.

We will start the analysis by assuming that the line voltage V_L is the reference phasor and that the line current lags V_L by an angle ϕ_L. Thus,

$$V_L = |V_L|\,\underline{/0}$$

$$I_L = |I_L|\,\underline{/-\phi_L}$$

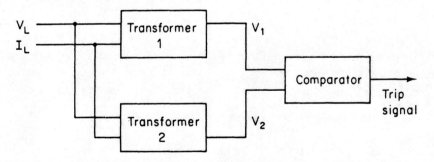

Figure 9-17. Schematic of Relay Comparator Circuit.

The impedance Z_L is thus

$$Z_L = \frac{V_L}{I_L}$$
$$= |Z_L|\underline{/\phi_L}$$

The transformers' output voltages V_1 and V_2 are assumed to be linear combinations of the input quantities

$$V_1 = k_1 V_L + Z_1 I_L \qquad (9.15)$$
$$V_2 = k_2 V_L + Z_2 I_L \qquad (9.16)$$

The impedances Z_1 and Z_2 are expressed in the polar form:

$$Z_1 = |Z_1|\underline{/\psi_1}$$
$$Z_2 = |Z_2|\underline{/\psi_2}$$

The comparator input voltages V_1 and V_2 are thus given by

$$V_1 = |I_L|\left(k_1|Z_L| + |Z_1|\underline{/\psi_1 - \phi_L}\right) \qquad (9.17)$$
$$V_2 = |I_L|\left(k_2|Z_L| + |Z_2|\underline{/\psi_2 - \phi_L}\right) \qquad (9.18)$$

Amplitude Comparison

The trip signal is produced for an amplitude comparator when

$$|V_2| \geq |V_1| \qquad (9.19)$$

The operation threshold condition $|V_2| = |V_1|$ is of interest and occurs when

$$\left|k_1|Z_L| + |Z_1|\underline{/\psi_1 - \phi_L}\right|^2 = \left|k_2|Z_L| + |Z_2|\underline{/\psi_2 - \phi_L}\right|^2 \qquad (9.20)$$

Both sides of the above equation are of the following form:

$$\left|A + B\underline{/\beta}\right|^2 = |A + B\cos\beta + jB\sin\beta|^2$$
$$= (A + B\cos\beta)^2 + (B\sin\beta)^2$$
$$= A^2 + B^2 + 2AB\cos\beta$$

Thus the operation condition (9.20) is obtained as

$$(k_1^2 - k_2^2)|Z_L|^2 + 2|Z_L|[k_1|Z_1|\cos(\psi_1 - \phi_L) - k_2|Z_2|\cos(\psi_2 - \phi_L)]$$
$$+ (|Z_1|^2 - |Z_2|^2) \leq 0 \qquad (9.21)$$

This is the general equation for an amplitude comparison relay. The choices of k_1, k_2, Z_1, and Z_2 provide different relay characteristics.

Ohm Relay

The following parameter choice is made:

$$k_1 = k \qquad k_2 = -k$$
$$Z_1 = 0 \qquad Z_2 = Z$$
$$\psi_2 = \psi$$

As a result, the general equation (9.21) reduces to

$$|Z_L|(\cos\psi\cos\phi_L + \sin\psi\sin\phi_L) \leq \frac{|Z|}{2k} \qquad (9.22)$$

The relay characteristic can be shown in the X-R plane by setting the line impedance in the rectangular form

$$Z_L = R_L + jX_L$$

Thus,

$$R_L = |Z_L|\cos\phi_L$$
$$X_L = |Z_L|\sin\phi_L$$

Thus the relay threshold equation (9.22) becomes

$$R_L\cos\psi + X_L\sin\psi = \frac{|Z|}{2k} \qquad (9.23)$$

This is a straight line in the $X_L - R_L$ plane as shown in Figure 9-18. The shaded area is the restrain area; an operate signal is produced in the nonshaded area.

Mho Relay

The mho relay characteristic is obtained with the choice

$$k_1 = -k \qquad k_2 = 0$$
$$Z_1 = Z_2 = Z$$
$$\psi_1 = \psi_2 = \psi$$

As a result, the general equation (9.21) reduces to

$$k^2|Z_L|^2 - 2k|Z_L||Z|\cos(\psi - \phi_L) \leq 0 \qquad (9.24)$$

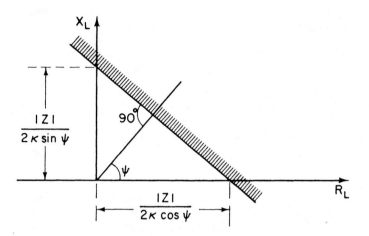

Figure 9-18. Ohm Relay Characteristic.

In terms of R_L and X_L, we thus have the characteristic

$$R_L^2 + X_L^2 - \frac{2|Z|}{k}(R_L\cos\psi + X_L\sin\psi) \leq 0$$

or

$$\left(R_L - \frac{|Z|}{k}\cos\psi\right)^2 + \left(X_L - \frac{|Z|}{k}\sin\psi\right)^2 \leq \frac{|Z|^2}{k^2} \quad (9.25)$$

The threshold condition with equality sign is a circle as shown in Figure 9-19.

Impedance Relay

Here we set

$$k_1 = -k \qquad k_2 = 0$$
$$Z_1 \neq Z_2$$

As a result, the relay characteristic is given by

$$k^2|Z_L|^2 - 2k|Z_L||Z_1|\cos(\psi_1 - \phi_L) + |Z_1|^2 - |Z_2|^2 \leq 0 \quad (9.26)$$

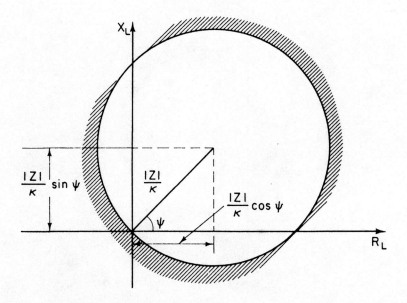

Figure 9-19. Mho Relay Characteristic.

or in the $R_L - X_L$ plane,

$$\left(R_L - \frac{|Z_1|\cos\psi}{k}\right)^2 + \left(X_L - \frac{|Z_1|\sin\psi}{k}\right)^2 \leq \frac{|Z_2|^2}{k^2} \tag{9.27}$$

The threshold condition is a circle with center at $|Z_1|/k\underline{/\psi}$ and radius $|Z_2|/k$ as shown in Figure 9-20.

Phase Comparison

Let us now consider the comparator operating in the phase comparison mode. Assume that

$$V_1 = |V_1|\underline{/\theta_1}$$
$$V_2 = |V_2|\underline{/\theta_2}$$

Then the phasor ratio is

$$\frac{V_1}{V_2} = \frac{|V_1|}{|V_2|}\underline{/\theta_1 - \theta_2}$$

Let the phase difference be defined as

$$\theta = \theta_1 - \theta_2$$

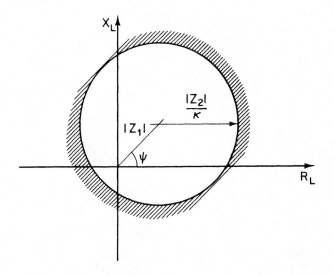

Figure 9-20. Impedance Relay Characteristic.

A criterion for operation of the $\pm 90°$ phase comparator such as

$$\frac{-\pi}{2} \leqslant \theta \leqslant \frac{\pi}{2}$$

is chosen. This implies that

$$\cos\theta \geqslant 0$$

For our manipulation purposes, we rewrite this condition as

$$\frac{|V_1||V_2|}{|I_L|^2}\cos(\theta_1 - \theta_2) \geqslant 0 \tag{9.28}$$

Expanding the right-hand side, we obtain

$$\frac{|V_1||V_2|}{|I_L|^2}\cos(\theta_1 - \theta_2) = \frac{Re(V_1)Re(V_2) + Im(V_1)Im(V_2)}{|I_L|^2} \tag{9.29}$$

Recall that

$$\frac{Re(V_1)}{|I_L|} = k_1|Z_L| + |Z_1|\cos(\psi_1 - \phi_L)$$

$$\frac{Im(V_1)}{|I_L|} = |Z_1|\sin(\psi_1 - \phi_L)$$

$$\frac{Re(V_2)}{|I_L|} = k_2|Z_L| + |Z_2|\cos(\psi_2 - \phi_L)$$

$$\frac{Im(V_2)}{|I_L|} = |Z_2|\sin(\psi_2 - \phi_L)$$

As a result, we have

$$k_2 k_2 |Z_L|^2 + |Z_L|[k_1|Z_2|\cos(\psi_2 - \phi_L) + k_2|Z_1|\cos(\psi_1 - \phi_L)]$$
$$+ |Z_1||Z_2|\cos(\psi_1 - \psi_2) \geq 0 \qquad (9.30)$$

This is the general equation for the $\pm 90°$ phase comparator. By assigning values to the parameters k_1, k_2, Z_1, and Z_2, different relay characteristics are obtained.

Ohm Relay

The following choice results in an ohm-relay characteristic:

$$k_1 = -k \qquad k_2 = 0$$
$$Z_1 = Z_2 = Z$$
$$\psi_1 = \psi_2 = \psi$$

These are exactly the values associated with the mho relay using the amplitude comparison mode. Substitution of these parameters results in

$$-k|Z_L||Z|\cos(\psi - \phi_L) + |Z|^2 \geq 0$$

or

$$|Z_L|\cos(\psi - \phi_L) \leq \frac{|Z|}{k} \qquad (9.31)$$

In the $(X_L - R_L)$ plane the result is the straight line and is shown in Figure 9-21.

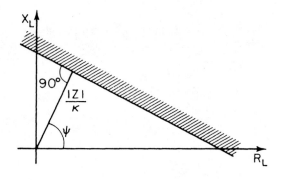

Figure 9-21. Ohm-Relay Characteristic Using Phase Comparison.

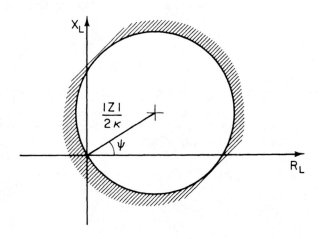

Figure 9-22. Mho-Relay Characteristic Using Phase Comparison.

Mho Relay

In this case we choose the circuit that would produce the ohm-relay characteristic in the amplitude comparison case:

$$k_1 = k \qquad k_2 = -k$$
$$Z_1 = 0 \qquad Z_2 = Z\underline{/\psi}$$

As a result, we have

$$-k^2|Z_L|^2 + k|Z_L||Z|\cos(\psi - \phi_L) \geq 0$$

or

$$\left[R_L - \frac{|Z|\cos\psi}{2k}\right]^2 + \left[X_L - \frac{|Z|\sin\psi}{2k}\right]^2 \leq \frac{|Z|^2}{4k^2} \qquad (9.32)$$

Again this is the equation of a circle with radius $|Z|/2k$, i.e., half of the radius of the amplitude comparator mho circle, and with its center at $(|Z|/2k)\underline{/\psi}$ as shown in Figure 9-22.

9.5 GENERATOR PROTECTION

There are a number of abnormal conditions that may occur with rotating equipment, including:

1. Faults in the windings.
2. Loss of excitation.
3. Motoring of generators.
4. Overload.
5. Overheating.
6. Overspeed.
7. Unbalanced operation.
8. Out-of-step operation.

Several of these conditions can be corrected while the unit is in service and should be detected and signaled by alarms. Faults, however, require prompt tripping and are the result of insulation breakdown or flashovers that occur across the insulation at some point. The result of a fault is a conducting path between points that are normally of different potential. If the path has a high resistance, the fault is accompanied by a noticeable voltage change in the affected area. If, on the other hand, the path resistance is low, a large current results, which may cause serious damage.

Figure 9-23 shows the various types of faults that may occur in the insulation system of a generator's windings. The faults shown are identified as:

1. Interphase short circuit.
2. Interturn fault.
3. Stator earth fault.
4. Rotor earth fault.
5. Interturn fault in rotor.

Note that A denotes the insulation of individual windings and B denotes the stator core.

A short circuit between parts of different phases of the winding, such as faults 1 and 2 above, results in a sever fault current within the machine. A consequence of this is a distinct difference between the currents at the

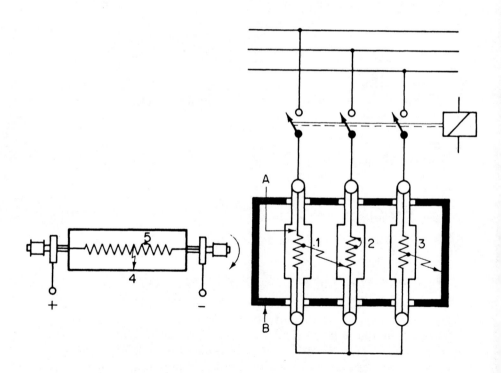

Figure 9-23. Various Types of Faults That May Occur in the Insulation System of the Generator Windings.

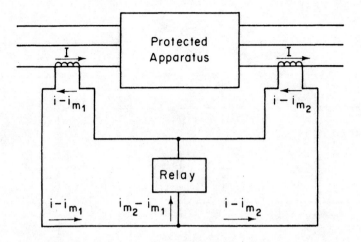

Figure 9-24. Basic Differential Connection.

neutral and terminal ends of the particular winding, which is detected by a differential protection system. Here the currents on each side of the protected apparatus for each phase are compared in a differential circuit. Any difference current will operate a relay. Figure 9-24 shows the relay circuit for one phase only. On normal operation, only the difference between the current transformer magnetizing currents i_{m_1} and i_{m_2} passes through the relay. This is due to the fact that with no faults within the protected apparatus, the currents entering and leaving are equal to I. If a fault occurs between the two sets of current transformers, one or more of the currents (in a three-phase system) on the left-hand side will suddenly increase, while that on the right-hand side may decrease or increase with a direction reversal. In both instances, the total fault current will flow through the relay, causing it to operate. In units where the neutral ends are inaccessible, differential relays are not used. In this case, reverse power relays are employed instead.

Current leakage can take place between turns of the same phase of a winding or between parallel coils of the same phase. This is referred to as an *interturn fault*. In generators with one winding per phase, a voltage transformer is connected between each phase terminal and the neutral of the winding. The secondary terminals are connected in an open delta to a polarized voltage relay as shown in Figure 9-25. In the event of an interturn

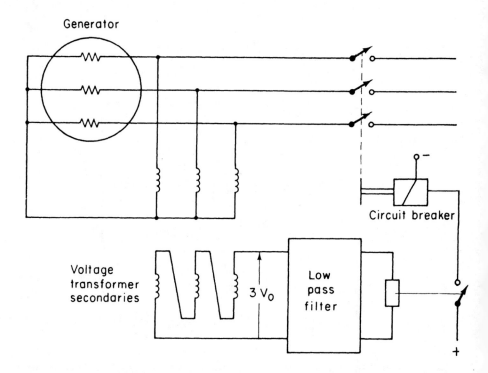

Figure 9-25. Interturn Fault Protection Scheme.

fault, a voltage appears at the terminals of the open delta and causes the relay to trip. In the phasor diagram of Figure 9-26, $3V_0$ is the resulting voltage at the terminals of the open delta. Thus,

$$3V_0 = V_1 + V_2 + V_3$$

Note that V_0 is essentially a zero sequence voltage.

To protect against stator earth faults, the neutral point of the generator is connected to ground through a high resistance. With an earth fault, current will flow through the resistance, producing a potential difference across the resistance between the neutral point and ground. This potential difference is picked up by a voltage transformer connected to a polarized voltage relay, which trips the generator circuit breakers as shown in Figure 9-27. This protection system is designed for a fault current exceeding 10 A.

If more than one ground fault in the rotor (field) circuit occurs, magnetic unbalances and hence machine vibrations take place. This makes

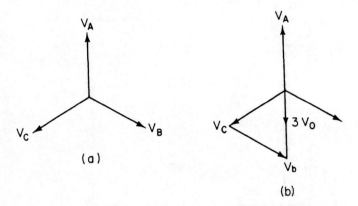

Figure 9-26. Phasor Diagrams for Normal and Fault Voltages in Interturn Fault Protection. (a) Normal Conditions. (b) Fault on Phase A.

the detection of earth faults on the rotor circuit of fundamental importance. A scheme utilizing dc such as that shown in Figure 9-28 is used. The voltage divider principle is used with R_1 and R_2 being fixed resistance (typically $R_1 = 5$ kΩ and $R_2 = 23$ kΩ for an exciter with 250-V rating). Note R_N is a resistance that varies with the voltage applied to it (typically 45 kΩ at 60 V and 4.7 kΩ at 150 V). When the field becomes grounded, a voltage (whose

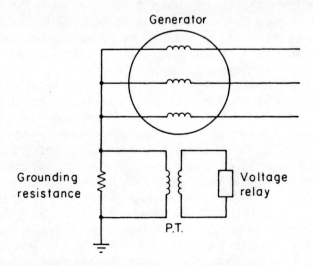

Figure 9-27. Schematic for Ground Fault Protection for a Machine Grounded Through a High Resistance.

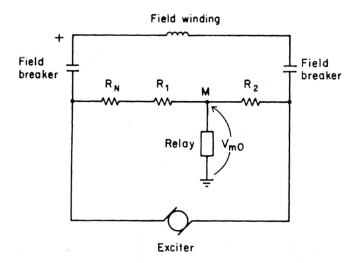

Figure 9-28. Field Ground Protection Scheme.

magnitude depends on the exciter voltage and the point of fault on the field) will appear between the point M and the ground. This voltage V_{M0} will be a maximum if the fault is at either end of the field. A point on the field where $V_{M0} = 0$ is called the *null point*. If $R_2 = R_1 + R_N$, then the null point is at the middle of the field winding. A dc instrument-type relay detects abnormal values of V_{M0} and hence rotor faults.

The protection mechanisms discussed above are shown in Figure 9-29 and are designed to react to faults that occur on the generator. Preventive measures taken include lightning arresters connected between the input phases and the star point of the winding and ground. This limits the stresses induced in the windings by atmospheric surges. In the event of turbine overspeeding, overvoltage protection should act to suppress the field. This is also necessary for regulator and exciter system faults. In hydro units, overvoltage time-lag relays with high-speed stage are used; whereas for turbogenerators, an instantaneous overvoltage relay is adequate.

Thermal relays are used as protection against overloads that cause inadmissibly high temperatures. These intervene when the protected winding has almost reached the designed temperature limits. The relay possesses a thermal replica whose time-constant is matched with that of the generator. The relay is capable of being overloaded for a short period so that the generator can be loaded to advantage up to its maximum thermal capacity.

Thermal overcurrent time-relays with modest time-constants are provided in the output circuits of the voltage regulator. For sustained positive or negative maximum excitation due to a fault in the regulator, the generator is switched over to manual regulation.

9.5 Generator Protection

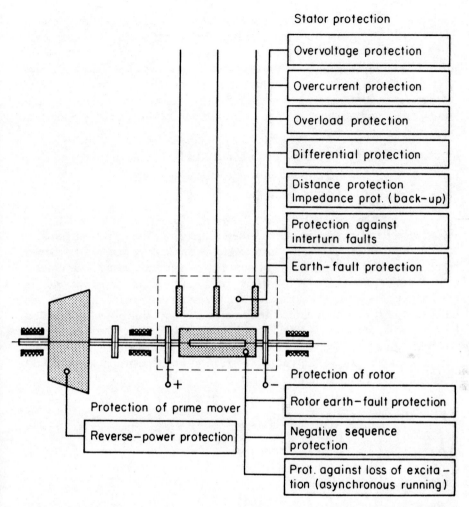

Figure 9-29. Allocation of Protective Devices for the Stator, Rotor, and Prime Mover.

Negative sequence relays are used to indicate whether the permissible single-phase loading of a three-phase generator has been exceeded. The relay affords a thermal protection of the rotor because, in the event of an unsymmetrical loading of the machine, out-of-phase currents with double the frequency are generated in the rotor, resulting in harmful overheating of the core.

Capacitive minimum-reactance relays are used for generators where a corresponding risk is present. This relay is combined with an undelayed overvoltage relay in such a way that when self-excitation commences, both relays pick up and trip the generator breaker immediately. Through a second output, a time-lag relay is switched on, which, on expiry of its time lag, causes hydrogenerators to be switched off or turbogenerators to have the field suppressed and a sustained signal of asynchronous running to be given. This relay also detects interruptions or short circuits in the excitation circuit. For such faults, the generator is either switched off immediately, or it remains in the network while resynchronization into the network is attempted. Figure 9-30 shows the X-R diagram for a capacitive minimum-reactance relay where

X_d = Synchronous reactance of the generator
X_d' = Transient reactance of the generator
a, b = Adjustment factors

Reverse-power relays are used to switch off the generator when it runs at full speed as a motor due to turbine failure when the generator remains connected to the network. In some systems, instead of the reverse-power relay, a very sensitive power directional relay with low setting is employed.

To eliminate currents in the bearings, at least one bearing and its auxiliary piping must be insulated from earth. A possible fault in this insulation or inadvertent short-circuiting by a conducting object is detected by the bearing current protection, which determines the current flowing in the bearing directly. It is normally provided as a two-stage facility.

Figure 9-31 shows the various protection functions their location and the faults covered for generator protection.

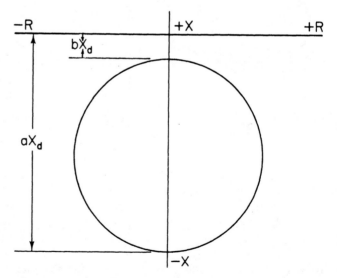

Figure 9-30. Tripping Locus of the Capacitive Minimum-Reactance Relay in the Negative Reactance Zone of the R-X Diagram.

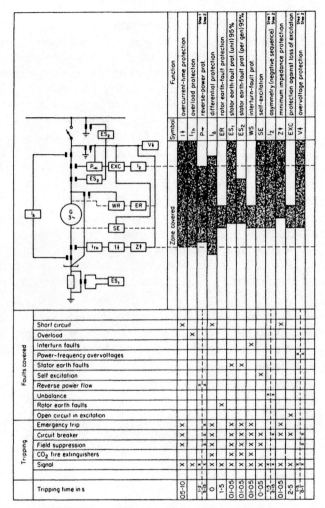

Figure 9-31. A General Protection Scheme for Generators.
(Courtesy Brown-Boveri Co.)

9.6 TRANSFORMER PROTECTION

A number of fault conditions can arise within a power transformer. These include:

1. *Earth faults*: A fault on a transformer winding will result in currents that depend on the source, neutral grounding impedance, leakage reactance of the transformer, and the position of the fault

576 System Protection

in the windings. The winding connections also influence the magnitude of fault current. In the case of a Y-connected winding with neutral point connected to ground through an impedance Zg, the fault current depends on Zg and is proportional to the distance of the fault from the neutral point. If $Zg = 0$, i.e., the neutral is solidly grounded, the fault current is controlled by the leakage reactance. It is clear that the leakage reactance is dependent on the fault location. The reactance decreases as the fault becomes closer to the neutral point. As a result, the fault current is highest for a fault close to the neutral point. Figure 9-32 compares the general variation of fault current with the location of the fault in the winding for Y-connected winding. In the case of a fault in a Δ-connected winding, the range of fault current is less than that for a Y-connected winding, with the actual value being controlled by the method of grounding used in the system. Phase fault currents may be low for a Δ-connected winding due to the high impedance to fault of the Δ winding. This factor should be considered in designing the protection scheme for such a winding.

2. *Core faults* due to insulation breakdown can permit sufficient eddy-current to flow to cause overheating, which may reach a magnitude sufficient to damage the winding.

3. *Interturn faults* occur due to winding flashovers caused by line surges. A short circuit of a few turns of the winding will give rise to high currents in the short-circuited loops, but the terminal currents will be low.

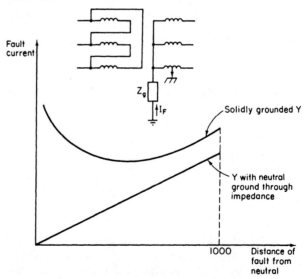

Figure 9-32. Earth Fault Current Variation with Location of Fault.

9.6 Transformer Protection

4. *Phase-to-phase faults* are rare in occurrence but will result in substantial currents of magnitudes similar to earth faults'.
5. *Tank faults* resulting in loss of oil reduce winding insulation as well as producing abnormal temperature rises.

In addition to fault conditions within the transformer, abnormal conditions due to external factors result in stresses on the transformer. These conditions include:

1. *Overloading*, which results in increased I^2R losses and an associated temperature rise.
2. *System faults* produce effects similar to but sometimes more severe than those due to overloading.
3. *Overvoltages* due to either transient surges or power frequency voltage increases produce insulation stresses and increased flux.
4. *Underfrequency operation* of the system results in an increased flux, causing increased core losses and a corresponding temperature rise.

When a transformer is switched in at any point of the supply voltage wave, the peak values of the core flux wave will depend on the residual flux as well as on the time of switching. The peak value of the flux will be higher than the corresponding steady-state value and will be limited by core saturation. The magnetizing current necessary to produce the core flux can have a peak of eight to ten times the normal full-load peak and has no equivalent on the secondary side. This phenomenon is called *magnetizing inrush current* and appears as an internal fault. Maximum inrush occurs if the transformer is switched in when the supply voltage is zero. Realizing this, is important for the design of differential relays for transformer protection so that no tripping takes place due to the magnetizing inrush current. A number of schemes based on the harmonic properties of the inrush current are used to prevent tripping due to large inrush currents.

Overheating protection is provided for transformers by placing a thermal-sensing element in the transformer tank. Overcurrent relays are used as a backup protection with time delay higher than that for the main protection. Restricted earth fault protection is utilized for Y-connected windings. This scheme is shown in Figure 9-33. The sum of the phase currents is balanced against the neutral current, and hence the relay will not respond to faults outside the winding.

Differential protection is the main scheme used for transformers. A number of considerations should be dealt with, including:

1. Transformer ratio: The current transformers should have ratings to match the rated currents of the transformer winding to which they are applied.

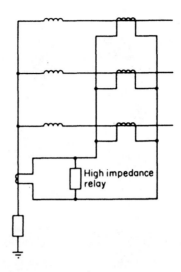

Figure 9-33. Restricted Ground Fault Protection for a Y Winding.

2. Due to the 30°-phase change between Y- and Δ-connected windings and the fact that zero sequence quantities on the Y side do not appear on the terminals of the Δ side, the current transformers should be connected in Y for a Δ winding and in Δ for a Y winding. Figure 9-34 shows the differential protection scheme applied to a Δ/Y transformer, and Figure 9-35 provides details for a three-winding Y/Δ/Y transformer differential protection scheme. When current transformers are connected in Δ, their secondary ratings must be reduced to $(1/\sqrt{3})$ times the secondary rating of Y-connected transformers.
3. Allowance should be made for tap changing by providing restraining coils (bias). The bias should exceed the effect of the maximum ratio deviation.

Example 9-3

Consider a Δ/Y-connected, 15-MVA, 33/11-kV transformer with differential protection applied, for the current transformer ratios shown in Figure 9-36. Calculate the relay currents on full load. Find the minimum relay current setting to allow 125 percent overload.

Solution

The primary line current is given by

$$I_p = \frac{15 \times 10^6}{(\sqrt{3})(33 \times 10^3)} = 262.43 \text{ A}$$

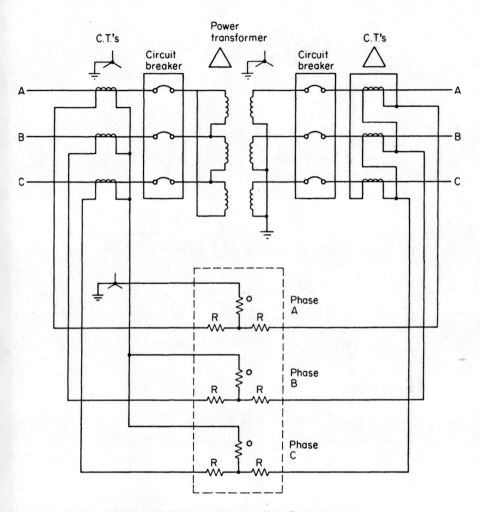

Figure 9-34. Differential Protection of a Δ/Y Transformer.

The secondary line current is

$$I_s = \frac{15 \times 10^6}{(\sqrt{3})(11 \times 10^3)} = 787.30 \text{ A}$$

The C.T. current on the primary side is thus

$$i_p = 262.43\left(\frac{5}{300}\right) = 4.37 \text{ A}$$

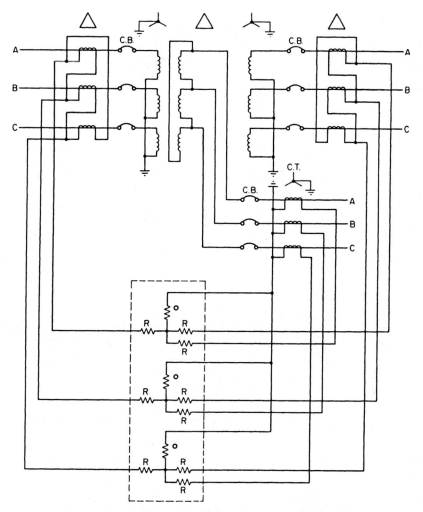

Figure 9-35. Transformer Differential Relay for a Three-Winding Transformer.

The C.T. current in the secondary side is

$$i_s = 787.30\left(\frac{5}{2000}\right)\sqrt{3} = 3.41 \text{ A}$$

Note that we multiply by $\sqrt{3}$ to obtain the values on the line side of the Δ-connected C.T.'s. The relay current on normal load is therefore

$$i_r = i_p - i_s = 4.37 - 3.41 = 0.9648 \text{ A}$$

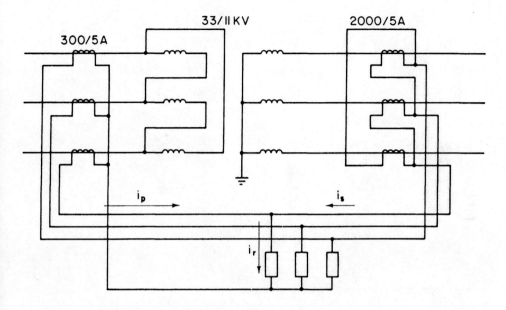

Figure 9-36. Transformer for Example 9-3.

With 1.25 overload ratio, the relay setting should be

$$I_r = (1.25)(0.9648) = 1.206 \text{ A}$$

Buchholz Protection

In addition to the above-mentioned protection schemes, it is common practice in transformer protection to employ gas-actuated relays for alarm and tripping. One such a relay is the Buchholz relay.

Faults within a transformer will result in heating and decomposing of the oil in the transformer tank. The decomposition produces gases such as hydrogen, carbon monoxide, and light hydrocarbons, which are released slowly for minor faults and rapidly for severe arcing faults. In the gas-activated relay named after its inventor, this phenomenon is utilized. With reference to Figure 9-37, the relay is connected into the pipe leading to the conservator tank. As the gas accumulates, the oil level falls and a float F is lowered and operates a mercury switch to sound an alarm. Sampling the gas and performing a chemical analysis provide a means for classifying the type of fault. In the case of a winding fault, the arc generates gas at a high

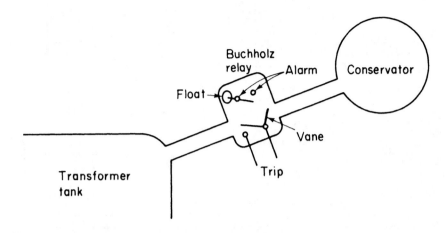

Figure 9-37. Principle of the Buchholz Relay.

release rate that moves the vane V to cause tripping through contacts attached to the vane.

Buchholz protection provides an alarm for a number of fault conditions including:

1. Interturn faults or winding faults involving only lower power levels.
2. Core hot spots due to short circuits on the lamination insulation.
3. Faulty joints.
4. Core bolt insulation failure.

Major faults resulting in tripping include severe earth or interphase winding faults and loss of oil through leakage.

9.7 BUS BAR PROTECTION

Bus bars are an essential link in the electric power system, and short circuits in their zone have to be interrupted in the shortest possible time. In distribution systems (6–20 kV) with supply through transformers, overcurrent time-relays provide an easy protection mechanism. The relays interrupt the supply to the bus bars if one or more supplies (but not feeders) are

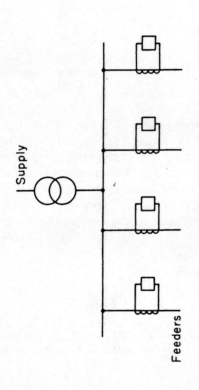

Figure 9-38. Single Bus Bar Protection Using Overcurrent Time-Relays.

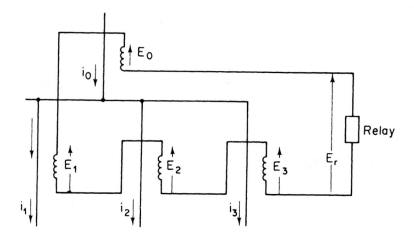

Figure 9-39. Linear Coupler Scheme for Bus Bar Protection.

conducting a fault current, as shown in Figure 9-38. Power direction relays are used in the transformer supply side to respond to a fault at the bus bars.

Differential protection is the most reliable method of protection for bus bars. Since a large number of circuits are involved, different current levels are encountered. The C.T.'s current on a faulted feeder must balance out the sum of C.T. currents on the nonfaulty feeders. This operates the C.T. at a high saturation level and may lead to false tripping. A number of techniques are used to overcome this. One such technique is to use linear couplers, which are simply ironless current transformers. The secondary voltage of the coupler is given by

$$E = Z_m I_p$$

where Z_m is the mutual impedance (typically 0.05 ohms). Figure 9-39 shows a linear coupler system, which is essentially a voltage differential scheme. The voltage on the relay is E_r and is given by

$$E_r = \sum_{j=1}^{n} (Z_m i_j - E_0)$$

Note that in the circuit shown with normal load or with fault outside the bus bar,

$$i_0 = \sum_{j=1}^{n} (i_j)$$

and thus $E_r = 0$. For a fault on the bus bar, this balance will be upset. The relay current is

$$I_r = \frac{E_r}{Z_r + \sum_{j=0}^{n}(Z_{c_j})}$$

where the Z_{c_j}'s are the coupler self-impedances and Z_r is the relay impedance.

Conventional current transformers are used in the multirestraint differential system for bus bar protection. In this system the inaccuracies resulting from current transformer saturation on heavy faults is compensated for by using a variable percentage differential relay. This relay consists of three induction-type restraint units and one induction-type operating unit. Two of the units operate on a common disc. The two discs

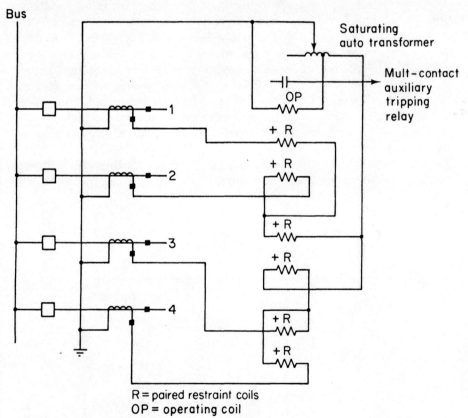

Figure 9-40. Schematic Connections of Multirestraint Differential Relay per Phase to Protect a Bus with Four Circuits. Connections for Only One Phase Are Shown.

have a common shaft with the moving contacts. Current flows through the windings of the four units, causing contact-opening torque for the restraint unit or contact-closing torque for the operating unit. The two windings on the restraint units are such that currents in the same direction provide restraint proportional to the sum, while currents in the opposite direction provide restraint proportional to the difference. When the currents in the two windings are equal and opposite, no restraining torque results. This relay is called the *variable percentage differential relay.*

At light fault currents, the C.T.'s performance is adequate, and the percentage is small for maximum sensitivity. For heavy external faults where the C.T.'s performance is likely to be poor, a large percentage is available. This characteristic is obtained by energizing the operating unit through a saturating autotransformer. Figure 9-40 shows a schematic connection of one phase for a multi-restraint differential system applied to a bus with four circuits.

Another scheme for bus bar protection is the high-impedance voltage differential system, which utilizes conventional current transformers. The current transformers are loaded with a high impedance relay unit to nullify the unequal current transformer performance problem. Bushing-type current transformers are used because they have a low secondary impedance. The relay unit shown in Figure 9-41 is an instantaneous voltage-plunger unit operated through a full-wave rectifier. The capacitance and inductance tune the circuit to fundamental frequency to reduce response to all harmonics. The impedance of this branch is around 3000 ohms, which means that C.T. secondaries and relay are subject to high voltages on a bus fault. A thyrite voltage-limiting unit is connected in parallel with the relay to limit the voltage. An instantaneous overcurrent unit is connected in series with

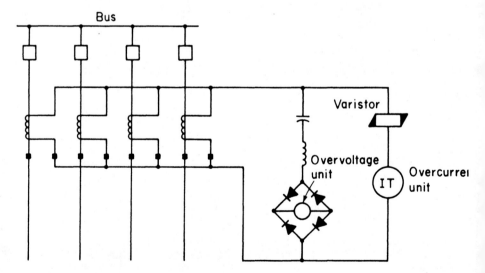

Figure 9-41. Schematic of the High-Impedance Voltage Differential System.

this combination and is set to operate at very high internal fault magnitudes. It must be set high to avoid operation on external faults. On internal faults, the voltage applied to the relay is high approaching the open-circuit voltage of the C.T. secondaries. On external faults the voltage will be as low as zero unless the C.T.'s saturate.

9.8 TRANSMISSION LINE OVERCURRENT PROTECTION

The earliest protective schemes evolved around the excessive currents resulting from a fault, which is the basis of overcurrent protection schemes. For transmission line protection in interconnected systems, it is necessary to provide the desired selectivity such that relay operation results in the least service interruption while isolating the fault. This is referred to as *relay coordination*. There are various possible methods to achieve the desired selectivity. Time/current gradings are involved in three basic methods discussed below for radial or loop circuits where there are several line sections in series.

Three Methods of Relay Grading

Time Grading

The purpose of time grading is to ensure that the breaker nearest to the fault opens first, by choosing an appropriate time setting for each of the relays. The time settings increase as the relay gets closer to the source. A simple radial system shown in Figure 9-42 will illustrate the point.

At each of the points 2, 3, 4, and 5, a protection unit comprising a definite time-delay overcurrent relay is placed. The time-delay of the relay provides the means for selectivity. The relay at circuit breaker 2 is set at the shortest possible time necessary for the breaker to operate (typically 0.25 second). The relay setting at 3 is chosen here as 0.5 second, that of the relay at 4 at 1 second, and so on. In the event of a fault at F in Figure 9-42, the relay at 2 will operate and the fault will be isolated before the relays at 3, 4, and 5 have sufficient time to operate. The shortcoming of the method is that the longest fault-clearing time is associated with the sections closest to the source where the faults are most severe.

Current Grading

The fact that fault currents are higher the closer the fault is to the source is utilized in the current-grading method. Relays are set to operate at a suitably graded current setting that decreases as the distance from the

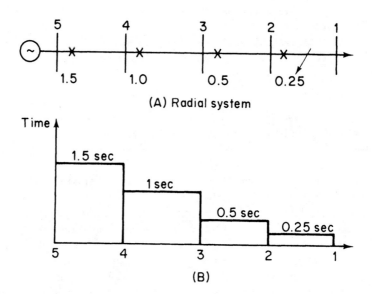

Figure 9-42. Principle of Time Grading.

source is increased. Figure 9-43 shows an example of a radial system with current grading. The advantages and disadvantages of current grading are best illustrated by way of examples.

Example 9-4

Consider the radial system shown in Figure 9-44. Calculate the fault currents for faults F_A, F_B, F_C, F_D, and F_E. Propose relay settings on the basis of current grading, assuming a 30 percent relay error margin.

Solution

The system voltage is 11 kV; hence the fault current is given by

$$I = \frac{V}{X_f} = \frac{11{,}000}{X_f\sqrt{3}}$$

where X_f is the reactance from the source to the fault point.

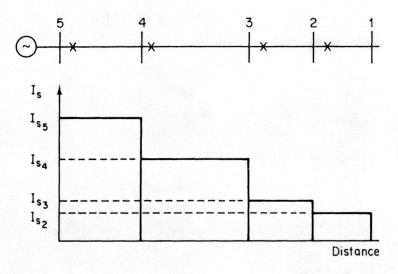

Figure 9-43. Current Grading for a Radial System.

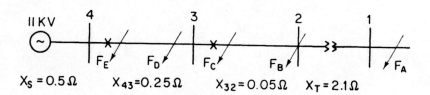

Figure 9-44. Radial System for Example 9-4.

For fault F_A we have
$$X_{F_A} = 0.5 + 0.25 + 0.05 + 2.1 = 2.9 \ \Omega$$
Thus the fault current is
$$I_{F_A} = \frac{11{,}000}{2.9\sqrt{3}} = 2189.95 \text{ A}$$
For fault F_B we have
$$X_{F_B} = 0.5 + 0.25 + 0.05 = 0.8 \ \Omega$$
Thus the fault current is
$$I_{F_B} = \frac{11{,}000}{0.8\sqrt{3}} = 7938.57 \text{ A}$$
For a fault at C we have
$$X_{F_C} = 0.5 + 0.25 = 0.75$$
Thus the fault current is
$$I_{F_C} = \frac{11{,}000}{0.75\sqrt{3}} = 8467.8 \text{ A}$$
Since F_D is very close to F_C, we conclude that
$$I_{F_D} = I_{F_C} = 8467.8 \text{ A}$$
For a fault at E we have
$$X_{F_E} = 0.5 \ \Omega$$
Thus,
$$I_{F_E} = \frac{11{,}000}{0.5\sqrt{3}} = 12701.71 \text{ A}$$

The relay at 1 should respond to faults F_B and F_C and should be set at 130 percent of fault current at F_A. Thus,
$$I_{S_1} = 1.3 I_{F_A} = 2846.93 \text{ A}$$
The relay at 2 should respond to faults F_E and F_D and should be set at
$$I_{S_2} = 1.3 I_{F_C} = 11008.14 \text{ A}$$
Note that relay 2 will not respond to F_A, F_B, and F_C.

In practice, there would be variations in the source fault level that result typically in a reduction of source apparent power by 50 percent. The apparent power reduction can be considered as an increase in source impedance (doubling of X_s). As a result, lower fault currents arise. The consequences of this are illustrated in the following example.

Example 9-5

Suppose that for the system of the above example, source level variations result in changing X_s from 0.5 Ω to 1 Ω. Find the resulting fault currents and study their effects on relay response.

Solution

The following are the revised currents:

$$I_{F_A} = \frac{11{,}000}{3.4\sqrt{3}} = 1867.90 \text{ A}$$

$$I_{F_B} = \frac{11{,}000}{1.3\sqrt{3}} = 4885.27 \text{ A}$$

$$I_{F_C} = I_{F_D} = \frac{11{,}000}{1.25\sqrt{3}} = 5080.68 \text{ A}$$

$$I_{F_E} = \frac{11{,}000}{1\sqrt{3}} = 6350.85 \text{ A}$$

Relay 1 will still respond to faults F_B and F_C. Relay 2 will not respond to any fault including F_E. Note the presence of the transformer with $X = 2.1$ is the main reason for relay 1 to operate properly.

Current grading is therefore not a practical proposition to protect the circuit between breakers 2 and 1. However, when there is a significant impedance between the breakers, the scheme is practical.

Inverse-Time Overcurrent Relaying

Each of the two methods considered so far has a disadvantage. Therefore the inverse-time overcurrent relay method has evolved because of the limitations imposed by the use of either current or time alone. With this third method, the time of operation is inversely proportional to the fault current level, and the actual characteristics are a function of both time and current settings. Figure 9-45 shows some typical inverse-time relay characteristics. Relay type CO-7 is in common use. Figure 9-46 shows a radial system with time-graded inverse relays applied at breakers 1, 2, and 3.

For faults close to the relaying points, the inverse-time overcurrent method can achieve appreciable reductions in fault-clearance times.

The operating time of the time-overcurrent relay varies with the current magnitude. There are two settings for this type of relay:

1. *Pickup current* is determined by adjusted current coil taps or current tap settings (C.T.S.). The pickup current is the current that causes the relay to operate and close the contacts.

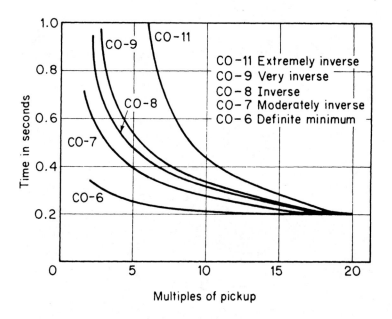

Figure 9-45. Comparison of CO Curve Shapes.

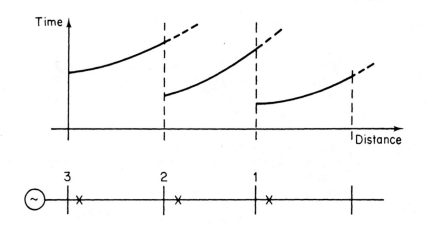

Figure 9-46. Time-Graded Inverse Relaying Applied to a Radial System.

2. *Time dial* refers to the reset position of the moving contact, and it varies the time of operation at a given tap setting and current magnitude.

The time characteristics are plotted in terms of time versus multiples of current tap (pickup) settings, for a given time dial position. There are five different curve shapes referred to by the manufacturer:

CO-11	Extreme inverse
CO-9	Very inverse
CO-8	Inverse
CO-7	Moderately inverse
CO-6	Definite minimum

These shapes are given in Figure 9-45. Figures 9-48 and 9-49 show detailed characteristics of two relay types.

Example 9-6

Consider the 11-kV radial system shown in Figure 9-47. Assume that all loads have the same power factor. Determine relay settings to protect the system assuming relay type CO-7 (with characteristics shown in Figure 9-48) is used.

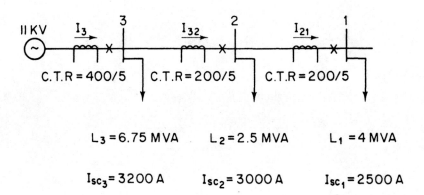

Figure 9-47. An Example Radial System.

594 System Protection

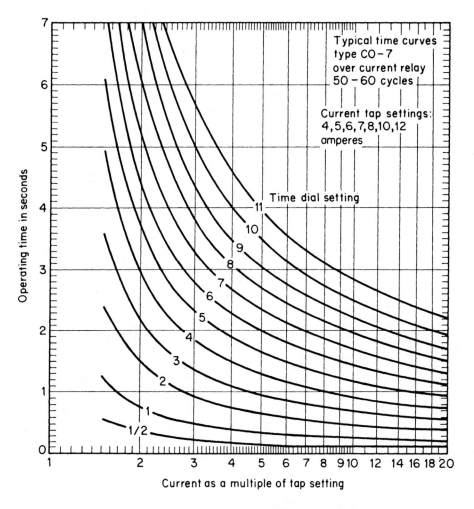

Figure 9-48. CO-7 Time-Delay Overcurrent Relay Characteristics.

Solution

The load currents are calculated as follows:

$$I_1 = \frac{4 \times 10^6}{\sqrt{3}\,(11 \times 10^3)} = 209.95 \text{ A}$$

$$I_2 = \frac{2.5 \times 10^6}{\sqrt{3}\,(11 \times 10^3)} = 131.22 \text{ A}$$

$$I_3 = \frac{6.75 \times 10^6}{\sqrt{3}\,(11 \times 10^3)} = 354.28 \text{ A}$$

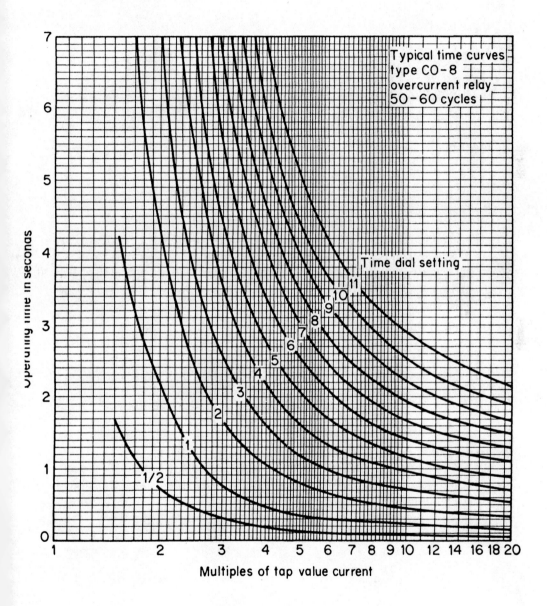

Figure 9-49. Type CO-8 Time-Current Curves.

The normal currents through the sections are calculated as

$$I_{21} = I_1 = 209.95 \text{ A}$$
$$I_{32} = I_{21} + I_2 = 341.16 \text{ A}$$
$$I_S = I_{32} + I_3 = 695.44 \text{ A}$$

With the current transformer ratios given, the normal relay currents are

$$i_{21} = \frac{209.95}{\frac{200}{5}} = 5.25 \text{ A}$$

$$i_{32} = \frac{341.16}{\frac{200}{5}} = 8.53 \text{ A}$$

$$i_S = \frac{695.44}{\frac{400}{5}} = 8.69 \text{ A}$$

We can now obtain the current tap settings (C.T.S.) or pickup current in such a manner that the relay does not trip under normal currents. For this type of relay, the current tap settings available are 4, 5, 6, 7, 8, 10, and 12 amperes. For position 1, the normal current in the relay is 5.25 A; we thus choose

$$(\text{C.T.S.})_1 = 6 \text{ A}$$

For position 2, the normal relay current is 8.53 A, and we choose

$$(\text{C.T.S.})_2 = 10 \text{ A}$$

Similarly for position 3,

$$(\text{C.T.S.})_3 = 10 \text{ A}$$

Observe that we have chosen the nearest setting higher than the normal current.

The next task is to select the intentional delay indicated by the time dial setting (T.D.S). We utilize the short-circuit currents calculated to coordinate the relays. The current in the relay at 1 on a short circuit at 1 is

$$i_{SC_1} = \frac{2500}{\left(\frac{200}{5}\right)} = 62.5 \text{ A}$$

Expressed as a multiple of the pickup or C.T.S. value, we have

$$\frac{i_{SC_1}}{(\text{C.T.S.})_1} = \frac{62.5}{6} = 10.42$$

We choose the lowest T.D.S for this relay for fastest action. Thus

$$(\text{T.D.S.})_1 = \frac{1}{2}$$

By reference to the relay characteristic, we get the operating time for relay 1 for a fault at 1 as

$$T_{1_1} = 0.15 \text{ s}$$

To set the relay at 2 responding to a fault at 1, we allow 0.1 second for breaker operation and an error margin of 0.3 second in addition to T_{1_1}. Thus,

$$T_{2_1} = T_{1_1} + 0.1 + 0.3 = 0.55 \text{ s}$$

The short circuit for a fault at 1 as a multiple of the C.T.S. at 2 is

$$\frac{i_{SC_1}}{(\text{C.T.S.})_2} = \frac{62.5}{10} = 6.25$$

From the characteristics for 0.55-second operating time and 6.25 ratio, we get

$$(\text{T.D.S.})_2 \cong 2$$

The final steps involve setting the relay at 3. For a fault at bus 2, the short-circuit current is 3000 A, for which relay 2 responds in a time T_{22} obtained as follows:

$$\frac{i_{SC_2}}{(\text{C.T.S.})_2} = \frac{3000}{\left(\frac{200}{5}\right)10} = 7.5$$

For the $(\text{T.D.S.})_2 = 2$, we get from the relay's characteristic,

$$T_{22} = 0.50 \text{ s}$$

Thus allowing the same margin for relay 3 to respond to a fault at 2, as for relay 2 responding to a fault at 1, we have

$$T_{32} = T_{22} + 0.1 + 0.3$$
$$= 0.90 \text{ s}$$

The current in the relay expressed as a multiple of pickup is

$$\frac{i_{SC_2}}{(\text{C.T.S.})_3} = \frac{3000}{\left(\frac{400}{5}\right)10} = 3.75$$

Thus for $T_3 = 0.90$, and the above ratio, we get from the relay's characteristic,

$$(\text{T.D.S.})_3 \cong 2.5$$

We note here that our calculations did not account for load starting currents that can be as high as five to seven times rated values. In practice, this should be accounted for. Problem 9-A-3 treats the question of finding the time of response of the relays to various faults as set in this example.

9.9 PILOT-WIRE FEEDER PROTECTION

The application of graded overcurrent systems to feeder protection has two disadvantages. The first is that the grading settings may lead to tripping times that are too long to prevent damage and service interruption. The second is that satisfactory grading for complex networks is quite difficult to attain. This led to the concept of "unit protection" involving the measurement of fault currents at each end of a limited zone of the feeder and the transmission of information between the equipment at zone boundaries. The principle utilized here is the differential (often referred to as Merz-price) protection scheme. For short feeders, pilot-wire schemes are used to transmit the information. Pilot-wire differential systems of feeder protection are classified into three types: (1) the circulating-current systems, (2) the balanced-voltage systems, and (3) the phase-comparison (Casson-Last) system. All depend on the fact that, capacitance current neglected, the instantaneous value of the current flowing into a healthy conductor at one end of the circuit is equal to the instantaneous current flowing out of the conductor at the other end, so that the net instantaneous current flowing into or out of the conductor is zero if the conductor is healthy. If, on the other hand, the conductor is short-circuited to earth or to another conductor at some point, then the net current flowing into or out of the conductor is equal to the instantaneous value of the current flowing out of or into the conductor at the point of fault.

Circulating-Current Systems.

The basic principle of operation of a circulating-current system is shown in Figure 9-50, which illustrates its application to a single-phase feeder. The two equal-ratio current transformers, one at each end of the protective circuit, have their secondary windings connected in series so that under load or external fault conditions their induced secondary voltages add together to produce a circulating current in the pilot-wire circuit. The relay R, shown connected at the midpoint of the pilot-circuit, carries the difference between the two C.T. secondary currents. This difference current is zero since the secondary currents of the two C.T.'s are identical. Under an internal short-circuit condition, however, this equality is no longer valid, and there is a resultant current in the relay that operates to trip the faulted feeder from the system. The simple system just described is not practical since it requires that tripping be initiated from a relay situated at the middle of the protective circuit.

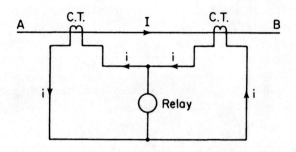

Figure 9-50. Circulating Current System.

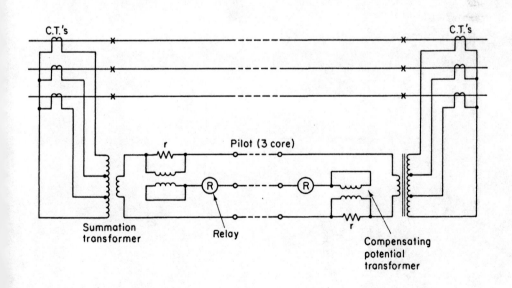

Figure 9-51. Self-Compensating Version of Figure 9-50.

To provide for the requirement that tripping be initiated from relays situated at the ends, the self-compensating circulating-current pilot-wire system evolved. In this system, two relays (one at each end of the protected circuit) are connected in series via a third pilot-wire as shown in Figure 9-51. The pilot wire and its two relays carry the difference current of the two summation transformer outputs. Shunts, r, in the pilot circuit and compensating potential transformers keep the third pilot at the midpotential of the two outer pilot cores under all through-current conditions. The pilot capacitance currents in the relays are thus kept to a tolerable value.

Balance-Voltages Systems.

The principle of operation of balance-voltage systems of differential feeder protection is illustrated in Figure 9-52 for a single-phase circuit. In this system, the two C.T.'s at the ends of the protected circuit have their secondary windings connected in series opposition around the pilot loop, so that under load or external fault conditions, there is no current in the relay that is connected in series with the pilot wires. Under internal fault conditions, however, the C.T. secondary voltages are no longer equal and opposite, so that the resultant voltage produces a current in the pilot wires and relay. Operation of the relay disconnects the faulted circuit from the system. Principles of the third type, phase comparison are discussed in Section 9-11.

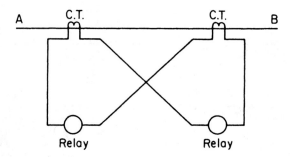

Figure 9-52. Principle of Balanced-Voltage System.

9.10 DISTANCE PROTECTION

Protection of lines and feeders based on comparison of the current values at both ends of the line can become uneconomical. Distance protection utilizes the current and voltage at the beginning of the line in a comparison scheme that essentially determines the fault position. Impedance measurement is performed using relay comparators whose theory was discussed in Section 9.4. A simple distance relay is offered by the balanced-beam relay shown in Figure 9-9, with one coil supplied by a current proportional to the fault current and the other supplied by a current proportional to the fault loop voltage. Under normal conditions, the pull developed by the voltage electromagnet exceeds that developed by the current electromagnet, and the contacts are open. For a short circuit within reach of the relay, the current in the current coil increases, while that in the voltage coil decreases, and thus the relay operates to trip the circuit breaker.

The relay as described is a plain impedance relay and has a characteristic such as that shown in Figure 9-20. It will thus respond to faults behind it (third quadrant) in the X-R diagram as well as in front of it. One way to prevent this is to add a separate directional relay that will restrain tripping for faults behind the protected zone. The reactance or mho relay with characteristics as shown in Figure 9-19 combines the distance-measuring ability and the directional property. The term *mho* is given to the relay where the circumference of the circle passes through the origin, and the term was originally derived from the fact that the mho characteristic (*ohm* spelled backward) is a straight line in the admittance plane.

Early applications of distance protection utilized relay operating times that were a function of the impedance for the fault. The nearer the fault, the shorter the operating time. This is shown in Figure 9-53. This has the same disadvantages as overcurrent protection discussed earlier. Present practice is to set the relay to operate simultaneously for faults that occur in the first 80 percent of the feeder length (known as the *first zone*). Faults beyond this point and up to a point midway along the next feeder are cleared by arranging for the zone setting of the relay to be extended from the first zone value to the second zone value after a time delay of about 0.5 to 1 second. The second zone for the first relay should never be less than 20 percent of the first feeder length. The zone setting extension is done by increasing the impedance in series with the relay voltage coil current. A third zone is provided (using a starting relay) extending from the middle of the second feeder into the third feeder up to 25 percent of the length with a further delay of 1 or 2 seconds. This provides backup protection as well. The time-distance characteristics for a three-feeder system are shown in Figure 9-54.

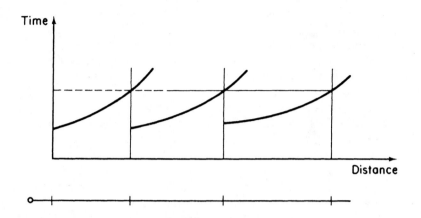

Figure 9-53. Principle of Time-Distance Protection.

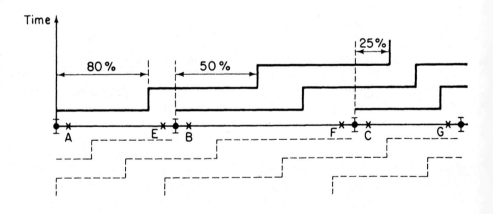

Figure 9-54. Time-Distance Characteristics of Distance Protection.

9.10 Distance Protection

Distance relaying schemes employ several relay units that are arranged to give response characteristics such as that shown in Figure 9-55. A typical system comprises:

1. Two offset mho units (with three elements each). The first operates as earth-fault starting and third zone measuring relay, and the second operates as phase-fault starting and third zone measuring relay.

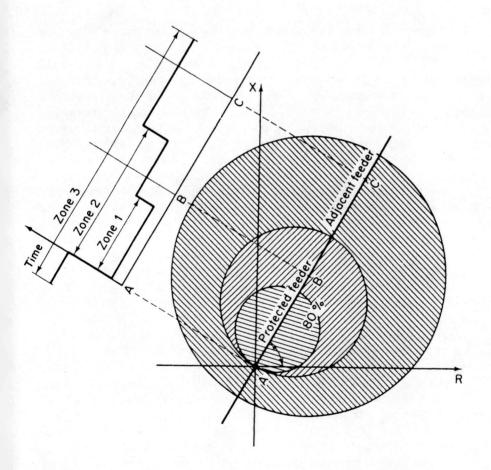

Figure 9-55. Characteristics of a Three-Zone Offset Mho-Relaying Scheme.

2. Two polarized mho units (with three elements each). The first unit acts as first and second zone earth-fault measuring relay, and the second unit acts as first and second zone phase-fault measuring relay.
3. Two time-delay relays for second and third zone time measurement.

The main difference between earth-fault and phase-fault relays is in the potential transformer (P.T.) and C.T. connections, which are designed to cause the relay to respond to the type of fault concerned.

9.11 POWER LINE CARRIER PROTECTION

Carrier-current protection systems utilize overhead transmission lines as pilot circuits. A carrier-frequency signal (30–200 kHz) is carried by two of the line conductors to provide communication means between ends of the line. The carrier signal is applied to the conductors via carrier coupling into units comprising inductance/capacitor circuits tuned to the carrier signal frequency to perform a number of functions. The carrier signals thus travel mainly into the power line and not into undesired parts of the system such as the bus bars. The communication equipment that operates at impedance levels of the order of 50–150 Ω is to be matched to the power line that typically has a characteristic impedance in the range of 240–500 Ω.

Figure 9-56 shows a typical arrangement of a power line carrier coupling system. The line trap L_1 and C_1 is tuned to the middle of the carrier band required and thus has a low impedance at power frequency and a high impedance to the carrier signals. Thus the carrier frequencies are prevented from entering the bus bars. The series tuning circuit L_2, C_2, and C_3 is tuned to the carrier midband and is the converse of the line trap in that it offers low impedance to the carrier signals but high impedance to the power frequency waves. T_1 is an isolating transformer that also serves as a grounding coil for C_3 so that the capacitor will have ground potential at power frequency. T_2 is also an isolating and impedance-matching transformer. C_4 is used to tune the shunt reactance of T_1 and T_2 at carrier midband frequency to minimize the losses. Modern line trap design for multichannel operation includes a series RLC (resistance-inductance-capacitance) network across the trap so that the trap presents an almost constant resistance over a wide band of frequencies to offset the effect of bus bar reactance. Typical values for C_2 and C_3 are 2000 pF, and for L_1, 200 μH corresponding to 150 kHz carrier frequency.

Power line carrier systems are used for two purposes. The first involves measurements, and the second conveys signals from one end of the line to the other with the measurement being done at each end by relays.

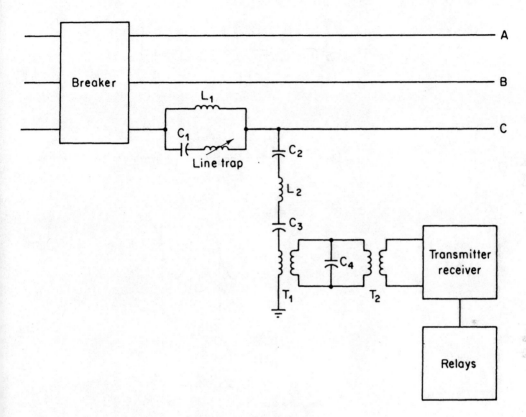

Figure 9-56. Power Line Carrier Coupling System.

When the carrier channel is used for measurement, it is not practical to transmit amplitude measurements from one end to the other since signal attenuation beyond the control of the system takes place. As a result, the only feasible measurement carrier system compares the phase angle of a derived current at each end of the system in a manner similar to differential protection as discussed below.

Phase-Comparison Protection

The principle of phase-comparison protection is illustrated in Figure 9-57 for an internal fault. The scheme is such that if the current at a relaying point is of sufficient magnitude, a carrier signal is transmitted to

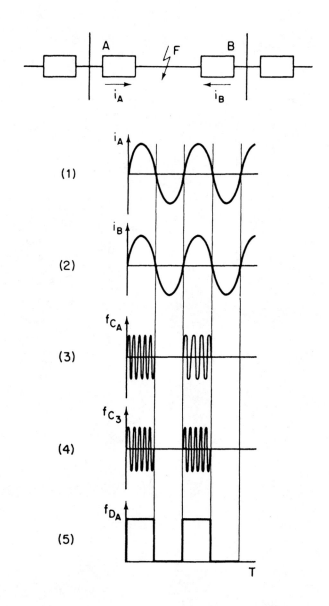

Figure 9-57. Principle of Phase Comparison with an Internal Fault.

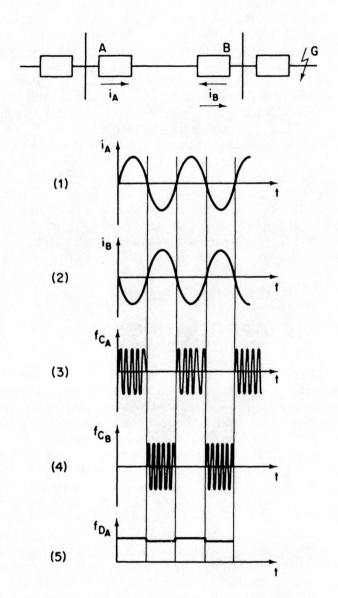

Figure 9-58. Principle of Phase Comparison with an External Fault.

the relaying systems at both ends of the line A and B. The carrier signal is transmitted only during the positive half-cycle of the current wave. Consider an internal fault at F; the current i_A flowing into line AB measured at A is assumed to be sinusoidal as shown in graph (1). The carrier signal generated at A and transmitted to relaying equipment at A and B is denoted by f_{C_A} and is shown is graph (3). The current i_B flowing into line AB measured at B will have a magnitude different than that of i_A but will be almost in phase with i_A as shown in graph (2). The corresponding carrier signal generated at B and transmitted to both A and B is f_{C_B} and is shown in graph (4). The detector at A sums f_{C_A} and f_{C_B} and rectifies the result to produce f_D shown in graph (5). When the signal f_{D_A} is zero for a specified time, the relaying equipment at A activates to trip breaker A.

In Figure 9-58, the situation with an external fault at G is illustrated. Observe now that i_B is in a direction opposite to that for an internal fault. The sequence of graphs is self-explanatory. Note f_{D_A} does not assume a zero value, and relaying equipment does not trip the breaker at A.

Radio and microwave links are now being applied in power systems to provide communication channels for teleprotection as well as for supervisory control and data acquisition.

9.12 COMPUTER RELAYING

The electric power industry has been one of the earliest users of the digital computer as a fundamental aid in the various design and analysis aspects of its activity. Computer-based systems have evolved to perform such complex tasks as generation control, economic dispatch (treated in Chapter 11), and load-flow analysis for planning and operation, to name just a few application areas. Research efforts directed at the prospect of using digital computers to perform the tasks involved in power system protection date back to the mid-sixties and were motivated by the emergence of process-control computers. A great deal of research is going on in this field, which is now referred to as *computer relaying*. Up to the early 1980s there had been no commercially available protection systems offering digital computer-based relays. However, the availability of microprocessor technology has provided an impetus to computer relaying. Microprocessors used as a replacement for electromechanical and solid-state relays can provide a number of advantages while meeting the basic protection philosophy requirement of decentralization.

There are many perceived benefits of a digital relaying system:

1. *Economics*: With the steady decrease in cost of digital hardware, coupled with the increase in cost of conventional relaying, it seems reasonable to assume that computer relaying is an attractive alter-

native. Software development cost can be expected to be evened out by utilizing economies of scale in producing microprocessors dedicated to basic relaying tasks.

2. *Reliability*: A digital system is continuously active providing a high level of self-diagnosis to detect accidental failures within the digital relaying system.
3. *Flexibility*: Revisions or modifications made necessary by changing operational conditions can be accommodated by utilizing the programmability features of a digital system. This would lead to reduced inventories of parts for repair and maintenance purposes.
4. *System interaction*: The availability of digital hardware that monitors continuously the system performance at remote substations can enhance the level of information available to the control center. Postfault analysis of transient data can be performed on the basis of system variables monitored by the digital relay and recorded by the peripherals.

The main elements of a digital computer-based relay are indicated in Figure 9-59. The input signals to the relay are analog (continuous) and digital power system variables. The digital inputs are of the order of five to ten and include status changes (on-off) of contacts and changes in voltage levels in a circuit. The analog signals are the 60-Hz currents and voltages. The number of analog signals needed depends on the relay function but is in the range of 3 to 30 in all cases. The analog signals are scaled down (attenuated) to acceptable computer input levels (± 10 volts maximum) and then converted to digital (discrete) form through analog/digital converters (ADC). These functions are performed in the block labeled "Analog Input Subsystem."

The digital output of the relay is available through the computer's parallel output port. Five-to-ten digital outputs are sufficient for most applications.

The analog signals are sampled at a rate between 240 Hz to about 2000 Hz. The sampled signals are entered into the scratch pad [random access memory (RAM)] and are stored in a secondary data file for historical recording. A digital filter removes noise effects from the sampled signals. The relay logic program determines the functional operations of the relay and uses the filtered sampled signals to arrive at a trip or no trip decision, which is then communicated to the system.

The heart of the relay logic program is a relaying algorithm that is designed to perform the intended relay function such as overcurrent detection, differential protection, or distance protection, etc. It is not our intention in this introductory text to pursue this interesting topic in detail. However, to give the reader a feel for what is involved in a relaying

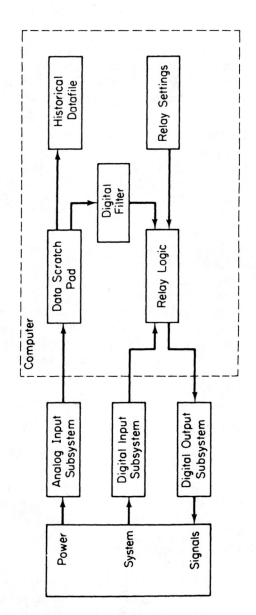

Figure 9-59. Functional Block Diagram of a Digital Relay.

610

algorithm, we discuss next one idea for peak current detection that is the function of a digital overcurrent relay.

Consider a pure sinusoidal current signal described by

$$i(t) = I_m \sin(\omega t + \phi)$$

The rate of change of $i(t)$ with respect to time is

$$i'(t) = \omega I_m \cos(\omega t + \phi)$$

Thus,

$$\frac{i(t)}{I_m} = \sin(\omega t + \phi)$$

$$\frac{i'(t)}{\omega I_m} = \cos(\omega t + \phi)$$

Squaring and adding, we get

$$I_m^2 = i^2(t) + \left[\frac{i'(t)}{\omega}\right]^2$$

We can also conclude that

$$\tan(\omega t + \phi) = \frac{\omega i(t)}{i'(t)}$$

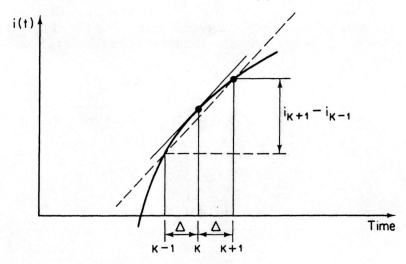

Figure 9-60. Approximating the Current Derivative for Digital Overcurrent Protection.

It is clear that all we need to detect the peak of the sinusoid is to know the value of the current and its derivative at one instant of time. In digital form, we need at least three sample points to define the current and its derivative. To see this, assume that the current is sampled every Δ seconds. With the three samples at $K-1$, K, and $K+1$, we conclude by reference to Figure 9-60 that

$$i'_K = \frac{i_{K+1} - i_{K-1}}{2\Delta}$$

Thus we get an estimate of I_m based on the three samples as the square root of

$$I_m^2 = i_K^2 + \frac{1}{4\omega^2 \Delta^2}(i_{K+1} - i_{K-1})^2$$

Clearly this enables the detection of any sudden increase in I_m. Actual implementation should take into account transient conditions and noise effects.

SOME SOLVED PROBLEMS

Problem 9-A-1

A three-phase, 15-MVA, 12-kV generator is provided with negative sequence protection employing a 1-A overcurrent relay and a negative sequence filter such as that discussed in Example 9-1. Assume that the maximum allowable negative sequence current is 10 percent of rated generator current. Find the most acceptable current transformer ratio from the given listing of C.T. ratios. Assume relay current to be related to filter's input current by

$$i_{\text{relay}} = 0.5 i_-$$

Standard C.T. ratios:

$$50/5, 100/5, 200/5, 400/5, 600/5, 800/5, 1200/5$$

Solution

The rated current is

$$I_r = \frac{15 \times 10^3}{\sqrt{3} \times 12} = 721.69 \text{ A}$$

The relay should respond to negative sequence current, which is 10 percent of rated. Thus,

$$I_- = 72.169 \text{ A}$$

The corresponding relay current is 1 A. Thus, the negative sequence current to the filter is given by

$$i_- = 2 \text{ A}$$

The current transformer ratio (C.T.R.) is therefore

$$\text{C.T.R.} = \frac{72.169}{2} = 36.08$$

The closest C.T.R. in the listing is 200/5, which is the desired answer.

Problem 9-A-2

The line of Example 9-2 is to be provided with 80 percent distance protection using either a resistance or a reactance ohm relay. Find the relay design parameters in each case, assuming that magnitude comparison is used.

Solution

The line impedance is

$$Z_L = 0.1 + j0.3 \text{ pu}$$

For 80 percent protection, the operating point's impedance is

$$Z_P = 0.8 Z_L = 0.08 + j0.24 \text{ pu}$$

An ohm relay based on magnitude comparison has the characteristics

$$V_1 = kV_L$$
$$V_2 = -kV_L + ZI_L$$

For a resistance-distance characteristic, the angle $\psi = 0$, and we have a characteristic as shown in Figure 9-61(a). To protect 80 percent of the line

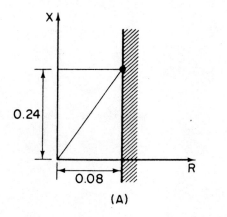
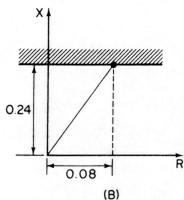

Figure 9-61. Relay $X - R$ Diagrams for Problem 9-A-2. (a) Resistance Distance. (b) Reactance Distance.

we thus have $Z = R$, with

$$\frac{R}{2k} = 0.08$$

For reactance distance, the angle $\psi = 90°$, and the characteristic is as shown in Figure 9-61(b). Thus $Z = jX$, with

$$\frac{X}{2k} = 0.24$$

Assuming $k = 1$ in each case gives

$R = 0.16$ p.u. for resistance-distance relay
$X = 0.48$ p.u. for reactance-distance relay

Problem 9-A-3

Consider the radial system of Example 9-6. It is required to construct the relay response time-distance characteristics on the basis of the design obtained as follows:

A. Assuming the line's impedance is purely reactive, calculate the source reactance and the reactances between bus bars 3 and 2, and 2 and 1.

B. Find the current on a short circuit midway between buses 3 and 2 and between 2 and 1.

C. Calculate the relay response times for faults identified in Example 9-6 and part (b) above and sketch the relay response time-distance characteristics.

Solution

A. The equivalent circuit of the system is shown in Figure 9-62. Assume

$$E = \frac{11 \times 10^3}{\sqrt{3}} = 6350.85 \text{ V}$$

For a short circuit at bus 3, we have

$$3200 = \frac{6350.85}{X_s}$$

Thus the source reactance is

$$X_s = 1.9846 \text{ ohms}$$

For a short circuit at bus 2, we have

$$3000 = \frac{6350.85}{X_s + X_{32}}$$

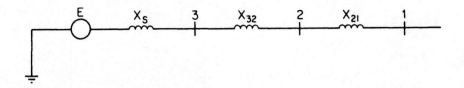

Figure 9-62. Equivalent Circuit for Problem 9-A-3.

Thus the reactance between buses 3 and 2 is
$$X_{32} = 0.1323 \text{ ohms}$$
For a short a circuit at bus 1, we have
$$2500 = \frac{6350.85}{X_s + X_{32} + X_{21}}$$
Thus the reactance between buses 2 and 1 is
$$X_{21} = 0.4234 \text{ ohms}$$

B. For a short circuit midway between buses 3 and 2 labeled A in Figure 9-63 we have
$$I_{SC_A} = \frac{6350.85}{X_s + 0.5 X_{32}} = 3096.77 \text{ A}$$
For a short circuit midway between buses 2 and 1 labeled B in Figure 9-63, we have
$$I_{SC_B} = \frac{6350.85}{X_s + X_{32} + 0.5 X_{21}} = 2727.27 \text{ A}$$

C. Consider the relay at 3, the current transformer ratio is 400/5, the current tap setting is 10 A, and the time dial setting is 2.5. The following list shows the steps to obtain the points on the time-distance curve desired:

Step 1 —For a fault at 3:
$$i_{relay} = \frac{3200}{\frac{400}{5}} = 40 \text{ A}$$

$$\frac{i_{relay}}{\text{C.T.S.}} = \frac{40}{10} = 4 \text{ A}$$

From the characteristic of CO-7:
$$T_{33} = 0.8 \text{ s}$$

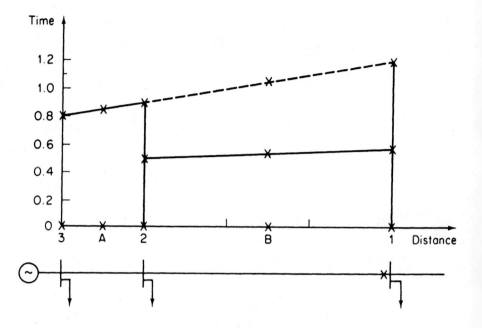

Figure 9-63. Time-Distance Curves for System of Problem 9-A-3.

Step 2 —For a fault at A:

$$i_{\text{relay}} = \frac{3096.77}{\frac{400}{5}} = 38.71 \text{ A}$$

$$\frac{i_{\text{relay}}}{\text{C.T.S.}} = \frac{38.71}{10} = 3.871 \text{ A}$$

From the characteristic of CO-7:

$$T_{3A} = 0.85 \text{ s}$$

Step 3 —For a fault at 2, we found in Example 9-6 that

$$T_{32} = 0.9 \text{ s}$$

Step 4 —For a fault at B; we get by a similar procedure,

$$T_{3B} = 1.05 \text{ s}$$

Step 5 —For a fault at 1, we get
$$T_{3_1} = 1.2 \text{ s}$$

Consider the relay at 2, the C.T. ratio is 200/5, C.T.S. = 10 A, and T.D.S. = 2. The following operating times are obtained:
$$T_{22} = 0.5 \text{ s}$$
$$T_{2B} = 0.53 \text{ s}$$
$$T_{21} = 0.55 \text{ s}$$

The relay operating time-variation with location of fault is shown in Figure 9-63.

Problem 9-A-4

It was pointed out in the text that motor starting and transformer inrush currents should be accounted for. It is required in this problem to treat the system of Example 9-6 assuming that the design is for starting currents 300 percent as much as the normal load currents. Assume that the same type of relay is used. However, current transformer ratios are still negotiable.

A. Calculate the new current transformer ratios.

B. Set the relays for this more realistic condition.

Solution

A. We will find the starting currents by simply multiplying the normal current by 3. Thus,
$$I_{21} = 629.85 \text{ A}$$
$$I_{32} = 1023.48 \text{ A}$$
$$I_s = 2086.32 \text{ A}$$

Using the C.T. ratios specified in Example 9-6, we have
$$i_{21} = 15.75 \text{ A}$$
$$i_{32} = 25.59 \text{ A}$$
$$i_s = 26.07 \text{ A}$$

For relay CO-7, the starting currents are all higher than the available current tap settings. We thus have to specify different current transformers.

Using C.T. at 1 with 400/5 ratio results in a relay current at starting of
$$i_{21} = 7.88 \text{ A}$$

We take a current tap setting of
$$\text{C.T.S.}_{\cdot 1} = 8$$

Using a C.T. at 2 with 400/5 ratio results in a relay current at starting, which is higher than the available current tap setting. We thus take a C.T. at 2 with 600/5 ratio. Thus,

$$i_{32} = \frac{1023.48}{600/5}$$
$$= 8.53 \text{ A}$$

The current tap setting is thus taken as

$$\text{C.T.S.}_2 = 10 \text{ A}$$

Taking a C.T. at 3 with 1200/5 ratio gives

$$i_s = \frac{2086.36}{\frac{1200}{5}}$$
$$= 8.69 \text{ A}$$

We thus take a current tap setting of 10:

$$\text{C.T.S.}_3 = 10 \text{ A}$$

B. To select the T.D.S., we have for relay 1 on a short circuit at 1.

$$i_{SC_1} = \frac{2500}{\frac{400}{5}}$$
$$= 31.25 \text{ A}$$

Expressed as a multiple of C.T.S. value, we have

$$\frac{i_{SC1}}{\text{C.T.S.}_1} = \frac{31.25}{8} = 3.91$$

We choose the lowest T.D.S., as

$$(\text{T.D.S.})_1 = 0.5$$

From the relay characteristic we get

$$T_{11} = 0.15 \text{ s}$$

To set the relay at 2, we allow the 0.4-second margin to get

$$T_{21} = 0.55 \text{ s}$$

The short-circuit current for a fault at 1 as a multiple of the C.T.S. at 2 is

$$\frac{I_{SC1}}{(\text{C.T.})_2 (\text{C.T.S.})_2} = \frac{2500}{10\left(\frac{600}{5}\right)} = 2.08$$

For 0.55-second operating time with 2.08 current as a multiple of the tap setting, we get

$$(\text{T.D.S.})_2 = 0.75$$

To set the relay at 3, we consider the fault at bus 2, the short-circuit current is 3000 A, to which relay 2 responds in time T_{22}. Thus,

$$\frac{i_{SC_2}}{\text{C.T.S.}_2} = \frac{3000}{\left(\frac{600}{5}\right)10} = 2.5$$

For T.D.S.$_2 = 0.75$, we get

$$T_{22} = 0.4 \text{ s}$$

Allowing the 0.4-second margin, we thus have

$$T_{32} = 0.4 + 0.4 = 0.8 \text{ s}$$

The short-circuit current at 2 as seen by the relay at 3 is

$$\frac{I_{2SC}}{(\text{C.T.})_3(\text{C.T.S.})_3} = \frac{3000}{\left(\frac{1200}{5}\right)10} = 1.25$$

For $T_{32} = 0.8$, we get from the relay's characteristic,

$$\text{T.D.S.}_3 = 0.5$$

Note that we needed to extrapolate the given characteristics. This completes the relay settings.

Problem 9-A-5

For the 13.8-kV system shown in Figure 9-64, determine the relay settings and the current transformer ratios to protect the system from faults. Assume the following data:

$$L_1 = 9 \text{ MVA} \qquad PF_1 = 0.9 \text{ lagging}$$
$$L_2 = 4 \text{ MVA} \qquad PF_2 = 0.85 \text{ lagging}$$

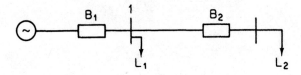

Figure 9-64. System for Problem 9-A-5.

620 System Protection

The maximum fault currents are
$$I_{sc_{max1}} = 3200 \text{ A}$$
$$I_{sc_{max2}} = 2800 \text{ A}$$

Assume type CO-8 relays are used and that the available current tap settings are the same as that for the CO-7 type.

Solution

The normal load currents are calculated as
$$I_1 = \frac{9 \times 10^6}{13.8 \times 10^3 \sqrt{3}} \underline{/-\cos^{-1} 0.9}$$
$$= 376.53 \underline{/-25.84°}$$
$$I_2 = \frac{4 \times 10^6}{13.8 \times 10^3 \sqrt{3}} \underline{/-\cos^{-1} 0.85}$$
$$= 167.35 \underline{/-31.79°}$$

The normal source current is thus
$$I_s = I_1 + I_2$$
$$= 543.26 \underline{/-27.67°} \text{ A}$$

The normal current through the line is thus 167.35 A, and we can choose a current transformer ratio of
$$\text{C.T.R.}_2 = 200/5$$

The normal source current is 543.26 A, and we choose a current transformer ratio of
$$\text{C.T.R.}_1 = 400/5$$

As a result, the normal relay currents are
$$i_2 = \frac{167.35}{200/5} = 4.18 \text{ A}$$
$$i_s = \frac{543.26}{400/5} = 6.79 \text{ A}$$

Therefore the current tap settings are
$$\text{C.T.S.}_2 = 5$$
$$\text{C.T.S.}_1 = 7$$

On a short circuit at 2, the relay current is
$$i_{SC_2} = \frac{2800}{200/5} = 70 \text{ A}$$

Choosing the lowest time dial setting,
$$T.D.S._2 = 0.5$$
The short-circuit current at 2 expressed as a multiple of the current tap setting is thus
$$\frac{i_{SC2}}{C.T.S._2} = \frac{70}{5} = 14$$
From the relay characteristics, we get
$$T_{22} = 0.05 \text{ s}$$
To set the relay at 1, we allow a 0.4-second margin to respond to the fault at 2. Thus
$$T_{12} = 0.45 \text{ s}$$
The short-circuit current at 2 seen at relay 1 is
$$i_{SC_{12}} = \frac{2800}{400/5} = 35 \text{ A}$$
This current is expressed as a multiple of the current tap setting, we get
$$\frac{i_{SC_{12}}}{C.T.S._1} = \frac{35}{7} = 5 \text{ A}$$
Corresponding to 5 A and 0.45 seconds, we obtain from the relay's characteristic,
$$T.D.S._1 = 1.2$$
This completes the relay settings.

The relay at 1 will respond to a short circuit at 1 as follows:
$$\frac{i_{SC_{11}}}{C.T.S._1} = \frac{3200}{\left(\frac{400}{5}\right)7} = 5.71 \text{ A}$$
Thus from the relay's characteristic,
$$T_{11} = 0.4 \text{ s}$$
This is naturally lower than T_{12}.

PROBLEMS

Problem 9-B-1

Use the relations between phase and sequence currents to prove that Eq. (9.1) leads to Eq. (9.3). Show that the choice
$$X_m = R_1/\sqrt{3}$$

leads to an output voltage containing only negative and zero sequence quantities.

Problem 9-B-2

A three-phase, 85-MVA, 13.8-kV generator is provided with negative sequence protection employing a 5-A overcurrent relay and a negative sequence filter. Assume that the maximum allowable negative sequence current is 10 percent of rated generator current and that the relay current is related to the filter's input current by

$$i_{\text{relay}} = 0.5 i_-$$

Find the most acceptable current transformer ratio from the standard ratios given in Problem 9-A-1.

Problem 9-B-3

Consider the system of Example 9-2. Assume that the load is reduced to

$$S_r = 1 + j0.4 \text{ p.u.}$$

Find Z_r, Z_s, and the angle δ for this condition.

Problem 9-B-4

The line of Example 9-2 is to be provided with 80 percent distance protection using an ohm relay with $\psi = 45°$. Find the relay's impedance Z assuming $k = 1$ and that magnitude comparison is used.

Problem 9-B-5

Repeat Example 9-3, for a transformer rating of 12-MVA.

Problem 9-B-6

Consider the radial system shown in Figure 9-44. Calculate the fault currents for faults F_A, F_B, F_C, F_D and F_E. Propose relay settings on the basis of current grading. Assume that

$$X_s = 0.45 \text{ ohm}$$
$$X_{43} = 0.20 \text{ ohm}$$
$$X_{32} = 0.1 \text{ ohm}$$
$$X_T = 2.1 \text{ ohms}$$

Problem 9-B-7

Suppose that for the system of Problem 9-B-6, source level variations result in changing X_s from 0.45 ohms to 0.9 ohms. Find the resulting fault currents.

Problem 9-B-8

Consider the system of Example 9-6. Assume now that the load at the far end of the system is increased to
$$L_1 = 6 \text{ MVA}$$
Determine the relay settings to protect the system using relay type CO-7.

Problem 9-B-9

Repeat Problem 9-B-8, assuming that the design is for starting currents of 250 percent as much as the normal load currents (Refer to Problem 9-A-4).

Problem 9-B-10

Repeat Problem 9-A-5, for
$$L_2 = 6 \text{ MVA}$$
with all other information given in the problem statement unchanged.

CHAPTER X

Power System Stability

10.1 INTRODUCTION

In this chapter we are concerned with the implications of a major network disturbance such as a short circuit on a transmission line, the opening of a line, or the switching on of a major load. We are mainly interested in the analysis of the behavior of the system immediately following such a disturbance. Studies of this nature are called *transient stability analysis*. The term *stability* is used here to denote the ability of the system machines to recover from small random perturbing forces and still maintain synchronism.

Stability considerations have been recognized to be among the essential tools in electric power system planning. The possible consequences of instability in an electric power system were dramatized by the Northeast power failure of 1965. This is an example of a situation that arises when a severe disturbance is not cleared away fast enough. This blackout began with a loss of a transmission corridor, which isolated a significant amount of generation from its load. More recently, a transmission tower in the Consolidated Edison system was hit by a severe lightning stroke in July 1977.

The events that followed this led to the shutdown of New York's power. Both events dramatized the consequences of an instability in an interconnected electric power system.

In the present chapter our intention is to give an introduction to the important topic of transient stability in electric power systems. We treat the case of a single machine operating to supply an infinite bus. The analysis of the more complex problem of large electric power networks with the interconnections taken into consideration is not treated in the present work. This requires methods of study that are beyond the scope of this book.

10.2 THE SWING EQUATION

In the power system engineer's terminology, the dynamic equation relating the inertial torque to the net accelerating torque of the synchronous machine rotor is called the *swing equation*. This simply states

$$J\left(\frac{d^2\theta}{dt^2}\right) = T_a \; \text{N} \cdot \text{m} \tag{10.1}$$

The left-hand side is the inertial torque, which is the product of the inertia (in kg. m^2) of all rotating masses attached to the rotor shaft and the angular acceleration. The accelerating torque T_a is in newton meters and can be expressed as

$$T_a = T_m - T_e \tag{10.2}$$

In the above, T_m is the driving mechanical torque, and T_e is the retarding or load electrical torque.

The angular position of the rotor θ may be expressed as the following sum of angles:

$$\theta = \alpha + \omega_R t + \delta \tag{10.3}$$

The angle α is a constant that is needed if the angle δ is measured from an axis different from the angular reference. The angle $\omega_R t$ is the result of the rotor angular motion at rated speed. The angle δ is time-varying and represents deviations from the rated angular displacements. This gives the basis for our new relation

$$J\left(\frac{d^2\delta}{dt^2}\right) = T_m - T_e \tag{10.4}$$

We find it more convenient to make the following substitution of the dot notation:

$$\ddot{\delta} = \frac{d^2\delta}{dt^2}$$

Therefore we have

$$J\ddot{\delta} = T_m - T_e \qquad (10.5)$$

Some Alternate Forms

Some alternative useful forms of Eq. (10.5) have been developed. The first is the power form that is obtained by multiplying both sides of Eq. (10.5) by ω and recalling that the product of the torque T and angular velocity is the shaft power. This results in

$$J\omega\ddot{\delta} = P_m - P_e$$

The quantity $J\omega$ is called the *inertia constant* and is truly an angular momentum denoted by M (Js/rad). Thus,

$$M = J\omega \qquad (10.6)$$

Thus the power form is

$$M\ddot{\delta} = P_m - P_e \qquad (10.7)$$

A normalized form of the swing equation can be obtained by dividing Eq. (10.5) by the rated torque T_R to obtain the dimensionless equation:

$$\left(\frac{J}{T_R}\right)\ddot{\delta} = \frac{T_m}{T_R} - \frac{T_e}{T_R}$$

The left-hand side of the above equation can be further manipulated to yield a form frequently used. Recall the definition of the kinetic energy of a rotating body. This gives the kinetic energy at rated speed as

$$W_k = \left(\frac{1}{2}\right)J\omega_R^2$$

Then

$$\frac{J}{T_R} = \frac{2W_k}{\omega_R^2 T_R}$$

We know further that the rated power is

$$P_R = \omega_R T_R$$

Thus,

$$\frac{J}{T_R} = \frac{2W_k}{\omega_R P_R}$$

Consequently, we have

$$\left(\frac{2W_k}{\omega_R P_R}\right)\ddot{\delta} = \frac{T_m}{T_R} - \frac{T_e}{T_R}$$

A constant that has proved very useful is denoted by H which is equal to the kinetic energy at rated speed divided by the rated power P_R. Thus,

$$H \triangleq \frac{W_k}{P_R} \tag{10.8}$$

The units of H are in seconds. As a result, we write the per unit or normalized swing equation as

$$\left(\frac{2H}{\omega_R}\right)\ddot{\delta} = T_{m\,pu} - T_{e\,pu} \tag{10.9}$$

Observing that $T_{pu} = P_{pu}$, we can then write

$$\left(\frac{2H}{\omega_R}\right)\ddot{\delta} = P_m - P_e \tag{10.10}$$

where the equation is in the per unit system.

Machine Inertia Constants

The angular momentum inertia constant M as defined by Eq. (10.6) can be obtained from manufacturer-supplied machine data. The machine kinetic energy in megajoules N may be written in terms of M as follows:

$$N = \left(\frac{1}{2}\right)M\omega_e \tag{10.11}$$

where ω_e is the angular speed in electrical degrees per second. This in turn is related to the frequency by

$$\omega_e = 360f$$

We can therefore conclude that

$$M = \frac{N}{180f} \text{ MJ/elec. deg.} \tag{10.12}$$

The value of N is obtained from the moment of inertia of the machine usually denoted by WR^2 and traditionally given in lb-ft^2. The conversion formula is derived now.

We have

$$N = \left(\frac{1}{2}\right)\left(\frac{WR^2}{32.2}\right)\omega^2 \text{ ft-lb}$$

In terms of the speed in r/min,

$$N = \left(\frac{1}{2}\right)\left(\frac{WR^2}{32.2}\right)\left(\frac{2\pi n}{60}\right)^2 \text{ ft-lb}$$

To convert to megajoules, we get

$$N = \left(\frac{1}{2}\right)\left(\frac{WR^2}{32.2}\right)\left(\frac{2\pi n}{60}\right)^2\left(\frac{1}{550}\right)(746)(10^{-6})$$

This reduces to our final desired formula:

$$N = 2.3097 \times 10^{-10}(WR^2)n^2 \tag{10.13}$$

The relation between H and M can be simply obtained using Eq. (10.8), which is rewritten as

$$H = \frac{N}{G} \tag{10.14}$$

Here G is the machine rating. Combining this with Eq. (10.12), we obtain

$$M = \frac{GH}{180f} \text{ MJs/elec. deg} \tag{10.15}$$

The quantity H does not vary greatly with the rated power and speed of the machine but instead has a characteristic value or set of values for each class of machine. In the absence of definite information, typical values of H may be used. The curves in Figures 10-1, 10-2, and 10-3 give the general characteristic variations of H for existing and future large turbogenerators.

In system studies where several machines with different ratings are used, the H constant for each machine, given to a base of the machine rating, must be converted to the common system base by multiplying H in Eq. (10.14) by the ratio (machine base MVA/system base MVA).

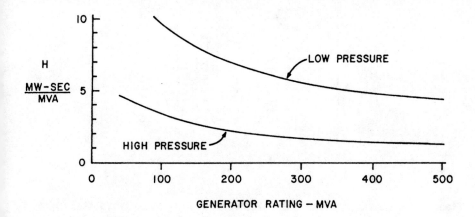

Figure 10-1. Inertia Constants for Large Turbogenerators Rated 500 MVA and Below.

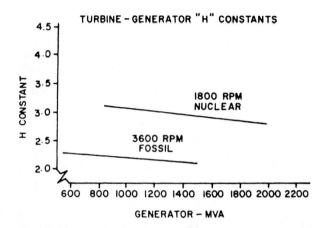

Figure 10-2. Expected Inertia Constants for Future Large Turbogenerators.

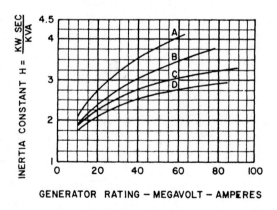

Figure 10-3. Inertia Constants of Large Vertical-Type Water-Wheel Generators, Including Allowance of 15 Percent for Water Wheels. A—450–514 r/min. B—200–400 r/min. C—138–180 r/min. D—80–120 r/min.

Example 10-1

Given a 60-Hz, four-pole turbogenerator rated 10 MVA, 13.2 kV, with an inertia constant of $H = 6.5$ kWs/kVA:

A. Calculate the kinetic energy stored in the rotor at synchronous speed.
B. Find the acceleration if the net mechanical input is 13,400 hp and the electric power developed is 8 MW.
C. Assume that the acceleration in part B is constant for a period of 10 cycles. Find the change in δ in that period.
D. If this generator is delivering rated MVA at 0.8 PF lag when a fault reduces the electric power output by 50 percent, determine the accelerating torque at the fault time.

Solution

A. Using the definition of Eq. (10.14), we obtain
$$N = (10)(6.5) = 65 \text{ MJ}$$

B. The accelerating power P_a is calculated as
$$P_a = P_m - P_e$$
$$= (13,400)(0.746) - 8000$$
$$= 1996.400 \text{ kW}$$

To calculate the acceleration, we need to calculate M. Thus,
$$M = \frac{N}{180f}$$
$$= \frac{65}{(180)(60)} = 6.0185 \times 10^{-3} \text{ MJs/elec. deg}$$

Therefore,
$$\frac{d^2\delta}{dt^2} = \frac{P_a}{M} = \frac{1996.400 \times 10^{-3}}{6.0185 \times 10^{-3}}$$
$$= 331.71 \text{ elec. deg/s}^2$$

C. Using the assumption of constant acceleration and zero initial conditions, we have
$$\delta(t) = \left(\frac{1}{2}\right) at^2$$

For ten cycles to elapse,
$$t = \frac{10}{60} = 0.167 \text{ s}$$

Therefore the required angle is

$$\delta = \left(\frac{1}{2}\right)(331.71)(0.167)^2$$
$$= 4.63 \text{ elec. deg}$$

D. The mechanical power input is equal to the electrical power output at equilibrium. Hence,

$$P_m = 0.8 \times 10 = 8 \text{ MW}$$

On fault, we have

$$P_e = (0.5)(8) = 4 \text{ MW}$$

Therefore the accelerating power is

$$P_a = 8 - 4 = 4 \text{ MW}$$

But we are required to find the accelerating torque. Therefore,

$$T_a = \frac{P_a}{\omega}$$
$$= \frac{4 \times 10^6}{\frac{2\pi(120) \times 60}{4 \times 60}}$$
$$= 21{,}220.66 \text{ N} \cdot \text{m}$$

10.3 ELECTRIC POWER RELATIONS

We have seen in Chapter 3 that an infinite bus is a node with fixed voltage and frequency. In practice, a major bus in an electric network of a large capacity compared to the machine under consideration is treated as an infinite bus.

Consider a simple system consisting of a machine connected to an infinite bus through a network represented by the *ABCD* parameters. Figure 10-4 shows a schematic representation of the system.

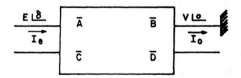

Figure 10-4. One Machine Connected to an Infinite Bus Through a Two-Port Network.

We will assume that the parameters of the network are given in polar form as

$$A = A\underline{/\alpha}$$
$$B = B\underline{/\beta}$$
$$D = D\underline{/\Delta}$$

The sending-end and receiving-end voltages are assumed as

$$E = E\underline{/\delta}$$
$$V = V\underline{/0}$$

We have the following expression for the receiving-end voltage (based on the inverse transmission form discussed earlier):

$$V = DE - BI_e \tag{10.16}$$

Using the definitions of the various quantities involved, we have

$$I_e = \frac{DE}{B}\underline{/\Delta + \delta - \beta} - \frac{V}{B}\underline{/-\beta}$$

The complex power at the machine (sending) side is thus

$$S_e = EI_e^*$$

Consequently, we write

$$S_e = \frac{DE^2}{B}\underline{/\beta - \Delta} - \frac{EV}{B}\underline{/\beta + \delta}$$

The real part gives

$$P_e = \frac{DE^2}{B}\cos(\beta - \Delta) - \frac{EV}{B}\cos(\beta + \delta) \tag{10.17}$$

Now define

$$\gamma = \frac{\pi}{2} - \beta \tag{10.18}$$

Then the electric power output of the machine is

$$P_e = \frac{DE^2}{B}\cos(\beta - \Delta) + \frac{EV}{B}\sin(\delta - \gamma) \tag{10.19}$$

This can further simplify to

$$P_e = P_c + P_M \sin(\delta - \gamma) \tag{10.20}$$

where

$$P_c = \left(\frac{DE^2}{B}\right)\cos(\beta - \Delta) \tag{10.21}$$

$$P_M = \frac{EV}{B} \tag{10.22}$$

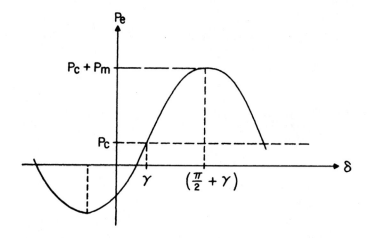

Figure 10-5. Power Angle Curve of a Synchronous Machine Connected to an Infinite Bus.

The relationship between P_e and δ in Eq. (10.20) is shown in Figure 10-5. We note that the power angle curve of a synchronous machine connected to an infinite bus is a sine curve displaced from the origin by an amount P_c vertically and by the angle γ horizontally. Since usually $\beta < \pi/2$, and in view of Eq. (10.18), we see that $\gamma > 0$.

The maximum value of the electric power delivered by the generator is achieved for

$$\frac{\partial P_e}{\partial \delta} = 0$$

This results in requiring

$$P_M \cos(\delta - \gamma) = 0$$

The angle δ at which this occurs is

$$\delta_{m_e} = \frac{\pi}{2} + \gamma \qquad (10.23)$$

The maximum electric power is thus

$$P_{e_{max}} = P_c + P_M \qquad (10.24)$$

It is of interest to consider the corresponding expressions for the electric power received at the infinite bus. For this we have

$$E = AV + BI_0$$

10.3 Electric Power Relations

which results in

$$I_0 = \frac{E}{B} - \frac{AV}{B}$$

The output apparent power is again given by

$$S_0 = VI_0^*$$

or

$$S_0 = \frac{VE^*}{B^*} - \frac{AV^2}{B}$$

Thus,

$$S_0 = \frac{VE}{B}\underline{/\beta - \delta} - \frac{A^*V^2}{B^*}\underline{/\beta - \alpha}$$

The real part of the above gives

$$P_0 = \frac{VE}{B}\cos(\beta - \delta) - \frac{AV^2}{B}\cos(\beta - \alpha) \qquad (10.25)$$

Using the definitions of Eqs. (10.18), (10.21), and (10.22), we get

$$P_0 = P_M \sin(\delta + \gamma) - \left(\frac{V}{E}\right)^2 P_c$$

Here we assume a symmetrical network with $A = D$. The maximum output power is obtained for

$$\delta_{m_0} = \frac{\pi}{2} - \gamma \qquad (10.26)$$

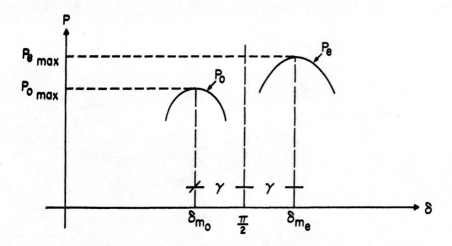

Figure 10-6. Illustrating the Relative Locations of Maximum Angular Shifts for Input and Output Powers for the System Shown in Figure 10-4.

The maximum output power is given by

$$P_{0_{max}} = P_M - \left(\frac{V}{E}\right)^2 P_c \qquad (10.27)$$

Let us observe that

$$\delta_{m_0} < \delta_{m_e}$$

from which we conclude that the maximum allowable output power is attained well before the maximum allowable generated power. The separation of the two angles is 2γ as shown in Figure 10-6. We can conclude that a design on the basis of the output power will be more conservative. Note that for $\gamma = 0$, the two angle limits are the same.

Example 10-2

Assume that the voltages at the sending and receiving terminals of a line are fixed at 110 kV. The **ABCD** parameters of the line are

$$\mathbf{A} = \mathbf{D} = 0.980 \underline{/0.3°}$$
$$\mathbf{B} = 82.5 \underline{/76.0°}$$

What are the steady-state power limits and the angles at which these occur?

Solution

The angle γ is obtained as

$$\gamma = \frac{\pi}{2} - \beta = 90 - 76.0 = 14°$$

Consequently, the angles for maximum power are

$$\delta_{m_e} = 90 + 14 = 104°$$
$$\delta_{m_0} = 90 - 14 = 76°$$

Here we have

$$\mathbf{E} = \mathbf{V} = 110 \text{ kv (line-to-line)}$$

Thus we calculate

$$P_c = \frac{DE^2}{B} \cos(\beta - \Delta)$$
$$= \frac{(0.98)(110)^2}{82.5} \cos(76 - 0.3)$$
$$= 35.5 \text{ MW}$$
$$P_M = \frac{EV}{B}$$
$$= \frac{(110)^2}{82.5}$$
$$= 146.67 \text{ MW}$$

As a result, the maximum generator side power limit is

$$P_{e_{max}} = P_c + P_M$$
$$= 182.17 \text{ MW}$$

For the output side we have

$$P_{0_{max}} = P_M - \left(\frac{V}{E}\right)^2 P_c$$
$$= 111.16 \text{ MW}$$

An alternate but equivalent formulation of the problem discussed above utilizes the admittance matrix representation of two-port networks. Here we have

$$\begin{bmatrix} I_e \\ I_0 \end{bmatrix} = \begin{bmatrix} Y_{11} & Y_{12} \\ Y_{12} & Y_{22} \end{bmatrix} \begin{bmatrix} E \\ V \end{bmatrix}$$

where clearly node (1) is the source bus and node (2) is the infinite bus. From our previous discussions of the load-flow problem, we can show that the power at the sending (generator) end is

$$P_e = E^2 Y_{11} \cos\theta_{11} + EVY_{12}\cos(\theta_{12} - \delta) \tag{10.28}$$

The driving point admittance at node 1 is

$$\mathbf{Y}_{11} = Y_{11}\underline{/\theta_{11}}$$

and the negative transfer admittance is

$$\mathbf{Y}_{12} = Y_{12}\underline{/\theta_{12}}$$

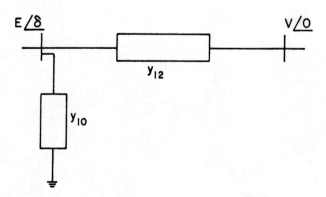

Figure 10-7. Equivalent Circuit for a System of One Machine against Infinite Bus.

Equation (10.28) is clearly of the form

$$P_e = P_c + P_M \sin(\delta - \gamma)$$

Here we have

$$\gamma = \theta_{12} - \frac{\pi}{2} \tag{10.29}$$

$$G_{11} = Y_{11} \cos \theta_{11} \tag{10.30}$$

$$P_c = E^2 G_{11} \tag{10.31}$$

$$P_M = EVY_{12} \tag{10.32}$$

Let us observe that for a network of the configuration shown in Figure 10-7 we have

$$\mathbf{Y}_{11} = \mathbf{y}_{12} + \mathbf{y}_{10}$$
$$\mathbf{Y}_{12} = -\mathbf{y}_{12}$$

Example 10-3

A synchronous machine is connected to an infinite bus through a transformer having a reactance of 0.1 pu and a double-circuit transmission line with 0.45 pu reactance for each circuit. The system is shown in Figure 10-8. All reactances are given to a base of the machine rating. The direct-axis transient reactance of the machine is 0.15 pu. Determine the variation of the electrical power with angle δ. Assume $V = 1.0$ pu.

Solution

An equivalent circuit of the above system is shown in Figure 10-9. From this we have the following equations:

$$\mathbf{y}_{12} = \frac{1}{j0.475} = -j2.1053$$

$$\mathbf{y}_{10} = 0$$

$$\mathbf{Y}_{12} = j2.1053$$

$$\mathbf{Y}_{11} = -j2.1053$$

$$\theta_{11} = -\frac{\pi}{2}$$

$$\theta_{12} = \frac{\pi}{2}$$

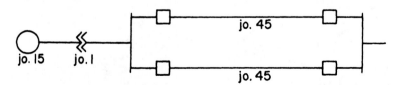

Figure 10-8. System for Example 10-3.

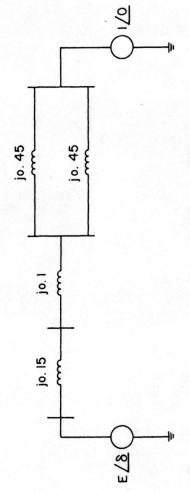

Figure 10-9. Equivalent Circuit of System for Example 10-3.

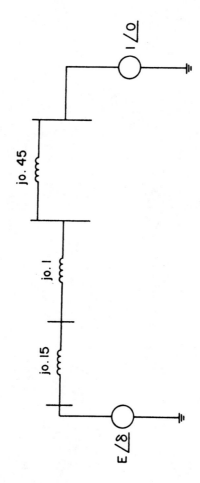

Figure 10-10. Equivalent Circuit of System in Example 10-4.

Therefore,
$$P_c = 0 \quad \text{and} \quad \gamma = 0$$
The electrical power is given by
$$P_e = EVY_{12}\sin(\delta - \gamma)$$
Therefore,
$$P_e = 2.1053E \sin \delta$$

Changes in the network configuration between the two sides (sending and receiving) will alter the value of Y_{12} and hence the expression for the electric power transfer. The following example illustrates this point.

Example 10-4

Assume for the system of Example 10-3 that only one circuit of the transmission line is available. Obtain the relation between the transmitted electrical power and the angle δ. Assume other variables to remain unchanged.

Solution

The network configuration presently offers an equivalent circuit as shown in Figure 10-10. For the present we have
$$y_{12} = \frac{1}{j0.7} = -j1.43$$
As a result,
$$Y_{12} = j1.43$$
$$P_e = 1.43E \sin \delta$$
Observe that the maximum value of the new curve is lower than the one corresponding to the previous example.

10.4 CONCEPTS IN TRANSIENT STABILITY

In order to gain an understanding of the concepts involved in transient stability prediction, we will concentrate on the simplified network consisting of a series reactance X connecting the machine and the infinite bus. Under these conditions our power expression given by Eq. (10.28), reduces to

$$P_e = \frac{EV}{X} \sin \delta \qquad (10.33)$$

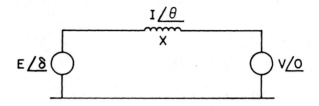

Figure 10-11. A Schematic Representation of a Single Machine Infinite Bus System.

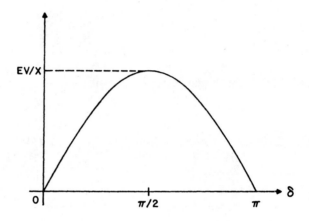

Figure 10-12. Power Angle Curve Corresponding to Eq. (10.33).

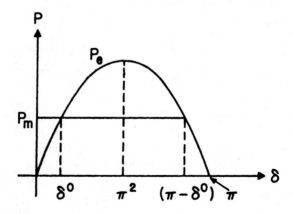

Figure 10-13. Stable and Unstable Equilibrium Points.

The situation is shown in Figure 10-11, and the corresponding power angle curve is shown in Figure 10-12.

An important assumption that we adopt is that the electric changes involved are much faster than the resulting mechanical changes produced by the generator/turbine speed control. Thus we assume that the mechanical power is a constant for the purpose of transient stability calculations. The functions P_m and P_e are plotted in Figure 10-13. The intersection of these two functions defines two values for δ. The lower value is denoted δ^0; consequently, the higher is $\pi - \delta^0$ according to the symmetry of the curve. At both points $P_m = P_e$; that is, $d^2\delta/dt^2 = 0$, and we say that the system is *in equilibrium*.

Assume that a change in the operation of the system occurs such that δ is increased by a small amount $\Delta\delta$. Now for operation near δ^0, $P_e > P_m$ and $d^2\delta/dt^2$ becomes negative according to the swing equation, Eq. (10.7). Thus δ is decreased, and the system responds by returning to δ^0. We refer to this as a *stable equilibrium point*. On the other hand, operating at $\pi - \delta^0$ results in a system response that will increase δ and moving further from $\pi - \delta^0$. For this reason, we call $\pi - \delta^0$ an *unstable equilibrium point*.

If the system is operating in an equilibrium state supplying an electric power P_{e_0} with the corresponding mechanical power input P_{m_0}, then

$$P_{m_0} = P_{e_0}$$

and the corresponding rotor angle is δ_0. Suppose the mechanical power P_m is changed to P_{m_1} at a fast rate, which the angle δ cannot follow as shown in Figure 10-14. In this case $P_m > P_e$, and acceleration occurs so that δ increases. This goes on until the point δ_1 where $P_m = P_e$, and the acceleration is zero. The speed, however, is not zero at that point, and δ continues to increase beyond δ_1. In this region, $P_m < P_e$ and rotor retardation takes place.

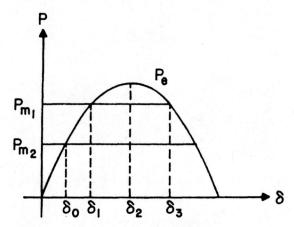

Figure 10-14. System Reaction to Sudden Change.

The rotor will stop at δ_2 where the speed is zero and retardation will bring δ down. This process continues on as oscillations around the new equilibrium point δ_1. This serves to illustrate what happens when the system is subjected to a sudden change in the power balance of the right-hand side of the swing equation.

The situation described above will occur for sudden changes in P_e as well. The system discussed in Examples 10-3 and 10-4 serves to illustrate this point, which we discuss further in the next example.

Example 10-5

The system of Example 10-3 is delivering an apparent power of 1.1 pu at 0.85 PF lagging with two circuits of the line in service. Obtain the source voltage (excitation voltage) E and the angle δ_0 under these conditions. With the second circuit open as in Example 10-4, a new equilibrium angle δ_1 can be reached. Sketch the power angle curves for the two conditions. Find the angle δ_0 and the electric power that can be transferred immediately following the circuit opening, as well as δ_1. Assume that the excitation voltage remains unchanged.

Solution

The power delivered is
$$P_0 = 1.1 \times 0.85 = 0.94 \text{ p.u.}$$

The current in the circuit is
$$\mathbf{I} = \frac{1.1}{1} \underline{/-\cos^{-1} 0.85}$$

We can thus write
$$E\underline{/\delta} = V\underline{/0} + I(jX)$$
$$= (1+j0) + (1.1\underline{/-31.79°})(0.475\underline{/90°})$$
$$= 1 + 0.28 + j0.44$$
$$= 1.35\underline{/19.20°}$$

Therefore,
$$E = 1.35$$
$$\delta_0 = 19.20°$$

The power angle curve for the line with two circuits according to Example 10-3 is
$$P_{e_0} = (2.1053)(1.35)\sin\delta$$
$$= 2.84 \sin\delta$$

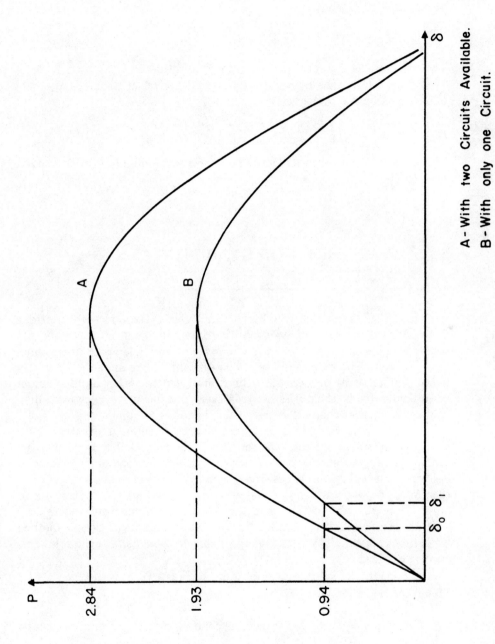

Figure 10-15. Illustrating the Power Curves for Example 10-5.

With one circuit open, the new power angle curve is obtained as given in Example 10-4. Thus,

$$P_{e_1} = (1.43)(1.35)\sin\delta$$
$$= 1.93\sin\delta$$

The two power angle curves are shown in Figure 10-15.

From inspection of the curves, we can deduce that the angle δ_1 can be obtained using the power relation

$$0.94 = 1.93\sin\delta_1$$
$$\delta_1 = 29.15°$$

We can obtain the value of electric power corresponding to δ_0 with one line open as

$$P_{e_{10}} = 1.93\sin 19.2°$$
$$= 0.63 \text{ p.u.}$$

10.5 A METHOD FOR STABILITY ASSESSMENT

In order to predict whether a particular system is stable after a disturbance it is necessary to solve the dynamic equation describing the behavior of the angle δ immediately following an imbalance or disturbance to the system. The system is said to be unstable if the angle between any two machines tends to increase without limit. On the other hand, if under disturbance effects, the angles between every possible pair reach maximum value and decrease thereafter, the system is deemed stable.

Assuming as we have already done that the input is constant, with negligible damping and constant source voltage behind the transient reactance, the angle between two machines either increases indefinitely or oscillates after all disturbances have occurred. In the case of two machines, therefore, the two machines either fall out of step on the first swing or never. Here the observation that the machines' angular differences stay constant can be taken as an indication of system stability. A simple method for determining stability known as the *equal-area method* is available. This we discuss here.

The Equal-Area Method

The swing equation for a machine connected to an infinite bus can be written as

$$\frac{d\omega}{dt} = \frac{P_a}{M} \tag{10.34}$$

where $\omega = d\delta/dt$ and P_a is the accelerating power. We would like to obtain an expression for the variation of the angular speed ω with P_a. We observe that Eq. (10.34) can be written in the alternate form

$$d\omega = \frac{P_a}{M}\left(\frac{d\delta}{d\delta}\right)dt$$

or

$$\omega\, d\omega = \frac{P_a}{M}(d\delta)$$

Integrating, we obtain

$$\int_{\omega_0}^{\omega} \omega\, d\omega = \frac{1}{M}\int_{\delta_0}^{\delta} P_a(d\delta)$$

Note that we may assume $\omega_0 = 0$; consequently,

$$\omega^2 = \frac{2}{M}\int_{\delta_0}^{\delta} P_a(d\delta)$$

or

$$\frac{d\delta}{dt} = \left[\frac{2}{M}\int_{\delta_0}^{\delta} P_a(d\delta)\right]^{1/2} \quad (10.35)$$

The above equation gives the relative speed of the machine with respect to a reference frame moving at a constant speed (by the definition of the angle δ).

If the system is stable, then the speed must be zero when the acceleration is either zero or is opposing the rotor motion. Thus for a rotor that is accelerating, the condition for stability is that a value of δ_s exists such that

$$P_a(\delta_s) \leq 0$$

and

$$\int_{\delta_0}^{\delta_s} P_a(d\delta) = 0$$

This condition is applied graphically in Figure 10-16 where the net area under the $P_a - \delta$ curve reaches zero at the angle δ_s as shown. Observe that at δ_s, P_a is negative, and consequently the system is stable. Also observe that area A_1 equals A_2 as indicated.

The accelerating power need not be plotted to assess stability. Instead, the same information can be obtained from a plot of electrical and mechanical powers. The former is the power angle curve, and the latter is assumed constant. In this case, the integral may be interpreted as the area between the P_e curve and the curve of P_m, both plotted versus δ. The area to be equal to zero must consist of a positive portion A_1, for which $P_m > P_e$, and an

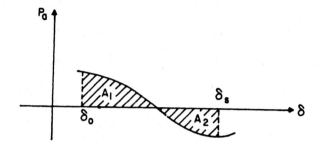

Figure 10-16. The Equal-Area Criterion for Stability for a Stable System.

equal and opposite negative portion A_2, for which $P_m < P_e$. This explains the term *equal-area criterion for transient stability*. This situation is shown in Figure 10-17.

If the accelerating power reverses sign before the two areas A_1 and A_2 are equal, synchronism is lost. This situation is illustrated in Figure 10-18. The area A_2 is smaller than A_1, and as δ increases beyond the value where P_a reverses sign again, the area A_3 is added to A_1.

Example 10-6

Consider the system of the previous three examples. Determine whether the system is stable for the fault of an open circuit on the second line. If the system is stable, determine δ_s.

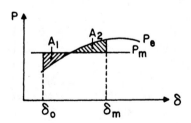

Figure 10-17. The Equal-Area Criterion for Stability.

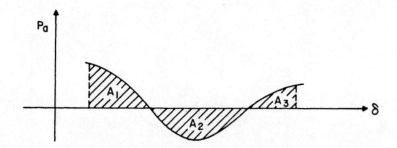

Figure 10-18. The Equal-Area Criterion for an Unstable System.

Solution

From the examples given above, we have

$$\delta_0 = 19.2°$$
$$\delta_1 = 29.15°$$

The geometry of the problem is shown in Figure 10-19. We can calculate the area A_1 immediately:

$$A_1 = 0.94(29.15 - 19.2)\frac{\pi}{180} - \int_{19.2°}^{29.15°} 1.93 \sin \delta \, (d\delta)$$

Observe that the angles δ_1 and δ_0 are substituted for in radians. The result is

$$A_1 = 0.0262$$

The angle δ_f is

$$\delta_f = 180° - \delta_1$$
$$= 150.85°$$

If the area enclosed by the power angle curve for one circuit and the fixed $P_m = 0.94$ line between δ_1 and δ_f denoted by A_{2_f}, is larger than A_1, then the system is stable. To ascertain this, we have

$$A_{2_f} = 2\int_{\delta_1}^{\pi/2} 1.93 \sin \delta \, (d\delta) - (0.94)(2)[90 - \delta_1]\frac{\pi}{180}$$
$$= 1.3745$$

This clearly gives

$$A_{2_f} > A_1$$

and the system is stable.

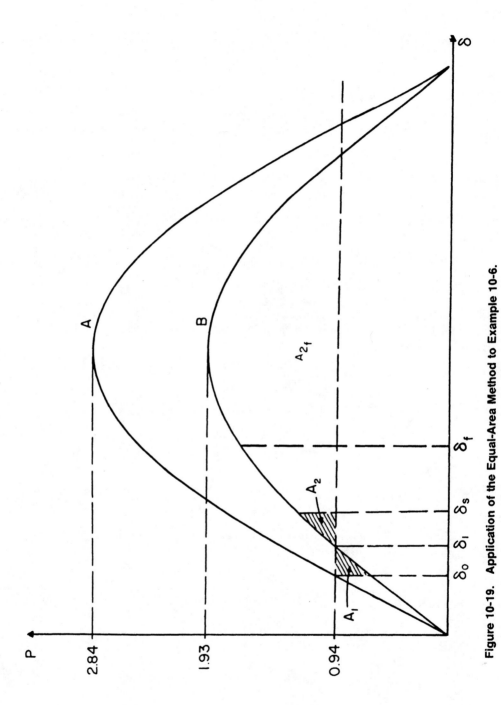

Figure 10-19. Application of the Equal-Area Method to Example 10-6.

10.5 A Method for Stability Assessment

The angle δ_s is obtained by solving for $A_2 = A_1$. Here we get

$$0.0262 = \int_{\delta_1}^{\delta_s} 1.93 \sin \delta \, (d\delta) - 0.94(\delta_s - \delta_1)\left(\frac{\pi}{180}\right)$$

This gives

$$0.0262 = 1.93(\cos 29.15° - \cos \delta_s) - 0.94(\delta_s - 29.15°)\left(\frac{\pi}{180}\right)$$

Further manipulations yield the following relation in δ_s (in degrees):

$$1.93 \cos \delta_s + 0.0164 \delta_s = 2.1376$$

The equation is nonlinear; we resort to iterative methods for its solution. The result is

$$\delta_s = 39.39°$$

The above example shows the application of the equal-area method to the case of a generator supplying power to an infinite bus over two parallel transmission lines. For the loading indicated above, the system is stable. The opening of one of the lines may cause the generator to lose synchronism even though the load could be supplied over a single line. The following example illustrates this point.

Example 10-7

Assume that the system in Example 10-5 is delivering an active power of 1.8 p.u. with the same source voltage E as before. Determine whether the system will remain stable after one circuit of the line is opened.

Solution

We have for the initial angle δ_0,

$$1.8 = 2.84 \sin \delta_0$$

Therefore,

$$\delta_0 = 39.33°$$

The angle δ_1 is obtained from

$$1.8 = 1.93 \sin \delta_1$$
$$\delta_1 = 68.85$$

The area A_1 is thus

$$A_1 = 1.8(68.85 - 39.33)\frac{\pi}{180} - 1.93 \int_{\delta_0}^{\delta_1} \sin \delta \, (d\delta)$$
$$= 0.93 + 1.93[\cos(68.85°) - \cos(39.33°)]$$
$$= 0.13$$

The area A_{2_f} is obtained as

$$A_{2_f} = 3.86 \int_{\delta_1}^{\pi/2} \sin\delta(d\delta) - 1.8(\delta_f - \delta_1)\left(\frac{\pi}{180}\right)$$

$$= 3.86\cos(68.85°) - 1.8(111.15° - 68.85°)\left(\frac{\pi}{180}\right)$$

$$= 0.06$$

We note that $A_1 > A_{2_f}$ and the system is therefore unstable.

If a three-phase short circuit took place at a point on the extreme end of the line, there is some impedance between the generator bus and the load (infinite) bus. Therefore, some power is transmitted while the fault is still on. The situation is similar to the ones analyzed above, and we use the following example to illustrate our point.

Example 10-8

A generator is delivering 0.25 of P_{max} to an infinite bus through a transmission line. A fault occurs such that the reactance between the generator and the bus is increased to two times its prefault value.

A. Find the angle δ_0 before the fault.
B. Show graphically what happens when the fault is sustained.
C. Find the maximum value of the δ swing in the case of a sustained fault.

Solution

A. Figure 10-20 illustrates the situation for this example. The amplitude of the power angle curve with the fault sustained is one-half of the original value.

Before the fault we have

$$0.25 = (1)\sin\delta_0$$

Hence

$$\delta_0 = 14.48°$$

B. At the fault instant, we get

$$0.25 = 0.5\sin\delta_1$$

Thus,

$$\delta_1 = 30°$$

$$A_1 = 0.25(\delta_1 - \delta_0) - \int_{\delta_0}^{\delta_1} 0.5\sin\delta(d\delta)$$

$$= \frac{(30 - 14.48)\pi(0.25)}{180} + 0.5(\cos 30° - \cos 14.48°)$$

$$= 0.0677 - 0.0511 = 0.0166$$

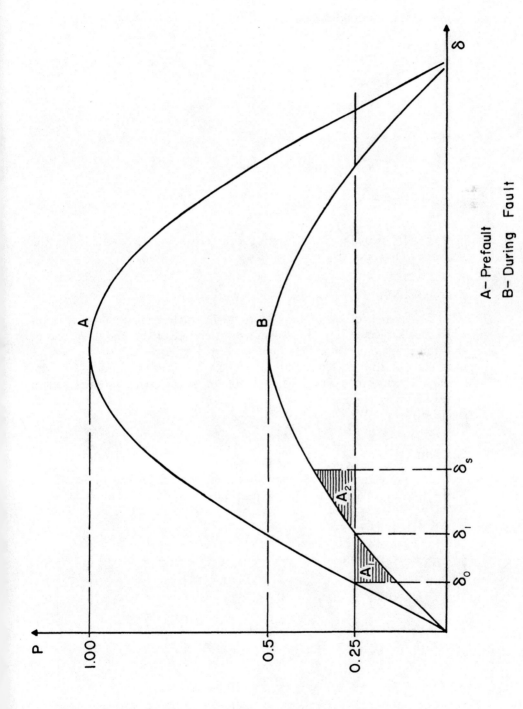

Figure 10-20. Power Angle Curves for Example 10-8.

C. As before, the condition for stability is
$$A_2 = A_1$$

Therefore,
$$0.0166 = \int_{\delta_1}^{\delta_s} 0.5 \sin \delta (d\delta) - 0.25(\delta_s - \delta_1)\left(\frac{\pi}{180}\right)$$

Hence
$$0.5473 = 0.5 \cos \delta_s + \frac{\delta_s}{720}\pi$$

By trial and error, we obtain
$$\delta_s = 46.3°$$

The following example illustrates the effects of short circuits on the network from a stability point of view.

Example 10-9

The system of Examples 10-3 to 10-7 is delivering a power of 1 pu when it is subjected to a three-phase short circuit in the middle of one of the transmission circuits. This fault is cleared by opening the breakers at both ends of the faulted circuit. If the fault is cleared for $\delta_c = 50°$, determine whether the system will be stable or not. Assume the same source voltage E is maintained as before. If the system is stable, find the maximum swing angle.

Solution

The power angle curves have been determined for the prefault network in Example 10-3 and for the postfault network in Example 10-4. In

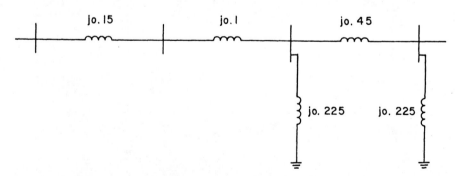

Figure 10-21. Equivalent Circuit for System of Example 10-3 with a Short Circuit in the Middle of One Transmission Circuit.

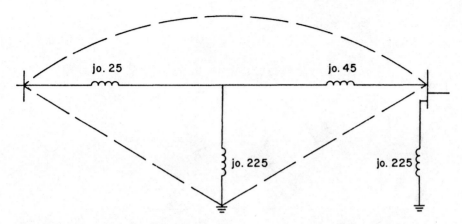

Figure 10-22. Reduced Equivalent Circuit for System of Example 10-9.

Example 10-5, we obtained

$$E = 1.35$$

Therefore,

Prefault: $P = 2.84 \sin \delta$
Postfault: $P = 1.93 \sin \delta$

During the fault, the network offers a different configuration, which is shown in Figure 10-21. We will need to reduce the network in such a way as to obtain a clear path from the source to the infinite bus. We do this by using a $Y - \Delta$ transformation as indicated in Figure 10-22. Consequently,

$$jX = j\left[0.45 + 0.25 + \frac{(0.45)(0.25)}{0.225}\right] = j1.2$$

and the fault power angle curve is given by

$$P = 1.13 \sin \delta$$

The three power angle curves are shown in Figure 10-23.

The initial angle δ_0 is given by the equation

$$1 = 2.84 \sin \delta_0$$

Therefore we have

$$\delta_0 = 20.62°$$

The clearing angle is

$$\delta_c = 50°$$

The area A_1 can thus be calculated as

$$A_1 = 1(50 - 20.62)\left(\frac{\pi}{180}\right) - 1.13 \int_{20.62°}^{50°} \sin \delta (d\delta)$$
$$= 0.51 + 1.13[\cos 50° - \cos 20.62°]$$
$$= 0.18$$

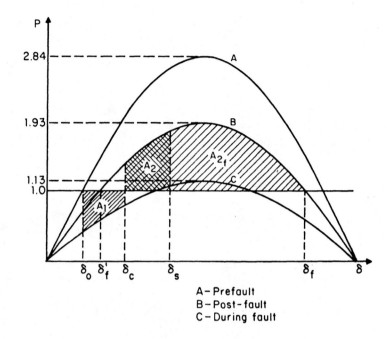

Figure 10-23. Power Angle Curves for Example 10-9.

The maximum area A_{2_f} is obtained using the angle δ_f. Thus,
$$1 = 1.93 \sin \delta_f'$$
$$\delta_f' = 31.21$$

Consequently
$$\delta_f = 180 - \delta_f' = 148.79°$$

Now
$$A_{2_f} = \left[1.93 \int_{\delta_c}^{\delta_f} \sin \delta (d\delta)\right] - (1)(\delta_f - \delta_c)\left(\frac{\pi}{180}\right)$$
$$= 1.93[\cos 50° - \cos 148.79°] - (148.79 - 50)\left(\frac{\pi}{180}\right)$$
$$= 1.17$$

Since $A_{2_f} > A_1$, the system is stable.

To calculate the angle of maximum swing, we have
$$A_2 = A_1$$

Thus,

$$0.18 = \left[\int_{\delta_c}^{\delta_s} 1.93 \sin \delta (d\delta)\right] - (1)(\delta_s - \delta_c)\left(\frac{\pi}{180}\right)$$

$$= 1.93(\cos 50 - \cos \delta_s) - (\delta_s - 50)\left(\frac{\pi}{180}\right)$$

$$1.93 \cos \delta_s + 1.7453 \times 10^{-2} \delta_s = 1.93324$$

Using iterations, we obtain

$$\delta_s \cong 66.3°$$

10.6 IMPROVING SYSTEM STABILITY

The stability of the electric power system can be affected by changes in the network or changes in the mechanical (steam or hydraulic) system. Network changes that adversely affect system stability can either decrease the amplitude of the power curve or raise the load line. Examples of events that decrease the amplitude of the power curve are: short circuits on tie lines, connecting a shunt reactor, disconnecting a shunt capacitor, or opening a tie line. Events that raise the load line include: disconnecting a resistive load in a sending area, connecting a resistive load in a receiving area, the sudden loss of a large load in a sending area, or the sudden loss of a generator in a receiving area. Changes in a steam or hydraulic system that can affect stability include raising the load line by either closing valves or gates in receiving areas or opening valves or gates in sending areas.

There are several corrective actions that can be taken in order to enhance the stability of the system following a disturbance. These measures can be classified according to the type of disturbance—depending on whether it is a loss of generation or a loss of load.

In the case of a loss of load, the system will have an excess power supply. Among the measures that can be taken are:

1. Resistor braking.
2. Generator dropping.
3. Initiation along with braking, of fast steam valve closures, bypassing of steam, or reduction of water acceptance for hydro units.

In the case of a loss of generation, countermeasures are:

1. Load shedding.
2. Fast control of valve opening in steam electric plants; and in the case of hydro, increasing the water acceptance.

The measures mentioned above are taken at either the generation or the load sides in the system. Measures that involve the interties (the lines) can be taken to enhance the stability of the system. Among these we have the switching of series capacitors into the lines, the switching of shunt capacitors or reactors, or the boosting of power on HVDC lines.

Resistor braking relies on the connection of a bank of resistors in shunt with the three-phase bus in a generating plant in a sending area through a suitable switch. This switch is normally open and will be closed only upon the activation of a control device that detects the increase in kinetic energy exceeding a certain threshold. Resistive brakes have short time ratings to make the cost much less than that of a continuous-duty resistor of the same rating. If the clearing of the short circuit is delayed for more than the normal time (about three cycles), the brakes should be disconnected and some generation should be dropped.

Generator dropping is used to counteract the loss of a large load in a sending area. This is sometimes used as a cheap substitute for resistor braking to counteract short circuits in sending systems. It should be noted that better control is achieved with resistor braking than with generator dropping.

To counteract the loss of generation, load shedding is employed. In this instance, a rapid opening of selected feeder circuit breakers in selected load areas is arranged. This disconnects the customer's premises with interruptible loads such as heating, air conditioning, air compressors, pumps where storage is provided in tanks, or reservoirs. Aluminum reduction plants are among loads that can be interrupted with only minor inconvenience. Load shedding by temporary depression of voltage can also be employed. This reduction of voltage can be achieved either by an intentional short circuit or by the connection of a shunt reactor.

The insertion of switched series capacitors can counteract faults on ac interties or permanent faults on dc interties in parallel with ac lines. In either case, the insertion of the switched series capacitor decreases the transfer reactance between the sending and receiving ends of the interconnection and consequently increases the amplitude of the sine curve and therefore enhances the stability of the system. It should be noted that the effect of a shunt capacitor inserted in the middle of the intertie or the switching off of a shunt reactor in the middle of the intertie is equivalent to the insertion of a series capacitor (this can be verified by means of a $Y - \Delta$ transformation).

To relieve ac lines of some of the overload and therefore provide a larger margin of stability, the power transfer on a dc line may be boosted. This is one of the major advantages of HVDC transmission.

SOME SOLVED PROBLEMS

Problem 10-A-1

A 60-Hz alternator rated at 20 MVA is developing electric power at 0.8 PF lagging with net mechanical input of 18 MW. Assume that acceleration is constant for a period of 15 cycles, in which δ attains a value of 15° electrical from zero initial conditions. Calculate the inertia constant H for the machine.

Solution

For 15 cycles to elapse,

$$t = \frac{15}{60} = 0.25 \text{ s}$$

With the assumption of constant acceleration and zero initial conditions, we have

$$\delta = \frac{1}{2} a t^2$$

But for $t = 0.25$ s, we have $\delta = 15°$ electrical degrees. Therefore the acceleration is

$$a = \frac{2 \times 15}{(0.25)^2}$$
$$= 480° \text{ elec./s}^2$$

The accelerating power P_a is

$$P_a = P_m - P_e$$
$$= 18 - (20)(0.8)$$
$$= 2 \text{ MW}$$

But

$$a = \frac{P_a}{M}$$
$$480 = \frac{2}{M}$$

Thus

$$M = \frac{1}{240}$$

We now use

$$M = \frac{GH}{180f}$$
$$\frac{1}{240} = \frac{(20)H}{(180)(60)}$$

This yields

$$H = 2.25 \text{ kWs/kVA}$$

Problem 10-A-2

Show that the speed of a generator subject to a constant decelerating power of 1 pu will be reduced from rated value to zero in $2H$ seconds.

Solution

The swing equation is

$$\left(\frac{2H}{\omega_R}\right)\ddot{\delta} = P_a$$

Integrating, we get

$$\dot{\delta}(t) = \dot{\delta}_0 + \left(\frac{\omega_R}{2H}\right)P_a t$$

With initial speed of

$$\dot{\delta}_0 = \omega_R$$

accelerating power of

$$P_a = -1$$

and final speed of

$$\dot{\delta}(t_1) = 0$$

we obtain

$$0 = \omega_R - \left(\frac{\omega_R}{2H}\right)t_1$$

From the above,

$$t_1 = 2H$$

Problem 10-A-3

The sending-end and receiving-end voltages of a short transmission link are fixed at 132 kV each. The line impedance is 100 ohms. The difference between the maximum power transfer at the generator side and the maximum power transfer at the receiving side at the receiving side of the link is 40 MW. Calculate the maximum power transfer at the generator side and the corresponding angle.

Solution

The difference in maximum power transfer is

$$P_{e_{max}} - P_{0_{max}} = P_c + P_M - \left[P_M - \left(\frac{V}{E}\right)^2 P_c\right]$$

$$= P_c\left[1 + \left(\frac{V}{E}\right)^2\right]$$

Using the information given, we get

$$40 = P_c\left[1 + \left(\frac{132}{132}\right)^2\right]$$

Thus,
$$P_c = 20 \text{ MW}$$

Now for a short line, $B = Z$. Thus,
$$P_M = \frac{EV}{B}$$
$$= \frac{(132)^2}{100}$$
$$= 174.24 \text{ MW}$$

As a result, the maximum power at the generator side is
$$P_{e_{max}} = P_c + P_M$$
$$= 20 + 174.24 = 194.24 \text{ MW}$$

We use
$$P_c = \frac{DE^2}{B} \cos(\beta - \Delta)$$

For this line,
$$D = 1$$
$$\Delta = 0.0$$
$$E = 132 \text{ kV}$$
$$B = 100 \text{ ohms}$$

Therefore,
$$20 = \frac{(1)(132)^2}{100} \cos\beta$$

Thus,
$$\cos\beta = 0.11$$
$$\beta = 83.41°$$

As a result, the angle δ_{m_e} is
$$\delta_{m_e} = 180 - 83.41$$
$$= 96.59°$$

Problem 10-A-4

The series impedance and shunt admittance of a transmission line are
$$\mathbf{Z} = 189.23 \underline{/78.2°}$$
$$\mathbf{Y} = 1.152 \times 10^{-3} \underline{/90°}$$

Calculate the maximum power transfer and the corresponding rotor angles when the sending-end and receiving-end voltages are 230 and 200 kV respectively. Assume that the line is long.

Solution

From given data,

$$Z = 189.23 \underline{/78.2°}$$
$$Y = 1.152 \times 10^{-3} \underline{/90°}$$

Given

$$E = 230 \text{ kV}$$
$$V = 200 \text{ kV}$$

We calculate for the long line assumption:

$$A = 0.895 \underline{/1.38°}$$
$$B = 181.85 \underline{/78.6°}$$

Thus,

$$P_M = \frac{EV}{B}$$
$$= \frac{(230)(200)}{181.85}$$
$$= 252.96 \text{ MW}$$

$$P_c = \frac{DE^2}{B} \cos(\beta - \Delta)$$
$$= \frac{(0.895)(230)^2}{181.85} \cos(78.6 - 1.38)$$
$$= 57.59 \text{ MW}$$

And

$$P_{e_{max}} = P_c + P_M$$
$$= 310.55 \text{ MW}$$

$$P_{0_{max}} = P_M - \left(\frac{V}{E}\right)^2 P_c$$
$$= 209.41$$
$$\gamma = 90 - \beta = 11.4°$$
$$\delta_{m_e} = 90 + 11.4 = 101.4°$$
$$\delta_{m_0} = 78.6°$$

Problem 10-A-5

Find the error involved in estimating the maximum power transfer for the line of Problem 10-A-4 if the short line assumption is used. Comment on the results.

Solution

For the short line assumption:

$$A = D = 1\underline{/0}$$
$$B = Z = 189.23\underline{/78.2°}$$

Thus,

$$P_M = \frac{EV}{B}$$
$$= \frac{(230)(200)}{189.23}$$
$$= 243.09 \text{ MW}$$
$$P_c = \frac{DE^2}{B}\cos(\beta - \Delta)$$
$$= \frac{(230)^2}{189.23}\cos(78.2)$$
$$= 57.17 \text{ MW}$$

And

$$P_{e_{max}} = P_c + P_M$$
$$= 300.26 \text{ MW}$$
$$P_{0_{max}} = P_M - \left(\frac{V}{E}\right)^2 P_c$$
$$= 199.86 \text{ MW}$$

Therefore, the error in sending-end estimate is

$$310.55 - 300.26 = 10.29 \text{ MW}$$

and the error in receiving-end estimate is

$$209.41 - 199.86 = 9.55 \text{ MW}$$

Thus the short line assumption gives conservative estimates.

Problem 10-A-6

Calculate the minimum value of sending-end voltage necessary to enable transmission of 320 MW for the line of Problem 10-A-4.

Solution

$$P_{e_{max}} = \frac{EV}{B} + \frac{DE^2}{B}\cos(\beta - \Delta)$$
$$320 = \frac{200E}{181.85} + \frac{0.895E^2}{181.85}\cos(78.6 - 1.38)$$
$$= 1.10E + 1.0887 \times 10^{-3}E^2$$

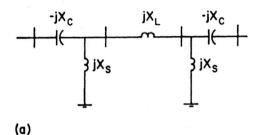

(a)

(b)

Figure 10-24. Systems for Problem 10-A-7.

Solving, we get

$$E = 235.88 \text{ kV}$$

Problem 10-A-7

Obtain the maximum power transfer capacity of the two compensated systems shown in Figure 10-24.

Solution

For the system of Figure 10-24(a), we perform a $\Delta - Y$ transformation to obtain the configuration shown in Figure 10-25.
A further transformation yields a π network with series reactance

$$X_{12} = 2\left(\frac{X_L X_s}{X_L + 2X_s} - X_c\right) + \frac{\left(\frac{X_L X_s}{X_L + 2X_s} - X_c\right)^2}{\frac{X_s^2}{X_L + 2X_s}}$$

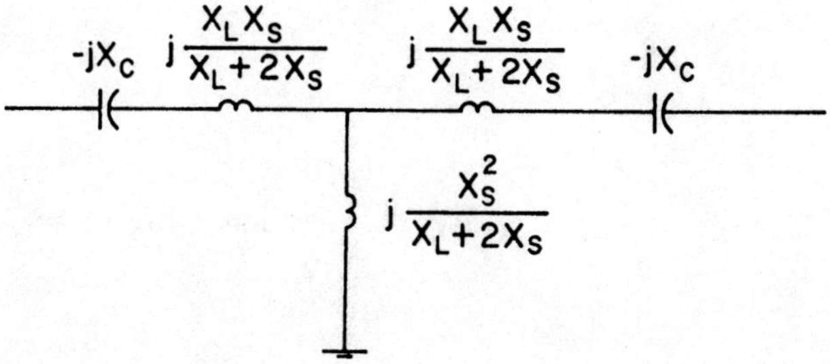

Figure 10-25. Reduced System of Figure 10-24(a) for Problem 10-A-7.

Thus the maximum power is given by

$$P_{e_{max_a}} = \frac{EV}{X_{12}}$$

For the system of Figure 10-24(b), we have

$$P_{e_{max_b}} = \frac{EV}{X_L - 2X_c}$$

Problem 10-A-8

Compare the maximum power transfer capacities for the two schemes of Problem 10-A-7 for

$$X_L = 1 \text{ p.u.}$$
$$X_c = 0.2 \text{ p.u.}$$
$$X_s = 0.1 \text{ p.u.}$$

Solution

For the scheme of Figure 10-24(a), we get

$$X_{12} = 2\left[\frac{(1)(0.1)}{1+2(0.1)} - 0.2\right] + \frac{\left(\frac{0.1}{1+0.2} - 0.2\right)^2}{\frac{(0.1)^2}{1+0.2}}$$

$$= 1.4 \text{ p.u.}$$

Thus,

$$P_{e_{max_a}} = 0.71 \, (E)(V)$$

For the scheme of Figure 10-24(b), we get

$$P_{e_{max_b}} = \frac{(E)(V)}{1-0.4} = 1.6\,(E)(V)$$

Thus the system for Figure 10-24(b) clearly has a higher transfer capacity.

Problem 10-A-9

A 760-kV, 50-Hz, 600-km, three-phase transmission line has the following parameters:

$$r = 0.0122 \text{ ohms/km}$$
$$x = 0.282 \text{ ohms/km}$$
$$g = 19 \times 10^{-9} \text{ siemens/km}$$
$$b = 4.04 \times 10^{-6} \text{ siemens/km}$$

A. Obtain the π equivalent circuit parameters of the line.
B. Calculate the sending-end power as a function of δ for equal sending-end and receiving-end voltages of 760 kV.

Solution

$$z = r + jx = 0.0122 + j0.282$$
$$= 0.2823\underline{/87.523°}\ \Omega/\text{km}$$
$$y = g + jb = 19 \times 10^{-9} + j4.04 \times 10^{-6}$$
$$= 4.04004 \times 10^{-6}\underline{/89.73°}\ \text{s/km}$$
$$\gamma = \sqrt{zy} = 1.0679 \times 10^{-3}\underline{/88.6267°}$$
$$\gamma l = 1.0679 \times 10^{-3} \times 600\underline{/88.6267°}$$
$$= 0.64073\underline{/88.6267°}$$
$$= 0.015356 + j0.64054$$
$$Z_c = \sqrt{\frac{z}{y}} = 264.34\underline{/-1.1°}$$

$$\cosh(\gamma l) = 0.80187 + j9.1775 \times 10^{-3} = 0.80192\underline{/0.65573°}$$
$$\sinh(\gamma l) = 1.2313 \times 10^{-2} + j0.5977 = 0.59783\underline{/88.82°}$$

The π equivalent circuit parameters are thus

$$\tilde{Z} = Z\frac{\sinh(\gamma l)}{\gamma l} = \frac{(0.2823\underline{/87.523°})(0.59783\underline{/88.82°})}{1.0679 \times 10^{-3}\underline{/88.6267°}}$$

$$= 158.04\underline{/87.72°}$$
$$= 6.29 + j157.91 \text{ ohms}$$

$$\frac{\tilde{Y}}{2} = \frac{\cosh(\gamma l) - 1}{Z_c \sinh(\gamma l)}$$

$$= \frac{0.80187 + j9.1775 \times 10^{-3} - 1}{(264.34\underline{/-1.1°})(0.59783\underline{/88.82°})}$$

$$= 1.2551 \times 10^{-3}\underline{/89.63°}$$

The admittances Y_{11} and Y_{12} are obtained as follows:

$$Y_{11} = \frac{\tilde{Y}}{2} + \frac{1}{\tilde{Z}}$$

$$= 5.0741 \times 10^{-3}\underline{/-87.065°}$$

$$Y_{12} = \frac{-1}{\tilde{Z}} = 6.3275 \times 10^{-3}\underline{/92.28°}$$

Consequently, we get

$$P_e = (760)^2(5.0741 \times 10^{-3})\cos(-87.065°)$$
$$+ (760)^2(6.3275 \times 10^{-3})\cos(92.28 - \delta)$$
$$= 150.07 + 3654.76 \sin(\delta - 2.28°) \text{ MW}$$

Problem 10-A-10

Series and shunt compensation of the line of Problem 10-A-9 are employed as shown in Figure 10-26. The degree of series compensation K_s is defined as

$$K_s = \frac{X_c}{lx}$$

where X_c is the total reactance of installed series capacitor bank, x is the line series reactance per unit length, and l is the line length. For shunt compensation, the degree of compensation K_d is

$$K_d = \frac{B_r}{lb}$$

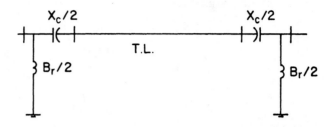

Figure 10-26. Compensation Scheme for System of Problem 10-A-10.

where B_r is the susceptance of shunt reactors connected, and b is the shunt susceptance per unit length of the line. Calculate the maximum sending-end power of the system with

$$K_s = 0.8$$
$$K_d = 0.8$$

Solution

Figure 10-27(a) shows the scheme using the π equivalent of Problem 10-A-9. We have

$$X_c = -0.8 \times 0.282 \times 600 = -135.36$$
$$B_r = -0.8 \times 4.04 \times 10^{-6} \times 600 = -1.9392 \times 10^{-3}$$

A $\Delta - Y$ transformation gives

$$Z_T = \frac{158.04 / 87.72° \cdot \dfrac{1.2551 \times 10^{-3} / 89.63°}{2}}{158.04 / 87.72° + \dfrac{1.2551 \times 10^{-3} / 89.63°}{2}}$$

$$= 87.708 / 87.43°$$
$$= 3.9353 + j87.62 \text{ ohms}$$

$$Z_M = \frac{\left(1.2551 \times 10^{-3} / 89.63°\right)^{-2}}{1.4356 \times 10^3 / -89.34°}$$

$$= 442.19 / -89.92 \text{ ohms}$$
$$= 0.62 - j442.19 \text{ ohms}$$

The reduced circuit is shown in Figure 10-27(b). A series combination gives

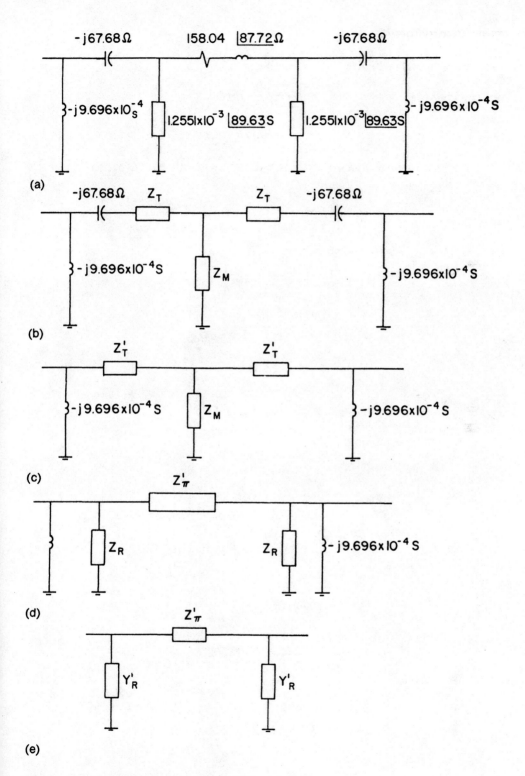

Figure 10-27. Reduction Steps for Problem 10-A-10.

the circuit of Figure 10-27(c) with

$$Z'_T = Z_T - j67.68$$
$$= 3.9353 + j19.94$$
$$= 20.325 \underline{/78.84°}$$

A Y − Δ transformation gives the circuit of Figure 10-27(d) with

$$Z'_\Pi = 2(3.9353 + j19.94) + \frac{\left(20.325\underline{/78.84°}\right)^2}{442.19\underline{/-89.92°}}$$
$$= 39.733\underline{/79.098°}$$
$$Y'_\Pi = \frac{1}{Z'_\Pi}$$
$$= 0.02517\underline{/-79.098°}$$
$$Z_R = 2(0.62 - j442.19) + 3.9353 + j19.94$$
$$= 864.46\underline{/-89.66°}$$
$$Y_R = 1.1568 \times 10^{-3}\underline{/89.66°}$$
$$= 6.9308 \times 10^{-6} + j1.1568 \times 10^{-3}$$

A parallel combination gives

$$Y'_R = Y_R - j9.696 \times 10^{-4}$$
$$= 6.9308 \times 10^{-6} + j1.8718 \times 10^{-4}$$
$$= 1.873 \times 10^{-4}\underline{/87.879°} \text{ siemens}$$

The final reduced network is shown in Fig. 10-27(e). We now have

$$Y_{11} = Y'_R + Y'_\Pi$$
$$= 2.4988 \times 10^{-2}\underline{/-79.000°}$$
$$Y_{12} = -Y'_\Pi$$
$$= 0.02517\underline{/100.902°}$$

We thus have

$$P_e = (760)^2(2.4988 \times 10^{-2})\cos(-79.000°)$$
$$+ (760)^2(0.02517)\cos(100.902 - \delta)$$
$$= 2753.96 + 14538.2\sin(\delta - 10.90) \text{ MW}$$

Problem 10-A-11

Obtain the expression for sending-end electric power for a scheme composed of two parallel lines. Each of the lines is similar to that of Problem 10-A-10. Transmission voltage is 760 kV.

Solution

It is clear that the power transmitted will be double that of the system given in Problem 10-A-10. Figure 10-28 shows the equivalent circuit

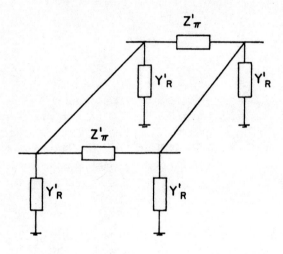

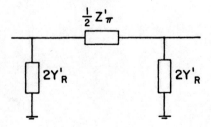

Figure 10-28. Equivalent Circuits for Problem 10-A-11.

of the system. The reduced circuit gives

$$Y_{11} = 2Y'_R + \frac{2}{Z'_\Pi}$$

$$Y_{12} = -\frac{2}{Z'_\Pi}$$

The numerical values are thus

$$Y_{11} = 4.9976 \times 10^{-2} \underline{/-79.000°}$$

$$Y_{12} = 0.05034 \underline{/100.902°}$$

Thus,

$$P_e = (760)^2 (4.9976 \times 10^{-2}) \cos(-79.000°)$$
$$+ (760)^2 (0.05034) \sin(\delta - 10.90°)$$
$$= 5507.9 + 29076.4 \sin(\delta - 10.90°)$$

Problem 10-A-12

A 760-kV, 1200-km, double-circuit transmission line is composed of two 600-km sections in tandem. The line parameters are identical to those of the line of Problem 10-A-9. Series and shunt compensations identical to that employed in Problem 10-A-10 are utilized. The scheme is shown in Figure 10-29. Obtain the expression for the sending-end electric power.

Solution

Each of the two sections is identical to the line in Problem 10-A-11. The equivalent circuit of the system as well as the reduction steps are shown in Figure 10-30. The notation employed is that of Problem 10-A-10.

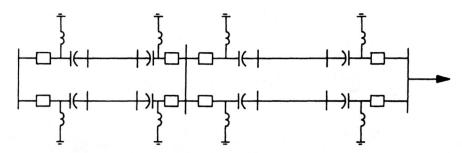

Figure 10-29. Transmission System for Problem 10-A-12.

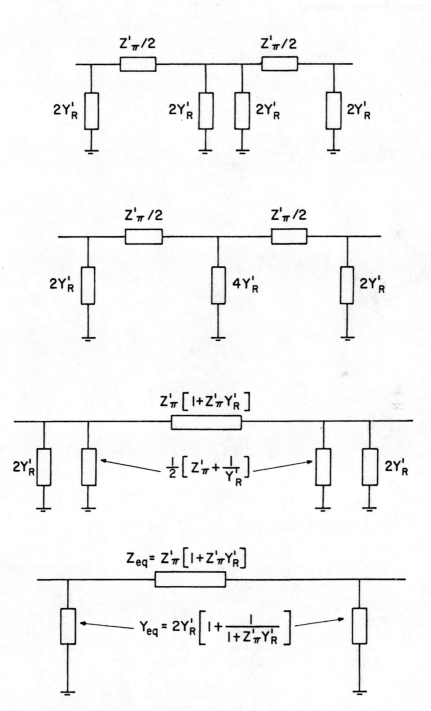

Figure 10-30. Reduction Steps for Problem 10-A-12.

We have

$$Z'_\Pi = 39.733 \underline{/79.098°} \ \Omega$$

$$Y'_R = 1.8731 \times 10^{-4} \underline{/87.879°} \ S$$

$$1 + Z'_\Pi Y'_R = 1 + 7.442 \times 10^{-3} \underline{/166.90°}$$

$$= 0.99275 \underline{/0.0968°}$$

$$Z_{eq} = Z'_\Pi (1 + Z'_\Pi Y'_R)$$

$$= 39.45 \underline{/79.19°}$$

$$Y_{eq} = 2Y'_R \left(1 + \frac{1}{1 + Z'_\Pi Y'_R}\right)$$

$$= 7.52 \times 10^{-4} \underline{/87.83°}$$

Thus,

$$Y_{12} = \frac{-1}{Z_{eq}}$$

$$= 0.02535 \underline{/100.81°}$$

$$Y_{11} = Y_{eq} + \frac{1}{Z_{eq}}$$

$$= 2.462 \times 10^{-2} \underline{/-78.802°}$$

The electric power is then

$$P_e = (760)^2 (2.462 \times 10^{-2}) \cos(78.802°)$$
$$+ (760)^2 (0.02535) \sin(\delta - 10.81°)$$
$$= 2761.66 + 14642.16 \sin(\delta - 10.81°)$$

Problem 10-A-13

A three-phase short circuit to ground takes place in the middle of one of the lines close to the load in the system of Problem 10-A-12. Evaluate the power formula in this case.

Solution

The equivalent circuit and the steps for its reduction are shown in Figure 10-31. Again we retain the same notation:

$$Z_{eq} = 2.5 \left(39.733 \underline{/79.098°}\right) \left[1 + 0.8 \left(39.733 \underline{/79.098°}\right) \right.$$
$$\left. \times \left(1.8731 \times 10^{-4} \underline{/87.879°}\right)\right]$$

$$= 98.756 \underline{/79.175°} \text{ ohms}$$

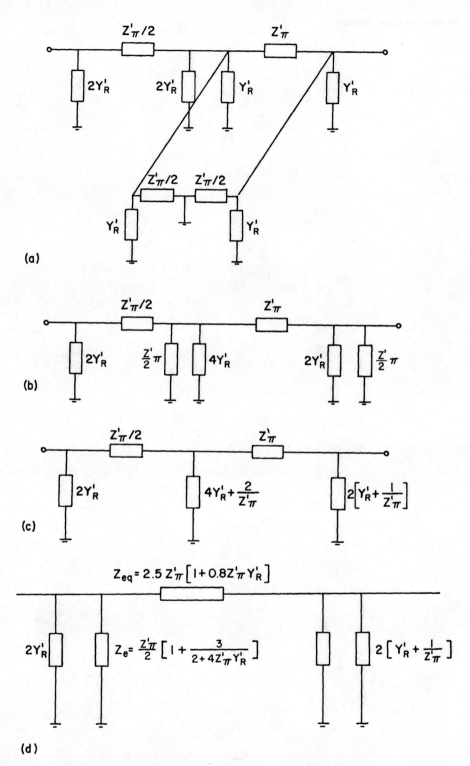

Figure 10-31. Steps in the Reduction of the Circuit for Problem 10-A-13.

$$Y_{12} = \frac{-1}{Z_{eq}}$$
$$= 10.126 \times 10^{-3}/100.83°$$
$$Y_e = \frac{1}{Z_e} = 0.01996/-78.98°$$
$$Y_{11} = 2Y'_R + Y_e + \frac{1}{Z_{eq}}$$
$$= 0.02972/-78.88°$$

We thus have

$$P_e = (760)^2(0.02972)\cos(78.88°)$$
$$+ (760)^2(10.126 \times 10^{-3})\sin(\delta - 10.83°)$$
$$= 3310.0 + 5848.77 \sin(\delta - 10.83°)$$

Problem 10-A-14

The short circuit in the line of Problem 10-A-13 is cleared by removing the affected section. Evaluate the power formula in this case.

Solution

Figure 10-32 shows the equivalent circuit for this case. We have from the previous problems:

$$1 + Z'_{\Pi}Y'_R = 0.99275/0.0968°$$
$$Z_{eq} = 1.5(39.733/79.098°)(0.99275/0.0968°)$$
$$= 59.17/79.195°$$
$$Y_{12} = \frac{-1}{Z_{eq}}$$
$$= 1.69 \times 10^{-2}/100.8°$$
$$Y_{11} = 2(1.8731 \times 10^{-4}/87.879°)\left(1 + \frac{1/-0.0968°}{0.99275}\right)$$
$$+ 1.69 \times 10^{-2}/-79.195°$$
$$= 0.01617/-78.596°$$

The electric power at the sending end is

$$P_e = (760)^2[0.01617 \cos(78.596°)$$
$$+ 1.69 \times 10^{-2}\sin(\delta - 10.8°)]$$

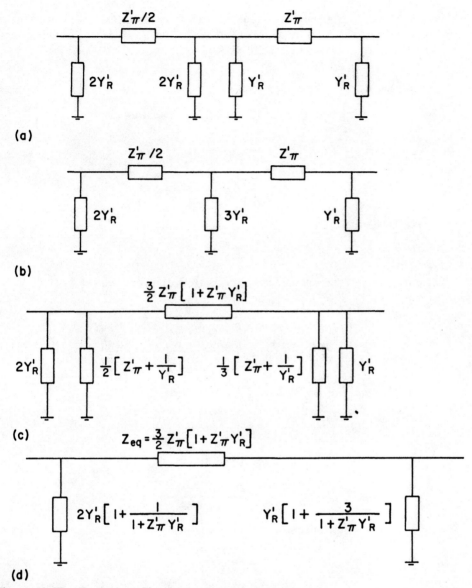

Figure 10-32. Equivalent Circuits for Problem 10-A-14.

This gives

$$P_e = 1846.65 + 9761.44 \sin(\delta - 10.8°) \text{ MW}$$

Problem 10-A-15

Derive an equivalent swing equation for an interconnection of two finite machines with inertia constants M_1 and M_2 and which have angles δ_1

and δ_2. Show that the equations for such a case are exactly equivalent to that of a single finite machine of inertia ($M = M_1 M_2 / M_1 + M_2$) with angle $\delta_{12} = \delta_1 - \delta_2$ connected to an infinite bus.

Solution

The swing equations of the two finite machines are

$$\frac{d^2 \delta_1}{dt^2} = \frac{P_{m_1} - P_{e_1}}{M_1}$$

$$\frac{d^2 \delta_2}{dt^2} = \frac{P_{m_2} - P_{e_2}}{M_2}$$

The relative angle is

$$\delta_{12} = \delta_1 - \delta_2$$

Subtracting the second swing equation from the first, we get

$$\frac{d^2 \delta_1}{dt^2} - \frac{d^2 \delta_2}{dt^2} = \frac{P_{m_1}}{M_1} - \frac{P_{m_2}}{M_2} - \left(\frac{P_{e_1}}{M_1} - \frac{P_{e_2}}{M_2} \right)$$

$$\frac{d^2 \delta_{12}}{dt^2} = \frac{P_{m_1}}{M_1} - \frac{P_{m_2}}{M_2} - \left(\frac{P_{e_1}}{M_1} - \frac{P_{e_2}}{M_2} \right)$$

Multiply each side by $M_1 M_2 / M_1 + M_2 = M$. Hence

$$M \frac{d^2 \delta_{12}}{dt^2} = \left(\frac{M_2 P_{m_1} - M_1 P_{m_2}}{M_1 + M_2} \right) - \left(\frac{M_2 P_{e_1} - M_1 P_{e_2}}{M_1 + M_2} \right)$$

Let us define the equivalent input:

$$P_m = \frac{M_2 P_{m_1} - M_1 P_{m_2}}{M_1 + M_2}$$

and equivalent output:

$$P_e = \frac{M_2 P_{e_1} - M_1 P_{e_2}}{M_1 + M_2}$$

Hence the equivalent swing equation is

$$M \frac{d^2 \delta_{12}}{dt^2} = P_m - P_e$$

And the equivalent inertia constant is

$$M = \frac{M_1 M_2}{M_1 + M_2}$$

Problem 10-A-16

Show that the equivalent power angle curve of two interconnected finite machines is given by

$$P_e = P_c + P_M \sin(\delta - \gamma)$$

where

$$P_c = \frac{M_2 E_1^2 Y_{11} \cos\Theta_{11} - M_1 E_2^2 Y_{22} \cos\Theta_{22}}{M_1 + M_2}$$

$$P_M = \frac{E_1 E_2 Y_{12}\sqrt{M_1^2 + M_2^2 - 2M_1 M_2 \cos 2\Theta_{12}}}{M_1 + M_2}$$

$$\gamma = -\tan^{-1}\left(\frac{M_1 + M_2}{M_1 - M_2} \tan\Theta_{12}\right) - 90°$$

Solution

The equations of a two-machine system are as follows:

$$P_{e1} = E_1^2 Y_{11} \cos\Theta_{11} + E_1 E_2 Y_{12} \cos(\Theta_{12} - \delta_1 + \delta_2)$$
$$P_{e2} = E_2 E_1 Y_{21} \cos(\Theta_{21} - \delta_2 + \delta_1) + E_2^2 Y_{22} \cos\Theta_{22}$$

Substituting these values of P_{e1} and P_{e2} into the expression for equivalent output and let $\delta = \delta_1 - \delta_2$. The result is

$$P_e = \frac{M_2 E_1^2 Y_{11} \cos\Theta_{11} - M_1 E_2^2 Y_{22} \cos\Theta_{22}}{M_1 + M_2}$$
$$+ \frac{E_1 E_2 Y_{12}[M_2 \cos(\delta - \Theta_{12}) - M_1 \cos(\delta + \Theta_{12})]}{M_1 + M_2}$$

The two cosine terms involving δ may be combined into a single cosine term as follows:

$$A = M_2 \cos(\delta - \Theta_{12}) - M_1 \cos(\delta + \Theta_{12})$$
$$= (M_2 - M_1)\cos\delta \cos\Theta_{12} + (M_1 + M_2)\sin\delta \sin\Theta_{12}$$

For simplicity, put

$$a = (M_1 - M_2)\cos\Theta_{12}$$
$$b = (M_1 + M_2)\sin\Theta_{12}$$

Then

$$A = b\sin\delta - a\cos\delta$$

$$A = \sqrt{a^2 + b^2}\left[\left(\frac{b}{\sqrt{a^2 + b^2}}\right)\sin\delta - \left(\frac{a}{\sqrt{a^2 + b^2}}\right)\cos\delta\right]$$

Now put

$$\cos\gamma = \frac{b}{\sqrt{a^2+b^2}}$$

$$\sin\gamma = \frac{a}{\sqrt{a^2+b^2}}$$

Then

$$A = \sqrt{a^2+b^2}\,(\cos\gamma\sin\delta - \sin\gamma\cos\delta)$$
$$= \sqrt{a^2+b^2}\,\sin(\delta-\gamma)$$

We have

$$a^2+b^2 = (M_1-M_2)^2\cos^2\Theta_{12} + (M_1+M_2)^2\sin^2\Theta_{12}$$
$$= M_1^2 + M_2^2 - 2M_1M_2(\cos^2\Theta_{12} - \sin^2\Theta_{12})$$
$$= M_1^2 + M_2^2 - 2M_1M_2\cos(2\Theta_{12})$$

Thus,

$$A = \sqrt{M_1^2 + M_2^2 - 2M_1M_2\cos 2\Theta_{12}}\,\sin(\delta-\gamma)$$

From the above, we have

$$P_e = \frac{M_2 E_1^2 Y_{11}\cos\Theta_{11} - M_1 E_2^2 Y_{22}\cos\Theta_{22}}{M_1+M_2}$$
$$+ \frac{E_1 E_2 Y_{12}\sqrt{M_1^2 + M_2^2 - 2M_1M_2\cos 2\Theta_{12}}}{M_1+M_2}\sin(\delta-\gamma)$$

Let

$$P_c = \frac{M_2 E_1^2 Y_{11}\cos\Theta_{11} - M_1 E_2^2 Y_{22}\cos\Theta_{22}}{M_1+M_2}$$

$$P_M = \frac{E_1 E_2 Y_{12}\sqrt{M_1^2 + M_2^2 - 2M_1M_2\cos(2\Theta_{12})}}{M_1+M_2}$$

Then

$$P_e = P_c + P_M\sin(\delta-\gamma)$$

Problem 10-A-17

High-voltage direct-current transmission systems have been used for long-distance high-power links or where bodies of water or other obstacles preclude the use of overhead alternating current lines. In addition to the main purpose of transporting large blocks of power, dc transmission systems

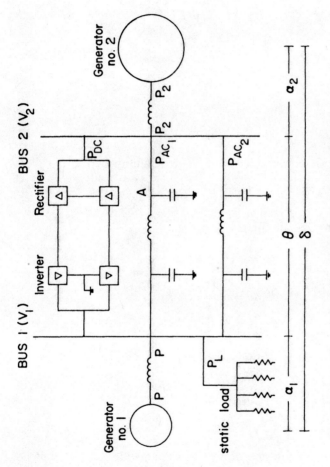

Figure 10-33. Single-Line Diagram of a Parallel ac-dc System.

have some fringe benefits. One of these is the possibility that a properly controlled dc transmission link in parallel with ac links may considerably increase the transient stability limit of the total system. Figure 10-33 shows such a scheme. Derive an equivalent swing equation for this system.

Solution

In the case of two machines, the single equivalent is

$$M\frac{d^2\delta}{dt^2} = P_{m_{eq}} - P_{e_{eq}}$$

With

$$\delta = \delta_2 - \delta_1$$

we have

$$P_{e_{eq}} = \frac{M_1 P_{e_2} - M_2 P_{e_1}}{M_1 + M_2}$$

The electric power delivered by machine number 1 is

$$P_{e_1} = P_L - P_{DC} - \frac{|V_1||V_2|}{X_{eq}}\sin\Theta$$

and by machine number 2 is

$$P_{e_2} = P_{DC} + \frac{|V_1||V_2|}{X_{eq}}\sin\Theta$$

As a result,

$$-P_{e_{eq}} = \frac{M_2}{M_1 + M_2}(P_L) - P_{DC} - \frac{|V_1||V_2|}{X_{eq}}\sin\Theta$$

Thus the required swing equation is

$$M\frac{d^2\delta}{dt^2} = P_{m_{eq}} + \frac{M_2}{M_1 + M_2}(P_L) - P_{DC} - \frac{|V_1||V_2|}{X_{eq}}\sin(\delta - \alpha_2 + \alpha_1)$$

Problem 10-A-18

Consider the case of an electric machine connected to an infinite bus through a reactive electric network such that the magnitude of the power angle curve is unity. A change in the network results in a new power angle curve with magnitude x. Suppose the machine is delivering a power p before the change occurs. Show that the maximum value of p such that the system remains stable satisfies

$$\frac{p}{x}\left[-\sin^{-1}\left(\frac{p}{x}\right) - \sin^{-1}p + \pi\right] = \sqrt{1 - \left(\frac{p}{x}\right)^2} + \sqrt{1 - p^2}$$

Verify that for $x = 0.5$, the maximum value of prefault power p is approximately 0.4245.

Solution

With reference to Figure 10-34, for a critically stable system the shaded areas A_1 and A_2 should be equal. From the geometry of the problem,

$$p = x \sin \delta_1$$

$$p = \sin \delta_0$$

The area A_1 is given by

$$A_1 = p(\delta_1 - \delta_0) - \int_{\delta_0}^{\delta_1} x \sin \delta \, (d\delta)$$

This reduces to

$$A_1 = p\left[\sin^{-1}\left(\frac{p}{x}\right) - \sin^{-1} p\right] + x(\cos \delta_1 - \cos \delta_0)$$

$$= p\left[\sin^{-1}\left(\frac{p}{x}\right) - \sin^{-1} p\right] + x\left[\sqrt{1 - \left(\frac{p}{x}\right)^2} - \sqrt{1 - p^2}\right]$$

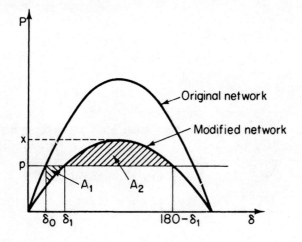

Figure 10-34. Swing Curves for Problem 10-A-18.

The area A_2 is

$$A_2 = \int_{\delta_1}^{\pi-\delta_1} x \sin\delta(d\delta) - \pi p + 2\delta_1 p$$

This reduces to

$$A_2 = 2x\cos\delta_1 - \pi p + 2\delta_1 p$$

$$= 2x\sqrt{1-\left(\frac{p}{x}\right)^2} - \pi p + 2p\sin^{-1}\frac{p}{x}$$

Equating A_1 to A_2 yields

$$p\left[-\sin^{-1}\left(\frac{p}{x}\right) - \sin^{-1}p\right] + x\left[\sqrt{1-\left(\frac{p}{x}\right)^2} - \sqrt{1-p^2}\right]$$

$$= 2x\sqrt{1-\left(\frac{p}{x}\right)^2} - \pi p$$

Rearranging, we get the required relation:

$$\frac{p}{x}\left[-\sin^{-1}\left(\frac{p}{x}\right) - \sin^{-1}p + \pi\right] = \sqrt{1-\left(\frac{p}{x}\right)^2} + \sqrt{1-p^2}$$

The left-hand side of the derived expression for $x = 0.5$ and $p = 0.4245$ is

$$\frac{0.4245}{0.5}\left(-\sin^{-1}\frac{0.4245}{0.5} - \sin^{-1}0.4245 + 180\right)\frac{\pi}{180} = 1.434$$

The right-hand side is

$$\left[1-\left(\frac{0.4245}{0.5}\right)^2\right]^{1/2} + \sqrt{1-(0.4245)^2} = 1.434$$

For this accuracy the two sides are identical; hence $p = 0.4245$ is the maximum initial power for stability with $x = 0.5$.

Problem 10-A-19

A generator is delivering 0.6 of P_{max} to an infinite bus through a transmission line. A fault occurs such that the reactance between the generator and the bus is increased to three times its prefault value. When the fault is cleared, the maximum power that can be delivered is 0.80 of the original maximum value. Determine the critical clearing angle using the equal-area criterion.

Solution

The initial angle δ_0 is obtained from

$$0.6 P_{max} = P_{max} \sin\delta_0$$

Thus

$$\delta_0 = 36.87°$$

The angle δ_m on the postfault curve is obtained from

$$0.80 P_{max} \sin \delta_m = 0.6 P_{max}$$

Thus

$$\delta_m = 131.41 \text{ or } 2.294 \text{ rad}$$

$$A_1 = P_{max}\left[0.6(\delta_c - \delta_0) - 0.33 \int_{\delta_0}^{\delta_c} \sin \delta (d\delta)\right]$$

$$= P_{max}(0.6\delta_c + 0.33 \cos \delta_c - 0.65)$$

$$A_2 = P_{max}\left[0.8 \int_{\delta_c}^{\delta_m} \sin \delta (d\delta) - 0.6(2.294 - \delta_c)\right]$$

$$= P_{max}(0.8 \cos \delta_c + 0.6 \delta_c - 0.847)$$

With reference to Figure 10-35, the following areas are obtained:

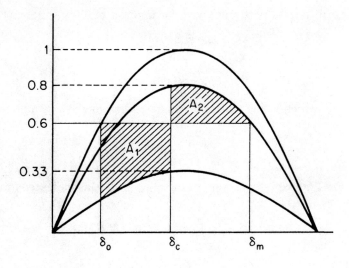

Figure 10-35. Power Angle Curve for Problem 10-A-19.

For the critical angle,
$$A_1 = A_2$$
We thus get
$$\delta_c = 65.05°$$

Problem 10-A-20

A generator is delivering 0.5 of P_{max} to an infinite bus through a transmission line. A fault occurs such that the new maximum power is 0.3 of the original. When the fault is cleared, the maximum power that can be delivered is 0.8 of the original maximum value.

A. Determine the critical clearing angle.
B. If the fault is cleared at $\delta = 75°$, find the maximum value of δ for which the machine swings around its new equilibrium position.

Solution

A. We start by finding the angle δ_0 that satisfies
$$0.5 = (1)\sin \delta_0$$
Thus,
$$\delta_0 = 30°$$
The angle δ_A is on the postfault curve for a power of 0.5 P_{max}; thus,
$$0.5 = 0.8 \sin \delta_A$$
Therefore,
$$\delta_A = 38.68°$$
The angle δ_m, which is the maximum angle permissible, is a complement of δ_A. Thus,
$$\delta_m = 180° - 38.68°$$
$$= 141.32°$$
We now find expressions for the areas A_1 and A_2 as indicated in Figure 10-36. We have
$$A_1 = 0.5(\delta_c - \delta_0) - \int_{\delta_0}^{\delta_c} 0.3 \sin \delta (d\delta)$$
Also
$$A_2 = \int_{\delta_c}^{\delta_m} 0.8 \sin \delta (d\delta) - 0.5(\delta_m - \delta_c)$$

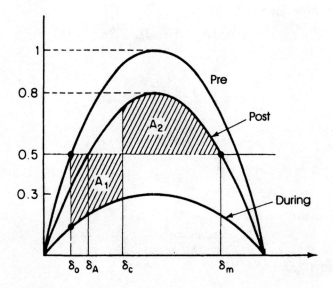

Figure 10-36. Swing Curves for Problem 10-A-20.

For critically stable operation, we have the condition $A_1 = A_2$. This gives

$$0.5\delta_c - 0.5\delta_0 - \int_{\delta_0}^{\delta_c} 0.3 \sin \delta (d\delta) = \int_{\delta_c}^{\delta_m} 0.8 \sin \delta \, d\delta - 0.5(\delta_m - \delta_c)$$

which reduces to

$$0.5(141.32 - 30)\left(\frac{\pi}{180}\right) = 0.3(\cos \delta_0 - \cos \delta_m) + 0.5(\cos \delta_c - \cos \delta_m)$$

Using the numerical values for the given angles, we have

$$\frac{\pi}{360}(111.32) = 0.3 \cos 30 - 0.8 \cos(141.32°) + 0.5 \cos \delta_c$$

which gives the desired critical clearing angle:

$$\delta_{c_c} = 79.97°$$

B. For fault clearing at 75°, we have

$$\delta_{c_1} = 75°$$

The area A_1 is given by
$$A_1 = 0.5(\delta_{c_1} - \delta_0) - 0.3 \int_{\delta_0}^{\delta_{c_1}} \sin\delta(d\delta)$$

This gives
$$A_1 = 0.5(75 - 30)\left(\frac{\pi}{180}\right) - 0.3(\cos 30° - \cos 75°)$$

Hence,
$$A_1 = 0.21054$$

The area A_2 is given by
$$A_2 = 0.8(\cos 75° - \cos \delta_{m_1}) - 0.5(\delta_{m_1} - \delta_{c_1})$$

This reduces to
$$A_2 = 0.862 - 0.8\cos\delta_{m_1} - 0.5\delta_{m_1}$$

For $A_1 = A_2$, we obtain
$$0.8\cos\delta_{m_1} + 0.5\delta_{m_1} - 0.651 = 0$$

We solve iteratively to get the required angle δ_{m_1} given by
$$\delta_{m_1} \cong 118.89°$$

Problem 10-A-21

The power angle curves for a single machine against an infinite bus system is
$$P = 2.8\sin\delta$$

Under fault conditions, the curve is described by
$$P = 1.2\sin\delta$$

Assume that the system is delivering a power of 1 pu prior to the fault and that fault clearing results in the system returning to the prefault conditions. If the fault is cleared at $\delta_c = 60°$, would the system be stable? Find the maximum angle of swing δ_s if the system is stable.

Solution

The angle δ_0 is given by
$$1 = 2.8\sin\delta_0$$

Thus,
$$\delta_0 = 20.925°$$

Also, we have

$$1 = 1.2 \sin \delta_1$$

Therefore,

$$\delta_1 = 56.443°$$

The area A_1 is thus

$$A_1 = (56.443 - 20.925)\frac{\pi}{180} - \int_{\delta_0}^{\delta_1} 1.2 \sin \delta (d\delta)$$

$$= 0.16237$$

The area A_{2_f} is made of two parts as shown in Figure 10-37. The first is the area between the fault curve and the input line; this extends from δ_1 to δ_c. The second extends from δ_c to δ_f, enclosed by the prefault curve and the input power line. Thus,

$$\delta_f = 180 - \delta_0$$
$$= 159.08°$$

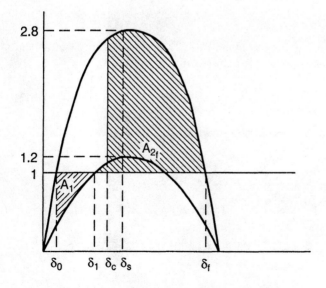

Figure 10-37. Swing Curves for Problem 10-A-21.

Therefore we have

$$A_{2_f} = \int_{\delta_1}^{\delta_c} 1.2 \sin \delta (d\delta) + \int_{\delta_c}^{\delta_f} 2.8 \sin \delta (d\delta) - (\delta_f - \delta_1) \frac{\pi}{180}$$

$$= 1.2(\cos \delta_1 - \cos \delta_c) + 2.8(\cos \delta_c - \cos \delta_f) - (\delta_f - \delta_1)\frac{\pi}{180}$$

$$= 2.2886$$

Clearly the system is stable.

To find δ_s, we have to check the area between δ_1 and δ_c. Thus,

$$A_{2_c} = \int_{\delta_1}^{\delta_c} 1.2 \sin \delta (d\delta) - (\delta_c - \delta_1)\frac{\pi}{180}$$

$$= 0.0012$$

Clearly $A_{2_c} < A_1$; hence the system will continue swinging beyond δ_c to δ_s, at which angle $A_1 = A_2$. Thus,

$$A_2 = A_{2_c} + \int_{\delta_c}^{\delta_s} 2.8 \sin \delta (d\delta) - (\delta_s - \delta_c)\frac{\pi}{180}$$

$$0.16237 = 0.0012 + 2.8(\cos \delta_c - \cos \delta_s) - \frac{\pi}{180}(\delta_s - \delta_c)$$

And,

$$2.2861 - 2.8 \cos \delta_s - 0.0175 \delta_s = 0$$

Solving, we get $\delta_s = 66°$.

Problem 10-A-22

If the fault of Problem 10-A-21 is not cleared, would the system be stable? If so, what is the angle of maximum swing?

Solution

The area A_1 from the previous problem is

$$A_1 = 0.16237$$

The area A_{2_f} is as shown in Figure 10-38.

$$A_{2_f} = 2\int_{\delta_1}^{90°} 1.2 \sin \delta (d\delta) - 2(90° - \delta_1)\frac{\pi}{180}$$

$$= 2.4 \cos \delta_1 - (90 - \delta_1)\frac{\pi}{90}$$

With

$$\delta_1 = 56.443°$$

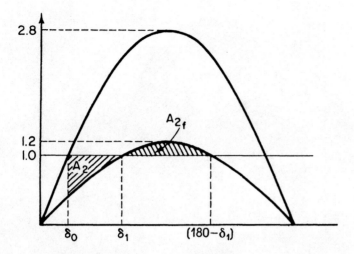

Figure 10-38. Swing Curves for Problem 10-A-22.

we get

$$A_{2_f} = 0.155$$

Since

$$A_{2_f} < A_1$$

therefore the system is unstable.

Problem 10-A-23

Find the critical clearing angle for the conditions of Problem 10-A-21.

Solution

We have from Problem 10-A-21,

$$A_1 = 0.16237$$

For critical stability,

$$A_1 = A_{2_f}$$

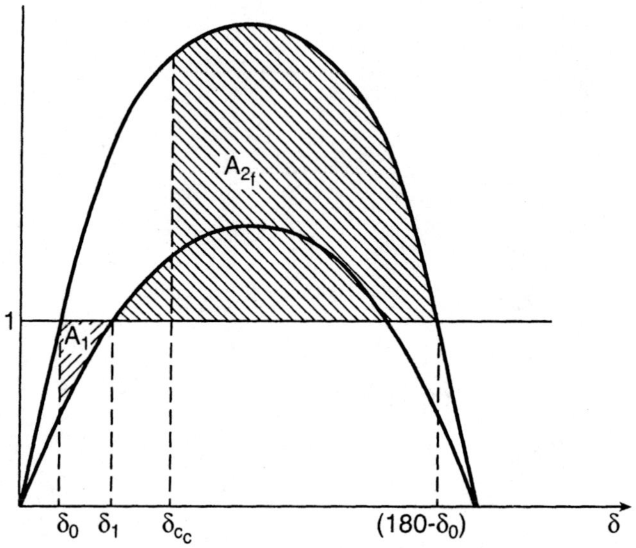

Figure 10-39. Swing Curves for Problem 10-A-23.

Thus with reference to Figure 10-39, we have
$$0.16237 = \int_{\delta_1}^{\delta_{cc}} 1.2 \sin\delta \, (d\delta) + \int_{\delta_{cc}}^{(180-\delta_0)} 2.8 \sin\delta \, (d\delta)$$
$$- (180 - \delta_0 - \delta_1)\frac{\pi}{180}$$

Thus,
$$1.3251 + 1.6 \cos \delta_{cc} = 0$$

As a result
$$\delta_{cc} = 145.913°$$

PROBLEMS

Problem 10-B-1

The speed of the rotor of a 60-Hz, 100-MVA generator subject to a constant decelerating power of 1 p.u. is reduced from rated value to zero in 12 seconds. If the net accelerating power is 20 MW, find the resulting acceleration.

Problem 10-B-2

The kinetic energy stored in the rotor of a 60-Hz, 20-MVA generator at synchronous speed is 100 megajoules. Assume that the rotor's acceleration is 400 elec. degree/s^2 when the machine is delivering rated MVA at 0.85 PF. Find the net mechanical hp input for this condition.

Problem 10-B-3

A four-pole, 60-Hz, 15-MVA generator develops an accelerating torque of 25,000 newton meters with a net mechanical power input of 16 MW while developing rated MVA. Calculate the unit's power factor.

Problem 10-B-4

A 60-Hz, three-phase transmission line has a total series impedance of $160 \underline{/75°}$ and a shunt admittance of $10^{-3} \underline{/90°}$. Determine the maximum receiving-end power transfer if the voltage is maintained at 220 kV at both ends of the line. Use:

A. The short line approximation.

B. The nominal π approximation.

Problem 10-B-5

Using the short line approximation only, determine whether or not a generated power of 400 MW can be transmitted over the line of Problem 10-B-4.

Problem 10-B-6

Find the minimum value of the voltage necessary to transmit 420 MW over the line of Problem 10-B-4. Assume equal sending-end and receiving-end voltages. Use the short line approximation.

Problem 10-B-7

Compare the maximum power transfer capacities of the two systems shown in Figure 10-40.

Problem 10-B-8

Compare the maximum power transfer capacities for the systems of Problem 10-A-7 for

$$X_L = 1.0 \text{ p.u.}$$
$$X_0 = 0.1 \text{ p.u.}$$
$$X_s = 0.2 \text{ p.u.}$$

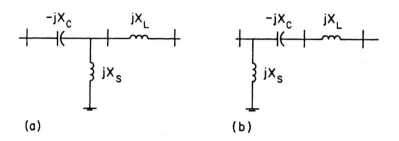

Figure 10-40. Systems for Problem 10-B-7.

Problem 10-B-9

Verify the results of Problem 10-A-9 using the $ABCD$ parameter method.

Problem 10-B-10

Repeat Problem 10-A-10 for
$$K_s = 0.8$$
$$K_d = 1.0$$

Problem 10-B-11

A 760-kV transmission line is modeled using an all-reactive, equivalent π circuit with series reactance $X_l = 160\ \Omega$, and shunt susceptance $B_l/2 = 1.25 \times 10^{-3}$ S. Series and shunt compensations of degrees K_s and K_d respectively are used as shown in Figure 10-41.

A. Obtain the sending-end electric power expression in terms of K_d and K_s. What effect does the degree of compensation have on power transfer for this all-reactive line?

B. If series compensation with $K_s = 0.5$ is used, calculate the steady-state angle δ for a load power of 3000 MW.

C. Find the degree of series compensation K_s required so that a power of 3000 MW is transmitted with an angle $\delta = 15°$.

Problem 10-B-12

A transmission line is composed of two identical sections in tandem. Each section consists of two identical parallel circuits with particulars similar to the line of Problem 10-B-11.

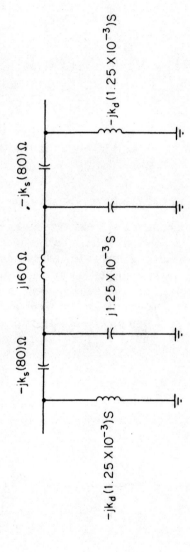

Figure 10-41. System for Problem 10-B-11.

A. Find the expression for the electric power in terms of K_s and K_d. Does K_d have any effect on the power transmission capability?
B. Find the maximum power with $K_s = 0.5$ for
  (i) $K_d = 0$.
  (ii) $K_d = 1$.

Problem 10-B-13

Repeat Problem 10-B-12 with a three-phase short circuit in the middle of one line on the load side.

Problem 10-B-14

Repeat Problem 10-B-13 with the fault cleared by removing the shorted line.

Problem 10-B-15

The system shown in Figure 10-42 has two finite synchronous machines, each represented by a constant voltage behind reactance, connected by a pure reactance. The reactance X includes the transmission line and the machine reactances. Write the swing equation for each machine and show that this system can be reduced to a single equivalent machine against an infinite bus. Find the inertia constant for the equivalent machine, the mechanical input power, and the amplitude of its power angle curve. The inertia constants of the two machines are H_1 and H_2 seconds.

Problem 10-B-16

Utilize the results of Problems 10-A-15 and 10-A-16 to reduce the two machine systems whose particulars are given below to a single equivalent

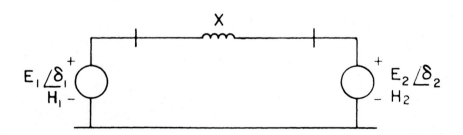

Figure 10-42. Circuit for Problem 10-B-15.

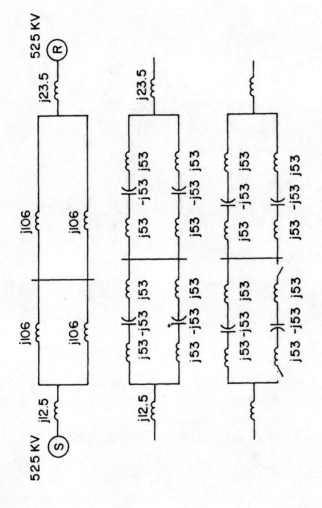

Figure 10-43. Network for Problem 10-B-17.
(a) Circuit with no compensation
(b) Compensated circuit
(c) Circuit with one section removed

machine against an infinite bus:

$$M_1 = 2.92 \times 10^{-4}$$
$$M_2 = 26.9 \times 10^{-4}$$
$$E_1 = 1.17$$
$$E_2 = 1.01$$
$$\delta_1 = 23.0$$
$$\delta_2 = 10.3°$$
$$P_{m_1} = 0.8$$
$$P_{m_2} = 3.2$$
$$Y_{11} = 1.84 \underline{/-90°}$$
$$Y_{22} = 13.5 \underline{/-88.1°}$$
$$Y_{12} = 0.172 \underline{/86.4°}$$

All quantities are in per unit.

Problem 10-B-17

Figure 10-43(a) shows a simplified diagram of a 360-mile, two-circuit, two-section, 525-kV ac transmission link interconnecting two regional power systems, each of which is represented by its Thévenin's equivalent circuit (EMF and impedance in series.)

A. Find the amplitude of the power angle curve for the circuit as indicated.

B. Series compensation of 50 percent is provided by capacitors in the individual line circuits as shown in Figure 10-43(b). Find the amplitude of the power angle curve for the circuit with the capacitors in.

C. A section of one circuit is removed for fault clearing as shown in Figure 10-43(c). Find the amplitude of the power angle curve.

D. Show that the transient stability limit for the circuit under the fault condition in part (c) is 2.215 Giga Watts.

Problem 10-B-18

The Gordon M. Shrum (G.M.S.) generating station is situated in the north central part of B.C. on the Peace River. Its power output is transmitted approximately 500 miles through two 500-kV transmission lines to the load centers in southern B.C. A schematic of the 500-kV system and its series capacitor banks is shown in Figure 10-44.

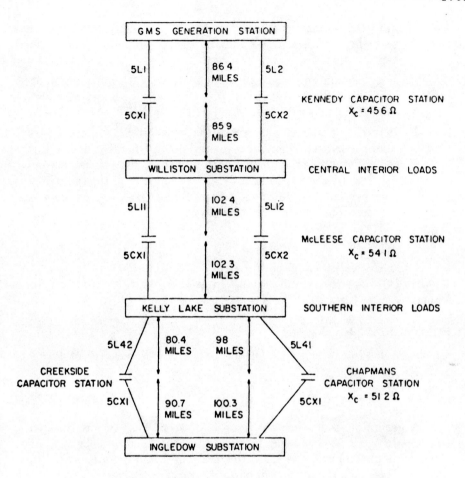

Figure 10-44. Schematic diagram for Problem 10-B-18.

A. Draw a labeled equivalent circuit of the system as given. (Do not reduce the circuit in so doing.)

B. Find the equivalent impedance Z_A between the G.M.S. generating station and the Ingledow substation prior to installing the capacitor banks at The Kennedy, McLeese, Chapmans, and Creekside stations.

C. Find the equivalent impedance Z_B in part (b) with capacitor banks installed. Neglecting resistances, find the ratio X_A/X_B.

D. Find the maximum power transfer P_{max} in each case given that both the G.M.S. station's and the Ingledow station's bus bars are kept at 500 kV.

E. If the maximum permissible design value for the angle difference δ is 30°, find the corresponding power transfer with and without the capacitor banks.

F. It turns out that an alternative that is $40 million more expensive than the capacitor option is a third 500-kV parallel line from the G.M.S. station to Ingledow with the three lines uncompensated. Assume the proposed line is 575.3 miles long with the same impedance per mile as the existing lines. Repeat part (e). As a systems planner would you choose this alternative? Why?

Problem 10-B-19

For the conditions of part (b) in Problem 10-B-11, a fault at the middle of the line occurs and is removed at angle $\delta = 70°$. Will the system be stable under these conditions?

Problem 10-B-20

Repeat Problem 10-B-19 for the conditions of part (c) of Problem 10-B-11.

Problem 10-B-21

A generator is delivering 0.55 of P_{max} to an infinite bus through a transmission line. A fault occurs such that the reactance between the generator and the bus is increased to three times its prefault value. When the fault is cleared, the maximum power that can be delivered is 0.75 of the original maximum value. Determine the critical clearing angle using the equal-area criterion.

Problem 10-B-22

Repeat Problem 10-A-21, with the power angle curve given by
$$P = 2.7 \sin \delta$$
All other information remains unchanged.

Problem 10-B-23

Repeat Problem 10-A-22 for the system described in Problem 10-B-22.

CHAPTER XI

Optimal Operation of Electric Power Systems

11.1 INTRODUCTION

A main objective in the operation of any of today's complex electric power systems is to meet the demand for power at the lowest possible cost, while maintaining safe, clean standards of environmental impact. Reliability and continuity of service are essential goals that the electric power systems engineer strives to meet at all times. Ordering these objectives and priorities is a very difficult task to perform since these generally change with the times and socioeconomic and political considerations.

It is our intent in this chapter to discuss a few relatively simple problems in the optimal economic operation of systems. The problems can be considered part of production scheduling activities. These activities are concerned with economic hourly scheduling of the available energy resources so that the lowest total production cost is achieved while meeting system loads within other system constraints. The activities require the availability of information regarding a forecast of load and resource availability, interchange possibilities, and current system limitations as well as energy resource limitations. We will begin by treating the case of systems with only thermal resources, followed by the case involving hydro and thermal resources in the system.

As a basic requirement we discuss certain aspects of modeling pertaining to thermal resources. It is emphasized that various models can be used to represent the same physical device or group of devices, depending on the purpose of the analysis. Our treatment is mainly concerned with models in general use for economic operation purposes.

11.2 MODELING OF FUEL COSTS FOR THERMAL GENERATION

Our purpose in this section is to briefly outline models for thermal generation for economic operational purposes. This is essentially a discussion of modeling the fuel cost variations with the active power generation for thermal generating plants.

Electric power is generated as a result of mechanical rotational energy produced by either steam or combustion turbines. Steam produced in the boiler or nuclear reactor is the medium of heat energy transfer to the turbines. Combustion turbines burn liquid or gaseous fuels, mostly light distillate oil or natural gas. No intermediate steps are needed in the latter case.

In fossil fuel-fired steam units, fuel is burnt and energy is released in the form of heat in the boiler, producing high temperature and pressure steam. The steam is led via the drum to the turbines where part of the thermal energy is transformed into mechanical form. The steam turbine drives the electric generator (alternator). The exhaust of the turbine is cooled in the condenser, and the resulting water is pumped back to the boiler. For the purpose of economic operation studies, our interest is in an input-output type of model. The input in this case is the fuel cost, and the output is the active power generation of the unit.

The input to the thermal plant is generally measured in MJ/h or in SI units (traditionally MBtu/h or kcal/h), and the output is measured in megawatts (MW). Although initially prepared on the basis of input versus main unit output, the input-output relationship must be converted to input versus net unit sendout. The total cost of operation includes the fuel cost and the cost of labor, supplies, and maintenance. The most common method to express these is to assume the cost of labor, supplies, and maintenance to be a fixed percentage of the incoming fuel costs.

11.2 Modeling of Fuel Costs for Thermal Generation

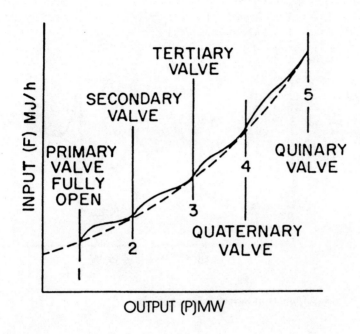

Figure 11-1. Typical Input-Output Curve for a Thermal Unit.

Figure 11-1 shows a typical thermal input-output curve $(F-P)$. The incremental heat rate characteristic is simply a plot of $\partial F/\partial P$ versus P, and a typical curve is shown in Figure 11-2. Heat rate units are MJ/kWh.

Typical heat rate data for sample unit sizes for steam units using coal, oil, and gas as primary sources of energy are given in Table 11-1.

Loading (output) levels at which a new steam admission valve is opened are called *valve points*. At these levels, discontinuities in the cost curves and in the incremental heat rate curves occur as a result of the sharp increases in throttle losses. As the valve is gradually lifted, the losses decrease until the valve is completely open. The shape of the input-output curve in the neighborhood of the valve points is difficult to determine by actual testing. Most utility systems find it satisfactory to represent the input-output characteristic by a smooth curve that can be defined by a polynomial.

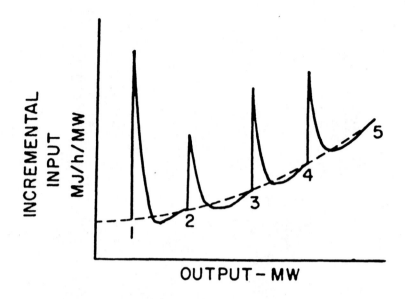

Figure 11-2. Typical Incremental Heat Curve.

TABLE 11-1
Typical Fossil Generation Unit Net Heat Rates

Fossil Fuel	Unit Rating	100% Output MJ/kWh	80% Output MJ/kWh	60% Output MJ/kWh	40% Output MJ/kWh	25% Output MJ/kWh
Coal	50	11.59	11.69	12.05	12.82	14.13
Oil	50	12.12	12.22	12.59	13.41	14.78
Gas	50	12.33	12.43	12.81	13.64	15.03
Coal	200	10.01	10.09	10.41	11.07	12.21
Oil	200	10.43	10.52	10.84	11.54	12.72
Gas	200	10.59	10.68	11.01	11.72	12.91
Coal	400	9.49	9.53	9.75	10.31	11.25
Oil	400	9.91	9.96	10.18	10.77	11.75
Gas	400	10.01	10.06	10.29	10.88	11.88
Coal	600	9.38	9.47	9.77	10.37	11.40
Oil	600	9.80	9.90	10.20	10.84	11.91
Gas	600	9.91	10.01	10.31	10.96	12.04
Coal	800/1200	9.22	9.28	9.54	10.14	
Oil	800/1200	9.59	9.65	9.92	10.55	
Gas	800/1200	9.70	9.75	10.03	10.67	

Note: For conversion: 1 Btu = 1054 joules.

11.2 Modeling of Fuel Costs for Thermal Generation

For economy operation problems treated here, the fuel cost curve is modeled as a quadratic in the active power generation. This we express as

$$F(P) = \alpha + \beta P + \gamma P^2 \qquad (11.1)$$

The determination of the parameters α, β, and γ requires the availability of data relating the cost $F(P_i)$ to the generation level P_i. We may then use a simple least-square estimation algorithm to do so.

Given n points where the cost $F(P_i)$ and the power P_i are known, the parameters are determined such that a least square error is involved. The problem is then to minimize J with respect to α, β, and γ, where J is given by

$$J = \sum_{i=1}^{n} \left[\alpha + \beta P_i + \gamma P_i^2 - F(P_i) \right]^2$$

The solution is obtained by setting the derivatives of J with respect to α, β, and γ to zero. The resulting relations are as follows:

$$\frac{\partial J}{\partial \alpha} = \sum_{i=1}^{n} 2\left[\alpha + \beta P_i + \gamma P_i^2 - F(P_i) \right] = 0$$

$$\frac{\partial J}{\partial \beta} = \sum_{i=1}^{n} 2 P_i \left[\alpha + \beta P_i + \gamma P_i^2 - F(P_i) \right] = 0$$

$$\frac{\partial J}{\partial \gamma} = \sum_{i=1}^{n} 2 P_i^2 \left[\alpha + \beta P_i + \gamma P_i^2 - F(P_i) \right] = 0$$

Rearranging, we have

$$(n)\alpha + \left(\sum_{i=1}^{n} P_i \right) \beta + \left(\sum_{i=1}^{n} P_i^2 \right) \gamma = \sum_{i=1}^{n} F(P_i) \qquad (11.2)$$

$$\left(\sum_{i=1}^{n} P_i \right) \alpha + \left(\sum_{i=1}^{n} P_i^2 \right) \beta + \left(\sum_{i=1}^{n} P_i^3 \right) \gamma = \sum_{i=1}^{n} P_i F(P_i) \qquad (11.3)$$

$$\left(\sum_{i=1}^{n} P_i^2 \right) \alpha + \left(\sum_{i=1}^{n} P_i^3 \right) \beta + \left(\sum_{i=1}^{n} P_i^4 \right) \gamma = \sum_{i=1}^{n} P_i^2 F(P_i) \qquad (11.4)$$

Solving the above linear set of equations in α, β, and γ yields the desired estimates. The following example illustrates the procedure.

Example 11-1

The data for the expected heat rate curve for a unit in a thermal station are shown below:

MW	70	75	112.5	150
Btu/kWh	8200	8150	7965	7955

A. Obtain the corresponding points on the input-output curve (input in Btu/h).

B. Obtain the parameters α, β and γ of the cost equation, Eq. (11.1).

Solution

A. We obtain the inputs $F(P_i)$ for various loadings P_i from the table by multiplying the heat rate value by the power output. Therefore, for $P_1 = 70$ MW, we have

$$F_1 = 8200 \times 70 \times 10^3 = 574 \times 10^6 \text{ Btu/h}$$

The other three points are similarly obtained:

For $P_2 = 75$ MW, $\quad F_2 = 611 \times 10^6$ Btu/h

For $P_3 = 112.5$ MW, $\quad F_3 = 896 \times 10^6$ Btu/h

For $P_4 = 150$ MW, $\quad F_4 = 1190 \times 10^6$ Btu/h

B. The following quantities are calculated

$$n = 4$$
$$\Sigma P_i = 407.50 \text{ MW}$$
$$\Sigma P_i^2 = 45.68125 \times 10^3$$
$$\Sigma P_i^3 = 5.5637 \times 10^6$$
$$\Sigma P_i^4 = 7.22 \times 10^8$$
$$\Sigma F_i = 3.271 \times 10^3$$
$$\Sigma P_i F_i = 3.65305 \times 10^5$$
$$\Sigma P_i^2 F_i = 4.43645 \times 10^7$$

We thus have to solve

$$\begin{bmatrix} 4 & (407.5) & (45.68125 \times 10^3) \\ (407.5) & (45.68125 \times 10^3) & (5.5637 \times 10^6) \\ (45.68125 \times 10^3) & (5.5637 \times 10^6) & (7.22 \times 10^8) \end{bmatrix} \begin{bmatrix} \alpha \\ \beta \\ \gamma \end{bmatrix} = \begin{bmatrix} 3.271 \times 10^3 \\ 3.65305 \times 10^5 \\ 4.43645 \times 10^7 \end{bmatrix}$$

for α, β, and γ. The solution is

$$\alpha = 69.23$$
$$\beta = 6.98$$
$$\gamma = 3.2828 \times 10^{-3}$$

Therefore the fuel cost expression required is given by

$$F(P) = 69.23 + 6.98P + 3.2828 \times 10^{-3} P^2 \text{ MBtu/h}$$

A Per Unit System

We observe that the elements of the relations leading to finding α, β, and γ are different by orders of magnitude. Certain numerical difficulties may arise due to this ill-conditioning. We speculate that normalizing the variables (which leads to the per unit system) can help in this regard.

As before, we need a base quantity for the power that we choose as the unit rating. Thus,

$$P_{\text{base}} = 150 \ MW$$

We also need a base for the fuel cost. Let us take this as the cost corresponding to the unit rating. Therefore,

$$F_{\text{base}} = 1190 \ \text{MBtu/h}$$

Consequently, we have in p.u. values,

$$P_1 = 0.4667 \qquad F_1 = 0.4824$$
$$P_2 = 0.5 \qquad F_2 = 0.5134$$
$$P_3 = 0.75 \qquad F_3 = 0.753$$
$$P_4 = 1.00 \qquad F_4 = 1.00$$

We therefore have

$$n = 4$$
$$\Sigma P_i = 2.7167$$
$$\Sigma P_i^2 = 2.0303$$
$$\Sigma P_i^3 = 1.6485$$
$$\Sigma P_i^4 = 1.4263$$
$$\Sigma F_i = 2.7488$$
$$\Sigma P_i F_i = 2.0466$$
$$\Sigma P_i^2 F_i = 1.657$$

Thus we have to solve

$$\begin{bmatrix} 4 & 2.7167 & 2.0303 \\ 2.7167 & 2.0303 & 1.6485 \\ 2.0303 & 1.6485 & 1.4263 \end{bmatrix} \begin{bmatrix} \alpha \\ \beta \\ \gamma \end{bmatrix} = \begin{bmatrix} 2.7488 \\ 2.0466 \\ 1.657 \end{bmatrix}$$

Clearly a marked improvement in the orders of magnitude is observed. The solution is

$$\alpha = 0.058$$
$$\beta = 0.880675$$
$$\gamma = 6.14534 \times 10^{-2}$$

The cost expression is therefore

$$F = 0.058 + 0.880675 P_{pu} + 6.14534 \times 10^{-2} P_{pu}^2$$

where both F and P are in p.u. values. To transform back to MW and MBtu/h, we have

$$F = 1190\left[0.058 + 0.880675\left(\frac{P_{MW}}{150}\right) + 6.14534 \times 10^{-2}\left(\frac{P_{MW}}{150}\right)^2\right]$$

or

$$F = 69.02 + 6.987P + 3.2502 \times 10^{-2}P^2$$

11.3 OPTIMAL OPERATION OF AN ALL-THERMAL SYSTEM: EQUAL INCREMENTAL COST LOADING

A simple yet extremely useful problem in optimum economic operation is treated here. Consider the operation of m thermal generating units on the same bus as shown in Figure 11-3. Assume that the variation of the fuel cost of each generator (F_i) with the active power output (P_i) is given by a quadratic polynomial. The total fuel cost of the plant is the sum of the individual unit cost converted to \$/h:

$$F = \sum_{i=1}^{m} \alpha_i + \beta_i P_i + \gamma_i P_i^2 \tag{11.5}$$

where α_i, β_i and γ_i are assumed available.

If one is interested in obtaining the power outputs so that F is a minimum, the first partial derivatives of F with respect to P_i are set to zero.

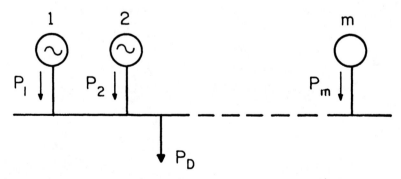

Figure 11-3. Units on the Same Bus.

11.3 Optimal Operation of an All-Thermal System: Equal Incremental Cost Loading

Thus,

$$\frac{\partial F}{\partial P_i} = 0 \quad (i = 1, \ldots, m) \tag{11.6}$$

This should hold for all m units in the system. For the example expression adopted for F, we obtain the optimal values for power generations as

$$P_{i_\ell} = \frac{-\beta_i}{2\gamma_i} \tag{11.7}$$

This expression for the optimal active power is guaranteed to result in minimizing F if the second partial derivative is positive. This condition for our case requires

$$\gamma_i > 0$$

Observe that Eq. (11.7) requires negative power generations as both β_i and γ_i are positive for practical systems.

Optimization problems in practice are seldom unconstrained. The usual situation is one where the cost is to be minimized subject to satisfying certain equations, which we refer to as *constraints*. One such case is when we wish to determine generation levels such that F is minimized while simultaneously satisfying an active power balance equation. This utilizes the principle of power flow continuity. Here the network is viewed as a medium of active power transfer from the generating nodes to the load node. Only one equation is needed.

The first active power balance equation model neglects transmission losses, and hence we can write

$$P_D = \sum_{i=1}^{m} (P_i) \tag{11.8}$$

with P_D being a given active power demand for the system.

The demand P_D is the sum of all demands at load nodes in the system. The model is useful in the treatment of parallel generating units at the same plant since in this case the negligible transmission losses assumption is valid.

We observe here that the results of the unconstrained minimization cited earlier in Eq. (11.7) lead to the sum

$$\sum_{i=1}^{m} (P_{i_\ell}) = -\frac{1}{2} \sum_{i=1}^{m} \left(\frac{\beta_i}{\gamma_i}\right)$$

The above sum is not equal to the power demand P_D as a general case. A popular method for solving constrained minimization problems uses the Lagrange multiplier technique. Here we write the constraint equation, Eq. (11.8), as

$$P_D - \sum_{i=1}^{m} (P_i) = 0 \tag{11.9}$$

The technique is based on including Eq. (11.9) in the original cost function by use of a Lagrange multiplier, say λ, which is unknown at the outset. Thus,

$$\hat{F} = F_T + \lambda \left[P_D - \sum_{i=1}^{m} (P_i) \right] \qquad (11.10)$$

where

$$F_T = \sum_{i=1}^{m} [F_i(P_i)]$$

Note that λ is to be obtained such that Eq. (11.9) is satisfied. The idea here is to penalize any violation of the constraint by adding a term corresponding to the resulting error. The Lagrange multiplier is in effect a conversion factor that accounts for the dimensional incompatibilities of the cost function (\$/h) and constraints (MW). The resulting problem is an unconstrained one, and we have increased the number of unknowns by one.

The optimality conditions are obtained by setting the partial derivatives of $\hat{F}$ with respect to P_i to 0. Thus,

$$\frac{\partial F_i}{\partial P_i} - \lambda = 0 \qquad (11.11)$$

Note that each unit's cost is independent of the generations of other units.

The expression obtained in Eq. (11.11) leads to the conclusion that

$$\lambda = \frac{\partial F_1}{\partial P_1} = \frac{\partial F_2}{\partial P_2} = \cdots \qquad (11.12)$$

The implication of this result is that for optimality, individual units should share the load such that their incremental costs are equal. We can see that the λ is simply the optimal value of incremental costs at the operating point. Equation (11.11) is frequently referred to as the *equal incremental cost-loading principle*. A graphical interpretation of the principle is shown in Figure 11-4.

Implementing the optimal solution is straightforward for the quadratic cost case where we have

$$F_i(P_i) = \alpha_i + \beta_i P_i + \gamma_i P_i^2$$

Our optimality conditions from Eq. (11.11) reduce to

$$\beta_i + 2\gamma_i P_i - \lambda = 0 \qquad (11.13)$$

The value of λ is determined such that Eq. (11.9) is satisfied. This turns out

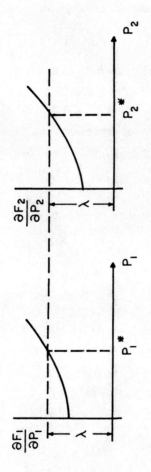

Figure 11-4. Illustrating the Principle of Equal Incremental Cost Loading.

to give

$$\lambda = \frac{2P_D + \sum_{i=1}^{m}\left(\frac{\beta_i}{\gamma_i}\right)}{\sum_{i=1}^{m} \gamma_i^{-1}} \qquad (11.14)$$

Finally the optimal generations are obtained using Eq. (11.13) as

$$P_i = \frac{\lambda - \beta_i}{2\gamma_i} \qquad (11.15)$$

We illustrate the use of the obtained results using the following example.

Example 11-2

The fuel-cost models for a two-unit plant are given by

$$F_1 = 462.28 + 8.28P_1 + 0.00053P_1^2 \text{ GJ/h}$$
$$F_2 = 483.44 + 8.65P_2 + 0.00056P_2^2 \text{ GJ/h}$$

where P_1 and P_2 are in MW.

The plant supplies a load of 1000 MW. Obtain the most economical loading of each unit and the incremental cost of power. Neglect losses.

Solution

Using the optimal equations—Eqs. (11.13) and (11.9)—we obtain

$$8.28 + 2(0.00053)P_1 - \lambda = 0$$
$$8.65 + 2(0.00056)P_2 - \lambda = 0$$
$$P_1 + P_2 = 1000$$

Solving the above, we obtain the optimal unit loadings:

$$P_1 = 683.49 \text{ MW}$$
$$P_2 = 316.51 \text{ MW}$$

The incremental cost of power is calculated as

$$\lambda = 9.00$$

The optimality conditions given in Eqs. (11.14) and (11.15) are straightforward to apply as can be judged from the preceding example. We find it instructive, however, to introduce a computer program written for this purpose.

Figure 11-5(a) shows the listing of a C language computer program designed to calculate the optimum power generation in a thermal system according to Eqs. (11.14) and (11.15). The dimension statements assume that up to five units are being scheduled. The program accepts the number

```
#include <stdio.h>
#include <stdlib.h>
#include <math.h>
int     M,N,T1[24],i,j,k,l;
float   P9[24],A[5],G4[5],G1[5],G5[5],P1[5],T,B;

main()
{
    printf("\n Please enter the number of units ");
    scanf("%d",&M);
                        printf("\n Please enter the number of time intervals ");
    scanf("%d",&N);

    for (i=1; i<=N; i++)
    {
    printf("\n Please enter the power demand for time interval %d ",i); scanf("%d",&P9[i]);
    }
    printf("\n Please input \n");
    T=0;
    B=0;
                        /* At this point we input the elements to the equations and
                        find the summations to calculate lambda */
```

Figure 11-5(a). A C Language Computer Program for Loss-Free All-Thermal Economic Operation Schedule.

```
    for (j=1; j<=M; j++)
    {
        printf("\n Elements to equation %d \n", j);
        printf("\n                                      Alpha ");
        scanf("%f", &A[j]);
        printf("\n                                      Beta ");
        scanf("%f", &G4[j]);
        printf("\n                                      Gamma ");
        scanf("%f", &G[j]);
        T=G4[j]/G[j]+T;
        B=(1/G[j])+B;
    }
    for (k=1; k<=N; k++)
    {
        G5[k]=(2*P9[k]+T)/B;
        printf("\n For Time Period %d ", k); printf("\n                   Lambda = %f ", G5[k]);
        /* calculate optimal power for each unit */
        for (l=1; l<=N; l++)
        {
            P1[l]=(G5[k]-G4[l])/(2*G[l]); printf("\n Pstar %d = %f \n", l, P1[l]);
        }
    }
}
```

Figure 11-5(a) (*Cont.*)

Please enter the number of units 2

Please enter the number of time intervals 2

Please enter the power demand for time interval 1 500

Please enter the power demand for time interval 2 300

Please input

Elements to equation 1

        Alpha 312.35

        Beta 8.52

        Gamma .00150

Elements to equation 2

        Alpha 483.44

        Beta 8.65

        Gamma .00056

For Time Period 1
Lambda = 9.022428
Pstar 1 =   167.475693

Pstar 2 =   332.524933

For Time Period 2
Lambda = 8.859321
Pstar 1 =   113.106728

Figure 11-5(b). Sample Output of Program Listed in Figure 11-5(a).

of units (M) as well as the number of times for which the schedules are to be obtained (N). This is followed by a request to input the power demand for each time instant. The fuel cost coefficients α_i, β_i, and γ_i are designated A, B, and G respectively and are entered next for each of the M units. Within the same loop the sums indicated in Eq. (11.14) are formed.

With the input data available, the program calculates the incremental cost of power λ denoted by D for each time instant, according to Eq. (11.14). The optimal power generations for each of the units is calculated using Eq. (11.15) as $P1(J)$ in the program.

A sample run of the program for a two-unit system is shown in Figure 11-5(b). Here we have three different levels of power demand and hence the number of time intervals is entered as 3. The optimal generations as well as the λ are obtained as shown.

11.4 ACCOUNTING FOR TRANSMISSION LOSSES

The problem of optimal operation considered in the preceding section utilized a power balance equation that neglects transmission losses in the interconnecting network. Although the problem is highly simplistic, it proves useful in the twin purposes of finding the optimum allocation of generation as well as obtaining single-unit equivalents for units on the same bus at a given power plant. A solution to the problem is also useful as a first approximation to more sophisticated problems as will become clear in the sequel.

Our intent presently is to outline an approach to modeling of the transmission losses in the system for economic operation purposes. To understand some of the basic principles involved, let us consider the single radial line system shown in Figure 11-6. Our purpose is to determine the dependence of the transmission losses P_L on the power generated P_G. From the equivalent circuit shown in Figure 11-7, we can deduce that

$$P_L = 3|I|^2 R$$

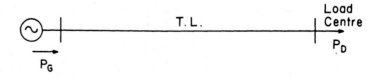

Figure 11-6. A Radial System.

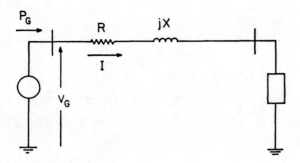

Figure 11-7. An Equivalent Circuit.

where R is the resistance of the line in ohms per phase. The current $|I|$ is obtained from

$$|I| = \frac{P_G}{(\sqrt{3})|V_G|\cos\phi_G}$$

In the above, P_G is the generated power, $|V_G|$ is the magnitude of the generated voltage (line-to-line), and $\cos\phi_G$ is the generator's power factor. Combining the above two relations, we conclude that

$$P_L = \frac{R}{|V_G|^2 \cos^2\phi_G}(P_G^2)$$

Assuming fixed generator voltage and power factor, we can write

$$P_L = BP_G^2 \qquad (11.16)$$

where in this case

$$B = \frac{R}{|V_G|^2 \cos^2\phi_G}$$

This shows that the losses may be approximated by a second-order function of the power generation.

Let us consider the case where a second generating plant is present—this time with the generation source very close to the power demand bus as shown in Figure 11-8. It is clear that the losses in this case are given by the following equation, which is identical to Eq. (11-16):

$$P_L = B_{11}P_1^2$$

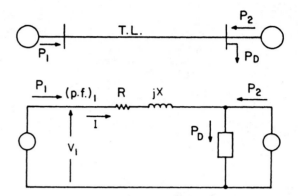

Figure 11-8. A Radial System with One Generation Source Close to the Demand Bus.

where

$$B_{11} = \frac{R}{(V_1)^2(PF_1)^2}$$

Here PF_1 denotes the power factor at bus 1.

As our second example we consider the system shown in Figure 11-9 where both generations are linked to the demand bus through lines of resistances R_{1D} and R_{2D} respectively. The following development is self-evident:

$$P_L = 3|I_1|^2 R_{1D} + 3|I_2|^2 R_{2D}$$
$$= \frac{R_{1D}}{(V_1)^2(PF)_1^2}(P_1^2) + \frac{R_{2D}}{(V_2)^2(PF)_2^2}(P_2^2)$$

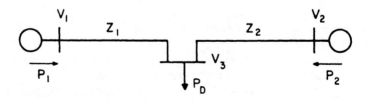

Figure 11-9. Two Radial Lines Feeding a Load.

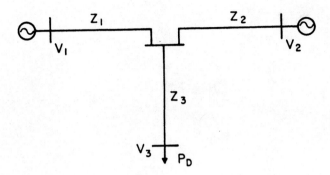

Figure 11-10. A Three-Line System.

This results in

$$P_L = B_{11}P_1^2 + B_{22}P_2^2 \tag{11.17}$$

We observe here that in Eqs. (11.16) and (11.17), no cross multiplication terms are present. This arises in cases such as for the system shown in Figure 11-10 where a third line of resistance R_{3D} feeds the demand bus. In this case, we have

$$P_L = B_{11}P_1^2 + 2B_{12}P_1P_2 + B_{22}P_2^2 \tag{11.18}$$

We now derive Eq. (11.18) as follows:

$$P_L = 3(I_1^2 R_{1D} + I_2^2 R_{2D} + I_3^2 R_{3D})$$

where we understand that all currents in the above expressions are magnitudes only. Thus,

$$|I_1| = \frac{P_1}{\sqrt{3}|V_1|(PF)_1}$$

$$|I_2| = \frac{P_2}{\sqrt{3}|V_2|(PF)_2}$$

$$|I_3| = \frac{P_D}{\sqrt{3}|V_3|(PF)_3}$$

Now we may approximately set

$$P_D \cong P_1 + P_2$$

Therefore,

$$|I_3| = \frac{P_1 + P_2}{\sqrt{3}|V_3|(PF)_3}$$

Substituting in the loss expression, we obtain

$$P_L = \frac{R_{1D}}{|V_1|^2(PF)_1^2}(P_1^2) + \frac{R_{2D}}{|V_2|^2(PF)_2^2}(P_2^2)$$
$$+ \frac{R_{3D}}{|V_3|^2(PF)_3^2}(P_1 + P_2)^2$$

Expanding and comparing with Eq. (1.18), we have

$$B_{11} = \frac{R_{1D}}{|V_1|^2(PF)_1^2} + \frac{R_{3D}}{|V_3|^2(PF)_3^2} \qquad (11.19)$$

$$B_{22} = \frac{R_{2D}}{|V_2|^2(PF)_2^2} + \frac{R_{3D}}{|V_3|^2(PF)_3^2} \qquad (11.20)$$

$$B_{12} = \frac{R_{3D}}{|V_3|^2(PF)_3^2} \qquad (11.21)$$

To illustrate the development of loss formulae from basic principles, we consider the following simple example.

Example 11-3

Consider a three-line, two-plant system such as the one shown in Figure 11-10. The following data are given in the per unit system:

$$|V_1| = 1.05$$
$$|V_2| = 1.03$$
$$|V_3| = 1.00$$
$$(PF)_1 = 0.95$$
$$(PF)_2 = 0.95$$
$$(PF)_3 = 0.85$$
$$R_{1D} = 0.04$$
$$R_{2D} = 0.05$$
$$R_{3D} = 0.03$$

Solution

Using eq. (11.19)–(11.21), we have

$$B_{11} = \frac{0.04}{(1.05)^2(0.95)^2} + \frac{0.03}{(1)^2(0.85)^2} = 0.0817$$

$$B_{22} = \frac{0.05}{(1.03)^2(0.95)^2} + \frac{0.03}{(1)^2(0.85)^2} = 0.0937$$

$$B_{12} = \frac{0.03}{(1)^2(0.85)^2} = 0.0415$$

Thus in per unit we have the loss formula given by

$$P_L = 0.0817 P_1^2 + 0.083 P_1 P_2 + 0.0937 P_2^2$$

The General Loss Formula

Ideally, the exact power flow equations should be used to account for the active power transmission losses in the power system. However, it is common practice to express the system losses in terms of active power generations only. This approach is commonly referred to as the *loss formula*, or *B-coefficient method*. The simplest form of the equation is called *George's formula* and is given by

$$P_L = \sum_{i=1}^{m} \sum_{j=1}^{m} (P_i B_{ij} P_j)$$

The coefficients B_{ij} are commonly referred to as the *loss coefficients*. A more general formula is

$$P_L = K_{L0} + \sum_{i=1}^{m} B_{i0} P_i + \sum_{i=1}^{m} \sum_{j=1}^{m} P_i B_{ij} P_j \tag{11.22}$$

Here a linear term and a constant have been added to the original quadratic expression. This expression is frequently called *Kron's loss formula*. Amazingly enough, this loss formula can be obtained by simply using the first three terms of the Taylor expansion of a function of several variables. This of course assumes dependence of the losses on the active power generations only.

When it is necessary to transmit electric energy over large distances or in the case of a relatively low load density over a vast area, the transmission losses are a major factor to be considered. The active power transmission losses may amount to 20 to 30 percent of the total load demand.

The ease of computation possible with the use of the loss formula is highly advantageous, especially if the complexity of calculating these coefficients can be reduced without loss of accuracy. It is emphasized that a number of approximations are involved in the loss formula, as noted below:

1. Assume a linear generator reactive characteristic such that

$$Q_{G_i} = Q_{G_{i_0}} + f_i P_{G_i}$$

2. Assume constant generator angular positions, δ_i.
3. Assume constant generator-bus voltage magnitudes.
4. Assume a fixed demand pattern.

Thus it is valid only for a certain range of loadings. In practice, however, the formula produces close answers with errors up to only a few percent. Very sophisticated methods for calculating the B-constants exist and are being used by utilities in connection with economic dispatch.

We will consider next the inclusion of loss in optimal operation studies for all-thermal systems.

11.5 OPTIMAL OPERATION OF AN ALL-THERMAL SYSTEM, INCLUDING LOSSES

The preceding section outlined how transmission losses are modeled for optimal operation studies. Including the losses in the active power balance equation leads to some modifications of the optimal solution obtained in Section 11.3. These are discussed here.

We are interested in minimizing the total cost given by Eq. (11.5), while satisfying the active power balance equation including losses. Thus,

$$P_D = \sum_{i=1}^{m} (P_i) - P_L \qquad (11.23)$$

Here P_L is the active power loss considered as a function of the active power generation alone as outlined in the previous section.

Following our treatment for the loss-free case, we form the augmented cost function:

$$\hat{F} = F_T + \lambda \left[P_D + P_L - \sum_{i=1}^{m} (P_i) \right] \qquad (11.24)$$

The optimality conditions are obtained using the same arguments as before

11.5 Optimal Operation of an All-Thermal System, Including Losses

and turn out to be

$$\frac{\partial F_i}{\partial P_i} + \lambda\left(\frac{\partial P_L}{\partial P_i} - 1\right) = 0 \qquad (11.25)$$

Note that with negligible transmission losses, the above expression reduces to Eq. (11.11) obtained in Section 11.3.

It is convenient to transform the obtained optimality expression into an equivalent form. This is done by defining the factors L_i:

$$L_i = \left(1 - \frac{\partial P_L}{\partial P_i}\right)^{-1} \qquad (11.26)$$

We can thus write Eq. (11.25) as

$$L_i \frac{\partial F_i}{\partial P_{s_i}} = \lambda \qquad (i = 1, \ldots, m) \qquad (11.27)$$

This is of the form of Eq. (11.12) except for the introduction of the new factors L_i, which account for the modifications necessitated by including the transmission loss. These are traditionally called the *penalty factors* to indicate that plant costs (F_i) are penalized by the corresponding incremental transmission losses ($\partial P_L/\partial P_i$).

Examination of Eq. (11.27) reveals that the optimal generations are obtained when each plant is operated such that the penalized incremental costs are equal. Let us consider an example that is primarily an extension of Example 11-2 to illustrate our new results.

Example 11-4

Consider the system of Example 11-2. Assume that losses are accounted for and are expressed as

$$P_L = B_{11} P_1^2$$

with

$$B_{11} = 1.5 \times 10^{-4} \text{ MW}^{-1}$$

Compute the optimal power generations in this case.

Solution

To start, we find the penalty factors L_1 and L_2. We have the incremental transmission losses evaluated as

$$\frac{\partial P_L}{\partial P_1} = 2B_{11}P_1 = 3 \times 10^{-4} P_1$$

$$\frac{\partial P_L}{\partial P_2} = 0$$

Therefore,
$$L_1^{-1} = (1 - 3 \times 10^{-4} P_1)$$
$$L_2 = 1$$

The optimal operation conditions of Eq. (11.27) are thus given by
$$8.28 + 0.00106 P_1 = \lambda(1 - 3 \times 10^{-4} P_1)$$
$$8.65 + 0.00112 P_2 = \lambda$$

We can eliminate λ between the above two equations to obtain
$$(8.28 + 0.00106 P_1) = (1 - 3 \times 10^{-4} P_1)(8.65 + 0.00112 P_2)$$

We also have the power balance equation
$$P_1 + P_2 = 1000 + (1.5 \times 10^{-4}) P_1^2$$

Care must be taken not to overlook the last term of the right-hand side representing the transmission losses. The last two equations can be combined to eliminate P_2. As a result, we have
$$\frac{8.28 + 0.00106 P_1}{1 - 3 \times 10^{-4} P_1} = 8.65 + 0.00112(1.5 \times 10^{-4} P_1^2 - P_1 + 1000)$$

Simplifying, the following third-order equation results:
$$x^3 - 10 x^2 + 101.4 x - 29.56 = 0$$
where
$$x = 10^{-3} P_1$$

The solution of this equation gives
$$x = 0.30014$$
or
$$P_1 = 300.14 \text{ MW}$$

The optimal value of P_2 is calculated according to the power balance equation:
$$P_2 = 1000 - 300.14 + 1.5 \times 10^{-4}(300.14)^2$$
$$= 713.3726 \text{ MW}$$

The corresponding loss is
$$P_L = 13.5126 \text{ MW}$$

The incremental cost of power delivered is obtained using the first optimality condition as
$$\lambda = \frac{8.28 + (0.00106)(300.14)}{1 - 3 \times 10^{-4}(300.14)}$$
$$= 9.44895$$

11.5 Optimal Operation of an All-Thermal System, Including Losses

To check the validity of our results, we use the second optimality condition to calculate P_2. This results in

$$P_2 = \frac{9.44895 - 8.65}{0.00112}$$
$$= 713.34896 \text{ MW}$$

This is close enough to the value computed earlier.

We make some observations on the results of the example. First, the optimal generation of the first plant is considerably less in the case accounting for losses than that for Example 11-2 neglecting losses. This is to be expected since the losses for the given model increase as the loading on plant 1 increases. It is natural to expect that more loading on plant 2 is advantageous. Indeed loss considerations seem to offset the cost ones! Our second observation is that the incremental cost of power delivered is higher in the present example than the corresponding value in the loss-free case of Example 11-3. One final observation is that even though our loss expression was a very simplistic one, we had to solve a cubic equation in P_1. This is no easy task and motivates us to consider use of a general iterative method for the solution.

A Computer Implementation

We assume that fuel costs are quadratic expressions as given in Eq. (11.5). Moreover the transmission losses are expressed by the general expression

$$P_L = K_{L_0} + \sum_{i=1}^{m} (B_{i0} P_i) + \sum_{i=1}^{m} \sum_{j=1}^{m} (P_i B_{ij} P_j) \tag{11.28}$$

In this case, our incremental loss expressions turn out to be

$$\frac{\partial P_L}{\partial P_i} = B_{i0} + 2 \sum_{j=1}^{m} (B_{ij} P_j) \tag{11.29}$$

We can conclude then that Eq. (11.27) requires that

$$f_i = \beta_i + 2\gamma_i P_i + \lambda \left[B_{i0} - 1 + 2 \sum_{j=1}^{m} (B_{ij} P_j) \right] = 0 \quad (i = 1, \ldots, m) \tag{11.30}$$

The multiplier λ is obtained such that the active power balance equation is satisfied. In our present case, this is given by

$$g = P_D + K_{L0} + \sum_{i=1}^{m} (B_{i0} - 1) P_i + \sum_{i=1}^{m} \sum_{j=1}^{m} P_i B_{ij} P_j = 0 \tag{11.31}$$

Equations (11.30) and (11.31) completely specify our desired optimal solution.

The Newton-Raphson method has proven successful in solving sets of algebraic equations. We evaluate the derivatives of the equations with respect to the unknowns. These turn out to be

$$\frac{\partial f_i}{\partial P_i} = 2(\gamma_i + \lambda B_{ii}) \qquad (i=1,\ldots,m) \tag{11.32}$$

$$\frac{\partial f_i}{\partial P_j} = 2\lambda B_{ij} \qquad (i, j=1,\ldots,m) \tag{11.33}$$

$$\frac{\partial f_i}{\partial \lambda} = \left[B_{i0} - 1 + 2 \sum_{j=1}^{m} (B_{ij} P_j) \right] \qquad (i=1,\ldots,m) \tag{11.34}$$

$$\frac{\partial g}{\partial P_i} = \left[B_{i0} - 1 + 2 \sum_{j=1}^{m} (B_{ij} P_j) \right] \qquad (i=1,\ldots,m) \tag{11.35}$$

Finally,

$$\frac{\partial g}{\partial \lambda} = 0 \tag{11.36}$$

Starting with estimates $P_i^{(0)}$ and $\lambda^{(0)}$ for the unknowns, new improved estimates are obtained according to

$$\lambda^{k+1} = \lambda^k + \Delta\lambda \tag{11.37}$$

$$P_i^{k+1} = P_i^k + \Delta P_i \tag{11.38}$$

The increments $\Delta\lambda$ and ΔP_i are obtained as the solution to the following set of linear equations:

$$\frac{\partial f_1}{\partial P_1}(\Delta P_1) + \frac{\partial f_1}{\partial P_2}(\Delta P_2) + \cdots + \frac{\partial f_1}{\partial P_m}(\Delta P_m) + \frac{\partial f_1}{\partial \lambda}(\Delta\lambda) = -f_1$$

$$\frac{\partial f_2}{\partial P_1}(\Delta P_1) + \frac{\partial f_2}{\partial P_2}(\Delta P_2) + \cdots + \frac{\partial f_2}{\partial P_m}(\Delta P_m) + \frac{\partial f_2}{\partial \lambda}(\Delta\lambda) = -f_2$$

$$\frac{\partial f_m}{\partial P_1}(\Delta P_1) + \frac{\partial f_m}{\partial P_2}(\Delta P_2) + \cdots + \frac{\partial f_m}{\partial P_m}(\Delta P_m) + \frac{\partial f_m}{\partial \lambda}(\Delta\lambda) = -f_m$$

$$\frac{\partial g}{\partial P_1}(\Delta P_1) + \frac{\partial g}{\partial P_2}(\Delta P_2) + \cdots + \frac{\partial g}{\partial P_m}(\Delta P_m) = -g$$

As the iterations progress, the solution to the problem is obtained to the desired degree of accuracy. It should be noted that the initial estimates for the solution are best obtained by assuming negligible losses.

We now discuss a computer program whose listing is given in Figure 11-11(a). The program obtains the optimum economic allocation of generation in an all-thermal system including losses. The fuel-cost model may be second-order to third-order. The second-order problem is denoted "IL2." The Newton-Raphson method is used. The Jacobian matrix is formed and inverted as needed to form the correction components D. The initial guess

```c
#include <stdio.h>
#include <stdlib.h>
#include <math.h>
#include "nrutil.h"
#include "matutil.h"
#define TINY 1.0e-20
#define NR_END 1
#define FREE_ARG char *
 void nrerror(char error_text[]) ;
 void ludcmp(float **a, int n, int *indx, float *d);
 void lubksb(float **a, int n, int *indx, float b[]) ;
 void zermat(float **mat, int n, int m);
 void zervec(float *vec, int n);
void gaussj(float **a, int n, float **b, int m);
int      M,N,O,p,i,j,k,l,C;
float    *P,*G1,*G2,*B,*S1,T;
float    *G8,*P1,**D1,n1;
float    **B1,**D,**F,**G3,L,E,X,K;
float    S3,S2,P2,P3,Z,L1,A1;
float    **a,d,*col,**mat,*vec;
int      m,n,order,*indx;

main()
{
        C=0;

        printf("\n Please enter the number of units \n");
    scanf("%d",&N);
        M=N+1;
        O=N-1;
        p=M+1;

/* need to call subroutine to allocate vectors and matrices */

        P=vector(M);
        G1=vector(M);
        G2=vector(M);
        B=vector(M);
        S1=vector(M);
        P1=vector(M);
        G8=vector(M);
        indx=ivector(M);
```

Figure 11-11(a). Computer Program Listing for Economic Operation of All-Thermal Including Losses.

```
B1=matrix(M,M);
D=matrix(p,p);
D1=matrix(p,p);
F=matrix(p,N);
G3=matrix(p,N);

printf("\n Please enter the power demand \n");
scanf("%f",&T);

printf("\n Please enter the order of the fuel cost model \n");
printf("\n either second (2) or third order (3) \n");
scanf("%d",&order);
/* initial guess for optimal power generation */

for (i=1; i<=N; i++)
{
        P[i]=T/(float)N;
}

printf("\n Please input approximate Lambda \n");
scanf("%f",&L);

printf("\n Please input Tolerance \n");
scanf("%f",&E);

printf("\n Please enter number of iterations \n");
scanf("%f",&X);

for (i=1; i<=N; i++)
{
        printf("\n For unit %d enter \n",i);

        printf("\n      Beta \n");
        scanf("%f",&G1[i]);
        printf("\n      Gamma \n");
        scanf("%f",&G2[i]);
                if (order == 3)
                {
                        printf("\n      Delta \n");
                        scanf("%f",&G8[i]);
                }
```

Figure 11-11(a). *(Cont.)*

```
        }
                printf("\n  Enter Loss coefficients \n");

printf("\n            KLO \n");
scanf("%f",&K);

 for (i=1; i<=N; i++)
{
/* enter B[i,0] elements */;

                printf("\n  B(%d,0) \n",i);
    scanf("%f",&B[i])
                /* enter B[i,j] elements */;

                for (j=1; j<=N; j++)
                {
                        if (i>j)
                        {
                                B1[i][j]=B1[j][i];
                        }
                        else
                        {
                                printf("\n  B(%d,%d) \n",i,j);
                    scanf("%f",&B1[i][j]);
                                B1[i][j]=B1[j][i];
                        }
                }
        }




        /* this is the start of the iterative loop */
A1=E+1;
while (C<X && A1>E)
{
        C=C+1;          /* Build the D (Jacobian) Matrix */
        S3=0;S2=0;P2=0;


        for (i=1; i<=N; i++)
```

Figure 11-11(a). (Cont.)

```
            {
                for (j=1; j<=N; j++)
                {
                    if (i==j & order==2)

                            D[i][j]=2*(G2[i]+(B1[i][j]*L));
                    else if (i==j & order!=2)
                            D[i][j]=2*(G2[i]+(B1[i][j]*L)+(3*(G8[i])*(P[i])));
                    else
                            D[i][j]=2*(B1[i][j]*L);
                }
            }
            for (i=1;i<=N;i++)
                {
                    S1[i]=0;
                }

            for (i=1; i<=N; i++)
        {
            for (j=1; j<=N; j++)
            {
                S1[i]=B1[i][j]*P[j]+S1[i];
            }
        }


            for(i=1;i<=N;++i)
            {
                j=N+1;
                D[i][j]=(B[i]-1)+(2*S1[i]);
            D[j][i]=D[i][j]; /* this gives us all other terms in  D*/

            }

    for (i=1;i<=M;i++)
    {
        for (j=1;j<=M;j++)
        {
        D1[i][j]=D[i][j]; /* make a copy of the original D matrix for future use */
        }
    }
```

Figure 11-11(a). *(Cont.)*

```
/* matrix inversion loop */

gaussj(D1,M,B1,0); /* call the gaussian-jordan elimination routine */
                   /* B1 is a matrix that determines inversion or solution of */
                   /* system of equations , D1 returns the inverse in any case */

/* matrix inversion loop end */

        for(i=1;i<=N;++i)
/* Here we create the "cost function" terms */
        {
                j=N+1;
                if (order != 2)
        F[i][1]=(G1[i]+(2*(G2[i])*(P[i]))+(3*(G8[i])*(P[i])*(P[i])))+(L*(D[i][j]));
                else
                F[i][1]=(G1[i]+(2*G2[i]*P[i])+(L*(D[i][j])));
        }

        for(i=1;i<=N;i++)
        {
                P2=P[i]+P2;
        }
        for(i=1;i<=N;++i)
        {
                S3=B[i]*P[i]+S3;
                for(j=1;j<=N;++j)
                {
                        S2=P[i]*B1[i][j]*P[j]+S2;
                }
        }
        P3=K+S3+S2;
        i=N+1;
        F[i][1]=-P2+T+P3;
        Z=F[i][1];

/* perform matrix multiplication here */

zermat(G3,M,M); /* zero the matrix first */

        for(j=1;j<=M;j++)
        {
                for(i=1;i<=M;i++)
```

Figure 11-11(a). (Cont.)

```
                        {
                        G3[j][1]=G3[j][1]+D1[j][i]*F[i][1];
                        }
        }
/* end  matrix mult */

        for(i=1;i<=N;++i)
        {
                P1[i]=G3[i][1];
        }
        j=N+1;
        L1=G3[j][1];
        if (C > X){ /* If we exceed number of iterations then stop program */
                printf("\n Does not converge in %d iterations \n",C);
                goto outside;
                }
        else
        {
                A1=fabs(Z); /* check if tolerence has been met */
                if (A1 <= E)
                {
                goto outside; /* if so then end program and print results */
                }
                else
                for(i=1;i<=N;++i)
                {
                P[i]=P[i]-P1[i]; /* otherwise reevaluate lambda and continue iterations */
                }
        }
        L=L-L1;


/*** goto statement ***/

outside:
printf("\n Lambda = %f",L);
for (i=1;i<=N;++i){
        printf("\n P[%d]=%f",i,P[i]);
        }
        printf("\n Power Demand = %6.3f",T);
        printf("\n Tolerance Was %.5f",A1);
```

Figure 11-11(a). *(Cont.)*

```c
        printf("\n Number of Iterations Was %d \n",C);
}
return;
} /* this is the end of the main function */


/*******************************************************************/
/* what follows here are utility routines needed in the main program */
/*******************************************************************/


void nrerror(char error_text[])
/* Numerical Recipes standard error handler */
{
        printf("Numerical Recipes run-time error...\n");
        printf("%s\n",error_text);
        printf("... now exiting to system ...\n");
        exit(1);
        return;
}

void zervec(float *vec, int n)

/* Zeros an input vector */
{
int i;

for(i=0;i<n;++i)
{
vec[i]=0;
}
return;
}

void zermat(float **mat, int n, int m)

/* Zeros an input matrix */
{
int i,j;

for(i=1;i<=n;i++)
```

Figure 11-11(a). (Cont.)

```
        {
                for(j=1;j<=m;j++)
                {
                        mat[i][j]=0;
                }
        }

        return;
}
```

Figure 11-11(a). (Cont.)

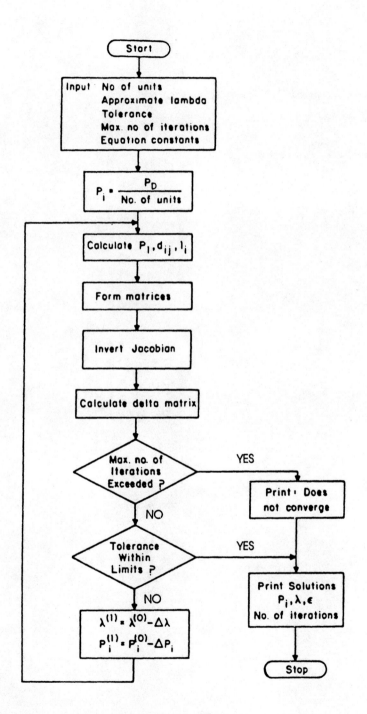

Figure 11-11(b). Computer Program Flow Chart.

Please enter the number of units
2

Please enter the power demand
1000

Please enter the order of the fuel cost model

either second (2) or third order (3)
2

Please input approximate Lambda
10

Please input Tolerance
.00005

Please enter number of iterations
20

For unit 1 enter

 Beta
7.74

 Gamma
.00107

For unit 2 enter

 Beta
7.72

 Gamma
.00072

Enter Loss coefficients

 KLO
0

B(1,0)
0

B(1,1)
.0005

B(1,2)
0

B(2,0)
0

B(2,2)
.0002

Figure 11-11(c). Sample Interactive Session Using Program in Figure 11-11(a).

Lambda = 12.869102
P[1]=304.328796
P[2]=841.044495
Power Demand = 1000.000
Tolerance Was 175.00000
Number of Iterations Was 1

Lambda = 13.432348
P[1]=367.839294
P[2]=838.367310
Power Demand = 1000.000
Tolerance Was 42.40588Tolerance Was 42.40588
Number of Iterations Was 2

Lambda = 13.470045
P[1]=367.072357
P[2]=842.133362
Power Demand = 1000.000
Tolerance Was 2.01828
Number of Iterations Was 3

Lambda = 13.470127
P[1]=367.077545
P[2]=842.133057
Power Demand = 1000.000
Tolerance Was 0.00308
Number of Iterations Was 4

Lambda = 13.470127
P[1]=367.077545
P[2]=842.133057
Power Demand = 1000.000
Tolerance Was 0.00000
Number of Iterations Was 5

Figure 11-11(c). *(Cont.)*

values are taken as the outcome of the division of the given demand among the available units. The approximate value of lambda is needed, however, as a user input. Figure 11-11(b) shows a flow chart of the program, and Figure 11-11(c) gives a sample output of an interactive session using the program. The student is encouraged to try this program in connection with problems such as the ones given at the end of this chapter.

11.6 OPTIMAL OPERATION OF HYDROTHERMAL SYSTEMS

The systems treated so far have included only thermal generation sources. In this section we intend to introduce a problem in optimal operation of systems with hydro as well as thermal generation. It is important to realize that in the case of a hydro unit, no variation in operating cost can be attributed to variations in output power. Consequently the criterion of minimum operating costs for thermal plants cannot be used for the hydro plants in the system. Instead it is usual to specify an allowable volume of water for release over a certain interval of time. This for the short range can vary from one day to a week.

Hydro-resource modeling is an important aspect of hydrothermal scheduling. We will now discuss the simplest models used for optimal operation studies purposes. As shown in Figure 11-12, a hydro unit's output (represented by its active power generation level) is a function of the rate of water discharge through the turbine q and the effective water head h. The conventional formula is given by

$$P_h = \frac{\eta h q}{c}$$

In the above, c is a dimensional conversion coefficient, and η is the efficiency that is dependent on h and q. We can thus state that

$$P_h = P_h(q, h) \tag{11.39}$$

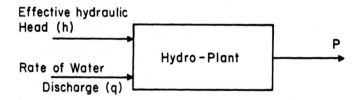

Figure 11-12. Input-Output Relationship for a Hydro Plant.

to denote the dependence of P_h on h and q. An assumption commonly made is that the power output is the product of two functions ψ and ϕ. In this case, ψ is a function of h only, and ϕ is a function of q only. Mathematically this states

$$P_h = \psi(h)\phi(q)$$

An important case is when the hydro plant has large storage reservoirs, for which, head variation is negligible. A popular model relating the active power generation to rate of water discharge is given by

$$q = \alpha + \beta P_h + \gamma P_h^2 \qquad (11.40)$$

The reason for its popularity is probably its similarity to the fuel-cost expression.

The Coordination Equations

Consider a system with hydro plants operating at constant head during the optimization interval $(0, T_f)$. It is desired to minimize the total fuel cost. Thus,

$$J = \int_0^{T_f} \sum_{i=1}^{m} (F_i) \, dt \qquad (11.41)$$

under the following conditions:

1. The total system generation matches the power demand $P_D(t)$ and the transmission losses $P_L(t)$. Thus,

$$\sum_{i=1}^{n} P_i(t) = P_D(t) + P_L(t) \qquad (11.42)$$

We have m thermal plants and $(n - m)$ hydro plants in this system.

2. The volume of water available for generation at each hydro plant is a specified amount b_i. Thus,

$$\int_0^{T_f} q_i(t) \, dt = b_i \quad (i = m+1, \ldots, n) \qquad (11.43)$$

The optimality conditions are obtained in a way similar to the methods of the previous sections. Presently we include the power balance given by, Eq. (11.42), in the cost expression using the multiplier functions $\lambda(t)$. The volume of water relation in Eq. (11.43) is accounted for using a new constant multiplier ν_j. Observe that the problem now is time-dependent, and consequently our incremental cost of power delivered λ is taken as a function of time. The yet unknown but constant multipliers ν_j are called *water conversion coefficients.* Our modified cost is thus

$$\tilde{J} = \int_0^{T_f} \left\{ \sum_{i=1}^{m} (F_i) + \lambda(t) \left[P_D(t) + P_L(t) - \sum_{i=1}^{n} (P_i) \right] \right.$$

$$\left. + \sum_{i=m+1}^{n} (\nu_i q_i) \right\} dt \qquad (11.44)$$

Taking derivatives with respect to the thermal generations, we have

$$\frac{dF_i}{dP_i} + \lambda(t) \left(\frac{\partial P_L}{\partial P_i} - 1 \right) = 0 (i = 1, \ldots, m) \qquad (11.45)$$

This is identical with the all-thermal equation, Eq. (11.25).
Next we take derivatives with respect to the hydro generation:

$$\lambda(t) \left(\frac{\partial P_L}{\partial P_i} - 1 \right) + \nu_i \left(\frac{\partial q_i}{\partial P_i} \right) = 0 (i = m+1, \ldots, n) \qquad (11.46)$$

This is new.
We can now introduce the penalty factors L_i as defined by Eq. (11.26). As a result, we now write Eqs. (11.45) and (11.46) as

$$L_i \left(\frac{dF_i}{dP_i} \right) = \lambda(t) (i = 1, \ldots, m) \qquad (11.47)$$

$$L_i \nu_i \left(\frac{\partial q_i}{\partial P_i} \right) = \lambda(t) (i = m+1, \ldots, n) \qquad (11.48)$$

It appears from Eq. (11.48) that an equivalent fuel cost F_e may be assigned to the hydro generation. This can be defined by

$$\frac{\partial F_{ei}}{\partial P_i} \triangleq \nu_i \left(\frac{\partial q_i}{\partial P_i} \right) (i = m+1, \ldots, n) \qquad (11.49)$$

Caution is advised here since the ν_i are unknown at the outset. Equations (11.47) and (11.48) are commonly referred to as the *coordination equations*. We will consider a number of examples to illustrate the application of the above results.

Example 11-5

Consider a hydrothermal electric power system with one thermal and one hydro plant. Assume that the thermal fuel cost expression is given by the usual quadratic:

$$F_1 = \alpha_1 + \beta_1 P_1 + \gamma_1 P_1^2$$

The hydro-discharge characteristic is assumed as

$$q_2 = \alpha_2 + \beta_2 P_2 + \gamma_2 P_2^2$$

11.6 Optimal Operation of Hydrothermal Systems

The transmission losses are expressed as

$$P_L = B_{11} P_1^2 + 2 B_{12} P_1 P_2 + B_{22} P_2^2$$

Write down the coordination equations for this system.

Solution

In the present case, $m = 1$ and $n = 2$. Our penalty factors are obtained as

$$L_1 = (1 - 2 B_{11} P_1 - 2 B_{12} P_2)^{-1}$$
$$L_2 = (1 - 2 B_{12} P_1 - 2 B_{22} P_2)^{-1}$$

The optimality conditions (coordination equations) are thus

$$\beta_1 + 2\gamma_1 P_1(t) = \lambda(t)[1 - 2 B_{11} P_1(t) - 2 B_{12} P_2(t)] \quad (11.50)$$
$$\nu_2[\beta_2 + 2\gamma_2 P_2(t)] = \lambda(t)[1 - 2 B_{12} P_1(t) - 2 B_{22} P_2(t)] \quad (11.51)$$
$$P_1(t) + P_2(t) - P_D(t) - B_{11} P_1^2(t) - B_{22} P_2^2(t) - 2 B_{12} P_1(t) P_2(t) = 0 \quad (11.52)$$
$$\int_0^{T_f} [\alpha_2 + \beta_2 P_2(t) + \gamma_2 P_2^2(t)] \, dt = b_2 \quad (11.53)$$

Observe that the result of Example 11-5 is quite complex. Under very restricted conditions we can solve directly for the powers $P_1(t)$ and $P_2(t)$ using Eq. (11.50)–(11.53). An example illustrates this concept.

Example 11-6

The system described in Example 11-5 has negligible losses. For $\beta_1 = \beta_2 = 0$, write down the optimality conditions.

Solution

With negligible losses and zero linear coefficients, our equations (11.50)–(11.53) reduce to

$$2\gamma_1 P_1(t) = \lambda(t) \quad (11.54)$$
$$\nu_2[2\gamma_2 P_2(t)] = \lambda(t) \quad (11.55)$$
$$P_1(t) + P_2(t) = P_D(t) \quad (11.56)$$
$$\int_0^{T_f} [\alpha_2 + \gamma_2 P_2^2(t)] \, dt = b_2 \quad (11.57)$$

We can now consider a numerical example.

Example 11-7

Obtain the optimal power generations $P_1(t)$, $P_2(t)$, the incremental cost of power delivered $\lambda(t)$, and the water conversion coefficient ν_2 for a

742 Optimal Operation of Electric Power Systems

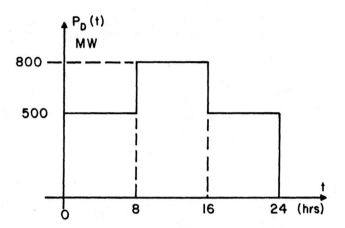

Figure 11-13. Power Demand Variation for System in Example 11-7.

system such as the one described in Example 11-6. The following data are available:

$$\gamma_1 = 0.003 \qquad \gamma_2 = 0.0024$$
$$\alpha_2 = 28 \qquad b_2 = 3500 \text{ Million Cubic Feet (MCF)}$$

The power demand is as shown in Figure 11-13.

Solution

With reference to the above example, we can obtain P_1 in terms of ν_2 and P_2 from Eqs. (11.54) and (11.55). This is given by

$$P_1(t) = \nu_2 \left(\frac{\gamma_2}{\gamma_1} \right) P_2(t) \tag{11.58}$$

Substituting in Eq. (11.56), we obtain

$$P_2(t) = \frac{P_D(t)}{1 + \nu_2 \left(\frac{\gamma_2}{\gamma_1} \right)} \tag{11.59}$$

Using the given data, we obtain

$$P_2(t) = \frac{P_D(t)}{1 + 0.8 \nu_2}$$

11.6 Optimal Operation of Hydrothermal Systems

Equation (11.57) gives

$$\int_0^{24} P_2^2(t)\,dt = \frac{3500-(28)(24)}{0.0024}$$

$$= 1.1783 \times 10^6$$

Using the expression for $P_2(t)$ in terms of $P_D(t)$, we attain

$$\int_0^{24} \frac{P_D^2(t)\,dt}{(1+0.8\nu_2)^2} = 1.1783 \times 10^6$$

Using the given power demand data, we get

$$\frac{(500)^2(16) + (800)^2(8)}{(1+0.8\nu_2)^2} = 1.1783 \times 10^6$$

Simplifying, we get

$$\nu_2 = 2.2275$$

Thus we have

$$P_2(t) = \frac{P_D(t)}{1+(0.8)(2.2275)}$$

or

$$P_2(t) = 0.35945 P_D(t)$$

Thus the optimal hydro generation is as follows:

$$P_2(t) = 179.73 \text{ MW for } P_D(t) = 500 \text{ MW}$$
$$P_2(t) = 287.56 \text{ MW for } P_D(t) = 800 \text{ MW}$$

The thermal generation is given by Eq. (11.58), which gives

$$P_1(t) = (2.2275)\left(\frac{0.0024}{0.003}\right) P_2(t)$$
$$= 1.7820 P_2(t)$$

This results in the following values:

$$P_1(t) = 320.27 \text{ MW for } P_D(t) = 500 \text{ MW}$$
$$P_1(t) = 512.44 \text{ MW for } P_D(t) = 800 \text{ MW}$$

The incremental cost of power delivered $\lambda(t)$ is obtained using Eq. (11.54). This gives

$$\lambda(t) = (2)(0.003) P_1(t)$$
$$= 0.006 P_1(t)$$

Thus,

$$\lambda(t) = 1.92 \text{ for } P_D(t) = 500 \text{ MW}$$
$$\lambda(t) = 3.07 \text{ for } P_D(t) = 800 \text{ MW}$$

The solution of problems including transmission losses involves the use of iterative techniques such as the Newton-Raphson method.

SOME SOLVED PROBLEMS

Problem 11-A-1

The fuel-cost models for a three-unit plant are given by

$$F_1 = 173.61 + 8.67 P_1 + 0.0023 P_1^2$$
$$F_2 = 180.68 + 9.039 P_2 + 0.00238 P_2^2$$
$$F_3 = 182.62 + 9.19 P_3 + 0.00235 P_3^2$$

The daily load curve for the plant is as shown in Figure 11-14. Obtain and sketch the optimal power generated by each unit and the plant's incremental cost of power delivered (λ).

Solution

From our given data we have

$$\beta_1 = 8.67 \quad \beta_2 = 9.039 \quad \beta_3 = 9.19$$
$$\gamma_1 = 0.0023 \quad \gamma_2 = 0.00238 \quad \gamma_3 = 0.00235$$

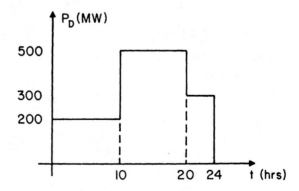

Figure 11-14. Load Curve for Problem 11-A-1.

We therefore form the summations

$$\sum \frac{\beta_i}{\gamma_i} = \frac{8.67}{0.0023} + \frac{9.039}{0.00238} + \frac{9.19}{0.00235} = 1.1478 \times 10^4$$

$$\sum \frac{1}{\gamma_i} = \frac{1}{0.0023} + \frac{1}{0.00238} + \frac{1}{0.00235} = 1.2805 \times 10^3$$

Using the formula for $\lambda(t)$ given by Eq. (11.14), we have

$$\lambda(t) = \frac{2P_D(t) + \sum\left(\dfrac{\beta_i}{\gamma_i}\right)}{\sum\left(\dfrac{1}{\gamma_i}\right)}$$

We thus have, using the given data,

$$\lambda(t) = \frac{2P_D(t) + 1.1478 \times 10^4}{1.2805 \times 10^3}$$

For $P_D = 200$ MW $\quad (0 < t < 10 \text{ h})$
$\lambda(t) = 9.2763$
For $P_D(t) = 500$ MW $\quad (10 < t < 20 \text{ h})$
$\lambda(t) = 9.7448$
For $P_D(t) = 300$ MW $\quad (20 < t < 24 \text{ h})$
$\lambda(t) = 9.4325$

Using the above, we get for $0 < t < 10$ h,

$$P_1 = 131.80 \text{ MW}$$
$$P_2 = 49.885 \text{ MW}$$
$$P_3 = 18.35 \text{ MW}$$

Similarly for $10 < t < 20$ h, we get

$$P_1 = 233.66 \text{ MW}$$
$$P_2 = 148.29 \text{ MW}$$
$$P_3 = 118.05 \text{ MW}$$

Finally for $20 < t < 24$ h, we get

$$P_1 = 165.75 \text{ MW}$$
$$P_2 = 82.66 \text{ MW}$$
$$P_3 = 51.59 \text{ MW}$$

Figure 11-15 shows sketches of λ, P_1, P_2, and P_3 time variations.

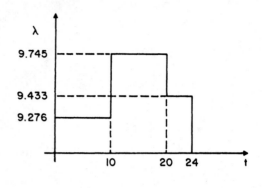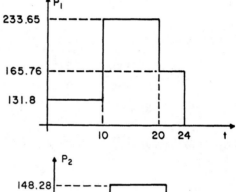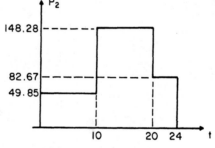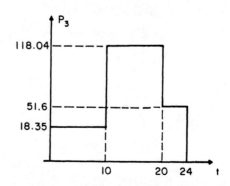

Figure 11-15. System λ and Optimal Generations for System of Problem 11-A-1.

Problem 11-A-2

Two oil-fired 400-MW thermal units are on the same bus. The unit cost models are given by

$$F_1 = 180 + 9P_1 + 0.002P_1^2$$
$$F_2 = 180 + \beta P_2 + \gamma P_2^2$$

A. For a power demand of 550 MW, the incremental cost of power λ is 10. Find the loading on both units assuming optimal operation.
B. For a power demand of 700 MW, the incremental cost of power λ is 10.25. Find the loading on both units assuming optimal operation.
C. Using data in parts A. and B. above, obtain β and γ, the coefficients of the second unit model.

Solution

A. For optimal operation we require that
$$9 + 0.004P_1 = \lambda$$
Given $\lambda = 10$,
$$P_1 = \frac{10 - 9}{0.004}$$
$$= 250 \text{ MW}$$
But $P_1 + P_2 = 550$. Therefore,
$$P_2 = 300 \text{ MW}$$

B. This is similar to part A:
$$9 + 0.004P_1 = 10.25$$
Thus
$$P_1 = 312.5 \text{ MW}$$
$$P_2 = 700 - 312.5$$
$$= 387.5 \text{ MW}$$

C. The second unit's loading satisfies
$$\beta + 2\gamma P_2 = \lambda$$
From parts A and B we have
$$\beta + 600\gamma = 10$$
$$\beta + 775\gamma = 10.25$$
Solving for β and γ, we get
$$\beta = 9.143$$
$$\gamma = 1.4286 \times 10^{-3}$$

Problem A-11-3

A. Assume that for two thermal generating units on the same bus, the cost models are given by

$$F_1 = \alpha_1 + \beta_1 P_1 + \gamma_1 P_1^2$$
$$F_2 = \alpha_2 + \beta_2 P_2 + \gamma_2 P_2^2$$

State the conditions for achieving minimum total cost when the two units supply a power demand P_D. Use the stated conditions to show that the minimum cost F_{min} is related to the incremental cost of power delivered λ by

$$F_{min} = F_0 + \left(\frac{\lambda}{2}\right)^2 \left(\frac{1}{\gamma_1} + \frac{1}{\gamma_2}\right)$$

with

$$F_0 = \left(\alpha_1 - \frac{\beta_1^2}{4\gamma_1}\right) + \left(\alpha_2 - \frac{\beta_2^2}{4\gamma_2}\right)$$

B. Two identical oil-fired 300-MW thermal units on the same bus have the following cost models:

$$F_1 = 180.68 + 9.039 P_1 + 0.00238 P_1^2 \text{ GJ/h}$$

At a certain load level, the minimum total cost for the overall plant is

$$F_{min} = 4205.16$$

Obtain the incremental cost of power delivered λ and the power demand P_D.

Solution

A. For optimal operation we have

$$\beta_1 + 2\gamma_1 P_1 = \lambda$$
$$\beta_2 + 2\gamma_2 P_2 = \lambda$$

The cost function F is

$$F = F_1 + F_2$$

where

$$F_1 = \alpha_1 + \beta_1 P_1 + \gamma_1 P_1^2$$
$$F_2 = \alpha_2 + \beta_2 P_2 + \gamma_2 P_2^2$$

The optimal cost in terms of λ is

$$F_1 = \alpha_1 + \beta_1 \left(\frac{\lambda - \beta_1}{2\gamma_1}\right) + \gamma_1 \left(\frac{\lambda - \beta_1}{2\gamma_1}\right)^2$$

Expanding, we have

$$F_1 = \alpha_1 - \frac{\beta_1^2}{2\gamma_1} + \frac{\beta_1 \lambda}{2\gamma_1} + \frac{\beta_1^2}{4\gamma_1} - \frac{\beta_1 \lambda}{2\gamma_1} + \left(\frac{\lambda}{2}\right)^2 (\gamma_1)^{-1}$$

This results in

$$F_{1_{min}} = F_{0_1} + \left(\frac{\lambda}{2}\right)^2 \left(\frac{1}{\gamma_1}\right)$$

with

$$F_{0_1} = \alpha_1 - \left(\frac{\beta_1^2}{4\gamma_1}\right)$$

F_2 can be treated in a similar way, which results in the required answer.

B. We have two identical units. Therefore,

$$\alpha_1 = \alpha_2 = 180.68$$
$$\beta_1 = \beta_2 = 9.039$$
$$\gamma_1 = \gamma_2 = 0.00238$$

We also have

$$F_{1_{min}} = \frac{F_{min}}{2} = 2102.58$$

Now

$$F_{0_1} = 180.68 - \left[\frac{(9.039)^2}{4 \times 0.00238}\right]$$
$$= -8401.62$$

Using

$$F_{1_{min}} = F_{0_1} + \left(\frac{\lambda}{2}\right)^2 \left(\frac{1}{\gamma_1}\right)$$

Then

$$2102.58 = -8401.62 + \left(\frac{\lambda}{2}\right)^2 \left(\frac{1}{0.00238}\right)$$

$$\left(\frac{\lambda}{2}\right)^2 = 25$$

$$\lambda = 10$$

For optimality

$$9.039 + 0.00476 P_1 = 10$$

Therefore,

$$P_1 = 201.89$$

But
$$P_2 = P_1$$
Hence,
$$P_D = 2P_1 = 403.78 \text{ MW}$$

Problem 11-A-4

Assume that m thermal units in a plant are operating in parallel to supply a power demand P_D. We are interested in obtaining the cost parameters α_E, β_E, and γ_E of the single equivalent unit with cost function defined by:

$$F_T(P_D) = \alpha_E + \beta_E P_D + \gamma_E P_D^2$$

This equivalent representation is to be based on the assumption that the units share the load optimally assuming quadratic cost models. Show that the equivalent parameters are given by

$$\gamma_E^{-1} = \sum_{i=1}^{m} \left(\frac{1}{\gamma_i}\right) \tag{11.60}$$

$$\frac{\beta_E}{\gamma_E} = \sum_{i=1}^{m} \left(\frac{\beta_i}{\gamma_i}\right) \tag{11.61}$$

$$\alpha_E = \sum_{i=1}^{m} \left[\alpha_i - \left(\frac{\beta_i^2}{4\gamma_i}\right)\right] + \frac{\beta_E^2}{4\gamma_E} \tag{11.62}$$

Solution

For any loading of the plant, the total cost is given by Eq. (11.5) as

$$F_T = \sum_{i=1}^{m} \left(\alpha_i + \beta_i P_i + \gamma_i P_i^2\right)$$

Our optimality conditions are given by Eqs. (11.14) and (11.15). Utilizing the definitions of β_E and γ_E, we can write Eq. (11.14) as

$$\lambda = \beta_E + 2\gamma_E P_D$$

The total fuel cost can thus be written as

$$F_T = \sum_{i=1}^{m} \left[\alpha_i + \beta_i \left(\frac{\beta_E - \beta_i + 2\gamma_E P_D}{2\gamma_i}\right)\right. $$
$$\left. + \gamma_i \left(\frac{\beta_E - \beta_i + 2\gamma_E P_D}{2\gamma_i}\right)^2\right]$$

It is clear that the total cost is a quadratic in the power demand. Expanding terms and rearranging, we have

$$F_T = \left\{ \sum_{i=1}^{m} (\alpha_i) + (\beta_E) \sum_{i=1}^{m} \left(\frac{\beta_i}{2\gamma_i}\right) \right.$$
$$\left. - \sum_{i=1}^{m} \left(\frac{\beta_i^2}{2\gamma_i}\right) + \sum_{i=1}^{m} \frac{(\beta_E - \beta_i)^2}{4\gamma_i} \right\}$$
$$+ \left\{ \sum_{i=1}^{m} \left(\frac{\gamma_E \beta_i}{\gamma_i}\right) + \left[\frac{(\beta_E - \beta_i)\gamma_E}{\gamma_i}\right] \right\} P_D$$
$$+ \left[\sum_{i=1}^{m} \left(\frac{\gamma_E^2}{\gamma_i}\right) \right] P_D^2$$

Using the definitions of β_E and γ_E, we see that the above reduces to

$$F_T = \left[\sum_{i=1}^{m} (\alpha_i) + \left(\frac{\beta_E^2}{2\gamma_E}\right) - \sum_{i=1}^{m} \left(\frac{\beta_i^2}{2\gamma_i}\right) \right.$$
$$\left. + \left(\frac{\beta_E^2}{4\gamma_E}\right) + \sum_{i=1}^{m} \left(\frac{\beta_i^2}{4\gamma_i}\right) - \left(\frac{\beta_E^2}{2\gamma_E}\right) \right]$$
$$+ \beta_E P_D + \gamma_E P_D^2$$

It is thus evident that the equivalent parameter expressions result in

$$F_T = \alpha_E + \beta_E P_D + \gamma_E P_D^2$$

An alternate method can be used to find γ_E and β_E. For optimal operation, the incremental cost is given by

$$\lambda = \frac{\partial F_T}{\partial P_D} = \beta_E + 2\gamma_E P_D$$

This is also given by

$$\lambda = \left[\frac{2P_D + \sum_{i=1}^{m} \left(\frac{\beta_i}{\gamma_i}\right)}{\sum_{i=1}^{m} \left(\frac{1}{\gamma_i}\right)} \right]$$

Equating the two expressions, we get

$$\beta_E + 2\gamma_E P_D = \left[\frac{\sum_{i=1}^{m} \left(\frac{\beta_i}{\gamma_i}\right)}{\sum_{i=1}^{m} \left(\frac{1}{\gamma_i}\right)} \right] + \left[\frac{2}{\sum_{i=1}^{m} \left(\frac{1}{\gamma_i}\right)} \right] P_D$$

The coefficients of P_D should be identical on both sides. Therefore,

$$\gamma_E = \left[\sum_{i=1}^{m}\left(\frac{1}{\gamma_i}\right)\right]^{-1}$$

The absolute coefficient on both sides gives

$$\beta_E = \frac{\sum_{i=1}^{m}\left(\frac{\beta_i}{\gamma_i}\right)}{\sum_{i=1}^{m}\left(\frac{1}{\gamma_i}\right)}$$

The above gives

$$\frac{\beta_E}{\gamma_E} = \sum_{i=1}^{m}\left(\frac{\beta_i}{\gamma_i}\right)$$

Problem 11-A-5

The fuel cost function for each of the two units at a thermal plant is given by $F(P) = 82.68 + 6.69P + 4.7675 \times 10^{-3}P^2$ MBtu/h where P is given in MW. Find the single unit equivalent of the plant. Assume that the two units are identical.

Suppose that the two units are not identical. In particular, the model given above applies to unit 1, and unit 2 has model parameters such that

$$F_2 = xF_1$$

with $x = 1.1$. Find the loss in economy if the units are assumed identical for a system load of 1000 MW.

Solution

Assuming the two units are identical, we can use results from Problem 11-A-4 to obtain

$$\gamma_E = \frac{\gamma}{2}$$
$$\beta_E = \beta$$
$$\alpha_E = 2\alpha$$

Using the given data, we obtain the following single unit equivalent:

$$F_{T_a}(P_D) = 165.36 + 6.69P_D + 2.38375 \times 10^{-3}P_D^2$$

With unidentical units,

$$\alpha_2 = 1.1\alpha_1$$
$$\beta_2 = 1.1\beta_1$$
$$\gamma_2 = 1.1\gamma_1$$

the equivalent single unit parameters are
$$\gamma_E^{-1} = \frac{1}{\gamma_1} + \frac{1}{1.1\gamma_1} = \frac{1.9091}{4.7675 \times 10^{-3}}$$
Thus
$$\gamma_{E_b} = 2.4973 \times 10^{-3}$$
$$\beta_{E_b} = \gamma_E \left(\frac{2\beta_1}{\gamma_1}\right) = 7.0087$$
$$\alpha_{E_b} = \alpha_1(1 + 1.1) - \frac{\beta_1^2}{4\gamma_1}(1 + 1.1) + \frac{\beta_{E_b}^2}{4\gamma_{E_b}}$$
$$= 2.1 \left[82.68 - \frac{(6.69)^2}{4 \times 4.7675 \times 10^{-3}}\right]$$
$$+ \frac{(7.0087)^2}{4 \times 2.4973 \times 10^{-3}}$$
$$= 162.5271$$

Hence we have
$$F_T(P_D) = 162.5271 + 7.008 P_D + 2.4973 \times 10^{-3} P_D^2$$
The total cost for the unidentical units operating optimally is
$$F_T = 162.5271 + (7.0087)(1000) + 2.4973 \times 10^{-3}(1000)^2$$
$$= 9668.53$$
If one erroneously assumes identical units, then loading will be equal. Therefore for
$$P_1 = P_2 = 500 \text{ MW}$$
the total operating cost in this case is
$$F_T = (82.68)(2.1) + 6.69(2.1)(500)$$
$$+ 4.7675 \times 10^{-3}(2.1)(500)^2$$
$$= 9701.07$$
Loss of economy is thus
$$\varepsilon = 9701.07 - 9668.53$$
$$= 32.54 \text{ MBtu/h}$$

Problem 11-A-6

The fuel costs of two coal-fired units are expressed in terms of third-order models as shown below:
$$F_{1c} = 119.7 + 10.33 P_1 - 0.012 P_1^2 + 0.039 \times 10^{-3} P_1^3 \text{ \$/h}$$
$$F_{2c} = 213.87 + 9.48 P_2 - 0.0044 P_2^2 + 0.0079 \times 10^{-3} P_2^3 \text{ \$/h}$$

754 Optimal Operation of Electric Power Systems

where P is in MW as usual in this type of problem. At a certain power demand, the incremental cost of power delivered is found to be $\lambda_c = 10$, for optimal economic operation of the plant.

A. Find the corresponding optimal power output of each of the two units and hence the power demand.

B. To reduce the computational complexity, we use quadratic cost models. For the two units of this problem, the corresponding models are

$$F_{1s} = 173.61 + 8.67P_1 + 0.0023P_1^2 \text{ \$/h}$$
$$F_{2s} = 300.84 + 8.14P_2 + 0.0015P_2^2 \text{ \$/h}$$

For the power demand of part A., find the incremental cost of power delivered λ_s, assuming quadratic models are used. Compute the optimal power output of each unit in this case.

C. The actual test cost figures for each of the two units in the neighborhood of the optimal solution are

$$F_1(160) = 1614.4$$
$$F_1(200) = 2002.00$$
$$F_2(320) = 3049.597$$
$$F_2(400) = 3796.00$$

Construct linear models of the cost of the form

$$F_i = \alpha_i + \beta_i P_i$$

for each of the two units in the ranges given.

D. Use the models obtained in part C. to compare the operating costs corresponding to the schedules of parts A. and B. What conclusions can you draw from this comparison?

Solution

A. For optimal operation we have the following conditions:

$$\frac{\partial F_{1c}}{\partial P_1} = 10.33 - 0.024P_1 + 0.117 \times 10^{-3}P_1^2 = \lambda = 10$$

$$\frac{\partial F_{2c}}{\partial P_2} = 9.48 - 0.0088P_2 + 0.0237 \times 10^{-3}P_2^2 = \lambda = 10$$

Solving the quadratic equations in P_1 and P_2, we obtain

$$P_1 = 190.31 \quad \text{or} \quad 14.82$$
$$P_2 = 423.16 \quad \text{or} \quad -51.85$$

We choose the first roots, and hence the power demand is obtained

as
$$P_D = P_1 + P_2 = 613.47 \text{ MW}$$

B. Following the same lines of reasoning as above, we have

$$\frac{\partial F_{1s}}{\partial P_1} = 8.67 + 0.0046 P_1 = \lambda$$

$$\frac{\partial F_{2s}}{\partial P_2} = 8.14 + 0.003 P_2 = \lambda$$

We do not have the value of λ available, but the demand is given. Hence we appeal to the following:

$$\lambda_s = \frac{2 P_D + \sum_{i=1}^{m}\left(\frac{\beta_i}{\gamma_i}\right)}{\sum_{i=1}^{m}\left(\frac{1}{\gamma_i}\right)}$$

This gives

$$\lambda_s = \frac{2 \times 613.4 + \left(\dfrac{8.67}{0.0023} + \dfrac{8.14}{0.0015}\right)}{\left(\dfrac{1}{0.0023} + \dfrac{1}{0.0015}\right)} = 9.463$$

From the optimality conditions, we thus get

$$P_1 = 172.39 \text{ MW}$$
$$P_2 = 441.01 \text{ MW}$$

C. For the first unit, we have

$$F_1 = \alpha_1 + \beta_1 P_1$$

Substituting the given data, we get

$$1614.4 = \alpha_1 + 160 \beta_1$$
$$2002.0 = \alpha_1 + 200 \beta_1$$

Therefore,

$$\beta_1 = 9.69$$
$$\alpha_1 = 64$$

Similarly for the second unit, we have

$$F_2 = \alpha_2 + \beta_2 P_2$$

Substituting the given data, we get

$$3049.557 = \alpha_2 + 320 \beta_2$$
$$3796.00 = \alpha_2 + 400 \beta_2$$

Thus,
$$\beta_2 = 9.33$$
$$\alpha_2 = 63.99$$

The required linear models are then
$$F_1 = 64 + 9.69 P_1$$
$$F_2 = 63.99 + 9.33 P_2$$

D. We use the models given above to calculate the cost of the schedule obtained in part (a):
$$F_1 = 64 + (9.69)(190.31) = 1908.1$$
$$F_2 = 63.99 + (9.33)(423.16) = 4012.07$$

The total cost is thus
$$F_T = 5920.18 \ \$/h$$

For the schedule of part (b), we have
$$F_1 = 64 + (9.69)(172.39) = 1734.46$$
$$F_2 = 63.99 + (9.33)(441.01) = 4178.61$$
$$F_T = 5913.07$$

We may conclude that the difference in cost between the two alternatives does not warrant the effort required for scheduling using the cubic model.

Problem 11-A-7

Consider a simple power system consisting of two generating plants and one load as shown in Figure 11-16.

A. Derive from basic principles the loss formula expression
$$P_L = B_{11} P_1^2 + B_{22} P_2^2$$

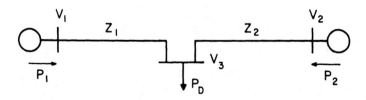

Figure 11-16. System for Problem 11-A-7.

with loss formula coefficients given by

$$B_{11} = \frac{R_1}{|V_1|^2 (PF_1)^2}$$

$$B_{22} = \frac{R_2}{|V_2|^2 (PF_2)^2}$$

B. Calculate the p.u. loss coefficients B_{11} and B_{22} for this system, given that

$$Z_1 = 0.04 + j0.16 \text{ p.u.}$$
$$Z_2 = 0.03 + j0.12 \text{ p.u.}$$
$$V_1 = 1 \text{ p.u.}$$
$$V_2 = 1.03 \text{ p.u.}$$
$$PF_1 = 0.85$$
$$PF_2 = 0.8$$

C. Given that the base MVA for this system is 100 MVA, write the active power balance equation for this system in terms of generations in MW.

Solution

A. For the given system we can write the transmission losses as

$$P_L = 3|I_1|^2 R_1 + 3|I_2|^2 R_2$$

where the current magnitudes are given by

$$|I_1| = \frac{P_1}{\sqrt{3}|V_1|(PF_1)}$$

$$|I_2| = \frac{P_2}{\sqrt{3}|V_2|(PF_2)}$$

Note that P_1 and P_2 are three-phase powers. As a result, we obtain the loss expression:

$$P_L = \frac{R_1}{|V_1|^2 (PF_1)^2}(P_1^2) + \frac{R_2}{|V_2|^2 (PF_2)^2}(P_2^2)$$

The loss expression stipulated is

$$P_L = B_{11} P_1^2 + B_{22} P_2^2$$

By comparison of the two expressions, we conclude that

$$B_{11} = \frac{R_1}{|V_1|^2(PF_1)^2}$$

$$B_{22} = \frac{R_2}{|V_2|^2(PF_2)^2}$$

B. Substituting the given parameters in the above expressions, we obtain

$$B_{11} = \frac{0.04}{(1)^2(0.85)^2} = 0.0554 \text{ p.u.}$$

$$B_{22} = \frac{0.03}{(1.03)^2(0.8)^2} = 0.0442 \text{ p.u.}$$

C. We consider first the conversion from the p.u. system to MW units. We have the conversion formula applied to losses:

$$P_{L_{MW}} = (P_{L_{pu}})(P_{\text{base}}) \qquad (11.63)$$

In the per unit system, we have

$$P_{L_{pu}} = B_{pu} P_{pu}^2$$

But

$$P_{pu} = \frac{P_{MW}}{P_{\text{base}}}$$

Therefore,

$$P_{L_{pu}} = B_{pu} \left(\frac{P_{MW}}{P_{\text{base}}}\right)^2$$

Substituting in Eq. (11.63), we thus have

$$P_{L_{MW}} = \left(\frac{B_{pu}}{P_{\text{base}}}\right)(P_{MW}^2)$$

We wish to have

$$P_{L_{MW}} = B_{MW^{-1}} P_{MW}^2$$

A simple comparison results in the conversion formula:

$$B_{MW^{-1}} = \frac{B_{pu}}{P_{\text{base}}} \qquad (11.64)$$

Application of the above formula to results of part (b), given that

$$P_{\text{base}} = 100 \text{ MVA}$$

results in

$$B_{11} = \frac{0.0554}{100} = 0.554 \times 10^{-3} \text{MW}^{-1}$$
$$B_{22} = \frac{0.0442}{100} = 0.442 \times 10^{-3} \text{MW}^{-1}$$

Problem 11-A-8

A simple power system consists of two generating plants and a load center. The power loss in the system is given by

$$P_L = 0.554 \times 10^{-3} P_1^2 + 0.442 \times 10^{-3} P_2^2$$

A. Obtain the penalty factors L_1 and L_2 for the two plants in terms of the generated powers.

B. Given a certain demand, plant 1 generates 100 MW for optimal operation. Find the incremental cost of power delivered λ, power generated by plant 2, power losses, and power demand. Assume the following cost coefficients:

$$\beta_1 = 8.21 \qquad \gamma_1 = 0.00225$$
$$\beta_2 = 8.56 \qquad \gamma_2 = 0.00235$$

Solution

A. The penalty factors L_i are defined by

$$L_i^{-1} = 1 - \frac{\partial P_L}{\partial P_i}$$

We have

$$\frac{\partial P_L}{\partial P_1} = 2(0.554 \times 10^{-3}) P_1$$

$$\frac{\partial P_L}{\partial P_2} = 2(0.442 \times 10^{-3}) P_2$$

Therefore we have

$$L_1^{-1} = 1 - 1.108 \times 10^{-3} P_1$$
$$L_2^{-1} = 1 - 0.884 \times 10^{-3} P_2$$

B. The optimality condition for plant 1 is

$$\frac{\beta_1 + 2\gamma_1 P_1}{1 - 1.108 \times 10^{-3} P_1} = \lambda$$

We have

$$\beta_1 = 8.21 \qquad \text{and} \qquad \gamma_1 = 0.00225$$

Also
$$P_1 = 100 \text{ MW}$$
Therefore,
$$\lambda = \frac{8.21 + 2(0.00225)(100)}{1 - 1.108 \times 10^{-1}}$$
This gives
$$\lambda = 9.7391$$
For plant 2, we have
$$\frac{\beta_2 + 2\gamma_2 P_2}{1 - 0.884 \times 10^{-3} P_2} = \lambda$$
Substituting
$$\beta_2 = 8.56 \quad \text{and} \quad \gamma_2 = 0.00235$$
we obtain
$$P_2 = 88.5918$$
The power loss is calculated as
$$P_L = 0.554 \times 10^{-3}(100)^2 + (0.442 \times 10^{-3})(88.5918)^2$$
$$= 9.0090$$
Consequently, the power demand is
$$P_D = P_1 + P_2 - P_L$$
Numerically we then have
$$P_D = 100 + 88.5918 - 9.0090$$
$$= 179.58$$

Problem 11-A-9

Two identical thermal plants have the following cost model parameters:
$$\beta = 6.7$$
$$\gamma = 4.8 \times 10^{-3}$$
The transmission network has the following B-coefficients:
$$B_{11} = 0.5 \times 10^{-3}$$
$$B_{22} = 0.2 \times 10^{-3}$$
Find the optimal values of P_1 and P_2 for a power demand of 500 MW.

Solution

The optimality conditions for this case give

$$\frac{\beta + 2\gamma P_1}{1 - 2B_{11}P_1} = \frac{\beta + 2\gamma P_2}{1 - 2B_{22}P_2}$$

Cross-multiplication and further manipulations yield

$$(\gamma + \beta B_{11})P_1 - (\gamma + \beta B_{22})P_2 - 2\gamma(B_{22} - B_{11})P_1 P_2 = 0$$

Substituting numerical values, we then have

$$[4.8 \times 10^{-3} + (6.7)(0.5 \times 10^{-3})]P_1$$
$$- [4.8 \times 10^{-3} + (6.7)(0.2 \times 10^{-3})]P_2$$
$$+ 9.6 \times 10^{-3}[0.3 \times 10^{-3}]P_1 P_2 = 0$$

This reduces further to

$$8.15 P_1 - 6.14 P_2 + 2.88 \times 10^{-3} P_1 P_2 = 0$$

or

$$P_2 = \frac{8.15 P_1}{6.14 - 2.88 \times 10^{-3} P_1}$$

The power balance equation for this system is

$$P_1 + P_2 - B_{11}P_1^2 - B_{22}P_2^2 - P_D = 0$$

Using the numerical values given and substituting for P_2 in terms of P_1, we obtain

$$P_1 + \left(\frac{8.15 P_1}{6.14 - 2.88 \times 10^{-3} P_1}\right) - 0.5 \times 10^{-3} P_1^2$$
$$- 0.2 \times 10^{-3} \left(\frac{8.15 P_1}{6.14 - 2.88 \times 10^{-3} P_1}\right)^2 - 500 = 0$$

Further manipulations yield

$$1.8850 \times 10^4 - 105.424 P_1 + 9.512 \times 10^{-2} P_1^2$$
$$- 2.5978 \times 10^{-5} P_1^3 + 4.1472 \times 10^{-9} P_1^4 = 0$$

The solution to the above equation is

$$P_1 = 219.903 \text{ MW}$$

Consequently, we obtain

$$P_2 = 325.461 \text{ MW}$$

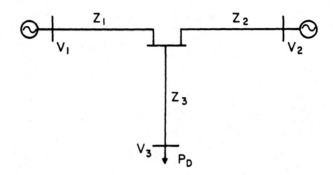

Figure 11-17. System for Problem 11-A-10.

Problem 11-A-10

Calculate the loss coefficients B_{11}, B_{12}, and B_{22} for the system shown in Figure 11-17. Assume a flat voltage profile.

$$V_1 = V_2 = V_3 = 1 \text{ p.u.}$$
$$R_1 = 0.0025 \text{ p.u.}$$
$$R_2 = 0.02 \text{ p.u.}$$
$$R_3 = 0.03 \text{ p.u.}$$
$$PF_1 = 0.85$$
$$PF_2 = 0.8$$
$$PF_3 = 0.75$$

The MW base for the system is 100 MW. Calculate the penalty factors L_1 and L_2 and the corresponding power loss for the two plants for an optimum loading of

$$P_1 = 120 \text{ MW}$$
$$P_2 = 100 \text{ MW}$$

Assume that the fuel-cost models for the two plants are given by

$$F_1 = \alpha + 6.69P + 4.7675 \times 10^{-3}P^2 \text{ \$/h}$$
$$F_2 = \alpha + 6.69P + \gamma P^2 \text{ \$/h}$$

If P is in MW, find the incremental cost of power delivered λ and the parameter γ.

Solution

$$B_{11} = \frac{R_1}{(V_1)^2(PF)_1^2} + \frac{R_3}{(V_3)^2(PF)_3^2}$$

$$= \frac{0.0025}{(1)^2(0.85)^2} + \frac{0.03}{(1)^2(0.75)^2}$$

$$= 0.0568 \text{ p.u.}$$

$$B_{22} = \frac{R_2}{(V_2)^2(PF)_2^2} + \frac{R_3}{(V_3)^2(PF)_3^2}$$

$$= \frac{0.02}{(1)^2(0.8)^2} + \frac{0.03}{(1)^2(0.75)^2}$$

$$= 0.0846 \text{ p.u.}$$

$$B_{12} = \frac{R_3}{(V_3)^2(PF)_3^2} = \frac{0.03}{(1)^2(0.75)^2} = 0.0533 \text{ p.u.}$$

In terms of MW units, we have

$$B_{11} = 5.679 \times 10^{-4}$$
$$B_{22} = 8.4583 \times 10^{-4}$$
$$B_{12} = 5.333 \times 10^{-4}$$

For

$$P_1 = 120 \text{ MW}$$
$$P_2 = 100 \text{ MW}$$

we have

$$\frac{\partial P_L}{\partial P_1} = 2B_{11}P_1 + 2B_{12}P_2$$

$$= (2)(5.679 \times 10^{-2})(1.2) + (2)(5.333 \times 10^{-2})(1)$$
$$= 0.24296$$

$$1 - \frac{\partial P_L}{\partial P_1} = 0.75704$$

$$L_1 = (0.75704)^{-1} = 1.3209$$

$$\frac{\partial P_L}{\partial P_2} = 2B_{12}P_1 + 2B_{22}P_2$$

$$= (2)(5.333 \times 10^{-2})(1.2) + (2)(8.4583 \times 10^{-2})(1)$$
$$= 0.29716$$

$$1 - \frac{\partial P_L}{\partial P_2} = 0.70284$$

$$L_2 = (0.70284)^{-1} = 1.4228$$

The optimality condition for the first unit is

$$\lambda = L_1 \frac{\partial F_1}{\partial P_1}$$

Substituting numerical values,

$$\lambda = (1.3209)[6.69 + (2)(4.7675 \times 10^{-3})(120)]$$
$$= 10.348$$

For the second unit,

$$\lambda = L_2 \left(\frac{\partial F_2}{\partial P_2} \right)$$

$$10.348 = (1.4228)[6.69 + 2\gamma(100)]$$

Solving for γ, we obtain

$$\gamma = 2.9149 \times 10^{-3}$$

Problem 11-A-11

A two-plant thermal system has the following loss coefficients:

$$B_{11} = 0.4 \times 10^{-3} \text{ MW}^{-1}$$
$$B_{22} = 0.3 \times 10^{-3} \text{ MW}^{-1}$$
$$B_{12} = 0$$

A. For a power demand of 600 MW, the transmission losses are 80 MW. Find the power generated by each plant under these conditions.

B. There are two solutions to the problem. Given that each plant is rated at 400 MW, which solution is feasible?

C. The two plants have fuel-cost expressions given by

$$F_1 = 2P_1 + 2 \times 10^{-3} P_1^2 \text{ \$/h}$$
$$F_2 = 2P_2 + \gamma_2 P_2^2 \text{ \$/h}$$

Assuming that the feasible solution of part B. is optimal, calculate γ_2 and the incremental cost of power λ.

D. Compare the total costs for the feasible and the nonfeasible schedules.

Solution

A. The power losses are given by the expression

$$P_L = B_{11} P_1^2 + B_{22} P_2^2$$

We use the given data to get

$$80 = 0.4 \times 10^{-3} P_1^2 + 0.3 \times 10^{-3} P_2^2$$

The total generated power is
$$P_1 + P_2 = P_D + P_L$$
Therefore,
$$680 = P_1 + P_2$$
We can thus eliminate P_2 from the loss expression to get
$$0.4 \times 10^{-3} P_1^2 + 0.3 \times 10^{-3}(680 - P_1)^2 - 80 = 0$$
Simplifying, we have
$$0.7 \times 10^{-3} P_1^2 - 0.41 P_1 + 58.72 = 0$$
The solution is obtained as
$$P_1 = \frac{582.86 \pm 64.65}{2}$$
The two roots are thus
$$P_1 = 323.75 \quad \text{or} \quad 259.10 \text{ MW}$$
The corresponding values of P_2 are
$$P_2 = 356.25 \quad \text{or} \quad 420.90 \text{ MW}$$

B. The second solution is not feasible since in this case $P_2 > 400$ MW.

C. For optimality, we have for the first plant,
$$\beta_1 + 2\gamma_1 P_1 = \lambda(1 - 2B_{11} P_1)$$
Using the given data,
$$2 + (2)(2 \times 10^{-3})(323.75) = \lambda[1 - (0.8 \times 10^{-3})(323.75)]$$
Thus,
$$\lambda = \frac{3.295}{0.741}$$
$$= 4.4467$$
For the second plant we have
$$2 + 2\gamma_2(356.25) = (4.4467)[1 - (0.6 \times 10^{-3})(356.25)]$$
This results in
$$\gamma_2 = 2.10 \times 10^{-3}$$

D. The cost corresponding to the feasible solution is calculated as follows:
$$F(P_1) = (2)(323.75) + (2 \times 10^{-3})(323.75)^2 = 857.13$$
$$F(P_2) = (2)(356.25) + (2.1 \times 10^{-3})(356.25)^2 = 979.02$$
Thus the total cost is
$$F_{T_A} = 857.13 + 979.02 = 1836.15 \text{ \$/h}$$

For the nonfeasible solution, we have

$$F(P_1) = (2)(259.10) + (2 \times 10^{-3})(259.10)^2 = 652.47$$
$$F(P_2) = 2(420.9) + (2.1 \times 10^{-3})(420.9)^2 = 1213.83$$

Thus the total cost is

$$F_{T_B} = 652.47 + 1213.83 = 1866.29 \ \$/h$$

The nonfeasible solution is more expensive than the feasible solution.

Problem 11-A-12

A hydrothermal electric power system consists of one hydro unit and a thermal unit. The daily load cycle is divided into three periods where the optimal water discharge q is found to be as shown in Figure 11-18. Assume that the thermal plant's cost model is given by

$$F_1 = 3.385 P_1 + 0.007 P_1^2$$

The hydro plant's discharge model is

$$q = 1.8 + 0.14 P_2 + 2.2 \times 10^{-4} P_2^2$$

where q is in MCF/h and P_2 is in MW.

A. Compute the active power generated by the hydro plant for each time interval. What is the available volume of water for the 24-hour period?

B. Assume that transmission losses are negligible and that the optimal water conversion coefficient is 21.00. Compute the incremental

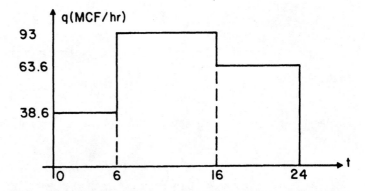

Figure 11-18. Variation of Discharge with Time for Problem 11-A-12.

cost of power delivered and the thermal power generation as well as the power demand for each of the time intervals.

Solution

A. Since we are given q over the three time intervals, we can solve the following quadratic in P_2:
$$q = 1.8 + 0.14 P_2 + 2.2 \times 10^{-4} P_2^2$$
For $0 < t \leq 6$, $q = 38.6$ MCF/h. Therefore we have
$$2.2 \times 10^{-4} P_2^2 + 0.14 P_2 + 1.8 = 38.6$$
$$P_2 = \frac{-0.14 \pm \sqrt{(0.14)^2 + (4)(36.8)(2.2 \times 10^{-4})}}{2 \times 2.2 \times 10^{-4}}$$
$$= 200 \text{ MW}$$
For $6 < t \leq 16$, $q = 93$ MCF/h, we have
$$2.2 \times 10^{-4} P_2^2 + 0.14 P_2 + 1.8 = 93$$
Thus
$$P_2 = \frac{-0.14 \pm \sqrt{(0.14)^2 + (4)(91.2)(2.2 \times 10^{-4})}}{2 \times 2.2 \times 10^{-4}}$$
$$= 400 \text{ MW}$$
For $16 < t \leq 24$, $q = 63.6$ MCF/h, we have
$$2.2 \times 10^{-4} P_2^2 + 0.14 P_2 + 1.8 = 63.6$$
$$P_2 = \frac{-0.14 \pm \sqrt{(0.14)^2 + (4)(61.8)(2.2 \times 10^{-4})}}{2 \times 2.2 \times 10^{-4}}$$
$$= 300 \text{ MW}$$
The total volume of water is obtained as
$$b = \Sigma q(i)(t_i)$$
where t_i is the duration of discharge $q(i)$. Thus we have
$$b = (38.6)(6) + (63.6)(8) + (93)(10)$$
$$= 1670.40 \text{ MCF}$$

B. For optimality we write for the hydro plant:
$$\nu [\beta_2 + 2\gamma_2 P_2] = \lambda$$
Thus we have
$$21(0.14 + 4.4 \times 10^{-4} P_2) = \lambda$$
This yields for each time interval the incremental cost of power

delivered λ as follows:

For $P_2 = 200$ MW, $\lambda = 4.788$
For $P_2 = 300$ MW, $\lambda = 5.712$
For $P_2 = 400$ MW, $\lambda = 6.636$

The optimality condition for the thermal plant is

$$\beta_1 + 2\gamma_1 P_1 = \lambda$$
$$3.385 + 0.014 P_1 = \lambda$$

This results in

$$P_1 = \frac{\lambda - 3.385}{0.014}$$

Consequently we obtain

$P_1 = 100.21$ MW $\lambda = 4.788$,
$P_1 = 166.21$ MW $\lambda = 5.712$,
$P_1 = 232.21$ MW $\lambda = 6.636$,

The power demands can be calculated now using

$$P_D = P_1 + P_2$$

Thus we get

$P_D = 300.21$ MW $(0 < t \leq 6)$
$P_D = 632.21$ MW $(6 < t \leq 16)$
$P_D = 466.21$ MW $(16 < t \leq 24)$

Problem 11-A-13

Losses are negligible in a hydrothermal system with characteristics as follows:

A. Fuel-cost model of the thermal plant is

$$F = 2P_1 + 0.001 P_1^2 \text{ \$/h}$$

B. The rate of water discharge as a function of active power generated is

$$q_2 = 50 P_2 + 0.01 P_2^2 \text{ cfs}$$

where P is in MW.

C. The optimal water conversion coefficient is found to be 12.01 $/MCF.

D. The load on the system is

Duration (h)	P_D (MW)
9	700
15	350

Compute the optimal active thermal and hydro power generations as well as the system's incremental cost of power delivered and the allowable volume of water.

Solution

The optimality equations for the thermal and hydro units are obtained as follows:

$$[2 + 0.002 P_1(t)] = \lambda(t)$$
$$(12.01)(3600 \times 10^{-6})[50 + 0.02 P_2(t)] = \lambda(t)$$

Observe that q is substituted in cf/h and that ν is substituted for in ($/cf). Eliminating $\lambda(t)$, we obtain

$$0.002 P_1(t) - 8.6472 \times 10^{-4} P_2(t) = 0.1618$$

or

$$P_1(t) - 0.43236 P_2(t) = 80.900$$

Now the power balance equation gives

$$P_1(t) + P_2(t) = P_D(t)$$

Upon combining the above two equations, we get

$$1.43236 P_2(t) = P_D(t) - 80.900$$

This gives P_2 for the specified demand as follows:

For $\quad P_D(t_1) = 700, \quad P_2(t_1) = 432.22$ MW

The power balance equation is now used to obtain

$$P_1(t) = 700 - 432.22$$
$$= 267.78 \text{ MW}$$

From the thermal plant's optimal equation:

$$\lambda(t_1) = 2.536$$

Similarly, for $P_D(t_2) = 350$ MW,

$$P_2(t_2) = 187.87 \text{ MW}$$
$$P_1(t_2) = 162.13 \text{ MW}$$
$$\lambda(t_2) = 2.324$$

The volume of water is obtained as the sum

$$b = (9)(3600)\left[(50)(432.22) + 0.01(432.22)^2\right]$$
$$+ (15)(3600)\left[(50)(187.87) + 0.01(187.87)^2\right]$$
$$= 1.287 \times 10^9 \text{ cf}$$

Problem 11-A-14

A two-plant hydrothermal system has the following characteristics:

A. Fuel cost as a function of active power generated at the thermal plant is

$$F = 2.7P_1 + 0.003P_1^2 \; \$/h$$

B. The transmission losses are given by

$$P_L = 1.43 \times 10^{-4} P_2^2$$

C. The rate of water discharge as a function of active power generated at the hydro plant is

$$q_2 = 2380 + 60P_2 \; \text{cfs}$$

D. The following table gives the system's power demand and optimal incremental costs.

Duration (h)	P_D (MW)	λ
14	700	3.911
10	450	3.627

It is required to calculate the optimal active power generated by each of the plants in each of the subintervals, the system power losses, the water conversion coefficient ν, and the allowable volume of water discharged at the hydro plant.

Solution

The optimality equations for the thermal plant are used to obtain

$$2.7 + 0.006P_1(t) = \lambda(t)$$

Note that we have used the fact that

$$\frac{\partial P_L}{\partial P_1} = 0$$

Therefore we have

$$P_1(t) = \frac{\lambda(t) - 2.7}{0.006}$$

Since λ is given for each time interval, we calculate the following: For $\lambda(t_1) = 3.911$

$$P_1(t_1) = 201.83 \; \text{MW}$$

and for $\lambda(t_2) = 3.627$

$$P_1(t_2) = 154.50 \; \text{MW}$$

The power balance equation is

$$P_D(t) + 1.43 \times 10^{-4} P_2^2(t) - P_1(t) - P_2(t) = 0$$

For $t = t_1$,

$$1.43 \times 10^{-4} P_2^2(t_1) - P_2(t_1) + 700 - 201.83 = 0$$

For $t = t_2$,

$$1.43 \times 10^{-4} P_2^2(t_2) - P_2(t_2) + 450 - 154.50 = 0$$

Solving for P_2, we get

$$P_2(t_1) = 539.84 \text{ MW}$$
$$P_2(t_2) = 309.17 \text{ MW}$$

The power losses are thus computed using the given loss expression as

$$P_L(t_1) = 41.67 \text{ MW}$$
$$P_L(t_2) = 13.67 \text{ MW}$$

The optimality condition for the hydro plant is

$$\nu(60)(3600) = \lambda(t)\left[1 - 2.86 \times 10^{-4} P_2(t)\right]$$

Using values at $t = t_1$, we obtain

$$\nu = 1.5311 \times 10^{-5} \text{ \$/cf}$$

The volume of water available is obtained as

$$b = T_1 q_2(t_1) + T_2 q_2(t_2)$$

This turns out to be

$$b = 14(3600)[2380 + (60)(539.84)]$$
$$+ 10(3600)[2380 + (60)(309.17)]$$
$$= 2.506 \times 10^9 \text{ cf}$$

Problem 11-A-15

A three-plant hydrothermal system has the following assumed characteristics:

$$F_1 = 47.84 + 9.50 P_1 + 0.01 P_1^2 \text{ \$/h}$$
$$F_2 = 50.01 + 9.94 P_2 + 0.01 P_2^2 \text{ \$/h}$$

The loss formula coefficients are given by

$$B_{11} = 1.6 \times 10^{-4}$$
$$B_{22} = 1.2 \times 10^{-4}$$
$$B_{33} = 2.2 \times 10^{-4}$$

The rate of water discharge at the hydroplant is given by
$$q_3 = 0.5087 + 0.1011 P_3 + 1.0 \times 10^{-4} P_3^2 \text{ Mcf/h}$$
The optimal schedule is given by
$$P_1 = 100.29 \text{ MW}$$
$$P_2 = 85.21 \text{ MW}$$
$$P_3 = 68.00 \text{ MW}$$

Calculate the volume of water available over a 24-hour period, the system incremental cost of power delivered λ, the system power demand, and the water conversion coefficient.

Solution

Given that
$$P_3 = 68 \text{ MW}$$
then using the discharge-power characteristic, we have
$$q_3 = 0.5087 + (0.1011)(68) + (1 \times 10^{-4})(68)^2$$
$$= 7.85 \text{ Mcf/h}$$
$$b = (T \times q_3)$$
$$= (24)(7.85) = 188.30 \text{ Mcf}$$

The first thermal plant's optimality equation is
$$[9.50 + 0.02 P_1(t)] = \lambda(t)[1 - 3.2 \times 10^{-4} P_1(t)]$$
Using $P_1(t) = 100.29$ MW, we obtain
$$\lambda(t) = 11.89$$
The second thermal plant's optimality equation is
$$[9.94 + 0.02 P_2(t)] = \lambda(t)[1 - 2.4 \times 10^{-4} P_2(t)]$$
Using $P_2(t) = 85.21$, we obtain the same value of $\lambda(t)$. The optimality equation for the hydroplant is
$$\nu[0.1011 + 2 \times 10^{-4} P_3(t)] = \lambda(t)[1 - 4.4 \times 10^{-4} P_3(t)]$$
This results in
$$\nu = 100.56 \text{ \$/MCF}$$
The system power demand is
$$P_D = 100.29 + 85.21 + 68 - (1.6 \times 10^{-4})(100.29)^2$$
$$- (1.2 \times 10^{-4})(85.21)^2 - (2.2 \times 10^{-4})(68)^2$$
$$= 250 \text{ MW}$$

Problem 11-A-16

A hydrothermal electric power system with three plants has the following transmission loss coefficients:

$$B_{11} = 1.6 \times 10^{-4}$$
$$B_{22} = 1.2 \times 10^{-4}$$
$$B_{33} = 2.2 \times 10^{-4}$$

Plants 1 and 2 utilize thermal generation, and plant 3 is hydraulic. The fuel-cost expressions for the thermal plants are

$$F_1 = 47.84 + 9.5P_1 + \gamma_1 P_1^2 \ \$/h$$
$$F_2 = 50.00 + \beta_2 P_2 + 0.01 P_2^2 \ \$/h$$

The values of γ_1 and β_2 are to be determined using available information specifying that for a given constant power demand, the optimal values are:

$$\lambda = 11.89$$
$$P_1 = 100.29 \text{ MW}$$
$$P_2 = 85.21 \text{ MW}$$
$$P_3 = 68.00 \text{ MW}$$

The hydro plant has discharge characteristics

$$q_3 = 0.5087 + 0.1011 P_3 + 1 \times 10^{-4} P_3^2 \ \text{Mcf/h}$$

A. Find γ_1 and β_2.
B. Compute the power loss and the power demand for the given data.
C. Compute the available volume of water over a 24-hour period.
D. Compute the water conversion coefficient $/Mcf.

Solution

A. Optimality conditions at the two thermal plants require

$$\frac{\partial F_1}{\partial P_1} = \lambda \left(1 - \frac{\partial P_L}{\partial P_1}\right)$$

$$\frac{\partial F_2}{\partial P_2} = \lambda \left(1 - \frac{\partial P_L}{\partial P_2}\right)$$

For the present data we have

$$9.5 + 2\gamma_1(100.29) = 11.89\left[1 - (2)(1.6 \times 10^{-4})(100.29)\right]$$

This results in

$$\gamma_1 = 1.0013 \times 10^{-2}$$

Moreover,
$$\beta_2 + 0.02(85.21) = 11.89[1 - (2)(1.2 \times 10^{-4})(85.21)]$$
which gives
$$\beta_2 = 9.9426$$

B. The power losses are obtained using
$$P_L = B_{11} P_1^2 + B_{22} P_2^2 + B_{33} P_3^2$$
With the given numerical values, we get
$$P_L = 3.50 \text{ MW}$$
As a result, we obtain the power demand using
$$P_D = P_1 + P_2 + P_3 - P_L$$
This turns out to be
$$P_D = 250 \text{ MW}$$

C. Since we know the hydro power, we obtain the rate of water discharge using the given characteristic
$$q_3 = 0.5087 + 0.1011(68) + (1 \times 10^{-4})(68)^2$$
This turns out to be
$$q_3 = 7.85 \text{ Mcf/h}$$
Over 24 hours we have
$$b = (7.85)(24) = 188.3 \text{ Mcf}$$

D. The optimality condition for the hydroplant is
$$\nu \frac{\partial q_3}{\partial P_3} = \lambda \left(1 - \frac{\partial P_L}{\partial P_3}\right)$$
With the given data we obtain
$$\nu = \frac{(11.89)[1 - (2)(2.2 \times 10^{-4})(68)]}{[0.1011 + (2)(1 \times 10^{-4})(68)]}$$
Hence the water conversion coefficient is obtained as
$$\nu = 100.56$$

PROBLEMS

Problem 11-B-1

Two thermal units at the same station have the following cost models:
$$F_1 = 793.22 + 7.74 P_1 + 0.00107 P_1^2$$
$$F_2 = 1194.6 + 7.72 P_2 + 0.00072 P_2^2$$

Find the optimal power generated P_1 and P_2 and the incremental cost of power delivered for power demands of 400, 600, and 1000 MW respectively.

Problem 11-B-2

Two thermal units have the following cost model parameters:

$\alpha_1 = 312.35 \qquad \alpha_2 = 483.44$

$\beta_1 = 8.52 \qquad \beta_2 = 8.65$

$\gamma_1 = 0.0015 \qquad \gamma_2 = 0.00056$

Evaluate the parameters of the expression for λ in terms of P_D given by Eq. (11.14). Sketch this variation and use it to obtain the incremental cost of power delivered and optimal generations for power demands of 300 and 500 MW.

Problem 11-B-3

Show that for the system given in Problem 11-B-2 positive power demand exists for which the optimal generations are negative.

Problem 11-B-4

The incremental fuel cost of two thermal units is given by

$$\frac{\partial F_1}{\partial P_1} = 2 + 0.012 P_1$$

$$\frac{\partial F_2}{\partial P_2} = 1.5 + 0.015 P_2$$

The variation of the power demand is as shown in Figure 11-19.

A. Express the power demand as a function of time.

B. For optimal economic operation, find the incremental cost of power delivered $\lambda(t)$, $P_1(t)$, and $P_2(t)$.

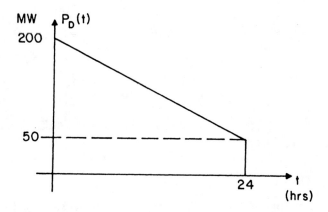

Figure 11-19. Power Demand for Problem 11-B-4.

Problem 11-B-5

The following data are available for the optimal operation of a three-unit thermal system:

P_D	λ	P_1	P_2	P_3
200	9.27627	131.798	?	18.3556
300	9.43246	?	?	51.5877
500	?	233.662	?	?

Complete the table assuming optimal operation neglecting losses. Evaluate the coefficients β_i and γ_i of the quadratic cost models.

Problem 11-B-6

Two thermal units have the following cost models:
$$F_1 = 366.0 + 9.85 P_1 - 0.003 P_1^2 + 3.6 \times 10^{-6} P_1^3$$
$$F_2 = 398 + 10.12 P_2 - 0.0034 P_2^2 + 2.7 \times 10^{-6} P_2^3$$

Assume that losses are negligible. The optimal power generation by unit 1 is 359.266 MW. Calculate the incremental cost of power delivered and the power demand under these conditions.

Problem 11-B-7

Two thermal units have the following cost models:
$$F_1 = 128 + 10.65P_1 - 0.012P_1^2 + 0.000038P_1^3$$
$$F_2 = 232 + 9.77P_2 - 0.004P_2^2 + 0.0000073P_2^3$$
For a power demand of 500 MW, find the optimal power generations P_1 and P_2.

Problem 11-B-8

The third-order polynomial cost model coefficients for three steam units are given as shown below:

$\beta_1 = 9.85$ $\qquad$ $\beta_2 = 10.12$ $\qquad$ $\beta_3 = 9.85$

$\gamma_1 = -0.003$ $\qquad$ $\gamma_2 = -0.0034$ $\qquad$ $\gamma_3 = -0.003$

$\delta_1 = 0.0000036$ $\qquad$ $\delta_2 = 0.0000027$ $\qquad$ $\delta_3 = 0.0000036$

If the incremental cost of power delivered is 9.01826, obtain the optimal power generations and the corresponding power demand.

Problem 11-B-9

An all-thermal two-unit system operates optimally with an incremental cost of power delivered of 9.8 for a power demand of 500 MW with unit 1 supplying 200 MW. The cost model for unit 1 is quadratic with
$$\gamma_1 = 3 \times 10^{-3}$$
Cost of unit 2 is modeled using a cubic with coefficients
$$\gamma_2 = -0.0002$$
$$\delta_2 = 2 \times 10^{-6}$$
Calculate the linear term coefficients β_1 and β_2.

Problem 11-B-10

The optimal power generation by unit 1 in a two-thermal unit system is 100 MW for a power demand of 250 MW. The cost model for the first unit is given by the quadratic.
$$F_1 = \alpha_1 + 12.00P_1 + 3 \times 10^{-3}P_1^2$$
The second unit is modeled using a cubic:
$$F_2 = \alpha_2 + 10.00P_2 - 1.5 \times 10^{-3}P_2^2 + \delta_2 P_2^3$$

Calculate the incremental cost of power delivered and the cubic coefficient δ_2.

Problem 11-B-11

A demand of 1000 MW is supplied by two thermal units that have the following cost models:

$$F_1 = 8.5P_1 + 0.0015P_1^2$$
$$F_2 = 9.00P_2 - 0.004P_2^2 + 9 \times 10^{-6}P_2^3$$

Obtain the optimal share of power generation by each unit to meet the demand. Calculate the incremental cost of power delivered λ.

Problem 11-B-12

The transmission loss equation for a two-plant thermal system is given by

$$P_L = 0.08P_2$$

The fuel-cost models are given by

$$F_1 = 9P_1 + 0.002P_1^2$$
$$F_2 = 8.7P_2 + 0.0018P_2^2$$

Calculate the optimal generations and incremental cost of power delivered λ for a power demand of 600 MW.

Problem 11-B-13

Repeat Problem 11-B-12 for a transmission loss equation given by

$$P_L = 10 + 0.02P_1 + 0.03P_2$$

All other data are unchanged.

Problem 11-B-14

An electric power system with two plants has the following transmission loss equation:

$$P_L = 0.5 \times 10^{-3}P_1^2 + 0.2 \times 10^{-3}P_2^2$$

The fuel-cost models are given by

$$F_1 = 8.52P_1 + 0.0015P_1^2$$
$$F_2 = 8.65P_2 + 0.00056P_2^2$$

Given that the incremental cost of power delivered is 10.9, obtain the optimal power generations and the corresponding power demand.

Problem 11-B-15

Use the Newton-Raphson method to find the optimal power generations and incremental power cost for the system of Problem 11-B-14 for a power demand of 500 MW.

Problem 11-B-16

The fuel-cost models for the two thermal units are as follows:

$$F_1 = 7.74 P_1 + 0.00107 P_1^2$$
$$F_2 = 7.72 P_2 + 0.00072 P_2^2$$

The loss expression is given by

$$P_L = 0.5 \times 10^{-3} P_1^2 + 0.2 \times 10^{-3} P_2^2$$

The optimal generation by plant #1 is 370 MW. Calculate the incremental cost of power delivered, the optimal generation by the second plant, the power loss, and the power demand.

Problem 11-B-17

Use the Newton-Raphson method to find the optimal power generations and incremental power cost for the system of Problem 11-B-16 for a power demand of 1000 MW.

Problem 11-B-18

A two-plant thermal system has transmission losses expressed as

$$P_L = 0.5 \times 10^{-3} P_1^2 + 0.2 \times 10^{-3} P_2^2$$

The cost models are given by

$$F_1 = 10.12 P_1 - 0.0034 P_1^2 + 0.0000027 P_1^3$$
$$F_2 = \beta_2 P_2 - 0.0016 P_2^2 + 0.0000009 P_2^3$$

For a certain power demand, the power generations for minimum cost are found to be

$$P_1 = 350 \text{ MW}$$
$$P_2 = 870 \text{ MW}$$

Calculate the value of λ, β_2, and the power demand.

Problem 11-B-19

The cost models for two plants are given by cubic models with coefficients given by

$$\beta_1 = 10.12 \quad\quad \beta_2 = 9.58$$
$$\gamma_1 = -0.0034 \quad\quad \gamma_2 = -0.0016$$
$$\delta_1 = 0.0000027 \quad\quad \delta_2 = 0.0000009$$

The transmission loss expression is

$$P_L = B_{11} P_1^2 + B_{22} P_2^2$$

Determine B_{11} and B_{22} given that for a given power demand the optimal power generations are

$$P_1 = 300 \text{ MW}$$
$$P_2 = 700 \text{ MW}$$

The corresponding incremental cost of power delivered is 13.5.

Problem 11-B-20

Assume that the transmission loss coefficients for the system described in Problem 11-B-19 are given by

$$B_{11} = 0.5 \times 10^{-3}$$
$$B_{22} = 0.2 \times 10^{-3}$$

Obtain the optimal generating schedule for a power demand of 1000 MW.

Problem 11-B-21

The following data pertain to a system similar to the one described in Problem 11-B-20:

$$\beta_1 = 10.12 \quad\quad \beta_2 = 9.85$$
$$\gamma_1 = -0.0034 \quad\quad \gamma_2 = -0.003$$
$$\delta_1 = 0.0000027 \quad\quad \delta_2 = 0.0000036$$

The transmission loss expression again is

$$P_L = 0.5 \times 10^{-3} P_1^2 + 0.2 \times 10^{-3} P_2^2$$

Given that the incremental cost of power delivered is 16.4291, calculate the corresponding power demand.

Problem 11-B-22

Plant #1 in the system of Problem 11-B-21 is modeled differently such that

$$\beta_1 = 9.77 \quad \text{and} \quad \delta_1 = 0.0000073$$

Note γ_1 is yet to be determined. Given that the incremental cost of power delivered is 17.5289 for a power demand of 1000 MW, determine P_1, P_2, and γ_1.

Problem 11-B-23

The power demand for a 24-hour period on a hydrothermal system is assumed constant at 120 MW. Assume that the thermal unit cost model is

$$F_1 = 50 + 9.5P_1 + 0.01P_1^2$$

The discharge power relationship for the hydro unit is

$$q_2 = 0.5 + 0.1P_2 + 10^{-4}P_2^2 \text{ Mcf/h}$$

Assume that the available volume of water is 200 Mcf for the 24-hour period. Calculate the optimal generations P_1 and P_2, the incremental cost of power λ, and the water conversion coefficient ν. Neglect losses.

Problem 11-B-24

The fuel-cost model for the thermal unit in a hydrothermal system is given by

$$F_1 = 3P_1 + 0.01P_1^2$$

The hydro-performance model is

$$q_2 = 0.03P_2 + 0.0005P_2^2 \text{ Mcf/h}$$

The power demand curve is given by

$$P_D(t) = 200 \text{ MW} \quad (0 \leqslant t \leqslant 12 \text{ h})$$
$$= 300 \text{ MW} \quad (12 \leqslant t \leqslant 24 \text{ h})$$

Assuming the water conversion coefficient is 90, obtain the optimal power generations, the volume of water available, and the incremental cost of power delivered. Neglect losses.

Problem 11-B-25

A hydrothermal system with two plants and negligible losses has the following performance models:

$$F_1 = 1.5P_1 + 0.002P_1^2$$
$$q_2 = 1.8 + 0.12P_2 + 0.0003P_2^2 \text{ Mcf/h}$$

The power demand over the first 12-hour period of the day is 600 MW, and the corresponding hydro plant's optimal output is 300 MW. The volume of water available over the 24-hour period is 2700 Mcf. Obtain the optimal generation by the hydro plant and the demand in the second 12 hours of the day assuming that the demand is constant over that interval. Find the incremental cost of power delivered over the two intervals as well as the water conversion coefficient. Neglect losses.

Problem 11-B-26

The following data pertain to models of performance of the thermal and hydro plants in a power system:

$$F_1 = 4P_1 + 0.0007P_1^2$$
$$q_2 = 0.6P_2 + 0.00035P_2^2$$

The water conversion coefficient is found to be 5. The power demand on the system has two values. The optimal incremental cost of power delivered for each of the two time intervals is found to be

$$\lambda(1) = 4.5$$
$$\lambda(2) = 4.7$$

Obtain the corresponding power demand values.

Problem 11-B-27

A two-plant hydrothermal system has the following optimal values:

$$P_1(1) = 350 \text{ MW} \quad P_1(2) = 500 \text{ MW}$$
$$P_2(1) = 400 \text{ MW} \quad P_2(2) = 500 \text{ MW}$$
$$\lambda(1) = 4.4 \quad \lambda(2) = 4.8$$

The water conversion coefficient is 5.5. Find the parameters of the cost model of the thermal plant as well as of the hydroperformance model. Neglect losses.

Problem 11-B-28

A hydrothermal electric power system with three plants has the following transmission loss coefficients:

$$B_{11} = 1.6 \times 10^{-4}$$
$$B_{22} = 1.2 \times 10^{-4}$$
$$B_{33} = 2.2 \times 10^{-4}$$

Plants 1 and 2 utilize thermal generation and plant 3 is hydraulic. The fuel cost expressions for the thermal plants are

$$F_1 = 47.84 + 9.5P_1 + \gamma_1 P_1^2 \ \$/h$$
$$F_2 = 50.00 + \beta_2 P_2 + 0.01P_2^2 \ \$/h$$

The values of γ_1 and β_2 are to be determined using available information specifying that for a given constant power demand, the optimal values are

$$\lambda = 11.89$$
$$P_1 = 100.29 \text{ MW}$$
$$P_2 = 85.21 \text{ MW}$$
$$P_3 = 68.00 \text{ MW}$$

The hydro plant has discharge characteristics

$$q_3 = 0.5087 + 0.1011 P_3 + 1 \times 10^{-4} P_3^2 \text{ Mcf/h}$$

A. Find γ_1 and β_2.
B. Compute the power loss and the power demand for the given data.
C. Compute the available volume of water over a 24-hour period.
D. Compute the water conversion coefficient $/Mcf.

Problem 11-B-29

A two-plant hydrothermal system has the following characteristics:

A. Fuel cost as a function of active power generated at the thermal plant is

$$F_1 = 10 P_1 + 0.0111 P_1^2 \text{ \$/h}$$

B. The transmission losses are given by

$$P_L = 1.43 \times 10^{-4} P_2^2$$

C. The rate of water discharge as a function of active power generated at the hydro plant is

$$q_2 = 2380 + 60 P_2 \text{ CFS}$$

D. The following table gives the system's power demand and the optimal incremental cost:

Duration (h)	P_D (MW)	λ
14	700	14.50
10	450	13.43

It is required to calculate the optimal active power generated by each of the plants in each of the subintervals, the system power losses, the water conversion coefficient ν, and the allowable volume of water discharge at the hydro plant.

Problem 11-B-30

A three-plant hydrothermal system has the following assumed characteristics:

$$F_1 = 28.50 P_1 + 0.03 P_1^2$$
$$F_2 = 29.82 P_2 + 0.03 P_2^2$$

The loss formula coefficients are given by

$$B_{11} = 1.6 \times 10^{-4}$$
$$B_{22} = 1.2 \times 10^{-4}$$
$$B_{33} = 2.2 \times 10^{-4}$$

The rate of water discharge at the hydroplant is given by

$$q_3 = 1.0174 + 0.2022 P_3 + 2.0 \times 10^{-4} P_3^2 \text{ Mcf/h}$$

The optimal schedule is given by

$$P_1 = 100.29 \text{ MW}$$
$$P_2 = 85.21 \text{ MW}$$
$$P_3 = 68.00 \text{ MW}$$

Calculate the volume of water available over a 24-hour period, the system incremental cost of power delivered λ, the system power demand, and the water conversion coefficients.

Index

ABCD Parameters:
    Definition, 157
    Exact Values for Transmission Lines, 168
    Relation to Admittance Parameters, 157
    For Cascaded Networks, 163
    Relation to Impedance Parameters, 158

Accelerating Torque, 626
Active Power, 13
Active Power Balance, 709
Air Gap Line, 57
All-Aluminum Conductors, 94
Aluminum Conductors, 94
Aluminum Conductor Steel Reinforced Conductor (ACSR), 94
American Electric Power System, 91
American National Standard Institute (ANSI), 31
American Wire Gage, 94
Amplitude Comparison, 560
Angular Momentum Inertia Constant, 628
Apparent Power, 13
Arc Drop, 361
Armature Reaction, 50
Armature Winding, 44
Autotransformers, 240

Index

Autotransformers, Three Winding, 252
Average Direct Voltage, 396
Average Power, 12

Balanced Beam Relay, 548
Balanced Three-Phase Systems: Current and Voltage Relations, 22
Balanced Voltage Systems, 600
Base Admittance, 296
Base Impedance, 296
B-Coefficients Method, 721
Bipolar Links, 363
Blackouts, 625
Bonneville Power Administration, 92
Break Even Between AC and DC, 373
Bridge Rectifier, Equivalent Circuit, 411
Brushes, 44
Buchholz Relay, 581
Bundle Circle, 100
Bundle Conductor Line, 100
Bundle Conductor Line, Capacitance of, 151
Bundle Conductor Line, Inductance for Double Circuit, 127
Busbar Protection, 582

Capacitive Reactance Spacing Factor, 132
Capacitive Susceptance, 132
Cascaded Circuits, 394
Cascaded Lines, 287
Casson-Last Scheme, 598
Central Stations, 2
Characteristic Impedance, 167
Circuit Breakers, 542
Circular Mils, 93
Circulating Current Systems, 598
Clapper Unit Relays, 545
Complex Power, 13

Complex Power in Three-Phase Circuits, 27
Compounding Curves, 59
Computer Relaying, 608
Conductance, 129
Conductor, 92
Conductor Resistance, 94
Conjugate, 13
Constant Extinction Angle Control, 418
Constant Ignition Angle Control, 417
Constant Current Control, 417
Control Centers, 2
Control of DC Links, 416
Control Transformers, 252
Coordination Equations, 739
Core Faults, 576
Core Loss, 220
Core Type Transformer, 219
Cumberland Station, 2
Current Grading, 587
Current Relays, 545
Current Transformers, 542
Cylinder Unit Relays, 548
Cylindrical Rotor Machine, 47

D'Arsonval Unit, 548
Delay Angle, 398
Delta Connection, 24
Delta-line Configuration, 114
Derivative Evaluation in Load Flow, 331
Differential Protection, 568
Differential Relays, 545
Direct Axis, 71
Direct Axis Magnetizing Reactance, 73
Directional Relays, 545
Distance Protection, 601
Distance Relays, 545
Double Circuit, Bundle Conductor Lines, 127

Double Circuit Lines, Capacitance of, 145
Double Circuit Lines, Inductance of, 122
Double Line-to-Ground Fault, 507
Driving Point Admittance, 157

Earth Faults, 575
Economic Operation of Power Systems, 701
Eddy Current Loss, 220
Edison Electric Illuminating Company, 2
Effect of Earth on Line Capacitance, 133
Effective Value, 12
Electrical Degrees, 47
Efficiency, Definition of, 27
Efficiency of Transformers, 229
Electromechanical Relay, 545
Equal Area Criterion, 646
Equal Incremental Cost, 708
Equivalent Circuits of Induction Motors, 256
Equivalent π-Model of Lines, 173
Errors in Line Models, 178
Exact Model of Transmission Lines, 168
Exciting Current, 221
Extinction Angle, 414
Extra High Voltage Lines, 7

Faraday, Michael, 43
Faults, Classification of, 469
Faults, Consequences of, 541
Ferranti Effect, 370
Frequency, 46
Frequency Relays, 545
Fuel Costs, 702

Generator Bus, 310
Generator Dropping, 657

Generator Protection, 567
George's Formula, 721
Grading Electrodes, 361
Graetz Circuit, 388

Heat Rates, 703
Horsepower, 27
HVDC, Applications of, 358
HVDC Circuit Breakers, 359
HVDC Converters, 360
HVDC, Economics of, 365
HVDC, Multi-terminal Systems, 359
HVDC, Systems in Operation, 356
H-type Line Configuration, 115
Homopolar Links, 364
Hoover Dam, 91
Hybrid Form of Load Flow Equations, 306
Hydro-thermal Systems, 738
Hydro-Quebec, 92
Hyperbolic Functions, 167
Hysteresis Losses, 220

Ideal Transformers, 220
Ignition Angle, 414
Image Charges, 133
Impedance Relay, 562
Impedance In Per Unit, 297
Incremental Cost of Power, 710
Incremental Transmission Losses, 723
Induction Disc Relays, 547
Induction Motors, 256
Induction Motor Classes, 265
Induction Motor, Maximum Torque, 263
Induction Motor, Rotor Current, 259
Induction Motor, Torque, 262
Inductive Reactance Spacing Factor, 100
Inductive Voltage Drop, 98
Inductorium, 2
Inertial Constant, 627

Index

Infinite Bus, 64, 632
Instantaneous Power, 12
Interconnections, 2
Interphase Short Circuit, 568
Interconnected System Reduction, 284
International Electrotechnical Commission, 31
Interturn Faults, 568, 576
Inverse-Time Overcurrent Relaying, 591
Inversion, 398

Kinetic Energy, 627

Lagrange Multipliers, 710
Lamm, U., 7
Lamme, B. G., 43
Linear Coupler, 584
Line Capacitance, 129
Line-Charging Current, 129
Line Inductance, 96
Line-to-Ground Fault, 495
Line-to-Line Fault, 512
Line Voltage:
    in Y connection, 22
    in Δ connection, 24
L-Network, 162
Load Bus, 310
Load Flow Equations, 304
Load Flow, Nonlinearity of, 312
Loss Formula, 721
Loss Load Factor, 432
Lumped Parameter Models for Transmission Lines, 173

Magnetizing Inrush Current, 577
Magnetizing Reactance, 52
Magnetomotive Force, 48
Maximum Efficiency of Transformer, 231

Maximum Power Transfer, 634
Maximum Torque in Induction Motor, 263
Mechanical Power Output of Induction Motor, 258
Mercury Arc Rectifier, 361
Mercury Arc Valve, 355, 361
Mho Relay, 561, 567
Mils, 93
Minimum Reactance Relays, 573
Monopolar Links, 363
Multiconductor Configuration:
    Capacitance of, 134
    Inductance of, 101
Multiconductor Three-Phase Systems, 119
Mutual Geometric Mean Distance, 105

Negative Sequence Current, 481
Negative Sequence Networks, 484
Negative Sequence Relays, 573
Negative Sequence Voltage, 476
Neutral Point, 22
Newton-Raphson Method for Optimal Operation, 726
Newton-Raphson Method, 318
Niagara Falls, 189
Nodal Admittance, 299
Nominal π-Model of a Line, 175

Offset Mho-relaying, 603
Ohm Relay, 561, 565
On-off Relay, 544
Ontario Hydro, 3
Open Circuit Characteristics, 55
Operational Objectives, 701
Overexcited Machines, 66
Overlap Angle, 402
Overlap, Voltage Drop Due to, 406

Parallel Connected Transformers, 235
Parallel Lines, Reduction, 285

Parallel-Series Connected Transformers, 233
Parameter Estimation, 705
P-pole Machines, 45
Peak Inverse Voltage, 378
Peak-to-Peak Ripple, 379
Pearl Street Station, 2
Penalty Factors, 723
Percentage Voltage Regulation, 227
Per Unit System, 296, 707
Phase Comparison, 563
Phase Comparison Protection, 605
Phase Sequence, 20
Phase Shift in Y/Δ Transformers, 244
Phase Voltage in Δ Connection, 24
Phase Voltage in Y Connection, 22
Pick-up Current, 591
Pilot Wire Protection, 598
Plunger Type Relay, 545
Polar Form of Load Flow Equations, 305
Polar Unit Relays, 547
Pole Faces, 44
Positive Sequence Current, 481
Positive Sequence Networks, 483
Positive Sequence Voltage, 476
Potential Transformers, 542
Power, 11
Power Angle Characteristics, 63
Power Angle Curves, 641
Power Factor, 12
Power Factor Correction, 18
Power Factor in HVDC Converters, 408
Power Line Carrier Protection, 604
Power Relays, 545
Power, Symmetrical Components, 482
Power Transformer, 217
Power Triangle, 15
Production Scheduling, 701
Propagation Constant, 166

Protective System Requirements, 542
Proximity Effect, 95
Pull-out Power, 65
Pulse Number, 381

Quadrature Axis, 71
Quadrature Axis Magnetizing Reactance, 73

Reactive Capability Curves, 61
Reactive Power, 13
Rectangular Form of Loadflow Equations, 305
Reference Values, 296
Regulating Transformers, 252
Relays, 544
Relay Comparators, 559
Resistor Braking, 657
Root-Mean-Square Value, 12
Rotor Earth Faults, 568

Saliency, 71
Salient Pole Machine, 47
    Analysis, 71
    Power Angle Characteristics, 75
Self Geometric Mean Distance, 105
Sequence Filters, 550
Series Connected Transformers, 232
Series-parallel connected transformers, 234
Sequence Impedance for Synchronous Machines, 490
Shell-type Transformer Construction, 219
Short-circuits, See Faults.
Short-circuit Characteristics, 55
Silicon-controlled Rectifiers, 362
Simultaneous Faults, 524
Single-line Diagrams, 31
Single-phase Circuit, Power in, 14

Single-phase Full Wave Rectifier, 376
Single-phase Two Wire Lines:
    Capacitance of, 130
    Capacitance Considering Earth, 136
    Inductance of, 96
Six-phase Diametrical Circuit, 395
Skin Effect, 95
Slack Bus, 310
Slip, 256
Slip Frequency, 256
Solid State Relays, 555
Speed Control of Induction Motors, 272
Squirrel Cage Rotor, 256
Stable Equilibrium Point, 643
Standard Symbols, 31
Stanley, W., 2
Star Point, 22
Stator Earth Fault, 568
Steady State Stability Limit, 65
Steel Conductors, 94
Stranded Conductor, 92
Subtransient Reactance, 474
Summation Rule for Power, 15
Swedish State Power Board, 91
Swing Equation, 626
Symmetrical Components, 474
Symmetrical Networks, 158
Symmetrical T-networks, 159
Synchronous Impedance, 53
Synchronous Machine, 43
    Construction, 44
    Fields in, 48
    Simple Equivalent Circuit, 52
Synchronous Reactance, 47, 53
Synchronous Reactance, Determination of, 55
Synchronous Speed, 256

T-network, 156
Tap-Changing Under Load (TCUL), 252
Temperature Coefficients, 95
Temperature Relays, 545
Thermal Relays, 572
Three-phase Faults, 515
Three-phase Power, 26
Three-phase Systems, 18
Three-phase One Way Circuit, 381
Three-phase Single Circuit Line:
    Capacitance of, 139
    Inductance of, 113
Three-phase Transformer Connections, 242
Three-Phase Two-way Circuit, 388
Three Winding Transformers, 236
Thyristors, 362
Time Dial, 593
Time-Distance Protection, 601
Time Grading, 587
Torque in Induction Motor, 258
Transfer Admittance, 157
Transfer Problems, 155
Transformer Equivalent Circuit, 221
Transformer Protection, 575
Transformer Symbols, 33
Transformer, Three-winding, 236
Transformer, Single-phase Symbols, 33
Transformer, Three-phase Symbols, 34
Transformer Zero Sequence Equivalents, 485
Transient Stability, 625
Transient Reactance, 474
Transmission Line Models, 164
Transmission Line Protection, 587
Transmission Lines:
    Equivalent π Models, 175
    Model Approximation Errors, 178
    Nominal π Models, 175
    Sequence Impedances, 493
    Short Line Model, 176
Transmission Losses, 716

Transmission Problems, 157
Transposition of Line Conductors, 116
Turns, Ratio, 218
Two-Port Networks, 154

Unconstrained Minimization, 709
Under-excited Machine, 66
Underground Distribution, 2
Unstable Equilibrium Points, 643

V Networks, 156
Valve Points, 703
Variable Percentage Differential Relay, 585
Voltage Angle Approximation, 315

Voltage Magnitude Approximation, 316
Voltage Relays, 545

Water Conversion Coefficients, 740
Water Discharge, 738
Water Head, 738
Wye Connection, 19

X-R Diagrams, 555

Zero Sequence Current, 481
Zero Sequence Networks, 484
Zero Sequence Voltage, 477
Zones of Protection, 542